PHYSICAL CONSTANTS

CONSTANT	SYMBOL	VALUE AND UNITS
Avogadro's number	N_A	6.0222×10^{26} (kg-mole)$^{-1}$
speed of light (vacuum)	c	2.9979×10^8 m/sec
universal gas constant	R	8.3143 J/mole-°K
Boltzmann's constant	$k = R/N_A$	1.3806×10^{-23} J/°K
mechanical equivalent of heat	J	4.1855 J/cal
vacuum permittivity	ε_0	8.8541×10^{-12} F/m
vacuum permeability	μ_0	$4\pi \times 10^{-7}$ H/m
gravitational constant	G	6.673×10^{-11} N-m^2/kg^2
Planck's constant	h	6.6262×10^{-34} J-sec
Dirac's constant	$\hbar = h/2\pi$	1.0546×10^{-34} J-sec
electronic charge	e	-1.6022×10^{-19} C
electron Compton wavelength	λ_0	2.4262×10^{-12} m
gyromagnetic ratio	$e/2m_e$	8.7940×10^{10} C/kg
electron Bohr magneton	$\mu_B = e\hbar/2m_e$	9.273×10^{-24} J/T
proton Bohr magneton	$e\hbar/2m_p$	5.050×10^{-27} J/T
nuclear magneton	μ_N	5.090×10^{-27} J/T
proton magnetic moment	μ_{pz}	$2.7928\mu_N$
neutron magnetic moment	μ_{nz}	$-1.1913\mu_N$
Rydberg constant	R_∞	1.0974×10^7 m^{-1}
Bohr radius	r_1	5.2917×10^{-11} m

Essentials of Modern Physics

Essentials of Modern Physics

VIRGILIO ACOSTA
UNITED STATES NAVAL ACADEMY

CLYDE L. COWAN
THE CATHOLIC UNIVERSITY OF AMERICA

B. J. GRAHAM
UNITED STATES NAVAL ACADEMY

HARPER & ROW, PUBLISHERS

New York, Evanston, San Francisco, London

We dedicate this text to our patient families

ESSENTIALS OF MODERN PHYSICS
Copyright © 1973 by Harper & Row, Publishers, Inc.

Printed in the United States of America. All rights reserved. No part of this book may be used or reproduced in any manner whatsoever without written permission except in the case of brief quotations embodied in critical articles and reviews. For information address Harper & Row, Publishers, Inc., 10 East 53rd Street, New York, N.Y. 10022

Standard Book Number: 06-040162-1

Library of Congress Catalog Card Number: 72-84326

Contents

Preface *xiii*

Part One SPACE AND TIME *1*

  1 Space and Time *3*
    1-1 The Physical Vacuum *4*
    1-2 The Mirror of Space–Time *7*
    1-3 The Measure of Space–Time *8*
    1-4 Matter and Space–Time *10*
    1-5 Summary *10*

  2 Conservation Laws *12*
    2-1 Conservation of Linear Momentum *13*
    2-2 Conservation of Angular Momentum *16*
    2-3 Conservation of Energy *18*
    2-4 Fields *21*

  3 Classical Relativity *26*
    3-1 Limits of "Common Sense" *27*
    3-2 Classical Principle of Relativity *28*
    3-3 Invariance of the Conservation of Linear Momentum *32*
    3-4 Invariance of Newton's Laws *33*

  4 The Michelson–Morley Experiment *40*
    4-1 Conflict Develops *41*
    4-2 Lorentz Transformations *45*
    4-3 Lorentz Composition of Velocities *49*

  5 Consequences of the Lorentz Transformations *55*
    5-1 Contraction of Length *56*
    5-2 Dilation of Temporal Intervals *59*
    5-3 Interpretation of the Michelson–Morley Experiment *63*
    5-4 Einstein's Solution to the Conflict *66*

Part Two PARTICLES AND WAVES *89*

  6 Relativistic Mechanics *71*
    6-1 Mass and Momentum *72*

6–2 Definition of Force   75
6–3 Relativistic Kinetic Energy   77
6–4 Total Energy   80
6–5 Schematic Review   84

7 Photoelectric Effect   91
7–1 Quanta of Electricity   92
7–2 Electron Emission   92
7–3 Photoelectric Effect   94

8 X rays   98
8–1 Roentgen   99
8–2 X Rays   99
8–3 Diffraction of X Rays   104
8–4 Diffraction of X Rays by Ruled Grating   107
8–5 Compton Effect   109

9 Pair Production   116
9–1 Radiation Interaction with Matter   117
9–2 Pair Production   117
9–3 Pair Annihilation   121
9–4 Absorption of Photons   122

10 Wave Nature of Particles   126
10–1 Wave-Particle Dilemma   127
10–2 De Broglie Waves   127
10–3 Experimental Confirmation of Particle-Waves   129
10–4 Wave Packets   131
10–5 The Uncertainty Principle   135
10–6 Another Form of the Uncertainty Principle   139

11 The Rutherford Experiment   143
11–1 The Nuclear Model of the Atom   144
11–2 Experimental Setup   145
11–3 Impact Parameter and Scattering Angle   146
11–4 Rutherford Scattering Formula   150

12 The Bohr Model I   157
12–1 Planetary Model   158
12–2 Atomic Spectra   161
12–3 The Bohr Model—Postulates   163
12–4 The Bohr Model—Energy States   164
12–5 The Rydberg Constant and Spectral Series   168
12–6 The Bohr Model and the Principle of Correspondence   169

Part Three THE ATOM   *189*

13 The Bohr Model II   *174*
13–1 Hydrogenlike Atoms   *175*
13–2 Correction for Nuclear Motion   *177*
13–3 The Franck–Hertz Experiment   *182*
13–4 The Franck–Hertz Experiment—Interpretation   *184*

14 The Schrödinger Equation I   *191*
14–1 Blackbody Radiation   *192*
14–2 Wave Functions   *195*
14–3 The Schrödinger Equation   *196*
14–4 The Time-Independent Schrödinger Equation   *198*

15 The Schrödinger Equation II   *202*
15–1 The Hamiltonian   *203*
15–2 Operators   *204*
15–3 The Potential Well   *206*

16 Some Applications of the Schrödinger Equation   *215*
16–1 The Classical Harmonic Oscillator   *216*
16–2 The Quantum Mechanical Harmonic Oscillator   *218*
16–3 The Tunnel Effect   *221*

17 Different Models of Mechanics   *229*
17–1 Models of Mechanics   *230*
17–2 Classical Mechanics   *230*
17–3 Relativistic Mechanics   *234*
17–4 Quantum Mechanics   *238*
17–5 Wave-Particle Duality   *240*
17–6 Uncertainty Principle   *241*

18 The Schrödinger Theory of the Hydrogen Atom   *244*
18–1 The Wave Equation: Separation of Variables   *245*
18–2 The Azimuthal Equation   *248*
18–3 The Polar Equation   *249*
18–4 The Radial Equation   *250*
18–5 The Complete Wave Function   *251*

19 Quantum Numbers I: Magnetic Moments   *255*
19–1 The Orbital Quantum Number l *256*
19–2 The Magnetic Quantum Number m_l *258*
19–3 The Magnetic Moment of the Hydrogen Atom   *262*

20 Quantum Numbers II: The Zeeman Effect   *266*
20–1 An Atom in an External Magnetic Field   *267*

20–2 The Normal Zeeman Effect 270
20–3 Total Number of States 273

21 The Wave Functions of the Hydrogen Atom 278
21–1 The Wave Functions of the Hydrogen Atom 279
21–2 The Radial Probability Distribution 280
21–3 Angular Probability Dependence 282

22 Electron Spin 287
22–1 Intrinsic Spin 288
22–2 Spin Angular Momentum 290
22–3 The Stern–Gerlach Experiment 292
22–4 Energy of The Spin-Orbit Interaction—Fine Structure 294

23 Atomic and Molecular Spectra 301
23–1 Total Angular Momentum 302
23–2 Atomic Spectra 304
23–3 Molecular Spectra 307
23–4 Lasers 309

24 The Exclusion Principle 316
24–1 The Exclusion Principle 317
24–2 Two-Electron Atoms 318
24–3 The Periodic Table 320

Part Four THE NUCLEUS 325

25 The Nucleus 327
25–1 The Nuclear Atom 328
25–2 Nuclear Forces 330
25–3 Some Properties of the Nucleus 333
25–4 Nuclear Binding Energy 335

26 Models of the Nucleus 339
26–1 Photodisintegration—Nuclear Stability 340
26–2 Spin Angular Momentum 343
26–3 Electrons in the Nucleus? 345
26–4 Liquid Drop Model 347
26–5 Shell Model 351

27 The Neutron 355
27–1 Discovery of the Neutron 356
27–2 Production of Neutrons 359

27–3 Detection of Neutrons   *360*
27–4 Neutron Capture   *361*

28 Nuclear Reactions I   *364*
28–1 Nuclear Reactions   *365*
28–2 *Q* Value of a Nuclear Reaction   *366*
28–3 *Q* Value and Binding Energy   *369*

29 Nuclear Reactions II   *373*
29–1 Kinetic Energies in the Laboratory and in the Center of Mass Frames   *374*
29–2 Threshold Energy of an Endoergic Reaction   *376*
29–3 Derivation of the Threshold Equation   *377*
29–4 Cross-Section Probability   *380*

30 Radioactivity I   *384*
30–1 Radioactivity   *385*
30–2 Disintegration Constant   *386*
30–3 Half-Life and Mean Life   *387*
30–4 Curve of Growth   *389*
30–5 Radioactive Series   *391*
30–6 Dating by Radioactive Decay   *393*

31 Radioactivity II   *399*
31–1 Alpha Decay   *400*
31–2 Positron Decay   *402*
31–3 Electron Decay   *405*
31–4 Electron Capture   *406*
31–5 Gamma Decay   *407*
31–6 Radiological Health Hazards   *407*

32 Fission and Fusion   *412*
32–1 Fission   *413*
32–2 Fusion   *416*
32–3 Nuclear Reactors   *418*

33 Particle Detectors   *423*
33–1 Properties of Particles   *424*
33–2 Nuclear Emulsions   *424*
33–3 Track Chambers   *427*
33–4 Electronic Detectors   *432*

34 Particle Accelerators   *442*
34–1 Accelerators   *443*
34–2 Cockroft–Walton Generator   *443*

34–3 Van de Graaff Generator   *445*
34–4 Cyclotron   *446*
34–5 Betatron   *450*
34–6 Linear Accelerator   *452*

35 Solid State I   *456*
35–1 Crystals   *457*
35–2 Metals   *460*
35–3 Band Theory   *465*

36 Solid State II   *468*
36–1 Fermi–Dirac Distribution   *469*
36–2 Semiconductors   *471*
36–3 Transistors   *474*

Part Five   ELEMENTARY PARTICLES   *477*

37 Elementary Particles   *479*
37–1 Charges and Forces   *480*
37–2 Quantum Numbers of Elementary Particles   *483*

38 Interactions of Elementary Particles   *493*
38–1 Antiparticles   *494*
38–2 Classes of Interactions   *496*
38–3 Interactions and Conservation Laws   *499*

39 The Family of Elementary Particles   *503*
39–1 Photons   *504*
39–2 Leptons   *505*
39–3 Hadrons   *508*

40 Origin of the Elements   *522*
40–1 The Puzzle of the Elements   *523*
40–2 Present Distribution of the Elements   *524*
40–3 Primordial Nucleosynthesis   *525*
40–4 Element Building in the Stars   *527*
40–5 Supernovae and the r Process   *533*
40–6 Explosions of Galactic Nuclei   *535*
40–7 Summary   *536*

41 Origin of the Universe   *538*
41–1 Age of the Universe   *539*
41–2 Size of the Universe   *542*
41–3 Expanding Universe   *545*
41–4 Birth of the Universe   *547*

Appendix   *551*

    Table 1 Periodic Table of the Elements   *552*
    Table 2 Useful Mathematical Aids   *554*
    Table 3 Natural Trigonometric Functions   *556*
    Table 4 Exponential Functions   *557*
    Table 5 Nobel Prizes in Physics   *558*
    Table 6 Table of Isotopes   *560*

Answers to Odd-Numbered Problems   *580*

Index   *590*

Preface

This book moves beyond the realm of classical physics to explore both the microscopic world of the atom, the nucleus and the elementary particles, and the macroscopic world of the cosmos. The coverage of quantum mechanics is more complete than that usually found in texts at this level because we feel that this topic is a natural and essential portion of modern physics. Likewise, the coverage of the elementary particles has been expanded to include some of the latest concepts.

The mathematical prerequisite for this text is two semesters of introductory calculus including the rudiments of vector calculus. We have purposely made all chapters of this book short and self-contained to encourage a feeling of achievement on the part of the student and, at the same time, to allow for greater organization flexibility by the instructor. If desired, rearrangement or deletion of chapters can thus be made more easily.

Such greats as Einstein and Dirac have contributed so much to the development of the concepts of physics that it is difficult to comprehend fully all that the fruits of their labor imply. Others have contributed small but significant steps toward greater ideas. These people and their contributions also make up a part of modern physics. A brief biography of a notable physicist is given at the beginning of each chapter to highlight the endeavors of some of these people.

The principle role of the exercises and questions at the end of each chapter is to help develop the student's skill in numerical solution of problems and to give the reader insights into understanding the nature of physics and its basic principles. The computer impact on physics has also been recognized by including a few computer-oriented problems throughout the book.

We wish to thank Lt. Comdr. Jack Kineke, USN, for patiently working out the problems and offering many helpful suggestions and Mary Hollywood Wilson for typing the manuscript. The cooperation given the authors and the attention given to our manuscript by the publisher, especially Jane Woodbridge and Ann B. Fox, has made the book a much better one and our task more pleasant.

Virgilio Acosta
Clyde L. Cowan
B. J. Graham

ONE
Space and Time

> The totality of physical phenomena is of such a character that it gives no basis for the introduction of the concept of "absolute motion"; or shorter but less precise: There is no absolute motion.
>
> ALBERT EINSTEIN
> *Out of My Later Years,* 1950

> It was Einstein who made the real trouble. He announced in 1905 that there was no such thing as absolute rest. After that there never was.
>
> STEPHEN LEACOCK

Einstein's theories of relativity did away with the need for concepts of absolute motion and absolute time. His theories, however, were more than exercises in mathematical abstraction because, with the strong evidence provided by the Michelson–Morley experiments on the luminiferous ether, they forced physicists to think through again all of the major concepts of physics. Some cherished principles had to be discarded, some had to be altered, and some stood the test presented to them by the new theories of relativity.

I
Space and Time

Georg Friedrich Bernhard Riemann
(1826–1866)

A native of Hanover, Germany, Riemann was a student of K. F. Gauss and later was professor of mathematics at the University of Göttingen where he had received his Ph.D. He expanded the geometry of Nikolai Lobachevsky and János Bolyai to develop a non-Euclidean system based on a postulate that permits no parallel lines. His theory that space is not necessarily uniform provided the basis for Riemannian geometry and was crucial to modern physical theories, including Einstein's theory of relativity.

1–1 THE PHYSICAL VACUUM
1–2 THE MIRROR OF SPACE-TIME
1–3 THE MEASURE OF SPACE-TIME
1–4 MATTER AND SPACE-TIME
1–5 SUMMARY

1-1 THE PHYSICAL VACUUM

The natural world in which we live appears to us to be a vast collection of objects and events, all of which are contained in a three-dimensional space. We perceive the events as being strung out in a continuing time sequence—each event is seen as the cause of another event, and this event becomes, in turn, the cause of yet another. Sometimes in the language of physics these observations that we have made are stated by saying that the natural world is contained in a three-plus-one-dimensional manifold called *space-time*. The purpose of this text is to examine the natural world in some detail and to discover a few of the laws of nature that help us to organize and describe space-time. In so organizing and defining space-time, we shall find that we understand the natural world better.

Before the objects and events of nature are studied directly, however, it is well to contemplate space-time itself. The concept of space-time contains the essence of the most profound questions that we as physicists attempt to answer. To the average person, a vacuum is a volume of space that contains absolutely nothing—no particles, no molecules. This is not, however, the way the physicist thinks of a vacuum. To illustrate a part of our understanding as physicists of the physical vacuum, we do an imaginary experiment. The various parts of this experiment have been in fact observed in the laboratory; so that even though this particular sequence of events has not been carried out as a single experiment, in principle it could be done. Let us begin with an absolute vacuum in an ideal container with perfectly white reflecting walls that are insulators of the best sort imaginable. There will be no detectable particles or radiation, since at first glance it appears to be the kind of vacuum that consists of absolutely nothing.

The experiment is begun by shining some light (electromagnetic radiation) into the vacuum through a very small window in one wall of the container. Since a small amount will be reflected back out through the window, more light is continually shone into the container. We must now begin to shine bluer and bluer light into the hole. Soon we would observe that the color of the escaping light indicates that the *temperature of the vacuum inside is rising*. As the temperature rises, the escaping light becomes bluer. Already, we have discovered that a vacuum can have a temperature.

To see just how "hot" this vacuum can become, let us continue to send more and more radiation into the container at a faster rate than it comes back out of the hole. At some point in this experiment, one photon of light will collide with another photon, and two electrons will appear (Figure 1-1). One of this pair of electrons will be negatively charged, and the other will be positively charged. The empty vacuum is no longer empty! *The vacuum contains two particles of matter —the two electrons.*

Where did these electrons come from? They were not in the beam of light, although the total energy that they contain did come in with the light. Electrons are very different particles from photons of light. The electrons are a part of that family of particles called *fermions*. They carry electrical charge as well as another sort of charge called *lepton number*, and they have a mass that continues to exist even if the electrons are brought to rest. A photon of light is very different. It is a *boson* and carries no charge of any sort; and a photon brought to rest ceases to exist.

As physicists we do not claim to know the complete answer to the origin of these electrons. We tend to think of the electrons as always being there in some sort of "virtual" state and as having been brought into detectable existence by the collision of light photons. The vacuum is thought of as a "state" of space-time that contains no detectable particles, and the following (or resulting) condition is thought of as a state that contains two electrons. In other words, we say that some sort of action applied to the vacuum state created from the vacuum two electrons in a "particle state."

Although the probability is small that these two electrons will ever collide with one another, it is possible that they might. One is positive and one is negative; *they are in some profound way totally different from one another and yet at the same time very much alike.* If they should collide, there will be a transition back to the vacuum state. That is, the two electrons will disappear, and two photons will appear in their places. This is commonly referred to as *matter–antimatter annihilation*. We

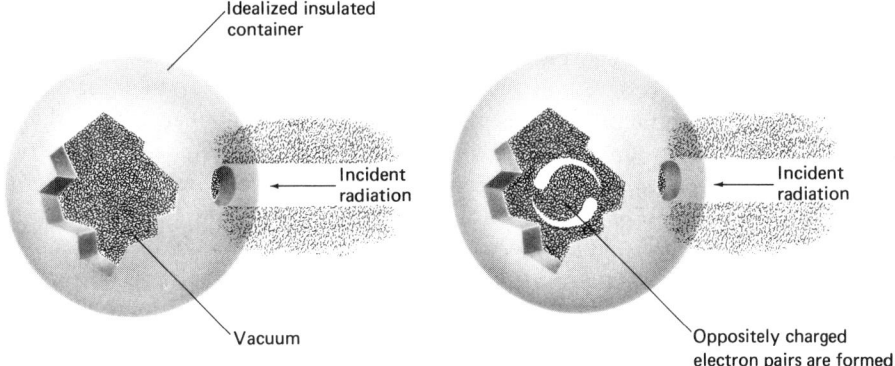

Figure 1-1 After a continuous influx of electromagnetic radiation into an insulated vacuum container, electron pairs are eventually formed.

may well ask: Where did they go? Are they still present in a nondetectable way?

Let us keep two detectable electrons in the container along with the radiation that we send in. Assume that they do not collide for a long time, during which time more radiation is poured in through the window. A continuing process of more photon collisions will produce more electron pairs, and collisions of photons with electrons will heat the electrons as well as producing even more electron pairs. The radiation keeps coming into the space, and it keeps getting hotter and hotter until finally, when a photon collides with an electron, a positive–negative muon pair is produced. Again, something new is found in the vacuum in the shape of these muons, and these muons are different from the electron pairs formed previously. For one thing, *the muons are radioactive.*

If the space is heated continuously by sending more and more radiation into the window faster than it can escape back through it, particles called *pi mesons* or *pions* will begin appearing. Another new item will be found in the container in the form of a very strong nuclear force that binds them together. The pions are very different both from muons and from electrons. With more heating, eventually proton-antiproton and neutron–antineutron pairs appear, and now we have the *material from which all atomic nuclei are made.*

Now we may ask: From where did these particles come? "From virtual states in the vacuum," is the reply of the physicist. Next we must ask: Was the vacuum truly empty? We might answer that if we consider the making of particles from the vacuum an observation, then it was not empty. Thus, if the appearance of particle–antiparticle pairs can be called evidence for a "detectable" vacuum, then we must conclude that the vacuum is jam-packed with electrons, muons, protons, and neutrons, as well as with other particles that appear as continued heating of the space takes place. So we may reason that the vacuum not only has a definite temperature but also contains an unimaginably dense assortment of all the particles in nature. It is certainly not a region of absolute nothingness!

As we have seen, with the appearance of the protons and neutrons as well as the electrons in space, we have the necessary materials with which to build all the known elements and compounds (or matter) found in nature. In addition to the element building continuously happening in our original container, there will also be particles colliding frequently with antiparticles and vanishing, leaving photons in their places. *When an equilibrium between matter and electromagnetic radiation has been established, all of the necessary components with which to build a real part of the universe are present.* Furthermore, the particles that have been produced are identical to their counterparts anywhere in the universe. The electrons and protons in the container are identical to the electrons and protons found in the oldest of rocks or the farthest of stars.

Our conclusion is that space in general contains a dense assortment of all known particles and that these particles are detectable with the aid of electromagnetic radiation (light). Thus, we say that the physical vacuum is something very real.

1-2 THE MIRROR OF SPACE-TIME

In our discussion of the physical vacuum, we mentioned the concepts of matter and antimatter. It is well to pause and investigate this phenomenon further. We have noted that a particle is just the opposite of its antiparticle, but that the two are very much alike. Let us consider an object standing in front of a plane mirror and suppose that we can see the object as well as its image in the mirror. In appearance, the object and its image are very much alike, but they are reversed with respect to each other in the same way a left hand is reversed with respect to the right hand. The image contains the same light and color distribution that the object contains but in a reversed order.

Now let there be an object with a distribution of electrical charges on it, and let the mirror be grounded, polished copper. Again there is a reversed optical image of the object, but now the image also has a charge distribution just like the one on the object, except that the distribution *is reversed in electrical sign*. If there is a concentration of *positive* charges on the top part of the object, there will be a similar concentration of *negative* charges on the top of the image. In this experiment, the object is a bit closer to being the same as its image, except that it is reversed (Figure 1-2).

In the ultimate case, space-time constitutes some sort of perfect mirror—one that reflects all aspects of every fundamental particle and also reverses each one in so doing. Every particle has a "reflection" in this perfect space-time mirror, and every property of the particle is faithfully contained in a reverse sense in its image. In this case, it makes little difference which is called object and which is called image. They are exactly "alike," but they are reversed in all senses with respect to one another.

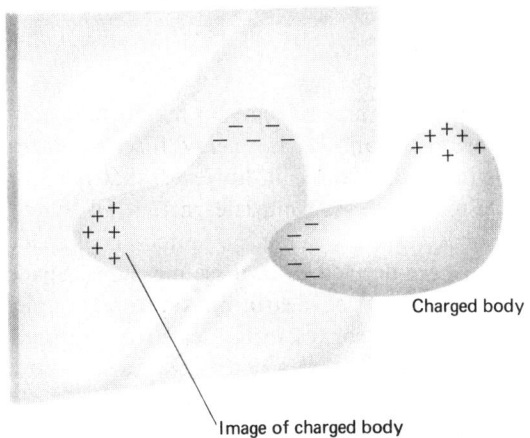

Figure 1-2 An object and its optical image are reversed in the same way the left hand is reversed with respect to the right, but by electrical induction the charge distribution on the image is reversed in sign.

Nature may thus be thought of as being composed of a vast number of particles and an equal number of antiparticles. As each is contained in the perfect mirror called space-time, one may be a great distance from the other, but both are "in" the mirror.

What happens when an object comes close to and "collides" with its image. We may go back to the case of optical images for an analogy. If we observe a leaf hanging from a branch of a tree and the image of the leaf in the surface of a quiet pond beneath the tree, we see both leaf and image. Now let the leaf fall toward the water. The image and the leaf "collide" as the leaf reaches the water surface. They both vanish as the leaf sinks. In their place is a ring of ripples expanding outward from the site of the collision.

This is an analogy but a very inadequate one. When a particle and its antiparticle combine in a collision, they both vanish completely, and some photons of electromagnetic radiation are produced or, in some cases pions are made, and these quickly leave the site of the collision.

We might ask: Where is the particular image of this particular electron on the point of my pencil? Does it have a particular corresponding single image? Further thought recalls that all negative electrons are identical to one another. Any positive electron may serve as an image for any negative electron, and vice versa.

Thus, all physical properties of matter are in some sense reflected in space-time, and these reflections are antimatter. We must put aside one property, however, for which this may not hold: the property of being alive. The property of life is apparently not "reflected" in space-time, and while being a perfectly evident property of many objects, life thus cannot be considered to be "in" space-time *in the sense that the physical properties are*. There is no evidence for "antilife," but only for the absence of life in particular cases.

1–3 THE MEASURE OF SPACE-TIME

We have learned in our prior studies of the natural sciences to regard nature in its many different aspects, which we variously name mass, energy, force, momentum, electrical charge, and so on. It is important to recall, however, that none of these qualities is ever measured in a direct sense. We must learn that all that is ultimately done in making a scientific observation is to measure intervals of space and intervals of time. All other quantities are derived from these measures. Space intervals may be measured directly with some sort of measuring rod (for example, a meter stick), or they may be indicated by some sort of spring scale (for example, by the varying positions of a needle on a scale).

Another method of performing a measure of a distance interval is to consider the time interval for a pulse of electromagnetic radiation to go out and be reflected back. Thus, we note that there exists a close relationship between time intervals and spatial ones. In an analogous way, the distances from one peak to another of waves in some medium may be used as a measure of temporal intervals. Most often, however, the measurement of a time interval is done by noting the successive positions of the hands on a clock. Just as intervals of time may be intimately related to intervals of space, so is the reverse true. In fact, all other quantities of which we speak in physics are derived from these two sorts of interval. Perhaps we should become more conscious of our methods in making these basic measurements.

Let us consider for a moment the direct measurement of an interval of space using measuring rods. A meter stick is placed along the interval so that the zero end of the stick is at one end of the interval, and then the number of centimeters that just falls short of filling out the interval is noted. Then the next centimeter on the stick is divided into, say, tenths. Counting the number of these just short of the interval gives us the next decimal. We may keep doing this, dividing the "next" interval into parts and counting the whole part, changing our instrument as we go from a meter stick to a microscope, then an interferometer, until we finally arrive at the limit of our ability to divide the next interval into parts. Our answer is a decimal number with, perhaps, eight or nine digits in it.

When the limit of our ability to divide the remaining space and still have a finite, rational decimal number for an answer has been reached, we assume that, in principle, if we could devise ever more sensitive instruments we could proceed to divide the remaining interval into yet smaller parts *ad infinitum*. We assume that the space that we have measured with a finite number of steps may be measured with an infinity of steps until the remaining bit of interval constitutes an "infinitesimal." How do we make this assumption? It is made immediately when our measurements are used as numerical values in differential and integral calculus expressions. The infinitesimal calculus is itself the mathematical model for the assumption we have just stated. Recall that in the calculus, the ratios of intervals approach a limit as an interval approaches zero. The analogy of this assumption, then, implies that space-time may be thought of as a continuous entity. Thus, we assume that logically we may consider any interval of space-time, no matter how small (or infinitesimal) that interval may be.

However, we should remind ourselves that there is no logical reason for considering that intervals are infinitesimal other than that the mathematics for describing such a continuum is much simpler than the mathematics would be if in fact there were some smallest-but-finite interval of space-time.* No coherent theory of physics has, as yet, been built on a noncontinuous, discrete, structure for space-time. It is well to remind ourselves, however, that as yet we have no compelling reason to consider space-time as a continuum other than the convenience of the mathematics. There is much for us to learn regarding the structure of space-time.

* See A. J. Greenberg et al., "Charged-Pion Lifetime and a Limit on a Fundamental Length," *Phys. Rev. Letters* **23,** 1267 (1969).

1-4 MATTER AND SPACE-TIME

We began this chapter by discussing material objects that are contained in a three-dimensional space, and it may be that perhaps we used the wrong words when we said that matter is "contained" in the space. This statement, if we examine it carefully, implies that the space-time manifold is merely a passive background into which all material objects are somehow embedded. Certainly this is the way space-time is treated in the classical physics and also in the classical philosophies. Space was considered to be a vast container of some sort, which provides the room for objects of the universe to move about in and to exert their mutual influences on one another.

However, it may be well to consider, on the other hand, if in some manner the existence of material objects could be due to the "geometry" of space-time itself, and further, that matter might be just one more property of an all-encompassing space-time. This was the view adopted by Albert Einstein at the beginning of the twentieth century in his efforts to build a general theory of relativity. He postulated that the existence of forces of any sort could be considered as manifestations of some particular curvatures in space-time that produce accelerations. All forces, whether gravitational, electrical, nuclear, and so on, are, perhaps, just convenient models for a more complex general situation in which a curvature produces the mass. The conception of general relativity as presented by Einstein is well summarized by Erwin Schrödinger*: "The ideal aspiration, the ultimate aim, of the theory is not more and nor less than this: A four-dimensional continuum endowed with a certain intrinsic geometric structure, a structure that is subject to certain inherent purely geometric laws, is to be an adequate model or picture of the real world around us in space and time with all it contains and including its total behavior, the display of all events going on in it."

This is, of course, a very different picture from the simple container concept of space-time with which we began our discussion. Which view of the relationship that exists between space-time and matter is correct, if either, is one of the most basic problems facing modern physics. It remains for future generations of physicists to solve. In this text, we can only assist those future scientists to make a start on it.

1-5 SUMMARY

We have seen that, in a very basic sense in physics, a vacuum in space-time is not an empty concept but contains a vast number of all the particles known. The manner in which these particles are to be detected involves the use of a high-energy light signal. It is also true that there are no basic reasons for assuming that space-time can be divided into smaller and smaller intervals indefinitely. Space-time may

* Erwin Schrödinger, *Space-Time Structure*, Cambridge University Press, London, 1950.

be a continuum, or it may be discrete, in that it consists of indivisible cells in some manner. In its interaction with matter, space-time acts in a fundamental way as a perfect mirror providing a complete and completely reversed image for each particle in the universe. It may also be true that space-time itself is either a passive container for the physical world or is itself the seat of all physical phenomena by and through its own intrinsic geometry.

QUESTIONS

1-1 In what sense can a vacuum be said to have a temperature?

1-2 In what sense can an "absolute" vacuum be considered as being absolutely empty, and in what sense can the same vacuum be considered as being very full? Use the concept of "detectability" in your answer.

1-3 In what sense can space-time be considered as a mirror?

1-4 Sketch in outline a figure of an "object" (it may be a human figure if you wish) and an "image" in a highly polished metallic mirror. Endow your object with a distribution of electrical charge, both negative and positive. Show the resultant electrical charge of the image.

1-5 In your opinion, can the property of "life" be classed as a purely physical property that should be reflected in the space-time mirror? Discuss your opinion, both pro and con.

1-6 Discuss the procedure by which we assume space-time to be a continuum when we apply the differential calculus to measurements taken in space-time. Use the fundamental definition of a differential in your answer.

1-7 What is meant, mathematically, by the term "continuum"? You may have to refer to texts on analysis found in the mathematics section of a library.

1-8 Discuss the difference between a universe in which all matter is simply embedded in space-time and one in which the geometry of space-time "produces" matter.

RECOMMENDED READING

DIRAC, P. A. M., "The Evolution of the Physicists' Picture of Nature," *Sci. Am.*, May 1963.
EINSTEIN, Albert, "On the Generalized Theory," *Sci. Am.*, April 1950.
FRISCH, David H., and THORNDIKE, Alan M., *Elementary Particles*, Van Nostrand Reinhold, New York, 1964.
GAMOW, George, "The Evolutionary Universe," *Sci. Am.*, September 1950.
SCHRÖDINGER, Erwin, *Space-Time Structure*, Cambridge University Press, London, 1950.

2
Conservation Laws

Sir Isaac Newton
(1642–1727)

Born in Lincolnshire, England, and educated at Trinity College, Cambridge, Newton held the Lucasian Chair in mathematics at Cambridge University (1669). In 1687, he published Philosophiae Naturalis Principia Mathematica, *one of the greatest works of all time. Newton, a most powerful influence in the realm of scientific thought, developed differential and integral calculus, the fundamental laws of classical mechanics, and the theory of gravitation; he also performed extensive research in optics and astronomy. He served as president of the Royal Society from 1703 until his death.*

2–1 CONSERVATION OF LINEAR MOMENTUM
2–2 CONSERVATION OF ANGULAR-MOMENTUM
2–3 CONSERVATION OF ENERGY
2–4 FIELDS

2-1 CONSERVATION OF LINEAR MOMENTUM

We know from our previous studies that the turn of the twentieth century marked the beginning of an era of unparalleled progress in the development of the science of physics. Even so, although the subject matter of classical mechanics is almost 400 years old, a knowledge of classical mechanics is still essential for a clear understanding of the basic principles of modern physics, for example, the theory of relativity and quantum mechanics. Let us review the development of classical physics before proceeding with our discussion of modern physics.

Kinematics, the study of motion, was developed mainly by GALILEO GALILEI (1564–1642), a brilliant Italian astronomer and mathematician. In the most basic sense, kinematics is just a study of geometry with the addition of a new parameter—time. The study of the causes of motion (dynamics) was developed by SIR ISAAC NEWTON (1642–1727), the great English astronomer, physicist, and mathematician. (In addition to establishing classical mechanics, Newton, independently of Leibnitz, developed infinitesimal calculus.)

Classical mechanics has been successful in solving a wide variety of problems in engineering, astronomy, and physics; however, the development of modern physics has shown that classical mechanics is not universal in its application. Investigation into the microscopic world of atoms, electrons, protons, and so on, has aided the development of new tools of modern physics—particularly relativity theory and quantum mechanics. We should note at this point that as physicists we are continually trying to establish a mathematical model to describe the space or universe around us. We note that: *A theory in physics is not considered as an ultimate truth, but merely as a model to be applied to solve problems and to yield solutions that are in close agreement with the experience offered by experimental determination.*

The most fundamental of these laws or models are the conservation laws. They divide themselves into two groups: the "extrinsic" elementary laws, consisting of the conservation of linear momentum, the conservation of angular momentum, and the conservation of energy; and "intrinsic" laws, consisting of the conservation of the total number of nucleons in a nuclear reaction, the conservation of lepton and baryon numbers, and so on. This latter group of conservation laws will be developed and discussed as we find the need arises, and here we will review the elementary conservation laws to establish a foundation for the subject matter of the remainder of the text.

Classical mechanics can be approached or studied either by beginning with Newton's laws as a foundation or by beginning with the principle of the conservation of linear momentum. We shall approach mechanics from the latter point of view, since the conservation of linear momentum is simpler and its applications are more general. Thus, we shall assume the principle of conservation of linear momentum to be the most fundamental law.

In discussing the relative motions of several bodies, we could use the several corresponding velocities: those velocities of each of the bodies with respect to each of the other bodies. This procedure quickly becomes very complicated, however. We shall find it to be a great simplification to use, instead, a three-dimensional set of orthogonal coordinates to describe a common "reference frame" in which all of the bodies move (or perhaps one or more remain at rest). By orthogonal we mean that the coordinates themselves are not dependent on one another. The (x, y, z) frame of mutually perpendicular linear coordinates is a most common example.

We shall also specify that this frame is an "inertial" frame of reference. By this we mean that in it, classical mechanics remains valid. We shall see later that "classical mechanics" includes the mechanics of special relativity. If we can specify such a frame of reference, then all other frames of reference that move with constant linear velocity with respect to the first are also inertial frames. The problem of the existence of a "fundamental reference frame" as the one in which Newton's laws are valid is a postulate of Newtonian mechanics and the theory of gravitation, known as Mach's principle.*

Inherently related to the concept of force, the cornerstone of mechanics, is what we call the *inertial mass*. Inertial mass represents a measure of the opposition a body exhibits to being accelerated. We know that for a given force, the larger the mass on which the force acts, the smaller the acceleration imparted to the body. Classically, the inertial mass is considered to be a universal constant and to be independent of all outside effects such as force, temperature, or velocity.

The linear momentum of a particle of inertial mass m moving with a velocity **v** is a vector defined by

$$\mathbf{p} = m\mathbf{v} \tag{2-1}$$

In terms of the unit vectors and components, we may write

$$\mathbf{p} = \hat{\mathbf{i}}mv_x + \hat{\mathbf{j}}mv_y + \hat{\mathbf{k}}mv_z$$

where $\hat{\mathbf{i}}$, $\hat{\mathbf{j}}$, and $\hat{\mathbf{k}}$ are vectors of unit length parallel to the coordinate axes x, y, and z, respectively, and where v_x, v_y, and v_z are the corresponding components of the velocity vector **v** referred to the three orthogonal axes.

---

* See, for example, the *Encyclopaedia Britannica*.

The principle of the *conservation of linear momentum* states: *For an isolated system of particles, the total linear momentum of the system will remain constant.* An isolated system is understood to be a system free from *any* external influence. For the isolated system of Figure 2–1,

$$m_A v_A + m_B v_B = \text{constant} \tag{2-2}$$

For a system composed of many particles, we have $m_A v_A + m_B v_B + \cdots + m_N v_N =$

$$\sum_{i}^{N} m_i v_i = \text{constant} \tag{2-3}$$

We shall now derive Newton's three laws of motion from the principle of conservation of momentum. For two isolated particles, differentiation of Equation (2–2) with respect to time yields

$$m_A \frac{dv_A}{dt} = -m_B \frac{dv_B}{dt}$$

Since $\mathbf{a} = dv/dt$, we have

$$m_A \mathbf{a}_A = -m_B \mathbf{a}_B \tag{2-4}$$

The accelerations are thus inversely proportional to the inertial masses, $a = F(1/m)$, where F is a constant of proportionality. Therefore, we have a definition of force:

$$\mathbf{F} = m\mathbf{a} \tag{2-5}$$

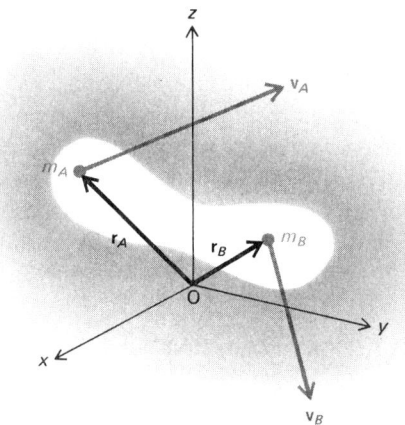

Figure 2–1 The principle of conservation of linear momentum for a system of two isolated particles requires that $m_A v_A + m_B v_B = \text{constant}$ throughout the interaction of the two particles, from $t = -\infty$ to $t = +\infty$.

This is *Newton's second law*. Now, for two isolated particles interacting only with each other by a force (e.g., an electrical or a gravitational force), $\mathbf{F}_A$ is the force particle B exerts on particle A and $\mathbf{F}_B$ is the force particle A exerts on particle B, or

$$\boxed{\mathbf{F}_A = -\mathbf{F}_B}$$

This is the principle of action and reaction referred to as *Newton's third law*.

Finally, for a single, free particle, since both $\mathbf{F} = 0$ and $\mathbf{a} = 0$, and we know that $\mathbf{a} = d\mathbf{v}/dt$, we conclude that

$$\boxed{\mathbf{v} = \text{constant}}$$

This is a statement of the law of inertia or *Newton's first law*.

Newton's second law can be written as

$$\mathbf{F} = \frac{d}{dt}(m\mathbf{v})$$

from which we obtain

$$\mathbf{F}\,dt = d(m\mathbf{v}) \tag{2-6}$$

When the force acts for a finite time t', we have

$$\boxed{\int_{t=0}^{t'} \mathbf{F}\,dt = m\mathbf{v} - m\mathbf{v}_0} \tag{2-7}$$

This integral is called the *impulse* of the force $\mathbf{F}$. We see that it equals the change in momentum that results from the application of this force during a time t'.

When an energetic particle makes a collision of short duration with a second particle, the forces between the particles are said to be *impulsive forces*. Although impulsive forces themselves are generally difficult to measure, the collisions can be analyzed through the conservation of linear momentum using Equation (2-7). Since impulsive forces are often large when compared to forces external to the system, and since they are applied for very short durations of time, we can often assume that forces external to the system are negligible. For these reasons, then, during any collision, whether elastic or inelastic, momentum can be assumed to be conserved.

2-2 CONSERVATION OF ANGULAR MOMENTUM

The *angular momentum* for a particle with linear momentum $\mathbf{p}$, located by position vector $\mathbf{r}$ with respect to a reference origin O, is a vector defined by

$$\boxed{\mathbf{L} = \mathbf{r} \times m\mathbf{v} = \mathbf{r} \times \mathbf{p}} \tag{2-8}$$

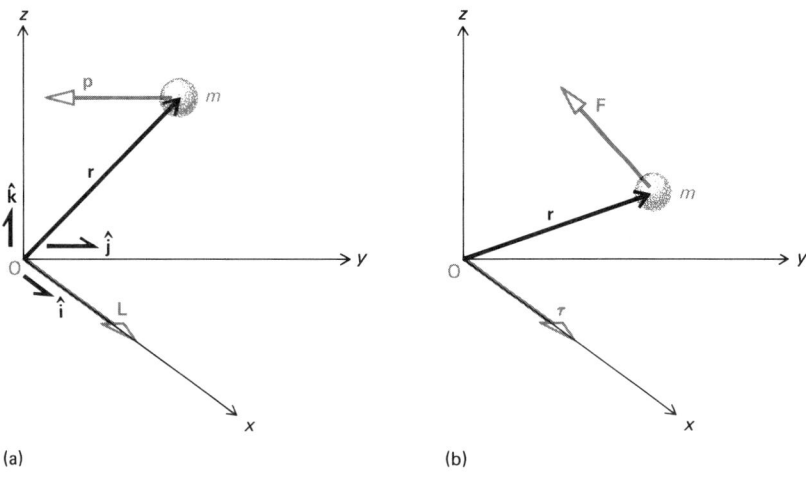

Figure 2–2 (a) A particle of mass *m* with a linear momentum **p** as shown in the $-y$ direction will have an angular momentum $\mathbf{L} = \mathbf{r} \times \mathbf{p}$. (b) A particle of mass *m* acted upon by a force **F** (in the *yz* plane) has a torque relative to the origin of $\boldsymbol{\tau} = \mathbf{r} \times \mathbf{F}$.

as illustrated in Figure 2–2(a). We should note that the angular momentum depends on the choice of location of a reference origin. Also, contrary to our expectations, the particle with respect to a given coordinate system need not be in any type of circular motion at all to possess an angular momentum. We can rewrite the angular momentum vector in terms of unit vectors and components of linear momentum as

$$\mathbf{L} = \begin{vmatrix} \hat{\mathbf{i}} & \hat{\mathbf{j}} & \hat{\mathbf{k}} \\ x & y & z \\ p_x & p_y & p_z \end{vmatrix}$$

$$= \hat{\mathbf{i}}(yp_z - zp_y) + \hat{\mathbf{j}}(zp_x - xp_z) + \hat{\mathbf{k}}(xp_y - yp_x) \tag{2-8a}$$

Recall that force may be considered the "cause" of linear motion. In the same manner torque, usually denoted by $\boldsymbol{\tau}$, may be considered the "cause" of rotational motion. In Figure 2–2(b) a force **F** applied to a particle located with position vector **r** from the reference origin produces a torque

$$\boxed{\boldsymbol{\tau} = \mathbf{r} \times \mathbf{F}} \tag{2-9}$$

To develop a relation between the angular momentum and the torque, we differentiate Equation (2–8) with respect to time, and we obtain

$$\frac{d\mathbf{L}}{dt} = \frac{d\mathbf{r}}{dt} \times m\mathbf{v} + \mathbf{r} \times \frac{d}{dt}(m\mathbf{v})$$

Since $d\mathbf{r}/dt = \mathbf{v}$, $(d\mathbf{r}/dt) \times m\mathbf{v} = 0$, and $\mathbf{F} = (d/dt)(m\mathbf{v})$, the equation can be simplified to

$$\frac{d\mathbf{L}}{dt} = \mathbf{r} \times \mathbf{F} = \boldsymbol{\tau} \tag{2-10}$$

In planetary motion, a body is continually acted on by the gravitational pull. This is always a force that is directed along the radius of the path of the body, given that the center of the body is the reference origin. Since the position vector $\mathbf{r}$ and the force $\mathbf{F}$ are in the same direction, $\boldsymbol{\tau} = \mathbf{r} \times \mathbf{F} = 0$, and from Equation (2–10) we conclude that the angular momentum $\mathbf{L}$ of such a system must be constant.

For a system of many bodies and forces, the resultant torque is

$$\boldsymbol{\tau}_R = \sum_{i=1}^{N} \boldsymbol{\tau}_i = \frac{d}{dt}\left(\sum_{i=1}^{N} \mathbf{L}\right) \tag{2-11}$$

Let us consider a system that is free from external torques. Our previous analysis has shown that the torques from internal forces between any pair of particles cancel out according to Newton's third law,

$$\frac{d}{dt}\left(\sum \mathbf{L}\right) = 0$$

and therefore

$$\sum \mathbf{L} = \text{constant} \tag{2-12}$$

This is the statement of the *conservation of angular momentum*.

2-3 CONSERVATION OF ENERGY

In Figure 2–3(a), a particle moving along the curvilinear path AB is acted on by a force $\mathbf{F}$ as it moves through a displacement $d\mathbf{r}$. The differential *work* of the force $\mathbf{F}$ is defined by

$$dW = \mathbf{F} \cdot d\mathbf{r} \tag{2-13}$$

If this force $\mathbf{F}$ is applied along the path AB, then the total work done is

$$W_{AB} = \int_A^B \mathbf{F} \cdot d\mathbf{r} = \int_A^B F \cos \alpha \, dr \tag{2-14}$$

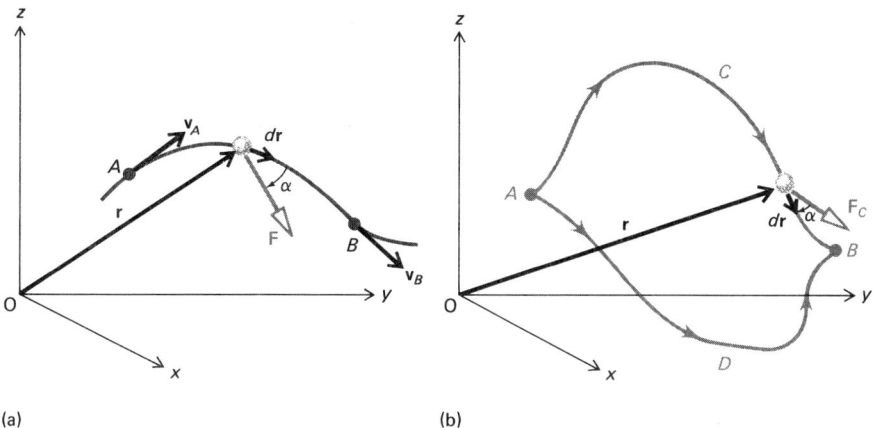

Figure 2–3 (a) The work done by the force **F** in moving the particle through a distance $d\mathbf{r}$ is $dW = \mathbf{F} \cdot d\mathbf{r}$. (b) For a conservative force $\mathbf{F}_c$, the work $W_{AB} = \int_A^B \mathbf{F}_c \cdot d\mathbf{r}$ is independent of the path connecting points A and B.

Let us assume that **F** is the resultant of all forces acting on the particle. Then

$$W_{AB} = \int_A^B \mathbf{F} \cdot d\mathbf{r} = \int_A^B m\frac{d\mathbf{v}}{dt} \cdot d\mathbf{r} = \int_{v_A}^{v_B} m\mathbf{v} \cdot d\mathbf{v}$$

because $d\mathbf{r}/dt = \mathbf{v}$. If we integrate, we obtain

$$W_{AB} = \int_{v_A}^{v_B} mv \, dv = \tfrac{1}{2}mv_B^2 - \tfrac{1}{2}mv_A^2 = K_B - K_A \qquad (2\text{--}15)$$

The quantity $K = \tfrac{1}{2}mv^2$ is defined as the kinetic energy. This is a statement of the *work-energy principle*: *The resultant work done by all the forces acting on a particle equals the corresponding change in kinetic energy.*

The force $\mathbf{F}_c$ in Figure 2–3(b) is said to be a *conservative force* if

$$W_{AB} = \int_{ACB} \mathbf{F}_c \cdot d\mathbf{r} = \int_{ADB} \mathbf{F}_c \cdot d\mathbf{r} = \text{constant}$$

We can state this in words as: *If the work done by $\mathbf{F}_c$ in moving the particle from point A to point B is independent of the path taken, then $\mathbf{F}_c$ is a conservative force.* As an example, let us review the work done by the gravitational force. Figure 2–4 shows a particle of mass m as it moves from position A to position B under the influence of the gravitational force $\mathbf{F}_g$. Since $\mathbf{F}_g = -\hat{\mathbf{j}}mg$, the work done by this force is

$$W_{AB} = \int_{h_1}^{h_2} (-\hat{\mathbf{j}}mg) \cdot (\hat{\mathbf{i}} \, dx + \hat{\mathbf{j}} \, dy) = \int_{h_1}^{h_2} -mg \, dy = mg(h_1 - h_2) = mgh$$

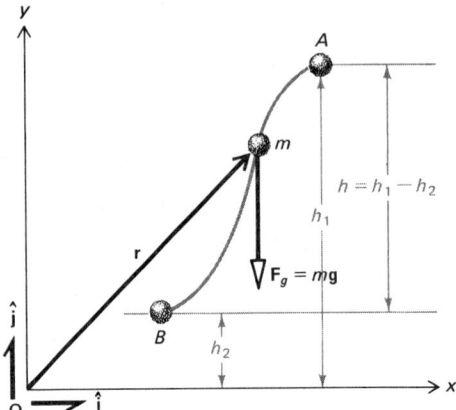

Figure 2–4 The work done by the conservative gravitational force is independent of the path between points A and B.

Since the work done by the gravitational force is independent of whatever path is taken between points A and B, it is a conservative force.

Potential energy is defined in terms of the work done by a conservative force:

$$U_{AB} = \int_A^B \mathbf{F}_c \cdot d\mathbf{r} = U_A - U_B \quad \text{(independent of path)} \tag{2-16}$$

The scalar function of position $U(x, y, z)$ is the potential energy function associated with the conservative force $\mathbf{F}_c$. The quantities U_A and U_B are just the values of the function $U(x, y, z)$ evaluated at the end points of the path. The potential energy at any given point is defined by Equation (2–16), in which position B may be chosen arbitrarily. Usually, B is chosen to be at infinity so that $U_B = 0$. Hence,

$$U_A = U_{AB} = \int_A^B \mathbf{F}_c \cdot d\mathbf{r} = -\int_B^A \mathbf{F}_c \cdot d\mathbf{r} \tag{2-17}$$

The *potential energy* at a point then is defined as *the work done by an equal but oppositely directed force used to move the particle from reference position B to the given position A.*

Let us recall the work-energy principle from Equation (2–15):

$$W_{AB} = K_B - K_A$$

This may now be rewritten to include both conservative and nonconservative forces:

$$W_{AB}(\text{conservative}) + W_{AB}(\text{nonconservative}) = K_B - K_A \tag{2-18}$$

We know from our previous discussion that

$$W_{AB}(\text{conservative}) = U_A - U_B$$

Now we shall rearrange the terms in Equation (2–18),

$$W_{AB}(\text{nonconservative}) = (K_B - K_A) - (U_A - U_B)$$

or

$$\boxed{W_{AB}(\text{nonconservative}) = (K_B + U_B) - (K_A + U_A)} \tag{2-19}$$

If all forces involved are conservative, so that $W_{AB}(\text{nonconservative}) = 0$, we obtain

$$\boxed{K_A + U_A = K_B + U_B = \text{constant}} \tag{2-20}$$

This is the statement of the *conservation of mechanical energy*. In words we may state that when all forces acting on a particle are conservative, the total energy at any one position is equal to a constant called the *total mechanical energy*.

When we consider all forces, both conservative and nonconservative, the work done by the nonconservative forces in Equation (2–19) will always be found to appear as some form of energy. For example, if the nonconservative force is a friction force, then the energy from this force will appear as heat energy. The *principle of conservation of energy*, a generalized statement that we deduce from experience, states that *the energy of an isolated system may be transformed from one kind of energy to another kind of energy; however, the total energy in its various forms can be neither created nor destroyed.*

2–4 FIELDS

One definition of physics is to state that it is the study of the different types of force interactions—gravitational, electromagnetic, weak, and strong. These interactions may be studied through the mechanism of fields. We briefly review the concepts of field here.

There are two simple categories of fields: vectors and scalars. We define a *field* as a region of space in which we can make a measurement of a physical quantity. A *scalar field of position* is defined by a position function $\phi(x, y, z)$ that assigns to every point in space a numerical scalar value. For example, let us consider a three-dimensional metal block containing a source of heat. The scalar field of temperature for this block might be given as

$$\phi(x, y, z) = 2x^2 - 3y^2 + z - 16 \, (C°)$$

The scalar value of temperature associated with a particular point $P(x = 2, y = 1, z = 0)$ is then

$$\phi(x, y, z) = \phi(2, 1, 0) = -11 C°$$

Many other examples of scalar fields exist, such as density distribution, pressure, and so on. In some cases, a fourth coordinate—time—is added, and the scalar field then becomes a function of both position and time. A simple case of this is when the temperature at a given point does not remain constant and does vary with time.

A vector field is defined by a vector function $\mathbf{F}(x, y, z)$ that assigns to every point in a given reference frame a vector. A good example of a vector field is the gravitational field of the earth, in which a vector $\mathbf{g}$ is assigned to every point in space. The magnitude of $\mathbf{g}$ depends on one parameter—the distance of the point from the center of the earth.

The interaction of the gravitational fields of two masses is illustrated in Figure 2–5. The gravitational fields at distances PA, PB, PC, and PD from mass

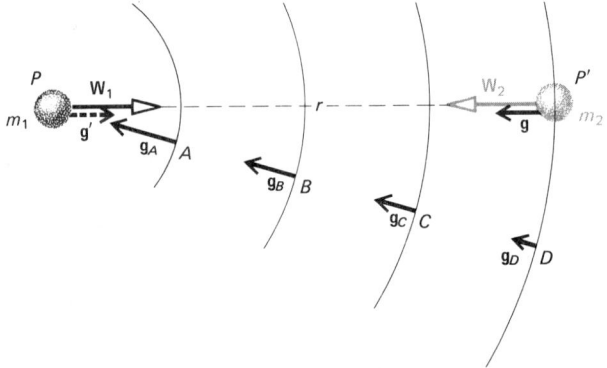

Mass m_1 sets up a gravitational field g in the space surrounding m_2	Gravitational field g acts on mass m_2	A force $W_2 = m_2 g$ will be exerted on m_2
Mass m_2 sets up a gravitational field g' in the space surrounding m_1	Gravitational field g' acts on mass m_1	A force $W_1 = m_1 g'$ will be exerted on m_1

Figure 2–5 Gravitational interaction between two masses.

m_1 are, respectively, $\mathbf{g}_A$, $\mathbf{g}_B$, $\mathbf{g}_C$, and $\mathbf{g}_D$. The test mass m_2 located at P' a distance r from mass m_1 experiences a gravitational field $\mathbf{g}$ set up at that point by mass m_1. This field acts on m_2 and produces a gravitational force $\mathbf{F}_2 = m_2\mathbf{g}$. This force is always an attractive force directed toward m_1. Following the same analysis, we see that m_2 exerts a gravitational pull $\mathbf{F}_1 = -m_1\mathbf{g}$ on m_1. The forces $\mathbf{F}_1$ and $\mathbf{F}_2$ are equal and opposite in direction, in agreement with the principle of action and reaction, $\mathbf{F}_1 = -\mathbf{F}_2$.

As we stated earlier, in addition to the gravitational force interactions, there are three other force interactions—electromagnetic, strong or nuclear, and weak. The relative strengths of these interactions are shown in Table 2–1.

Table 2–1 Force interactions

INTERACTION	RELATIVE STRENGTH
gravitational	1
weak (nuclear)	10^{27}
electromagnetic	10^{38}
strong (nuclear)	10^{40}

Although gravitational interactions are the weakest, their peculiar property of total gravitational force increasing without limit as the attracting masses increase makes gravitational forces among the most obvious of everyday life. They were built into a universal theory of gravitation by Newton in the seventeenth century. Electromagnetic forces were known to the ancients in the attraction of the lodestone for magnetic materials such as iron and in the attraction or repulsion of small bits of materials by glass or resin that has been rubbed with silk. Augustus Augustinus (St. Augustine) was the first to note the difference between magnetic and electrical forces in these examples. Many centuries later, Faraday, Maxwell, Lorentz, and others quantified the concept of the electromagnetic field. The nuclear field was discovered by Rutherford in his historic experiments using gold foils to scatter α particles from radioactive sources. The weak nuclear field is involved in the β decay of the elementary particles and of atomic nuclei and was first described quantitatively by Fermi in his theory of β decay developed in the 1930s.

Gravitational and electromagnetic interactions explain most of the phenomena taking place in the macroscopic world. Thus, these interactions were the first to be understood. On the other hand, the strong and weak interactions can be thought of as the proper working models for phenomena in the microscopic world.

In brief, then, we again state that the subject of physics can be defined as the study of the different types of interactions between particles and the conservation laws. The elementary conservation laws discussed in this chapter form the foundations of theoretical physics. The intrinsic conservation laws, such as conservation of parity, conservation of nucleons, and so on, will be developed and studied in later chapters as we need them.

We shall state without proof that every conservation law arises from some unique symmetry property of a field or of space-time itself.

PROBLEMS

2–1 A particle of mass $m_1 = 2.0$ kg has a velocity $v_1 = 3\hat{i} + 5\hat{j}$ m/sec, and a second particle of mass $m_2 = 6.0$ kg has a velocity $v_2 = 4\hat{i} + 2\hat{j}$ m/sec. What is the total momentum of the system composed of these two particles?

2–2 A neutron with a velocity of $8.0 \times 10^6 \hat{i}$ m/sec makes a head-on elastic collision with a helium nucleus initially at rest. Determine the momentum and velocity of the helium nuclei after the collision.

2–3 Show that for a planet orbiting the sun under the influence of radial forces only, the angular momentum of the planet is conserved.

2–4 An α particle with a speed of 6.0×10^5 m/sec makes an elastic collision with a carbon atom initially at rest. The α particle is scattered at an angle of $60°$ from the original direction and the carbon atom at an angle of $30°$ on the other side of the initial direction. The mass of the carbon atom is three times that of the α particle. Find the speed of the carbon atom after the collision.

2–5 A 3.0-kg mass with a velocity of $v = 9.6\hat{i} + 12.8\hat{j}$ m/sec strikes a wall perpendicular to the x axis. Assume that this is a perfectly elastic collision and determine the impulse given to the mass as a result of the collision.

2–6 A 5.0-kg mass with a velocity of $v_1 = 20\hat{i}$ m/sec collides with a 4.0-kg mass traveling with a velocity $v_2 = -65\hat{i}$ m/sec. If these masses become embedded and stick together as one piece, find the velocity of the combination after the collision. What is the impulse given the 5.0-kg mass?

2–7 A particle of mass $m = 2.0$ kg moves with a constant velocity $v = 20\hat{i}$ m/sec. If it passes the point P (0, 10 m) at time $t = 0$, find its angular momentum with respect to the origin when $t = 1.0$ sec and when $t = 3.0$ sec.

2–8 An electron revolves in a circular path of radius 5.3×10^{-11} m with a speed of 2.2×10^6 m/sec. What is the magnitude of the linear momentum of the electron? What is the magnitude of the angular momentum?

2–9 An astronaut uses a unicycle as an exerciser. If the astronaut and the unicycle are "floating" as he exercises, describe the resulting motion. What happens when he suddenly stops the turning wheel?

2–10 How much work is required to accelerate a 0.012-kg mass from a speed of 200 m/sec to a speed of 380 m/sec?

2–11 Binary stars of equal mass rotate about their center of mass. Discuss the conservation of momentum, linear and angular, for this system.

RECOMMENDED READING

COHEN, I., "Isaac Newton," *Sci. Am.*, December 1963.

FEINBERG, G., and GOLDHABER, M., "The Conservational Laws of Physics," *Sci. Am.*, October 1963.

FEYNMAN, R. P., LEIGHTON, R. B., and SANDS, L. M., *The Feynman Lectures on Physics*, Addison-Wesley, Reading, Mass., 1963, Vol. I, Chaps. 5 and 8.

HESSE, M., "Resource Letter on the Philosophical Foundation of Classical Mechanics," *Am. J. Phys.* **32,** 905 (1964).

LINDSAY, R. B., and MARGENAU, H., *Foundations of Physics*, Wiley, New York, 1936.

MCCUSKEY, S. W., *An Introduction to Advanced Dynamics*, Addison-Wesley, Reading, Mass., 1962, Chap. 1.

MAGIE, W. F., *A Source Book on Physics*, Harvard University Press, Cambridge, Mass., 1963, pp. 212–220, 228–236.

RABINOWICZ, E., "Resource Letter on Friction," *Am. J. Phys.* **31,** 897 (1963).

YOUNG, Hugh D., *Fundamentals of Mechanics and Heat*, McGraw-Hill, New York, 1964.

3
Classical Relativity

Galileo Galilei
(1564–1642)

Born in Pisa, Italy, and educated at the University of Pisa, Galileo built a mathematical physics that would be valid on a moving earth. His most famous treatise, Dialogues Concerning Two New Sciences *(1638) contains a detailed study of motion. By expressing his results in concise mathematical language, he set the example for all future scientists. Galileo was sentenced to permanent house arrest at his home near Florence in 1633 because he actively endorsed the Copernicanism that was strongly opposed by church leaders.*

3–1 LIMITS OF "COMMON SENSE"
3–2 CLASSICAL PRINCIPLE OF RELATIVITY
3–3 INVARIANCE OF THE CONSERVATION OF LINEAR MOMENTUM
3–4 INVARIANCE OF NEWTON'S LAWS

3-1 LIMITS OF "COMMON SENSE"

In the previous chapter, we began our study of physics, and hence of the physical world, by considering the effect that mechanical forces have on the objects in the universe. How forces act in making objects move or in changing their motion is the subject of that branch of physics called *mechanics*. We have all observed these phenomena from earliest childhood; we have all experienced forces and accelerations of objects. For example, we have learned to "lean in" to a curve when we are riding or skating or running. We have experienced momentum as we coast down and then up a long hill on a bicycle. We have observed the effects of forces generated against a surface: for example, when a rock or a ball strikes a surface and its momentum vector is very suddenly changed, especially if the surface happens to be a glass window. These events constitute our common experience, and so they become elements of what we refer to as our common sense.

Physicists and other scientists are trying to stretch these common sense lessons and apply them to the very large (the macroscopic), the very small (the microscopic), the very fast, and even the very distant objects. However, true progress in these attempts has been made only in recent centuries, as scientists learned to codify common experience into a set of general laws that could be expressed as equations. These equations can be applied to describe a vast amount of our commonly experienced universe.

The measurements by the early astronomer, Tycho Brahe, of the motions of the planets through the sky furnished Johannes Kepler with sufficient data in the sixteenth century so that he could describe the paths of the planets as being orbits around a common center. Then, toward the end of the seventeenth century, Newton developed his theory of mechanics using some of the concepts of kinematics originated by Galileo at the turn of the sixteenth century. Newton was able to show that the empirical laws of planetary motion formulated by Kepler had a physical basis in the laws of gravitational force.

The physicists of the seventeenth century were greatly encouraged to find that, evidently, the same laws of mechanics that describe the path of a rock thrown through the air could also describe the swing of the planets around the sun. They were then able to stretch legitimately their "common senses" to vast distances.

As twentieth century physicists, however, as we go to greater distances or to larger masses, or if we consider *very* small objects and objects that travel at high speeds, we shall find that our common senses no longer apply! Thus, we shall discover that the "laws" governing the world around us are, in truth, only approximations to a larger set of laws that cover a wider realm of nature. We shall find even then that this larger set of laws is still far from being a truly universal set of laws with which to describe the vast universe in all of its details. However, our common senses do apply with excellent precision to a large portion of nature, and a special term is used to cover this realm of physics: *classical*.

The classical mechanics, like other branches of mechanics that have been developed, depends on the type of conservation statements that assert that some quantity remains the same throughout a change in the motion of an object. For instance, the mass of a ball remains the same after it has been hit as it was before it was hit. Common sense dictates that nature should be the same, fundamentally, for the man riding by in a train as well as for the man standing by the side waiting for the train to pass. A distance, say of 1 m, on the train should be the same as an equivalent distance measured on the ground, and the watch in the pocket of the man on the train should keep the same time as does the watch belonging to the man standing by the side. These are the conclusions of the common sense or classical approach to nature. Let us now derive a fundamental statement of this approach.

3-2 CLASSICAL PRINCIPLE OF RELATIVITY

The definition of the term *relativity* provides the basic concept underlying much of physics, and yet we find that it seems so simple as to be almost trite. By relativity we mean the appearance of nature to one observer *and* its relationship to the appearance of nature to another observer, who may be moving with respect to the first one. We feel that it seems to be just common sense that the relative state of motion of an observer should not be allowed to change the laws of nature. If an observer's state of motion could change the laws, then we must ask ourselves if an infinite set of laws exists, or if no law at all exists? So we state our faith in our own common sense and in the stability of nature, and we state the *classical principle of relativity*: All the laws of nature must be the same for all observers who move relative to one another with constant velocity. If the relative motion is not constant, then it is accelerated, and the situation becomes more complicated, falling into the realm of *general relativity*.

We shall now derive the classical principle of relativity in somewhat more formal terms. In Figure 3-1 the two frames of reference or systems, S_1 and S_2, are moving relative to one another. For simplicity, they are oriented so that their respective x_i, y_i, and z_i axes are parallel, and the relative velocity vector **v** is parallel to the x_1 and the x_2 axes. Also, it is assumed that the clocks in S_1 and S_2 run at the same rate, and that they are synchronized to read $t = 0$ when the origins of the two systems coincide. Thus, we realize that it will not be necessary to write t_1 or t_2, since time is the same in each frame of reference, and it is sufficient to write t for the time in either system.

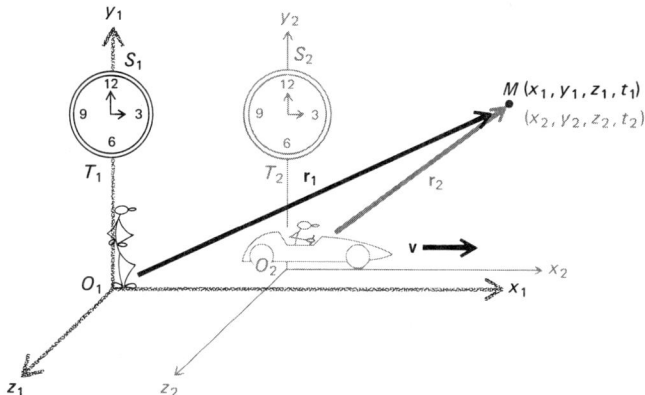

Figure 3–1 A point M, moving in space and time, is observed from a stationary system S_1 and from a system S_2 that is moving with a velocity **v** relative to S_1.

It is important that we understand in this context this seemingly trivial matter of reading the time from clocks. We shall discover later that perhaps this matter may be a bit more complicated than our original analysis. In the classical relativity, then, an observer O_1 in system S_1 reads the same time on his clock T_1 as he reads on the clock belonging to observer O_2 in system S_2. Conversely, observer O_2 reads the same time on the clock of observer O_1 as he does on his own clock T_2.

The "sameness" of the time as read in either system from the other is a basic assumption. At first this assumption might appear to be ascribed to mere "common sense." We should make some additional common sense assumptions about the space that S_1 and S_2 occupy in common. We assume that vectors of unit length, namely $\hat{\mathbf{i}}_1$ and $\hat{\mathbf{i}}_2$, may be layed off on the axis x_1 and on the axis x_2, and that $\hat{\mathbf{i}}_1$ always equals $\hat{\mathbf{i}}_2$ regardless of the value of t or of **v**. Our assumption is that a unit vector is a unit vector, regardless of the frame in which it is seen or measured, and that a unit vector always remains a unit vector. Initially, this statement may sound like pure common sense until we examine the way in which the "length" of a vector is measured. However, for the present we shall bypass this question and rest on our common sense assumption.

In order that we may treat all three axes equally, let us assume that a unit vector $\hat{\mathbf{j}}_1$ lies on the y_1 axis and $\hat{\mathbf{j}}_2$ lies on the y_2 axis, and that $\hat{\mathbf{j}}_1 = \hat{\mathbf{j}}_2$ for all t no matter which observer does the measuring. Finally, let unit vectors $\hat{\mathbf{k}}_1$ and $\hat{\mathbf{k}}_2$ lie along the z_1 and z_2 axes, respectively, with the same equality relationship as stated for the other unit vectors. Now consider the two frames of reference S_1 and S_2 as pictured in Figure 3–1, and let us drop the subscripts on the unit vectors since the vectors are the same in both systems.

Let us now imagine that an event is happening at M, a point in space and time that can be observed from both S_1 and S_2. This happening occurs at a time t as read on any clock in either of the two systems. Then, looking at Figure 3-1, we may write the vector equation

$$\mathbf{r}_1 = (O_1 O_2)\hat{\imath} + \mathbf{r}_2 \tag{3-1}$$

where $(O_1 O_2)$ is the distance from the origin of S_1 to the origin of S_2 at the time t of the happening. Since all of our clocks were started when the origins coincided, we may write this as

$$(O_1 O_2)\hat{\imath} = vt\hat{\imath} \tag{3-2}$$

Furthermore, the position vectors in S_1 and S_2 may be written in terms of their components as

$$\mathbf{r}_1 = x_1\hat{\imath} + y_1\hat{\jmath} + z_1\hat{k} \tag{3-3}$$

and

$$\mathbf{r}_2 = x_2\hat{\imath} + y_2\hat{\jmath} + z_2\hat{k} \tag{3-4}$$

where (x_1, y_1, z_1) are the coordinates of M in S_1 at t, and (x_2, y_2, z_2) are the coordinates of the same point M but in S_2 at t. Now, substitution of Equations (3-2), (3-3), and (3-4) into Equation (3-1) gives

$$x_1\hat{\imath} + y_1\hat{\jmath} + z_1\hat{k} = (x_2 + vt)\hat{\imath} + y_2\hat{\jmath} + z_2\hat{k} \tag{3-5}$$

Since $\hat{\imath}$, $\hat{\jmath}$, and $\hat{k}$ are orthogonal (or functionally independent objects), Equation (3-5) may be written as three simultaneous equations with the addition of a trivial one that was defined at the outset:

$$\boxed{\begin{aligned} x_1 &= x_2 + vt \\ y_1 &= y_2 \\ z_1 &= z_2 \\ t_1 &= t_2 \end{aligned}} \tag{3-6}$$

We note that only the coefficients in Equation (3-5) appear in the Set (3-6). The vector components on each side have been cancelled out. The Set (3-6) is the first example of a *coordinate transformation* that we have encountered. Let us examine the meaning behind this transformation. It tells the observer in S_1 how to relate the S_1 coordinates of M to the S_2 coordinates of M that he, the S_1 observer, measures in both reference systems. If the S_2 observer is to relate the coordinates in his frame to the coordinates that he measures in the S_1 frame, then the same transformation holds, but in reverse. The inverse of the Set (3-6) is

$$\boxed{\begin{aligned} x_2 &= x_1 - vt \\ y_2 &= y_1 \\ z_2 &= z_1 \\ t_2 &= t_1 \end{aligned}} \tag{3-7}$$

These two Sets (3–6) and (3–7) of simultaneous equations represent part of what is known as the *Galilean transformation group*. If we considered all the possible different ways that the two systems could be related to one another, we would include linear displacements along the y and z axes of the same sort that we have just described along the x axis. In addition, we would consider rotations about the different axes of varying angles and we would also consider reflections through the origin and in each direction. Taken all together, these relationships form a *group*. The properties of the group, when displayed in an algebraic fashion, are a representation of what is generally referred to as the *Galilean group*. However, we shall not deal with rotations and reflections here, but we shall deal with the transformations, sometimes referred to as mappings, corresponding to linear translations due to a constant vector velocity **v**.

We now extend our theory of Galilean transformations to include dynamic as well as static effects; we shall find out how velocities are to be understood when observed from the different frames. Let us imagine that our event at point M is now in motion away from M at time t. Then the event's velocity relative to S_1 is $\mathbf{v}_1$, where

$$\mathbf{v}_1 = \frac{d\mathbf{r}_1}{dt} \tag{3-8}$$

and the velocity of the event relative to S_2 is

$$\mathbf{v}_2 = \frac{d\mathbf{r}_2}{dt} \tag{3-9}$$

Now, substituting Equation (3–2) into (3–1) and differentiating with respect to t, and then using Equations (3–8) and (3–9), we obtain the vector equation

$$\boxed{\mathbf{v}_1 = \mathbf{v}_2 + \mathbf{v}} \tag{3-10}$$

We recommend that as an exercise Equation (3–10) be written in its component form [compare Equations (3–5) and (3–6)]. In either case, that is, as a vector equation such as the given form of Equation (3–10) or the same equation as a set of simultaneous equations in an equivalent component form, Equation (3–10) may be called the *Galilean (or classical) composition of velocities*. Equation (3–10), of course, has an inverse [compare Equations (3–6) and (3–7)],

$$\boxed{\mathbf{v}_2 = \mathbf{v}_1 - \mathbf{v}} \tag{3-11}$$

Let us differentiate with respect to time once again, remembering that system S_2 is moving with constant velocity **v** with respect to S_1. The same answer is obtained from either Equation (3–10) or Equation (3–11), so for both observers

$$\frac{d\mathbf{v}_1}{dt} = \frac{d\mathbf{v}_2}{dt} \tag{3-12}$$

or

$$\boxed{a_1 = a_2} \qquad (3\text{--}13)$$

Thus, accelerations appear to be the same as seen from either frame. We say that acceleration is an *invariant* with respect to a Galilean transformation. Since mass is also an invariant in these types of transformations, the product of mass and acceleration, or force, is also an invariant with respect to a Galilean transformation.

We have been using some new terminology in the past few pages, and we should note that hidden within that terminology are some very new thoughts and concepts. The conservation laws stated that certain quantities such as energy or momentum remain constant in total "quantity" before, during, and following a given interaction. Such interactions are examples of translations in time, and the conservation laws are statements about the invariance of some quantity under these translations. In this chapter, on the other hand, we have discussed invariance under a complete change of spatial frame. In the former case, the events happened in a single frame. In this section, we broadened our scope to include the relationship of two or more such frames moving relative to one another. In the next few chapters, we shall consider transformations between two such frames in more detail. In the course of this process, we shall expand on our "common sense."

3-3 INVARIANCE OF THE CONSERVATION OF LINEAR MOMENTUM

In Figure 3-2, the two particles of masses m and m' form an isolated system with no external forces. Let S_1 be an inertial frame of reference and S_2 another frame moving relative to S_1 with constant velocity $\mathbf{v}$. Then for system S_1, the conservation of momentum law states that

$$m\mathbf{v}_1 + m'\mathbf{v}_1' = \text{constant} \qquad (3\text{--}14)$$

where $\mathbf{v}_1$ and $\mathbf{v}_1'$ are the velocities of m and m', respectively. Thus, the value of the number given by the sum $m\mathbf{v}_1 + m'\mathbf{v}_1'$ at time t remains unchanged for any subsequent time, provided that no external forces appear.

Now let $\mathbf{v}_2$ and $\mathbf{v}_2'$ be the respective velocities of the same two particles relative to S_2. We know that according to the Galilean composition of velocities

$$\begin{aligned}\mathbf{v}_1 &= \mathbf{v} + \mathbf{v}_2 \\ \mathbf{v}_1' &= \mathbf{v} + \mathbf{v}_2'\end{aligned} \qquad (3\text{--}15)$$

Substituting Equation (3-15) in Equation (3-14) shows that

$$m(\mathbf{v} + \mathbf{v}_2) + m'(\mathbf{v} + \mathbf{v}_2') = \text{constant}$$

or

$$m\mathbf{v}_2 + m'\mathbf{v}_2' = \text{constant} - (m + m')\mathbf{v}$$

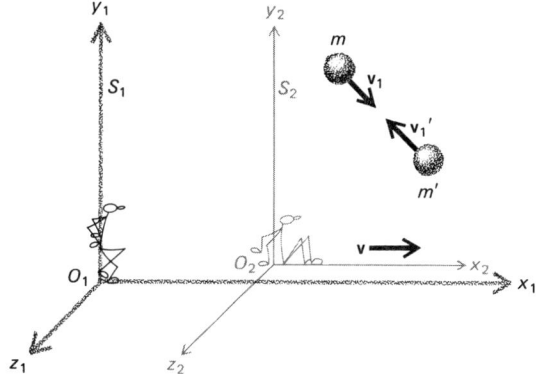

Figure 3-2 The total momentum of particles m and m' is invariant in form when transformed to inertial system S_2.

Finally,

$$m\mathbf{v}_2 + m'\mathbf{v}_2' = \text{constant} \tag{3-16}$$

since $(m + m')\mathbf{v} = $ constant. Thus, by comparing Equations (3-14) and (3-16) we see that *conservation of linear momentum remains invariant for all inertial systems moving relative to each other with constant velocity.*

3-4 INVARIANCE OF NEWTON'S LAWS

Let us consider again a particle of mass m with velocities $\mathbf{v}_1$ and $\mathbf{v}_2$ as seen from reference frames S_1 and S_2, respectively, where $\mathbf{v}$ is the constant velocity with which S_2 moves relative to S_1 (Figure 3-3).

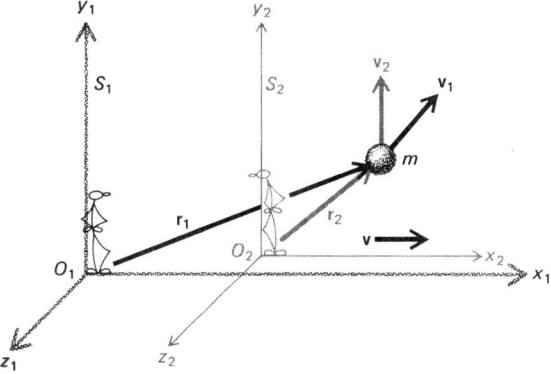

Figure 3-3 A particle of mass m moving with a velocity $\mathbf{v}_1$ in system S_1 and velocity $\mathbf{v}_2 = \mathbf{v}_1 - \mathbf{v}$ in system S_2.

We recall that according to the classical principle of relativity, the Galilean composition of velocities is again

$$\mathbf{v}_1 = \mathbf{v} + \mathbf{v}_2 \tag{3-17}$$

Since $d\mathbf{v}/dt_1 = 0$,

$$\frac{d\mathbf{v}_1}{dt_1} = \frac{d\mathbf{v}_2}{dt_2}$$

or

$$\mathbf{a}_1 = \mathbf{a}_2 \tag{3-18}$$

Thus, $m\mathbf{a}_1 = m\mathbf{a}_2$ and the two forces,

$$\mathbf{F}_1 = m\mathbf{a}_1$$
$$\mathbf{F}_2 = m\mathbf{a}_2$$

are the same in each system. Thus, we have shown that *Newton's second law of mechanics is invariant for all inertial frames moving relative to each other with constant velocity.*

Repeating the same reasoning, it can be shown that the other fundamental laws of mechanics—the conservation of angular momentum and the conservation of energy—also remain invariant for all inertial frames moving relative to each other with constant velocity. Before stating our conclusion, let us give one useful definition: An *inertial observer* is an observer at rest relative to an inertial frame. Hence, the classical principle of relativity can be stated: *All laws in mechanics remain invariant for all inertial observers moving relative to each other with constant velocity.*

EXAMPLE 3–1: A bomb is dropped from an airplane that is flying at an altitude of $h = 2000$ m and with a horizontal and constant speed of $v = 150$ m/sec (see Figure 3–4). Obtain the equations of (a) motion, (b) velocity, and (c) acceleration of the bomb as seen by a terrestrial observer O_1 in a stationary reference frame $S_1(x_1, y_1)$ and by the pilot O_2 in a moving frame $S_2(x_2, y_2)$.

SOLUTION

(a) *Equations of motion.* The acceleration of the bomb as seen by the terrestrial observer is just $g = 9.80$ m/sec^2 (the acceleration of gravity). As the bomb is released from the airplane, its horizontal velocity remains constant at $v = 150$ m/sec as measured by the terrestrial observer.

After t sec, the airplane has moved from O_2 to O_2' [Figure 3–4(b)], and the bomb will be at A just below it. If x_1 and y_1 are the coordinates of the bomb as

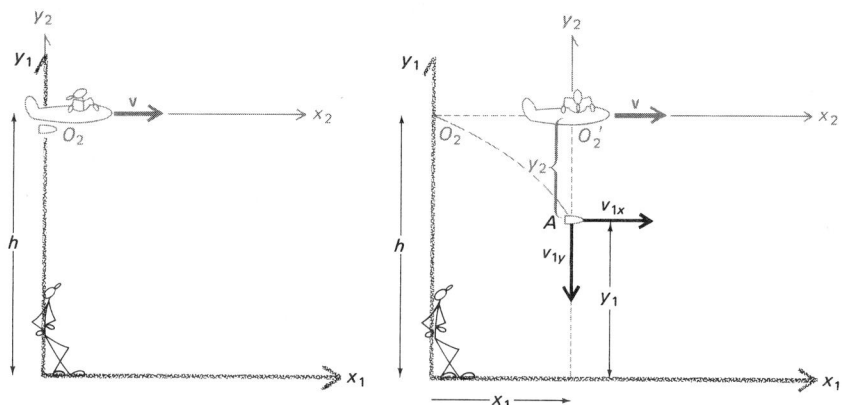

(a) Bomb is dropped (b) Bomb after t sec

Figure 3-4 Bomb dropped from an airplane as seen by a stationary observer and by the pilot.

measured from S_1, the motion of the bomb viewed by the terrestrial observer is

$$x_1 = v_{1x}t = 150t$$
$$y_1 = h - \tfrac{1}{2}gt^2 = 2000 - 4.9t^2$$

The *pilot* sees

$$x_2 = 0$$
$$y_2 = -\tfrac{1}{2}gt^2 = -4.9t^2$$

(b) *Velocity.* By differentiating the above equations of motion, we obtain for the terrestrial observer

$$\frac{dx_1}{dt} = v_{1x} = v = 150 \text{ m/sec}$$

$$\frac{dy_1}{dt} = v_{1y} = -gt = -9.8t$$

and for the pilot

$$\frac{dx_2}{dt} = v_{2x} = 0$$

$$\frac{dy_2}{dt} = v_{2y} = -gt = -9.8t$$

These are the respective rectangular components of the velocity as measured by each observer.

(c) *Acceleration.* Similarly, the components of acceleration are, for O_1,

$$\frac{d^2x_1}{dt^2} = a_{1x} = 0$$

$$\frac{d^2y_1}{dt^2} = a_{1y} = -g = -9.8 \text{ m/sec}^2$$

and for the pilot O_2,

$$\frac{d^2x_2}{dt^2} = a_{2x} = 0$$

$$\frac{d^2y_2}{dt^2} = a_{2y} = -g = -9.8 \text{ m/sec}^2$$

These accelerations are in agreement with the Galilean transformation.

PROBLEMS

3-1 An earthbound radar station monitors two very fast rocket ships that approach each other at speeds of $0.60c$ and $0.80c$, respectively, where c is the speed of light. According to Galilean transformations, what is the speed of approach of the two ships as seen by an astronaut on one of the ships? (See also Problem 4-15.)

3-2 A particle in a stationary system S_1 has a position given by

$$x_1 = 30t_1 + 10t_1^2$$

where t_1 is in seconds and x_1 is in meters. Find expressions for the position, velocity, and acceleration as measured by an observer moving in the positive x direction at a speed of 10 m/sec. Assume that $t_1 = t_2 = 0$ when the systems S_1 and S_2 coincide.

3-3 Prove that the conservation of angular momentum remains invariant under a Galilean transformation.

3-4 Two balls of masses m_a and m_b move parallel to the x axis in a system $S_1(x_1, y_1, z_1)$ with speeds v_a and v_b, respectively. For an elastic collision between these balls, show that the kinetic energy is also conserved in a second system $S_2(x_2, y_2, z_2)$ moving with constant velocity v in the x_1 direction.

3-5 An elevator is moving vertically upwards at a constant speed of 5.0 m/sec. When the elevator is 10 m above the ground, a person on the ground throws a ball vertically upwards at 20 m/sec. Write the expressions representing the position, velocity, and acceleration of the ball relative to the person on the ground and to a person riding the elevator.

3-6 At $t_1 = 0$ a ball is projected from O_1 in stationary system S_1 with an initial speed $v_0 = 30$ m/sec at an angle of 60° as shown in Figure 3-5. Systems S_1 and S_2 at $t_1 = 0$ coincide, and system S_2 moves in the positive x_1 direction with a speed of 10 m/sec. Write expressions for position and for rectangular components of the velocity and of the acceleration of the ball as seen from systems S_1 and S_2.

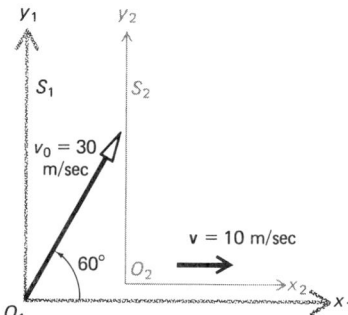

Figure 3-5

3-7 Two points A and B are located 2.0 km apart on the same bank of a river. Of two men making the round trip from A to B and back to A, the first rows a boat at 8.0 km/h relative to the river water, while the second walks along the shore at 8.0 km/h. (a) If the velocity of the river is 4.0 km/h from A to B, what is the time for each man to complete the trip? (b) What is the velocity of the walking man relative to the man in the boat on the trip from A to B?

3-8 A man who can row a boat 5.0 km/h in still water wishes to cross a river that is 1.0 km wide and that flows at a speed of 3.0 km/h. (a) At what angle with the shore must he direct the boat in order to reach the shore directly opposite him? (b) Compute the speed of the boat relative to the shore. (c) What is the time required to cross the river?

3-9 In Figure 3-6 a river of width L flows with a constant speed v. Swimmer A makes a round trip SRS parallel to the shore, and swimmer B makes a round trip STS perpendicular to the shore. If the speed of each swimmer relative to the water is c, show that (a) the traveling time for round trip SRS is

$$t_\| = \frac{2Lc}{c^2 - v^2}$$

(b) the traveling time for round trip STS is

$$t_\perp = \frac{2L}{\sqrt{c^2 - v^2}}$$

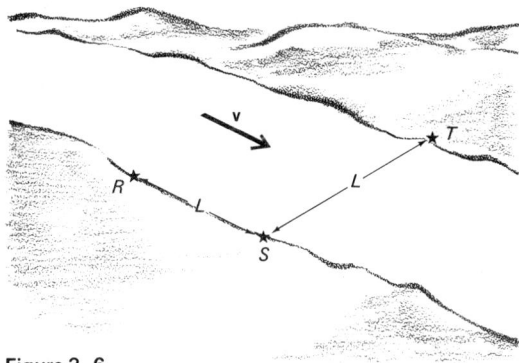

Figure 3–6

3–10 Two boys are tossing identical balls, each of mass 0.080 kg, in the aisle of an airplane traveling at 150 m/sec. Each boy tosses a ball to the other at speeds of 20 m/sec relative to the plane. While the balls are in flight, determine the total momentum and kinetic energy as measured by (a) a passenger on the plane, and (b) an observer on the earth. Are the momentum and kinetic energy invariant? Explain.

3–11 A radioactive atom emits an α particle at a speed of 5.0×10^6 m/sec relative to the atom. If the atom moves in the opposite direction with a velocity of 3.0×10^5 m/sec relative to the laboratory frame, determine the kinetic energy and momentum of the α particle as observed from (a) the moving atom, and (b) a stationary laboratory observer.

3–12 A system $S_2(x_2, y_2)$ is moving with a uniform translational motion relative to system $S_1(x_1, y_1)$ with a constant velocity of 30 m/sec parallel to the x axis. The corresponding axes in both systems are, respectively, parallel to each other. Two balls of masses $m_1 = 2.0$ kg and $m_1' = 3.0$ kg are moving relative to the S_1 frame with velocities $\mathbf{v}_1 = 3\hat{\imath}_1 + 4\hat{\jmath}_1$ (m/sec) and $\mathbf{v}_1' = 5\hat{\imath}_1 + 12\hat{\jmath}_1$ (m/sec). Compute (a) the velocities of the two balls relative to S_2, (b) the total linear momentum relative to S_1 and to S_2, respectively, and (c) the total kinetic energy relative to the systems S_1 and S_2.

RECOMMENDED READING

ALONSO, M., and FINN, E. J., *Fundamental University Physics*, Addison-Wesley, Reading Mass., 1968, Vol. 1.
Text with a good section on relativity and problems to go with it.

BONDI, H., *Relativity and Common Sense*, Doubleday, New York.
A readable introduction to the special theory of relativity.

BUCHDAHL, G., "Science and Logic: Some Thoughts on Newton's Second Law of Motion," *Brit. J. Phil. Sci.* **2,** 217 (1951).

DRAKE, S., "Galileo and the Law of Inertia," *Am. J. Phys.* **32,** 601 (1964).

DURELL, Clement V., *Readable Relativity*, Harper & Row, New York, 1960.

EINSTEIN, Albert, and INFELD, Leopold, *The Evolution of Physics*, Simon and Schuster, New York, 1938.
Good background reading for an introduction to relativity.

GALILEI, Galileo, *Dialogue on Two New Sciences*, H. Crewe, Trans., Macmillan, New York, 1939.

LANDAU, L. D., and RUMER, G. B., *What is Relativity?*, Oliver and Boyd, Edinburgh and London, 1960.
Small, readable, introductory book on the subject.

SEARS, Francis W., and BREHME, Robert W., *Introduction to the Theory of Relativity*, Addison-Wesley, Reading, Mass., 1968.
Clearly written text with examples and many problems.

Special Relativity Theory, Selected Reprints, American Institute of Physics, New York, 1963.
Contains many references and several excellent representative papers on the special theory of relativity.

4
The Michelson-Morley Experiment

Albert Abraham Michelson
(1852–1931)

Born in Strelno, Germany, Michelson emigrated to the United States. In 1869 he was appointed to the U.S. Naval Academy. As an instructor there (1875–1879), he performed his initial experiments on the speed of light. At the Case School of Applied Science (1883–1889), he determined the speed of light with an extremely high degree of accuracy. In 1920 Michelson first measured the diameter of a star. For his precision optical instruments and the investigations he made with them, he was awarded the 1907 Nobel Prize in physics.

4–1 CONFLICT DEVELOPS
4–2 LORENTZ TRANSFORMATIONS
4–3 LORENTZ COMPOSITION OF VELOCITIES

4-1 CONFLICT DEVELOPS

In the latter part of the nineteenth century, Maxwell and Hertz proposed the conception of light as electromagnetic radiation. Since then, physicists have investigated the many properties of light. Once it was known that light had wave properties, physicists naturally felt that it was necessary to propose a medium that would propagate this wave motion, that is, something in which the waves of light would travel. This medium was generally known as the *luminiferous ether*. To qualify as the carrier of light waves, it would be necessary for the ether to have some very strange properties. The ether was postulated to be a substance lighter than any known gas or vapor, and at the same time to have a rigidity comparable to that of steel.

In 1887 ALBERT A. MICHELSON and E. W. MORLEY devised and performed an experiment to test the nature of the luminiferous ether and to attempt to determine the velocity of light with respect to the ether. Physicists realized that if this ether existed, it must permeate all space and it must be the primary and absolute reference system for light, just as a pond of water is the reference system for waves traveling on its surface. They concluded that the earth then must either be at rest in this ether or moving relative to the ether, and that consequently the inertial reference frame for light is either at rest or is moving relative to the earth.

To perform such an experiment, a precise optical instrument was needed. The interferometer* is an instrument that had been developed to measure the phase, or positions, of wave peaks along a beam of light, and from these measurements the distance from one peak to the next could be deduced. This instrument is also capable of performing many other precise and interesting measurements. Figure 4–1 shows a schematic of the interferometer. Notice that a half-silvered mirror M divides the incident beam of light into two component beams that are then traveling 90° apart. These two beams are said to be *coherent* because they originate from the same original beam, and each portion of one beam's light waves has a constant phase difference with respect to the light waves making up the other beam. These two beams are next reflected from the fully silvered front surface mirrors M_1 and M_2 and are then returned to the observer via mirror M. If the two beams travel

---

* See A. A. Michelson, *Studies in Optics*, University of Chicago Press (Phoenix Books), Chicago, 1962.

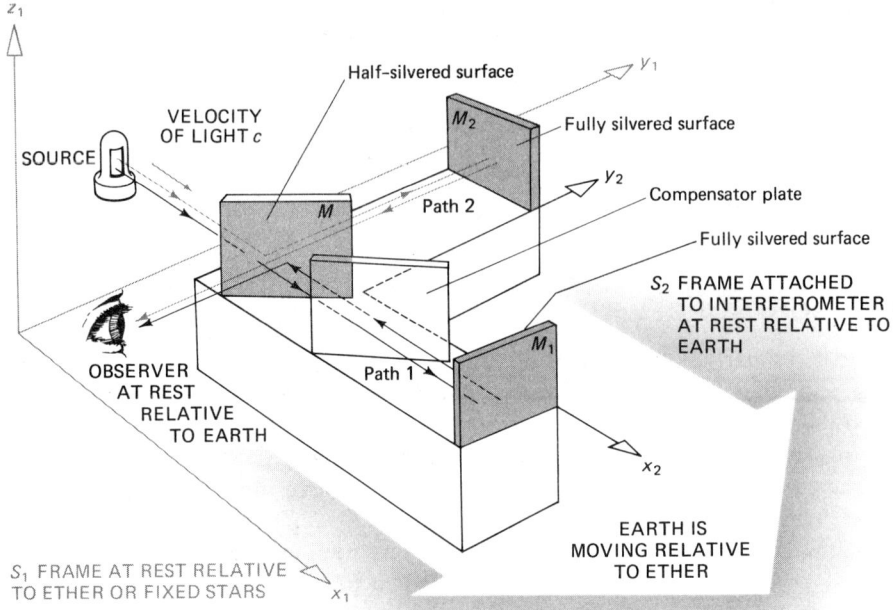

Figure 4–1 Schematic of Michelson interferometer as used to determine the velocity of light relative to earth.

equal optical paths, they will arrive in phase and produce a bright field by constructive interference. If the optical path of one beam is increased by translating either mirror M_1 or M_2 slightly, the beams begin to arrive at the observer more and more out of phase, with a resulting decrease in intensity due to destructive interference. If one mirror is moved a distance of $\lambda/4$ from its original position, the two beams are completely out of phase and interfere destructively to produce a dark field. Notice that a piece of glass, called the compensator plate, has been introduced into the beam traveling path 1. Both beams of light then will travel through a thickness of glass three times before arriving at the observer.

When Michelson and Morley decided to perform an experiment to test the properties of the ether, they concluded that an interferometer would serve their purposes. They wished to design an experiment that would in fact determine if an ether existed and if the ether was moving with respect to the earth. Like the waves on the surface of a moving river, light waves should appear to be moving at different speeds to an observer depending on whether or not the waves are traveling with the river of ether, against it, or across it. If the earth is moving through the ether (or, which is the same thing, the ether is streaming past the earth), an observer should then be able to detect a difference in the speed of light in different directions. To do this, Michelson and Morley constructed a large interferometer, which they floated in a pool of mercury. Then they looked for changes in the speed of light along path 1 relative to path 2 as they changed the interferometer's direction by rotating it in its mercury pool. A relative difference in the speed of light would be indicated by changes in the brightness of fringes in the final beam.

Let us repeat their experiment in our imagination; however, we shall eliminate the many side difficulties that had to be overcome by Michelson and Morley. Let us build a large interferometer with the paths MM_1 (no. 1) = MM_2 (no. 2) = L and float the apparatus on mercury, orienting the axis SM_1 in the direction which the earth is traveling relative to the distant fixed stars. We choose this orientation as a reasonable guess of the direction in which we are traveling through the ether (if it should exist).

The speed of light relative to the ether is c, and from the Galilean transformations the speed of light relative to the earth along the arm of the interferometer parallel to the velocity **v** of the earth is then

$$\begin{aligned} c - v \quad &\text{(from } M \text{ to } M_1\text{)} \\ c + v \quad &\text{(from } M_1 \text{ to } M\text{)} \end{aligned} \tag{4-1}$$

The time involved for each trip of a light wave will be

$$\begin{aligned} t_{MM_1} &= \frac{L}{c - v} \\ t_{M_1 M} &= \frac{L}{c + v} \end{aligned} \tag{4-2}$$

Thus, the time for one complete round trip, $MM_1 M$, in the direction *parallel* to the earth's motion is

$$\boxed{ t_\| = \underbrace{\frac{L}{c - v}}_{\substack{\text{time for} \\ MM_1}} + \underbrace{\frac{L}{c + v}}_{\substack{\text{time for} \\ M_1 M}} = \frac{2L/c}{1 - (v/c)^2} } \tag{4-3}$$

The traveling time for light to make the round trip, $MM_2 M$, a direction perpendicular to the motion of the earth, is

$$\boxed{ t_\perp = \underbrace{\frac{L}{\sqrt{c^2 - v^2}}}_{\substack{\text{time for} \\ MM_2}} + \underbrace{\frac{L}{\sqrt{c^2 - v^2}}}_{\substack{\text{time for} \\ M_2 M}} = \frac{2L}{\sqrt{c^2 - v^2}} = \frac{2L/c}{\sqrt{1 - (v/c)^2}} } \tag{4-4}$$

This equation results from the classical composition of velocities as shown in Figure 4–2. If c is the speed of light relative to the ether in reference frame S_1, then the speed of light relative to the earth (reference frame S_2) in both trips, MM_2 and $M_2 M$, is always $\sqrt{c^2 - v^2}$.

Equations (4–3) and (4–4) give the traveling times for trips $MM_1 M$ and $MM_2 M$ as measured by ourselves, the terrestrial observers. We note that since we

44 · PART ONE: SPACE AND TIME

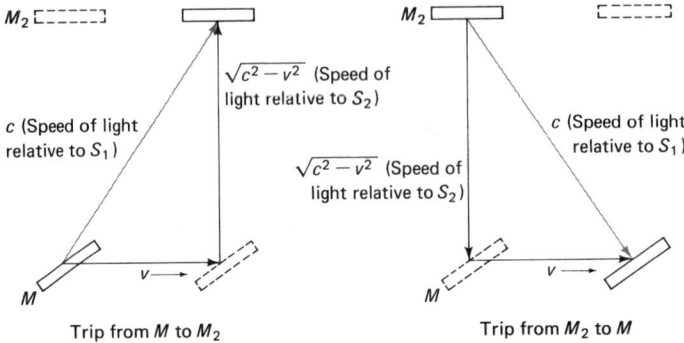

Figure 4–2 Relative motion of light according to classical composition of velocities as it reflects between mirrors M and M_2 in Figure 4–1.

have analyzed the experiment using the classical Galilean transformations, these times must be independent of our motion (i.e., the observer's motion). From Equations (4–3) and (4–4) then,

$$\frac{t_\parallel}{t_\perp} = \frac{(2L/c)/\sqrt{1-(v/c)^2}}{(2L/c)/[1-(v/c)^2]} = \frac{1}{\sqrt{1-(v/c)^2}} \tag{4-5}$$

Thus, $t_\parallel > t_\perp$, and the two portions of the coherent beam should produce an interference pattern when they are brought together.

When Michelson and Morley performed this experiment very carefully in 1887, they expected to observe a shift of at least 0.40 fringe. However, their efforts showed that, at most, the shift was only 0.005 fringe. Thus, they questioned if there was, in fact, an effect to be observed. Since that time, many other very careful experiments to measure the relative speed of light have been performed, but *none* of these have been able to demonstrate the existence of a luminiferous ether. The results always yield experimentally that

$$t_\parallel = t_\perp \tag{4-6}$$

In other words, Equation (4–6) is the experimental answer of nature to the question of whether or not the ether exists, which Michelson and Morley attempted to answer with their experiment.

A conflict arises, however, because—according to the Galilean analysis—an observer performing the experiment should observe that $t_\parallel > t_\perp$, and this result was not observed. On the other hand, if the Galilean composition of velocities is then rejected, and we accept that *the speed of light is the same for both inertial systems S_1 and S_2*, then we have

$$t_\parallel = \frac{2L}{c}$$

$$t_\perp = \frac{2L}{c}$$

and thus

$$t_\parallel = t_\perp$$

This result is in agreement with the results of many experiments. The results of the Michelson–Morley experiment, therefore, forced physicists to accept the invariance of the speed of light. Thus, we conclude that *the speed of light is the same, regardless of whether the speed is measured by an observer in a stationary system or by an observer in a system moving at a constant velocity relative to the light source.*

The Michelson–Morley experiment was a crucial one, because the "negative" results that it produced started a revolution in conceptual thinking in physics. A demand for deeper insight into the nature of space and time was created. Space and time are, after all, the framework into which all of nature is fitted. Perhaps much or even most of the events that we observe around us and that we call "natural" are just manifestations of different properties of space and time. As physicists, we shall sum up these properties and study them under the title of "transformations." That is, one question that a physicist usually asks is, "How will *this* particular natural event look if I observe it to happen from some other frame of reference—some situation in which I might be traveling or accelerating or rotating relative to the laboratory in which I am presently at rest?" This question is often difficult and, at times, even impossible to answer. By seeking answers to these problems of transformations, physicists have made great progress in their understanding and defining of physics in the past few decades.

The conclusions, particularly the invariance of the speed of light, resulting from the Michelson–Morley experiment were the experimental support for Einstein's relativity theory.* The results of this experiment and Einstein's work started a trend of investigation of the transformation properties of all of nature. The efforts of scientists to attain a more complete understanding of the nature of space and time are still at the forefront of physics. This quest was soundly established with the motion equations of Galileo and Newton and began to expand still further with those of Lorentz.

4–2 LORENTZ TRANSFORMATIONS

At this point, in connection with experiments dealing with light, we are forced to reject the use of Galilean transformations except as an approximation to the truth and to look for other more general and more compatible equations. We remind

---

*For an engrossing discussion of the genetic link between Michelson's experiments and Einstein's theory, see Gerald Holton, "Einstein and the 'Crucial' Experiment," *Am. J. Phys.* 37, 968 (1969).

ourselves that if $v/c \to 0$ (i.e., if v is small), then Equation (4–5) becomes $t_{\parallel} = t_{\perp}$. On the other hand, for large speeds (i.e., $v/c \to 1$), we are forced to reject the Galilean transformations. They can still be considered a good approximation, however, in the world of slower motion. Consider Figure 4–3, where an inertial system S_1 is at rest and an inertial system S_2 is moving with a uniform translational motion (i.e., $\mathbf{v}$ = constant). At time $t_1 = t_2 = 0$ both frames coincide, and the clocks are perfect and synchronized.

At time $t_1 = t_2 = 0$ a flash of light is emitted from the common origin of S_1 and S_2. Let M represent a point of advance of the light beam that has space and time coordinates (x_1, y_1, z_1, t_1) and (x_2, y_2, z_2, t_2) in systems S_1 and S_2, respectively. Now, according to the results of the Michelson–Morley experiment, the speed of light c must be the same for both inertial systems S_1 and S_2. The distances r_1 and r_2 from their respective origins to the point M (the point reached by the flash) are given by

$$\boxed{\begin{aligned} r_1 &= ct_1 \\ r_2 &= ct_2 \end{aligned}} \tag{4-7}$$

Therefore, we are forced to accept the fact that the two traveling times t_1 and t_2 (as measured by observers O_1 and O_2) are different, although this is contrary to what we might "ordinarily" experience.

From Equation (4–7),

$$\begin{aligned} x_1^2 + y_1^2 + z_1^2 &= c^2 t_1^2 \\ x_2^2 + y_2^2 + z_2^2 &= c^2 t_2^2 \end{aligned} \tag{4-8}$$

and from the symmetry conditions, $y_1 = y_2$ and $z_1 = z_2$. Equation (4–8) is now combined to yield

$$x_2^2 - c^2 t_2^2 = x_1^2 - c^2 t_1^2 \tag{4-9}$$

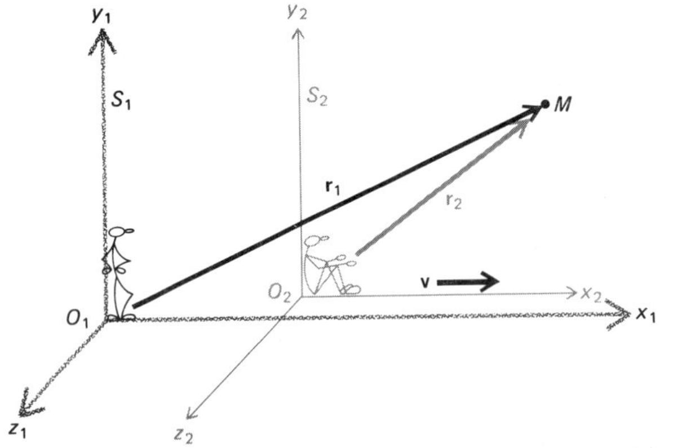

Figure 4–3 System S_2 is moving at a constant velocity relative to stationary system S_1.

At this point, we digress from our discussion to point out that we are making the following assumption: There exists a set of equations that interprets the description of a set of events as seen in one frame into a description of the same set of events as seen from another frame. We can think of many instances when such a set of equations can be applied. This method must work if all observers are to see the same nature in the same universe. The set of equations used for such an interpretation is called a transformation. We may think of this method as just a relabeling of the coordinates of an event as seen in one frame to another set of coordinates as seen in another frame. Thus, we are stating that we do not believe that our choice of labels should have any effect whatsoever on what we observe to be happening in nature.

We remind ourselves that in this discussion, we are considering only frames of reference that are moving with steady speed (or uniform velocity) with respect to one another. The treatment of transformations between frames that are being accelerated with respect to one another is a whole field of inquiry and beyond the scope of this text. This subject is usually called the study of general relativity.

We emphasize that we are concerned here with steadily moving frames. They are referred to as inertial frames because there is an especially simple relationship between certain vectors (such as the vectors of momentum) as seen from these different frames. We assume that the equations that relabel the coordinates from one inertial frame to another are linear equations of the following form

$$\begin{aligned} x_2 &= \gamma(x_1 - vt_1) \\ y_2 &= y_1 \\ z_2 &= z_1 \\ t_2 &= a(t_1 - bx_1) \end{aligned} \quad (4\text{-}10)$$

in which γ, a, and b are constants which we shall evaluate in the next few paragraphs. There are several additional requirements that we place on the assumed format of the transformations of Equation (4-10). We emphasize that the equations must be linear in form, because an event described in one system must transform into only one event in the second system. (A transformation of quadratic form could conceivably produce two solutions, which would imply that an event in one system could be interpreted as two events in the second system, an impossible situation.) Also, for velocities small compared to c ($v/c \to 0$), the new transformations must take the form of Galilean transformations. Equation (4-10) also holds for coincident systems, that is, when $t_1 = 0$ and $x_1 = 0$, then $x_2 = 0$ and $t_2 = 0$.

Now let us substitute Equation (4-10) into Equation (4-9) to get

$$x_1^2(\gamma^2 - a^2b^2c^2 - 1) + x_1 t_1(-2\gamma^2 v + 2a^2 bc^2)$$
$$+ t_1^2(\gamma^2 v^2 - a^2 c^2 + c^2) = 0 \quad (4\text{-}11)$$

Since this expression is identically zero,

$$\gamma^2 - a^2b^2c^2 - 1 = 0$$
$$-2\gamma^2 v + 2a^2bc^2 = 0 \qquad (4\text{-}12)$$
$$\gamma^2 v^2 - a^2c^2 + c^2 = 0$$

We solve these equations for the constants γ, a, and b and obtain

$$\boxed{\gamma = a = \frac{1}{\sqrt{1 - (v^2/c^2)}}} \qquad (4\text{-}13)$$

and

$$\boxed{b = \frac{v}{c^2}} \qquad (4\text{-}14)$$

The expression $1/\sqrt{1 - (v^2/c^2)}$, is known as the Lorentz factor and is usually represented by γ. Also, the expression v/c is usually denoted by β.

The transformation Equation (4–10) now becomes

$$\boxed{\begin{aligned} x_2 &= \frac{x_1 - vt_1}{\sqrt{1 - \beta^2}} = \gamma(x_1 - vt_1) \\ y_2 &= y_1 \\ z_2 &= z_1 \\ t_2 &= \frac{t_1 - (v/c^2)x_1}{\sqrt{1 - \beta^2}} = \gamma\left(t_1 - \frac{\beta}{c} x_1\right) \end{aligned}} \qquad (4\text{-}15)$$

The inverse transformations of Equation (4–15) is

$$\boxed{\begin{aligned} x_1 &= \frac{x_2 + vt_2}{\sqrt{1 - \beta^2}} = \gamma(x_2 + vt_2) \\ y_1 &= y_2 \\ z_1 &= z_2 \\ t_1 &= \frac{t_2 + (v/c^2)x_2}{\sqrt{1 - \beta^2}} = \gamma\left(t_2 + \frac{\beta}{c} x_2\right) \end{aligned}} \qquad (4\text{-}16)$$

These equations can be obtained either by algebraic manipulations or, practically, by interchanging subscripts in Equation (4–15) and replacing v with $-v$. These transformations are known as *Lorentz transformations* after H. A. LORENTZ (1853–1928), the Dutch physicist who derived them in 1890.

In 1923 Niels Bohr proposed a *correspondence principle*. This states that *any new theory in physics must reduce to the well-established corresponding classical theory when the new theory is applied to the special situation in which the less general theory is known to be valid.* Let us look at Equation (4–15) to see if the correspondence principle holds. When $\beta = v/c \to 0$, we see that Equation (4–15) reduces to

$$x_2 = x_1 - vt_1$$
$$y_2 = y_1$$
$$z_2 = z_1$$
$$t_2 = t_1$$

which are the Galilean transformations [Equation (3–7)]. Thus, we summarize:

Lorentz transformations → Galilean transformations

when $\beta = \dfrac{v}{c} \to 0$

4–3 LORENTZ COMPOSITION OF VELOCITIES

We differentiate Equation (4–15) to get

$$dx_2 = \frac{dx_1 - v\, dt_1}{\sqrt{1-\beta^2}} = \frac{(v_{1x} - v)\, dt_1}{\sqrt{1-\beta^2}}$$

$$dy_2 = dy_1$$

$$dz_2 = dz_1 \qquad (4\text{–}17)$$

$$dt_2 = \frac{dt_1 - (v/c^2)\, dx_1}{\sqrt{1-\beta^2}} = \frac{(1 - vv_{1x}/c^2)\, dt_1}{\sqrt{1-\beta^2}}$$

where $v_{1x} = dx_1/dt_1$. Thus, the Lorentz transformation equations for velocity are

$$\boxed{\begin{aligned} v_{2x} &= \frac{dx_2}{dt_2} = \frac{v_{1x} - v}{1 - (v/c^2)v_{1x}} \\ v_{2y} &= \frac{dy_2}{dt_2} = \frac{v_{1y}\sqrt{1-\beta^2}}{1 - (v/c^2)v_{1x}} \\ v_{2z} &= \frac{dz_2}{dt_2} = \frac{v_{1z}\sqrt{1-\beta^2}}{1 - (v/c^2)v_{1x}} \end{aligned}} \qquad (4\text{–}18)$$

Note that now, with the Lorentz transformations, even though the velocity **v** is along the x axis, the y and z components of $\mathbf{v}_2$ also depend on v_{1x}.

When $\beta = v/c \to 0$, these equations become

$$v_{2x} = v_{1x} - v$$
$$v_{2y} = v_{1y}$$
$$v_{2z} = v_{1z}$$

These are the Galilean composition of velocities. Thus, the correspondence principle does apply.

Interchanging subscripts 1 and 2 and replacing v by $-v$ gives the inverse velocity transformation

$$\boxed{\begin{aligned} v_{1x} &= \frac{v_{2x} + v}{1 + (v/c^2)v_{2x}} \\ v_{1y} &= \frac{v_{2y}\sqrt{1 - \beta^2}}{1 + (v/c^2)v_{2x}} \\ v_{1z} &= \frac{v_{2z}\sqrt{1 - \beta^2}}{1 + (v/c^2)v_{2x}} \end{aligned}} \quad (4\text{-}19)$$

Let us consider a particle M moving parallel to the x axis with a velocity $v_2 = v_{2x}$ in system S_2, which is itself moving with a velocity **v** relative to the inertial system S_1. Then, according to the Galilean transformation, the velocity components of M as measured in the inertial system S_1 are

$$\begin{aligned} v_{1x} &= v_{2x} + v = v_2 + v \\ v_{1y} &= v_{2y} = 0 \\ v_{1z} &= v_{2z} = 0 \end{aligned} \quad (4\text{-}20)$$

According to the Lorentz transformations, the velocity components are

$$\begin{aligned} v_{1x} &= \frac{v_{2x} + v}{1 + (v/c^2)v_{2x}} = \frac{v_2 + v}{1 + (v/c^2)v_{2x}} \\ v_{1y} &= \frac{v_{2y}\sqrt{1 - \beta^2}}{1 + (v/c^2)v_{2x}} = 0 \\ v_{1z} &= \frac{v_{2z}\sqrt{1 - \beta^2}}{1 + (v/c^2)v_{2x}} = 0 \end{aligned} \quad (4\text{-}21)$$

In particular, if we let $v_2 = c$, the Galilean transformation gives

$$\boxed{\begin{aligned} v_{1x} &= c + v \\ v_{1y} &= 0 \\ v_{1z} &= 0 \end{aligned}}$$

This result is incompatible with the observational results of the Michelson–Morley experiment. However, the Lorentz transformations yield

$$v_{1x} = \frac{c + v}{1 + (v/c^2)c} = c$$

$$v_{1y} = 0$$

$$v_{1z} = 0$$

which does agree with the observational results from the Michelson–Morley experiment.

EXAMPLE 4–1: Show that if (x_1, y_1, z_1, t_1) and (x_2, y_2, z_2, t_2) are the coordinates of one event in S_1 and the corresponding event in S_2, respectively, then the expression

$$ds_1^2 = dx_1^2 + dy_1^2 + dz_1^2 - c^2 dt_1^2$$

is invariant under a Lorentz transformation of coordinates (i.e., $ds_1^2 = ds_2^2$).

SOLUTION: Differentiation of the expressions in Equation (4–16) gives

$$dx_1 = \frac{dx_2 + v\, dt_2}{\sqrt{1 - \beta^2}}$$

$$dy_1 = dy_2$$

$$dz_1 = dz_2$$

$$dt_1 = \frac{dt_2 + (\beta/c)\, dx_2}{\sqrt{1 - \beta^2}}$$

where we have assumed that $v = $ constant. It is evident that

$$ds_1^2 = \left(\frac{dx_2 + v\, dt_2}{\sqrt{1 - \beta^2}}\right)^2 + dy_2^2 + dz_2^2 - c^2 \left(\frac{dt_2 + (\beta/c)\, dx_2}{\sqrt{1 - \beta^2}}\right)^2$$

which simplifies to

$$ds_1^2 = dx_2^2 + dy_2^2 + dz_2^2 - c^2\, dt_2^2 = ds_2^2$$

EXAMPLE 4–2: Show that two events that occur at the same time t_1 (i.e., simultaneously) at two different points $A(x_{1_A}, y_{1_A})$ and $B(x_{1_B}, y_{1_B})$ are not simultaneous when they are referred to the system S_2.

SOLUTION: Equation (4–15) shows the Lorentz transformation for time to be

$$t_2 = \gamma\left(t_1 - \frac{\beta}{c}x_1\right)$$

Using this equation for this case gives

$$t_{2A} = \gamma\left(t_1 - \frac{\beta}{c}x_{1A}\right) \quad \text{and} \quad t_{2B} = \gamma\left(t_1 - \frac{\beta}{c}x_{1B}\right)$$

Hence,

$$t_{2A} - t_{2B} = \gamma\frac{\beta}{c}(x_{1B} - x_{1A})$$

Thus, the two events cannot be simultaneous when referred to the system S_2 unless $x_{1A} = x_{1B}$.

PROBLEMS

4–1 Start with Lorentz transformation Equation (4–15) and solve algebraically for x_1, y_1, z_1, and t_1 to show that the inverse Lorentz transformation Equation (4–16) can be obtained by interchanging subscripts 1 and 2 for coordinates and by replacing v with $-v$.

4–2 Repeat Problem 4–1 for the Lorentz transformation of velocities in Equation (4–18) to show that the inverse equations can be obtained by interchanging subscripts 1 and 2 for coordinates and by replacing v with $-v$.

4–3 Use the Lorentz transformation of velocities to show that if $v_{1_x}^2 + v_{1_y}^2 + v_{1_z}^2 = c^2$ in inertial system S_1, then $v_{2_x}^2 + v_{2_y}^2 + v_{2_z}^2 = c^2$ in inertial system S_2. (This again shows that the speed of light is the same for all inertial systems according to the Lorentz transformations.)

4–4 Consider an inertial system S_2 that moves with a velocity $v = c$ relative to the inertial system S_1. An observer in system S_2 monitors a particle moving with a velocity with rectangular components $v_{2_x} = c$ and $v_{2_y} = c/2$. Compute the magnitude and direction of the velocity of the particle as measured (a) by Lorentz transformations of velocity, and (b) by Galilean composition of velocities and compare your results.

4–5 Two ion propulsion vehicles approach each other in opposite and parallel directions with velocities of $0.80c$ and $0.70c$ relative to an observer at rest along the line of action. Compute the relative velocity of the two vehicles (a) as measured by classical mechanics, and (b) as measured by relativistic mechanics and compare the results.

4-6 When a clock moves past us at a speed of $v = c/2$, it reads $t_2 = 0$ just as our clock reads $t_1 = 0$. Use the Lorentz transformation to determine the reading of our clock when the moving clock reads $t_2 = 10$ sec.

4-7 A man in a car moving at a velocity of 60 km/h throws a ball in the same direction in which the car is traveling. If the speed of the ball relative to the car is 80 km/h, compute the velocity of the ball relative to the ground using (a) relativistic, and (b) Galilean approaches and then compare the results.

4-8 The captain of a space vehicle traveling at a speed of $0.80c$ as monitored by a stationary radar system uses an electron gun to fire electrons in the same direction of travel at a speed of $0.90c$ relative to the space vehicle. Compute the speed of the electron relative to the radar station (a) as measured by relativistic mechanics, and (b) as measured by classical mechanics.

4-9 Show that the relativistic formula $v_{1_x} = (v_{2_x} + v)/[1 + (v/c^2)v_{2_x}]$ gives (a) $v_{1_x} < c$ when $v < c$ and $v_{2_x} < c$, and (b) $v_{1_x} = c$ when v_{2_x} or $v = c$.

4-10 An event that occurs in system S_1 has coordinates $x_1 = 1.0 \times 10^5$ m, $y_1 = 0$, $z_1 = 1.0 \times 10^5$ m, $t_1 = 1.0 \times 10^5$ sec. What would be the coordinates of this event as measured by an inertial observer attached to S_2 and moving at a relative velocity in the x_1 direction of $\tfrac{1}{2}c$?

4-11 An electron is projected at an angle of 37° with respect to the x_1 axis at a speed of $\tfrac{1}{2}c$. Determine the magnitude and direction of the velocity of this electron as measured from an inertial system moving as shown in Figure 4-4 at a speed of $\tfrac{1}{2}c$.

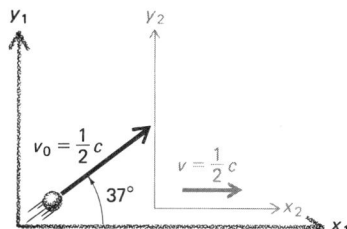

Figure 4-4

4-12 An experiment is initiated from the earth (assumed at rest) in which at $t_1 = 1.000$ sec a laser pulse is fired toward the moon and at $t_1{}^1 = 2.210$ sec a detector on the moon's surface detects the arrival of the pulse. What will be the time of travel of this pulse as measured by an observer moving in the same direction as the pulse at a speed of $0.800c$?

RECOMMENDED READING

BREHME, Robert W., "A Geometric Representation of Galilean and Lorentz Transformations," *Am. J. Phys.* **30,** 489 (1962).
  A clear and handy method of manipulating transformations is outlined.

FEYNMAN, R. P., LEIGHTON, R. B., and SANDS, L. M., *The Feynman Lectures on Physics*, Addison-Wesley, Reading, Mass., 1964, Vol. I, Chaps. 15, 18, and 20.

FRENCH, A. P., *Special Relativity*, Norton, New York, 1968.
  Interestingly written book with many problems on relativity. The account of the Michelson–Morley experiment is good reading.

JAFFE, Bernard, *Michelson and the Speed of Light*, Doubleday (Anchor Books), New York, 1960.
  An elementary description of the Michelson interferometer as well as an excellent biography of this famous scientist.

MICHELSON, A. A., "On a Method of Measuring the Velocity of Light," *Am. J. Sci.* **15,** 394–395 (1878).

MICHELSON, A. A., "The Relative Motion of the Earth and the Luminiferous Ether," *Am. J. Sci.* **22,** 120–129 (1881).

MICHELSON, A. A., *Light Waves and Their Uses*, University of Chicago Press, Chicago, 1903.

MICHELSON, A. A., *Studies in Optics*, University of Chicago Press (Phoenix Books), Chicago, 1962.
  A small, concise book with a good description of the Michelson interferometer.

MILLER, D. C., "Ether-Drift Experiments and the Determination of the Absolute Motion of the Earth," *Rev. Mod. Phys.* **5,** 203–242 (1933).

RUSH, J. H., "The Speed of Light," *Sci. Am.*, August 1965.

SHANKLAND, R. S., "The Michelson–Morley Experiment," *Sci. Am.*, November 1964.

SHANKLAND R. S., et al., "New Analysis of the Interferometer Observation of Dayton C. Miller," *Rev. Mod. Phys.* **27,** 167 (1955).
  A summary of many experimental efforts to substantiate (or disprove) the existence of the luminiferous ether.

Special Relativity Theory, Selected Reprints, American Institute of Physics, New York, 1963.
  Contains many good references and several excellent representative papers on the special relativity theory.

(See also those references at the end of Chapter 3.)

5
Consequences of the Lorentz Transformations

Hendrik Antoon Lorentz
(1853–1928)

A native of Arnheim, Holland, Lorentz received his Ph.D. from Leyden University in 1875. He was Director of Research at Teyler Laboratory in Haarlem and honorary professor at Leyden. In 1903 he developed the famous Lorentz transformations, which aided Einstein in formulating the theory of relativity. Lorentz also actively studied electromagnetism, gravitation, thermodynamics, radiation, and kinetic theory. For his theoretical explanation of the Zeeman effect, he shared the Nobel Prize with Pieter Zeeman in 1902.

5–1 CONTRACTION OF LENGTH
5–2 DILATION OF TEMPORAL INTERVALS
5–3 INTERPRETATION OF THE MICHELSON–MORLEY EXPERIMENT
5–4 EINSTEIN'S SOLUTION TO THE CONFLICT

5–1 CONTRACTION OF LENGTH

Let us consider for a moment the length of a meter stick. This might appear at first to be a rather silly exercise since by its very name the length of a meter stick is 1 m. But let us qualify this statement by adding that 1 m is the value of the meter stick *as seen in the rest frame of the stick*, say the frame S_2 (see Figure 5–1). If the meter stick lies parallel to the x axis in this frame, then the distance from end A, at x_{A_2}, to end B, at x_{B_2}, is 1 m. The length of the stick in S_2 is then defined as the difference between these two numbers on the x axis:

$$L_2 = x_{B_2} - x_{A_2} \tag{5-1}$$

Furthermore, these two numbers will remain the same with the passage of time, since S_2 is the rest frame of the stick. Their difference L_2 will also remain constant in time.

Now let us look at this same stick as observers in frame S_1. Let frame S_2 move with velocity **v** in a direction parallel to the x axis of S_1. End A lies at x_{A_1} in frame S_1, and the number x_{A_1} is constantly changing as the frame S_2 moves. The number x_{B_1}, which marks the other end of the meter stick, will also be changing with time. Looking at the stick as observers in S_1, we again define length as the difference between the two numbers that mark its ends:

$$L_1 = x_{B_1} - x_{A_1} \tag{5-2}$$

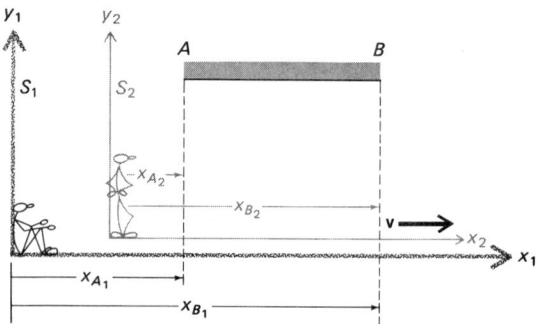

Figure 5–1

CHAPTER 5: CONSEQUENCES OF THE LORENTZ TRANSFORMATIONS · 57

It appears reasonable to require that the value of L_1 be constant in time. However, we must investigate to see if this is possible, since the two numbers that yield its value by their difference are changing. We wish that if the length of an object is constant in one frame (and in this case it is in S_2), then the length must also be constant as observed from any other frame. If this were not true, then the same object might appear rigid to one observer but nonrigid (or rubbery) to another observer moving relative to the first.

The Lorentz transformation of coordinate values provides the solution to the problem of maintaining a constant length of an object as seen from different frames. Let us apply the Lorentz transformation to the two numbers on the right-hand side of Equation (5-2). We obtain the following equivalent numbers in frame S_2:

$$x_{A_2} = \gamma(x_{A_1} - vt_1) \tag{5-3}$$

$$x_{B_2} = \gamma(x_{B_1} - vt_1) \tag{5-4}$$

where γ is the Lorentz factor [see Equation (4-13)], and t_1 is the instant in S_1 when we measured the length of the stick by noting the values of the coordinates of A and B. We are due for a surprise, however, when we subtract Equation (5-3) from Equation (5-4)! The expression for time drops out of the expression for length (as we said it must). Let us examine what is left:

$$x_{B_2} - x_{A_2} = \gamma(x_{B_1} - x_{A_1}) \tag{5-5}$$

or

$$\boxed{L_2 = \gamma L_1} \tag{5-6}$$

Since v must always be less than c, γ must always be greater than 1. Thus, we discover the surprising conclusion that *the stick observed from any frame moving with respect to the inertial frame (of the stick) appears to be shorter.* Thus, for any length of any object, we have the relation

$$L_1 < L_2 \tag{5-7}$$

or

$$\boxed{L_1 \begin{pmatrix} \text{observed length when} \\ \text{bar is in motion} \\ \text{relative to observer} \end{pmatrix} = L_2 \begin{pmatrix} \text{observed length when} \\ \text{bar is at rest} \\ \text{relative to observer} \end{pmatrix} \times \sqrt{1 - \beta^2}}$$

This law has even more general applications than this, since it can be applied to *any* object. The law is thus independent of the nature of the object being considered, and must therefore apply *to the space itself,* whether or not an object is, in fact, located in the interval between the coordinates being measured.

Albert Einstein proposed that the Lorentz transformation be a *fundamental law of nature,* superseding the Galilean transformation group, when the velocity becomes large enough to be measured in terms of c. The statement that the length of an object is dependent on the observer's state of motion so shocked physicists in the early years of this century that many of them questioned the validity of the experimental results of the Michelson–Morley experiment. These results, however, have withstood the test of time by being checked and thoroughly substantiated by many other and different experiments.

With our hindsight it is interesting for us to discover that Lorentz and an Irish physicist, G. F. FITZGERALD (1851–1901), thought that the shortening of an object in motion was due to some sort of force applied to the object by its passage through a stationary ether. Considerable thought was given at that time to the nature of this force. Einstein adopted the completely opposite view that the shortening is a *property of the space* itself, and that an absolute, or preferred, frame of reference over all others does not exist. Einstein rejected the idea of an *absolute* motion in nature for anything other than light as being without meaning. He believed that the speed of light in a vacuum *is* absolute. He also believed that the speed has the same value, called c (for the same light), as seen from *any* frame (regardless of the frame's velocity). It is important that we emphasize this point: *The speed of light in a vacuum is the same for every observer.* This statement includes the observer who is holding the source of the light or who is traveling at a high speed relative to the source.

The central theme to realize here is that we cannot use the speed of a beam of light to specify a preferred frame of reference. When we understand this point and realize that every observer views the universe from his own unique inertial frame (which is moving relative to someone else's inertial frame), we realize that we have a system that makes the universe understandable to *every* observer in the *same* terms. Thus, several corollary conclusions can be stated:

1. For small relative velocity (and especially as $\mathbf{v} \to 0$), L_1 becomes essentially equal to L_2 as is true in "classical" mechanics. (This is an example of Bohr's correspondence principle.)
2. The contraction of length occurs only for lengths measured parallel to the direction of relative motion.
3. If the Lorentz factor γ is to have a real rather than an imaginary value, then v must always be smaller than c.

EXAMPLE 5–1: A rigid bar of length $L_2 = 1.5$ m is at rest relative to system S_2 (Figure 5–2). If the bar makes an angle $\theta_2 = 45°$ with respect to the x_2 axis, what is the length L_1 and orientation of the bar θ_1 relative to S_1 when $v = 0.98c$?

SOLUTION: When the length of the bar is resolved into the two components parallel to the x_2 and y_2 axes, respectively, the corresponding lengths measured by S_2 will be

$$L_{2x} = L_2 \cos \theta_2$$
$$L_{2y} = L_2 \sin \theta_2$$

CHAPTER 5: CONSEQUENCES OF THE LORENTZ TRANSFORMATIONS · 59

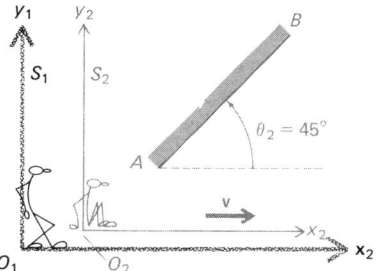

Figure 5–2

The vertical component is perpendicular to v and will not experience any contraction when viewed from S_1. Hence,

$$L_{1y} = L_{2y} = L_2 \sin \theta_2$$

The horizontal component is parallel to v and will be contracted according to Equation (5–5) to give

$$L_{1x} = L_{2x}\sqrt{1 - \beta^2} = L_2\sqrt{1 - \beta^2} \cos \theta_2$$

The length of the bar as measured by O_1 will be

$$L_1 = \sqrt{L_{1x}^2 + L_{1y}^2} = \sqrt{L_2^2(1 - \beta^2) \cos^2 \theta_2 + L_2^2 \sin^2 \theta_2}$$
$$= L_2\sqrt{1 - \beta^2 \cos^2 \theta_2}$$

The orientation relative to S_1 will be given by

$$\tan \theta_1 = \frac{L_{1y}}{L_{1x}} = \frac{L_2 \sin \theta_2}{L_2\sqrt{1 - \beta^2} \cos \theta_2}$$
$$= \frac{\tan \theta_2}{\sqrt{1 - \beta^2}}$$

Replacing numerical values, we get

$$L_1 = 1.08 \text{ m}$$
$$\theta_1 = 78.7°$$

5–2 DILATION OF TEMPORAL INTERVALS

Imagine the simplest sort of "event" in nature as a kind of happening that occurs at a point A in space and at the instant t_A. The space coordinates of this point as seen from a given reference frame may be designated by x_A, y_A, and z_A. Now let us also consider another event taking place at the same point A, but at a different time t_B. Both events are recorded in frame S_2, in which A is at rest. Naming the rest frame S_2 as before, the time interval between the two events is just

$$T_2 = t_{B_2} - t_{A_2} \tag{5-8}$$

where we have used the subscript 2 on the clock times t_A and t_B to remind us that these times were read from a clock at rest in the frame S_2.

Now consider the same pair of events at the same point but as seen from a frame S_1 that is moving parallel to the x axis of S_2 and with relative velocity $-\mathbf{v}$ (see Figure 5–3). We note that from the relativistic point of view, these two situations are equivalent: S_2 is moving relative to S_1 with velocity $\mathbf{v}$, or S_1 is moving relative to S_2 with velocity $-\mathbf{v}$. Obviously, the time interval observed from S_1 is given by

$$T_1 = t_{B_1} - t_{A_1} \tag{5-9}$$

but the coordinate values of the point for the first event will no longer be the same for the second event as they were in the rest frame S_2.

The value of a time interval in these frames must not be dependent on the space coordinate values of x, y, or z in any of the frames, otherwise we could change the rate of a clock merely by observing it from different places in the same frame. So the space coordinates that appear in the Lorentz transformation Equation (4–15) must drop out if they are to be used to consider *intervals* of time. We recall that the transformation for times is

$$t_{A_1} = \gamma \left(t_{A_2} + \frac{v x_A}{c^2} \right) \tag{5-10}$$

for event A, and

$$t_{B_1} = \gamma \left(t_{B_2} + \frac{v x_A}{c^2} \right) \tag{5-11}$$

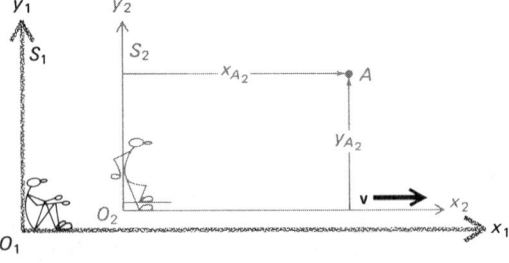

Figure 5–3   Point A is at rest relative to system S_2. Two events happen at point A at times t_2 and t_2' according to O_2.

for event B, and again γ is the Lorentz factor given by Equation (4–13). The role of x_A becomes clear when these equations are examined; it serves merely to *set* the clocks in S_1 relative to the clock at the stationary point in S_2. It does not affect the rates at all.

Subtracting Equation (5–10) from Equation (5–11) gives

$$t_{B_1} - t_{A_1} = \gamma(t_{B_2} - t_{A_2}) \tag{5-12}$$

Using Equations (5–8) and (5–9), this reduces to

$$\boxed{T_1 = \gamma T_2} \tag{5-13}$$

Since $\gamma > 1$, we arrive at another surprising conclusion, namely that

$$T_1 > T_2 \tag{5-14}$$

Thus, the relativistic time dilation is

$$\boxed{T_1 \begin{pmatrix} \text{measured time interval} \\ \text{between two events taking} \\ \text{place at a point in motion} \\ \text{relative to observer} \end{pmatrix} = T_2 \frac{\begin{pmatrix} \text{measured time interval} \\ \text{between two events taking} \\ \text{place at a point at rest} \\ \text{relative to observer} \end{pmatrix}}{\sqrt{1 - \beta^2}}}$$

Thus, an interval of time that separates two successive events is longer in any frame moving relative to the rest frame than it is in the rest frame! Now, since the only way a measured time interval can be made longer is to slow down the clock used to measure the interval, this statement becomes the assertion that *moving clocks run slower than do stationary ones*!

We now find ourselves saying that for every observer, his *own* clocks in his *own* laboratory run faster than do any other clocks if these other clocks are moving relative to him. We note that every observer may consider himself to be at rest and consider all that moves as moving relative to him. This privilege is stated for each observer by the *principle of special relativity*: *Every observer is equivalent to every other observer.* It does this by giving every observer the right to claim that *he* stands at the center of the universe—and that *his* rest frame is the stationary one in all creation! He may do this, but at the same time, he must recognize and respect the right of *every other* observer to do likewise. Only in this manner may all persons understand one another when they describe what they see in nature. They do so by means of the Lorentz transformation equations.

It is interesting for us to speculate with our hindsight on what thoughts Galileo might have had if he had known about these equations when he insisted that the earth moves around the sun, and not vice versa. It would seem that the study of nature now teaches us that the road to the most unique and meaningful individuality lies in constantly recognizing the complete equivalence of every other observer to ourselves. And here "equivalence" means equality in a deeper sense than is usually used.

We shall not obtain here the *inverse* transformation equations from those given, but we suggest that the reader do so and thus prove what has just been said. An additional exercise is most instructive: that of using the inverses to Equations (5–10) and (5–11) to show how one makes sure that clocks in different frames are truly equivalent clocks and not just poorly built ones.

EXAMPLE 5–2: A typical situation in which length intervals appear shortened and clocks appear to run faster may be found in the beam of pi mesons that is produced by a modern giant accelerator. In such a machine, protons are accelerated from near standstill to extreme energies and are then allowed to fall into a metal target. One of the products of these collisions is a beam of very fast pi mesons. These are particles that produce nuclear forces that hold atomic nuclei together.

In some cases, these pi mesons (or pions) are slowed down by sending them through a thick wall of concrete or iron, and then they are stopped in another target. Here the positive pions will decay to other particles because they are radioactive. These daughter particles are muons and neutrinos. In such cases as this, the time when the pion stopped can be found by means of a counter placed just before the last target. Another counter can time the appearance of the muon from the decay, and so the lifetime of the pion at rest is measured. When many such cases are recorded, the mean lifetime for pions that are at rest is found to be 2.60×10^{-8} sec.

In other cases, the fast pions are directed down a long corridor filled with air or into a pipe containing a vacuum. Many of them now decay in flight. One can measure the number of pions that start their journey down the corridor and the number that make it to the end. The difference is just the number that decayed enroute and while moving very fast.

It is not unusual for such pions to have a total energy that is 20 times their rest mass, or a Lorentz factor $\gamma = 20$. The speed of the pion can be calculated from the definition of the Lorentz factor. This speed is very close to the speed of light, c. If N_0 such pions start down a 100-m corridor with this velocity, they will make the trip in 100 m/3.00×10^8 m/sec $= 3.33 \times 10^{-7}$ sec if they do not decay on the way.

The decay equation for N, the number of pions that survive the trip, is

$$N = N_0 \exp(-\lambda t) = N_0 \exp(-t/\overline{T})$$

where λ is the decay constant and $\overline{T}$ is the average life. Thus, when N is calculated from the decay rate,

$$N = N_0 \exp[-(3.33 \times 10^{-7}/2.60 \times 10^{-8})] = N_0 \exp(-12.8)$$
$$= 2.76 \times 10^{-6} N_0$$

it would seem that less than 0.00028% of the pions reach the end of the corridor.

However, this is incorrect. The Lorentz factor $\gamma = 20$ must be used to slow down the pion clock, and its in-flight mean life is then $20 \times 2.60 \times 10^{-8}$ or 5.20×10^{-7} sec. The number that reach the end is then

$$N = N_0 \exp[-(3.33 \times 10^{-7}/5.20 \times 10^{-7})] = N_0 \exp(-0.642)$$
$$= 0.52 N_0$$

or 52% survive. The internal timing mechanism of the pions appears to run much slower as seen from the laboratory at the end of the corridor.

How does the laboratory "appear" to the pion? Certainly, an observer traveling along with the pion would say that the pion clock is running normally and that its mean lifetime is 2.6×10^{-8} sec. The corridor would appear shortened, however, by the Lorentz factor to one-twentieth of its length, or only 5.00 m long. The trip, according to the pion, would take only 5 m/3×10^8 m/sec $= 1.66 \times 10^{-8}$ sec. The number that reach the end is then

$$N = N_0 \exp[-(1.66 \times 10^{-8}/2.6 \times 10^{-8})]$$
$$= N_0 \exp(-0.642)$$
$$= 0.52 N_0$$

Thus, the observer in the laboratory counts the same number at the end of the corridor as would an observer traveling with the pion beam. The same Lorentz factor affects both, but in complementary ways.

5-3 INTERPRETATION OF THE MICHELSON–MORLEY EXPERIMENT

The Lorentz transformations can be used to show that time dilation and length contraction are just direct consequences of the invariance of the speed of light for all inertial frames moving relative to each other with uniform translational motion. This statement, which is the result of the Michelson–Morley experiment, is known as the *principle of special relativity*.

Let us examine the notion of *time dilation* by considering an example. Consider again the Michelson interferometer (see Figure 4–1) located as before in a reference frame S_2 attached to the interferometer and hence to the earth. In Figures 5–4(a) and 5–4(b), the mirror M_2 is at rest relative to S_2, which is moving with a speed v (the translational speed of the earth) relative to S_1. Reference frame S_1 is in turn attached to the fixed stars or the "ether." At time $t_1 = t_2 = 0$ [Figure 5–4(a)], S_2 coincides with S_1 and a flash of light is sent on its way from O_1 to mirror M_2, where it will be reflected from M_2 and then reach O_2 after some time interval. Call T_1 and T_2 the traveling times as measured by observers O_1 and O_2, respectively.

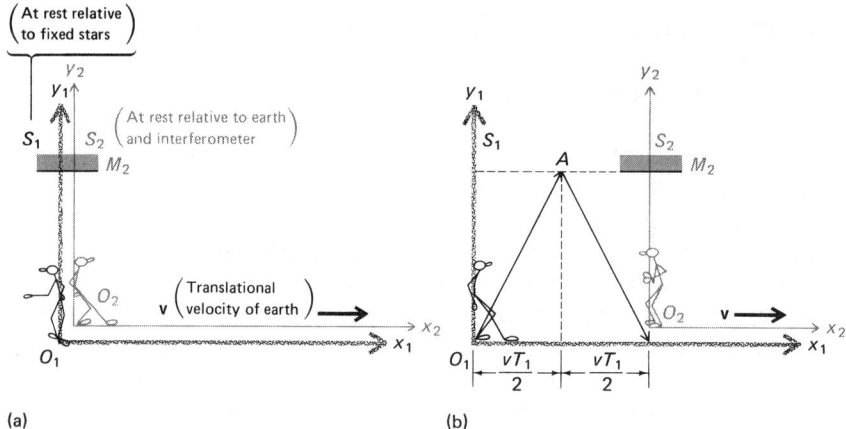

Figure 5–4

Evidently (since c, the speed of light, is invariant), Figure 5–4(a) shows the travel times to be

$$T_2 = \frac{2L}{c} \tag{5-15}$$

and

$$T_1 = \frac{O_1A + AO_2}{c} = \frac{2\sqrt{L^2 + (v^2/4)T_1^2}}{c} \tag{5-16}$$

From Equation (5–16),

$$L = \frac{c}{2} T_1 \sqrt{1 - \frac{v^2}{c^2}} \tag{5-17}$$

and from combining Equations (5–15) and (5–17),

$$\frac{c}{2} T_1 \sqrt{1 - \frac{v^2}{c^2}} = \frac{c}{2} T_2$$

which simplifies to

$$\boxed{T_1 = \frac{T_2}{\sqrt{1 - (v^2/c^2)}}} \tag{5-18}$$

the same result obtained by direct application of the Lorentz transformations.

Now let us consider the companion notion of *length contraction*. When S_2 coincides with S_1, a flash of light is sent from the common origins toward M_1, which is a distance L_2 away as measured by observer O_2. The light is reflected by mirror M_1 and comes back to O_2. As before, T_1 and T_2 are the respective times measured for the round trip of the light by observers O_1 and O_2, respectively. Now, for the round trip $O_2 M_1 O_2$,

$$T_2 = 2 \frac{L_2}{c} \tag{5-19}$$

Now, if t_1 is the traveling time from O_2 to M_1 as measured by observer O_1, Figure 5-5(b) shows that

$$ct_1 = vt_1 + L_1 \tag{5-20}$$

in which L_1 is the distance from M_1 to O_2, as measured from S_1. If the travel time from M_1 to O_2 as measured by O_1 is t_2, then

$$ct_2 = L_1 - vt_2 \tag{5-21}$$

Hence, from Equations (5-20) and (5-21),

$$T_1 = t_1 + t_2 = \frac{L_1}{c-v} + \frac{L_1}{c+v} = \frac{2L_1/c}{1-(v^2/c^2)} \tag{5-22}$$

and Equations (5-19) and (5-22) then give

$$\frac{T_1}{T_2} = \frac{(2L_1/c)/[1-(v^2/c^2)]}{2L_2/c} = \frac{L_1}{L_2}\left(\frac{1}{1-(v^2/c^2)}\right) \tag{5-23}$$

According to time dilation,

$$\frac{T_1}{T_2} = \frac{1}{\sqrt{1-(v^2/c^2)}}$$

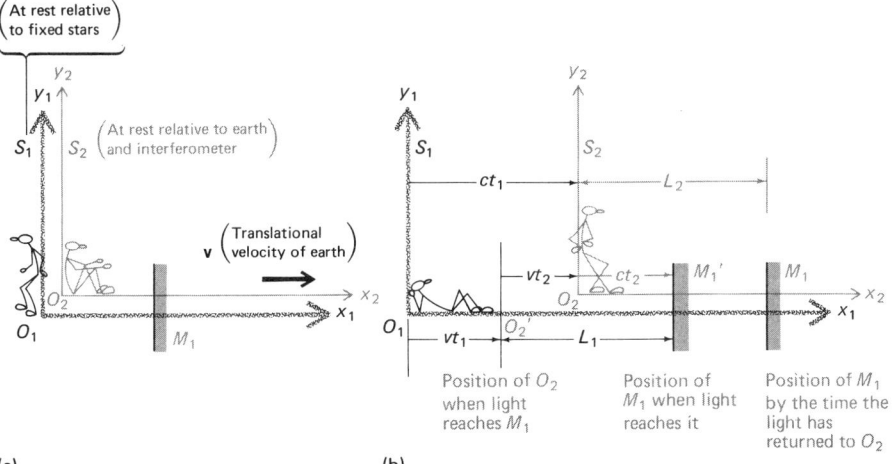

(a)   (b)

Figure 5-5

which then makes Equation (5–23)

$$\frac{1}{\sqrt{1-(v^2/c^2)}} = \frac{L_1}{L_2}\left(\frac{1}{1-(v^2/c^2)}\right)$$

and finally, we have

$$\boxed{L_1 = L_2\sqrt{1-\frac{v^2}{c^2}}} \qquad (5\text{–}24)$$

which is the formula for *length contraction.*

5–4 EINSTEIN'S SOLUTION TO THE CONFLICT

For an experimental setup like the Michelson–Morley experiment (Figure 4–1), the round trip travel times, as we have already seen, of the light were found to be for trip MM_1M,

$$t_{\parallel} = \frac{2L/c}{1-(v^2/c^2)} \qquad (5\text{–}25)$$

and for trip MM_2M,

$$t_{\perp} = \frac{2L/c}{\sqrt{1-(v^2/c^2)}} \qquad (5\text{–}26)$$

where $L = MM_1 = MM_2$, the distance from M to the mirrors M_1 and M_2, as measured by the terrestrial observer. Evidently, then,

$$\frac{t_{\parallel}}{t_{\perp}} = \frac{1}{\sqrt{1-(v^2/c^2)}}$$

Hence, according to the Galilean approach,

$$t_{\parallel} > t_{\perp}$$

On the other hand, the experimental results yielded the relationship

$$t_{\parallel} = t_{\perp}$$

A possible explanation of this experimental result to be suggested was that the speed of light is invariant regardless of the motion of the observer. This need for the rejection of the Galilean or classical composition of velocities, as we have already seen, was difficult for many physicists to accept, since it was a principle that at that time was considered as a dogma in physics.

Of several attempts made not to violate the ideas of classical physics, G. F. Fitzgerald proposed a clever solution. He suggested that all objects moving through the ether do experience a real contraction along the direction of motion and that the contracted length L_{motion} be given by

$$L_{motion} = L\sqrt{1 - \frac{v^2}{c^2}}$$

where $L = L_{rest}$ is the length of the same object when it is at rest relative to the ether (the reference system S_1 in the Michelson–Morley experiment). Hence, if L is replaced by L_{motion} in Equation (5–25), then

$$t_{\parallel} = \frac{2L\sqrt{1 - (v^2/c^2)}/c}{1 - (v^2/c^2)} = \frac{2L/c}{\sqrt{1 - (v^2/c^2)}}$$

and hence $t_{\parallel} = t_{\perp}$, which agrees with the results of the experiment.

The contraction cannot be detected by observer O_2 (the terrestrial observer), who is traveling with the object, because his meter stick is also contracted by the same ratio.

Einstein's solution to the problem was to reject the classical principle of composition of velocities and assume as a valid result that the speed of light is invariant regardless of the motion of the observer. This conclusion led, as we have shown before, to the Lorentz transformations and to the immediate conclusion of length contraction and time dilation.

It is important for us to stress that the length contraction is not a real contraction but a contraction in the "measured length," which is the only length that can be discussed. We must not use the words "observe" and "see" loosely. The act of "seeing" an object involves the finite amount of time required for the transit of light. Victor Weisskopf* shows that a rather distant object moving at relativistic speeds will not appear to be distorted in shape, but will appear to be rotated somewhat away from the position that it had at rest.

The solution provided by Einstein has proved to be the valid one, and there is much experimental evidence to support his theory. Hence, according to his interpretation:

1. The Galilean transformations should be rejected and considered not a valid approximation when $v/c \to 1$.
2. The Lorentz transformations should be considered to be valid (in agreement with the results of the Michelson–Morley experiment).
3. The postulation of the existence of the "ether" is rejected as unnecessary.
4. The concepts of an absolute space and an absolute time are rejected. Space and time are considered to depend on the reference frame or, in other words, they are relative.

---

*V. F. Weisskopf, "The Visual Appearance of Rapidly Moving Objects," *Phys. Today* **13**, 24–27 (September 1960).

In 1905 Einstein went a step further and stated the *special principle of relativity* in the following way: *All the laws in physics must be the same for all inertial frames moving relative to each other with uniform translational motion (constant velocity)*.

We realize that this implies that the laws of dynamics will remain invariant or have the same form when they are referred to different inertial frames of reference. This principle may be considered as the starting point of the special theory of relativity.

We have seen that measuring sticks are longest and clocks run fastest when they are seen from their own rest frames. These statements must be qualified in two important ways. In one, they will be generalized greatly, and in the other they will be restricted severely!

We first generalize by stating that the "observers" that we use in the various reference frames need not be persons, nor even animals or other living things. The effects that have been found here are impressed on every object in nature, from the largest to the smallest. Somehow, every particle has within itself the "measuring stick" and the "clock" which we have been talking about. Perhaps the property called length and the one called time—these properties that grow shorter or longer as we experience motion—are really a property of space (or space-time) itself, in which all of observable nature is embedded.

Now let us restrict this statement most severely. S_2 has been chosen to represent any frame in general that moves relative to S_1. The relative velocity vector **v** has always been assumed to be constant in time and in direction. The results do not hold necessarily when the velocity is changing—*there must be no acceleration.* Motion must be constant and linear! This condition is rarely, if ever, met in the real world. It may be *almost* met in small regions of space for short times, so the theory is an approximation only. It takes the title of the *restricted* or *special* theory of relativity. The real world contains accelerations, of course, and curved trajectories, and changing forces are found almost everywhere.

The problem of obtaining a single, unified description of the real world with its many different sorts of force, its accelerations, and its variety of particles, remains an unsolved problem today. It is a problem for the general relativity.

PROBLEMS

5-1 A rigid bar, 1 m in length, is measured by two observers, one at rest relative to the bar and the second moving in the direction of its length relative to the first. At what speed must the observer be moving so that he measures the bar as having a length of 0.999 m? 0.500 m?

5-2 Determine the dimensions and shape of a 1-m square plate that moves away from an observer in a straight line along its base at a relative speed of $0.80c$. Compare the area of the plate when at rest with the area as measured when the plate is in motion.

5-3 A meter stick that is moving parallel to its length is measured when its velocity is 0.98c. What is the length of this meter stick compared to its rest length?

5-4 An earth-referenced radar monitor observes space ship A, traveling at 0.80c, being overtaken by a second ship B, 10,000 m away and traveling at a speed of 0.98c. How long does it take for ship B to overtake ship A as measured by ship B? As measured by the radar monitor?

5-5 A "seconds" pendulum requires 2 sec to complete 1 cycle (1 sec to swing each direction). What will be the period of this pendulum measured by an observer moving at 0.80c?

5-6 How fast will a space ship have to travel in order that a 1-yr interval as measured by the observer in the ship be measured as a time interval of 2 yr by a stationary terrestrial observer?

5-7 A commuter is riding a train that is moving at a speed of 0.75c. When the train passes a station platform, a clerk at the station picks up a watch and sets it down. If the commuter observes that the clerk held the watch for 8.0 sec, how long does the clerk think that he held it?

5-8 The mean life of a charged pi meson when measured at rest is 2.6×10^{-8} sec. If the particle is traveling at a speed of 0.98c relative to the earth, what will be its mean life as measured by the terrestrial observer?

5-9 The distance from a given star to the earth is about 10^5 light years. Assuming that the lifetime of a person is 70 yr, at what speed must he travel to reach the star while he is still alive?

5-10 An earthbound astronomer observes a bright object in the northern sky 20 light years away approaching the earth at a speed of 0.80c. Assume that the earth is a stationary inertial system and calculate (a) the time required for the object to reach the earth as measured by the astronomer, (b) this time as measured by an astronomer on the bright object, and (c) the distance to the earth as measured by the astronomer on the bright object.

5-11 A rigid rod makes an angle $\theta_2 = 37°$ with respect to the x_2 axis. At what speed must the rod move parallel to the x_1 axis such that it will appear to be at an angle $\theta_1 = 45°$?

5-12 Show that the volume of a cube moving with a velocity v in the direction parallel to one edge is

$$V = V_0 \sqrt{1 - \left(\frac{v}{c}\right)^2}$$

where V_0 is the rest volume.

5-13 An astronomer on earth flashes a pulsed laser, and 1.3 sec later the pulse reaches the moon 3.9×10^8 m away. An observer traveling in the same direction as the pulse sees the two events (i.e., the flash and the arrival at the moon) as one event. What is the speed of this observer?

RECOMMENDED READING

ERBER, T., and MALHIST, R. J., "Transformation of Acceleration in Special Relativity," *Am. J. Phys.* **27,** 607 (1959).

FEYNMAN, R. P., LEIGHTON, R. B., and SANDS, L. M., *The Feynman Lectures on Physics*, Addison-Wesley, Reading, Mass., 1963, Vol. 1, Chaps. 15, 18, and 20.

FRENCH, A. P., *Special Relativity*, Norton, New York, 1968.
 Interestingly written book with many problems on relativity. Good chapter with discussion on moving clocks and other objects.

GOLD, T., "The Arrow of Time," *Am. J. Phys.* **30,** 137 (1963).

LIEBER, L. H., *The Einstein Theory of Relativity*, Holt, Rinehart & Winston, New York, 1945.
 A very elementary introduction to the special theory of relativity.

LOWRY, E. S., "The Clock Paradox," *Am. J. Phys.* **31,** 59 (1963).

SCOTT, G., and VINER, M., "The Geometric Appearance of Large Objects Moving at Relativistic Speeds," *Am. J. Phys.* **33,** 534 (1965).

SEARS, F. W., "Length of a Moving Rod," *Am. J. Phys.* **33,** 266 (1963).

Special Relativity Theory, Selected Reprints, American Institute of Physics, New York, 1963.
 Contains many good references and several excellent representative papers on the special theory of relativity.

WEISSKOPF, V. F., "The Visual Appearance of Rapidly Moving Objects," *Phys. Today* **13,** 24–27 (September 1960).
 This article is included in the *Selected Reprints* above. It corrects some common misconceptions of physicists about the shape of objects moving at relativistic velocities.

6
Relativistic Mechanics

Albert Einstein
(1879–1955)

Born in Ulm, Germany, Einstein studied at the University of Zurich. From 1914 to 1933 he taught at the University of Berlin and served as director of the Kaiser Wilhelm Institute. After being driven out of Nazi Germany in 1933, he continued his research at the Institute for Advanced Study at Princeton, N.J. Einstein, having published more than 300 scientific papers and books, is considered the outstanding physicist of the twentieth century. For his work on the photoelectric effect and his many profound contributions to theoretical physics, he was awarded the 1921 Nobel Prize.

6–1 MASS AND MOMENTUM
6–2 DEFINITION OF FORCE
6–3 RELATIVISTIC KINETIC ENERGY
6–4 TOTAL ENERGY
6–5 SCHEMATIC REVIEW

6–1 MASS AND MOMENTUM

Einstein's postulates of relativity forced physicists to reevaluate their views of mechanics. The classical expressions for momentum and energy must now be replaced by their relativistic expressions before they are incorporated into the laws of conservation of momentum and conservation of energy. In a sense, the ease with which the relativistic expressions fit into these conservation laws is a tribute to the great generality of these laws of physics. From classical mechanics, the linear momentum of a body with an inertial mass m and velocity $\mathbf{v}$ is defined by the equation

$$\mathbf{p} = m\mathbf{v} \tag{6-1}$$

In Chapter 2 we learned that the conservation of linear momentum for an isolated system of particles was presented as the most fundamental law of physics. For an isolated system of particles $m_1, m_2, \ldots, m_n$ with no external forces acting on it, the system will evolve in space and time in such a way that

$$\boxed{\sum_i m_i \mathbf{v}_i = m_1 \mathbf{v}_1 + m_2 \mathbf{v}_2 + \cdots + m_n \mathbf{v}_n = \text{constant}} \tag{6-2}$$

This conservation law expressed by Equation (6–2) is a consequence of the homogeneity of space in which all nature appears to be embedded. When a collision is observed from different moving frames of reference, then, there is no reason to expect space to become suddenly inhomogeneous! We must now find out how Equation (6–2) holds up under Lorentz transformations between moving coordinate systems. In anticipation of complications that may arise with regard to mass when the Lorentz transformations are made, the special symbol m_0 is assigned to the mass. The mass m_0 is measured for a body at rest in our own frame of reference and is known as the *rest mass* of the body.

Consider in a moving system S_2 two identical and perfectly elastic spheres each with a rest mass m_0 (Figure 6–1). In this moving system S_2, spheres A and B move with respective velocities

$$\begin{aligned} \mathbf{v}_{A_2} &= \mathbf{V} \\ \mathbf{v}_{B_2} &= -\mathbf{V} \end{aligned} \tag{6-3}$$

CHAPTER 6: RELATIVISTIC MECHANICS · 73

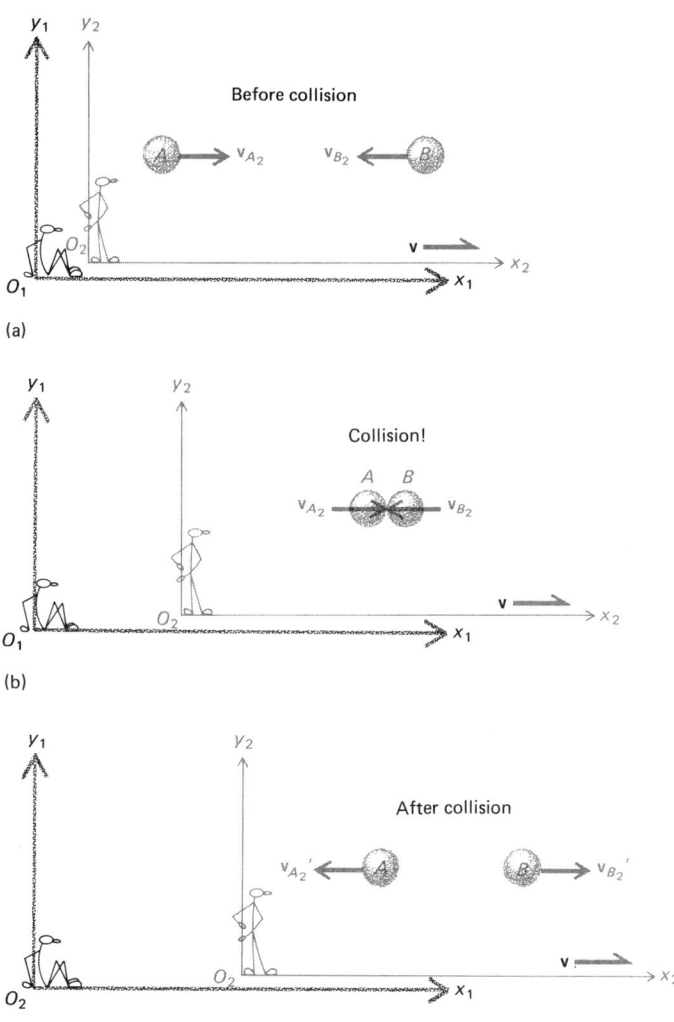

Figure 6–1 (a) Observer O_2 sees two spheres approach each other at equal speeds. (b) Here Observer O_2 sees two spheres just at the moment of impact, where $\mathbf{v}_{A_2} = \mathbf{v}_{B_2} = 0$. The spheres are momentarily at rest as far as observer O_2 is concerned. (c) Observer O_2 will see the sphere rebound with equal but oppositely directed speeds.

such that the spheres will make a head-on collision. Recalling the equation for the transformation of velocities, the Lorentz transformation is used to relate these two views of the same event. The Lorentz transformation of velocities shows that the speeds of the two spheres as seen by observer O_1 are

$$v_{A_1} = \frac{v_{A_2} + v}{1 + \beta(v_{A_2}/c)} = \frac{V + v}{1 + \beta(V/c)} \tag{6-4}$$

$$v_{B_1} = \frac{v_{B_2} + v}{1 + \beta(v_{B_2}/c)} = \frac{-V + v}{1 - \beta(V/c)} \tag{6-5}$$

where $\beta = v/c$.

If the sum of the masses as seen from S_1 is M, this total mass will remain constant throughout the collision and when they collide

$$m_{A_1} v_{A_1} + m_{B_1} v_{B_1} = M\mathbf{v} \tag{6-6}$$

in which

$$M = m_{A_1} + m_{B_1} \tag{6-7}$$

Thus, while observer O_2 sees the masses instantaneously at rest, observer O_1 sees them moving together with a speed v. From Equations (6-6) and (6-7), then,

$$\begin{aligned} m_{A_1}(v_{A_1} - v_{B_1}) &= M(v - v_{B_1}) \\ m_{B_1}(v_{B_1} - v_{A_1}) &= M(v - v_{A_1}) \end{aligned} \tag{6-8}$$

By using transformation Equations (6-4) and (6-5) and simplifying, the ratio of Equation (6-8) yields

$$\frac{m_{A_1}}{m_{B_1}} = \frac{1 + \beta(V/c)}{1 - \beta(V/c)} \tag{6-9}$$

Now, from Equation (6-4),

$$1 - \frac{v_{A_1}^2}{c^2} = 1 - \frac{(V + v)^2}{c^2[1 + \beta(V/c)]^2}$$

which can be rearranged algebraically to give

$$\sqrt{1 - \frac{v_{A_1}^2}{c^2}} = \frac{\sqrt{1 - \beta^2}\sqrt{1 - (V^2/c^2)}}{\sqrt{1 + \beta(V/c)}}$$

and similarly, from Equation (6-5),

$$\sqrt{1 - \frac{v_{B_1}^2}{c^2}} = \frac{\sqrt{1 - \beta^2}\sqrt{1 - (V^2/c^2)}}{1 - \beta(V/c)}$$

Factors $[1 + \beta(V/c)]$ and $[1 - \beta(V/c)]$ can now be extracted from these expressions and substituted in Equation (6–9). This now gives the ratio of the two masses as seen from S_1 as

$$\frac{m_{A_1}}{m_{B_1}} = \frac{\sqrt{1 - (v_{B_1}^2/c^2)}}{\sqrt{1 - (v_{A_1}^2/c^2)}} \tag{6-10}$$

Thus, the mass as seen from a moving frame of reference is not m_0 but is inversely proportional to the Lorentz factor $\gamma = 1/\sqrt{1 - (v^2/c^2)}$. Note that γ is always greater than 1 but it does approach unity as the velocity becomes very small compared to the velocity of light c. This allows us to write the general statement

$$m_{A_1}\sqrt{1 - \frac{v_{A_1}^2}{c^2}} = m_{B_1}\sqrt{1 - \frac{v_{B_1}^2}{c^2}} = m_0$$

or simply

$$\boxed{m = \gamma m_0} \tag{6-11}$$

The mass of a body then is not, in general, a constant nor the same for all observers but is a quantity that

1. depends on the reference from which the body is being observed, and
2. is less than or equal to m_0 when the body is at rest in the frame of reference from which the body is being observed.

The properties of the Lorentz factor γ cause the mass to become very large and finally to approach infinity as the relative velocity approaches c.

In agreement with the mass formula, the relativistic expression for the linear momentum is

$$\boxed{p = mv = \gamma m_0 v} \tag{6-12}$$

and the conservation of linear momentum for an isolated system is

$$\boxed{\sum_{i=1}^{n} m_i \mathbf{v}_i = \sum_{i=1}^{n} \gamma_i m_0 \mathbf{v}_i = \text{constant}} \tag{6-13}$$

6–2 DEFINITION OF FORCE

Although classical mechanics laws are not universal enough to include relativistic effects, the statement of Newton's second law,

$$\mathbf{F} = \frac{d}{dt}(\mathbf{p}) = \frac{d}{dt}(m\mathbf{v}) \tag{6-14}$$

is generally applicable, even to relativistic mechanics. By differentiating, Equation (6–14) can now be expressed as

$$\mathbf{F} = m \frac{d\mathbf{v}}{dt} + \mathbf{v} \frac{dm}{dt} \tag{6–15}$$

where m is now recognized as γm_0.

For a force acting in the positive x direction, we can write

$$F_x = \frac{d}{dt}(mv_x) = \frac{d}{dt}\left[\frac{m_0 v_x}{\sqrt{1 - (v^2/c^2)}}\right]$$

Differentiating gives

$$F_x = \frac{m_0}{\sqrt{1 - (v_x^2/c^2)}} \frac{dv_x}{dt} + \frac{m_0(v_x^2/c^2)}{[1 - (v_x^2/c^2)]^{3/2}} \frac{dv_x}{dt}$$

which simplifies further to

$$F_x = \frac{m_0}{[1 - (v_x^2/c^2)]^{3/2}} \frac{dv_x}{dt} \tag{6–15a}$$

or

$$F_x = \gamma^3 m_0 a_x \tag{6–15b}$$

where a_x is the acceleration observed in the laboratory.

EXAMPLE 6–1: Determine the relativistic force acting on a body moving with uniform circular motion.

SOLUTION: In this case, the magnitude of the velocity remains constant and

$$\mathbf{F} = \frac{d}{dt}\left[\frac{m_0}{\sqrt{1 - (v^2/c^2)}} \mathbf{v}\right] = \frac{m_0}{\sqrt{1 - (v^2/c^2)}} \frac{d\mathbf{v}}{dt}$$

Note that $m = m_0/\sqrt{1 - (v^2/c^2)}$ and $d\mathbf{v}/dt = \mathbf{a}_R$, which is the centripetal acceleration. Hence, in magnitude we can write

$$F = ma_R = m \frac{v^2}{R}$$

where R is the radius of the circle. Thus, Newton's second law covers the case of relativistic circular motion.

6-3 RELATIVISTIC KINETIC ENERGY

When the speed of a particle approaches relativistic values, the expression for classical kinetic energy must be changed to a relativistic form. To find an expression for the relativistic kinetic energy, we shall calculate the work done in increasing the speed of a particle from 0 to a final value v. To simplify the problem, the force and displacement are assumed to be in the same direction.

The kinetic energy, then, which is the net work done on the particle, is

$$\text{KE} = \int_0^r \mathbf{F} \cdot d\mathbf{r} \tag{6-16}$$

From Equation (6-15a) this becomes

$$\text{KE} = \int_0^r \frac{m_0}{[1 - (v^2/c^2)]^{3/2}} \frac{d\mathbf{v}}{dt} \cdot d\mathbf{r}$$

Since $d\mathbf{r} = \mathbf{v}\, dt$, $m = m_0/\sqrt{1 - (v^2/c^2)}$, and $\mathbf{v} \cdot d\mathbf{v} = v\, dv$

$$\text{KE} = \int_0^v \frac{m_0}{[1 - (v^2/c^2)]^{3/2}} v\, dv$$

which integrates to give

$$\text{KE} = \frac{m_0 c^2}{\sqrt{1 - (v^2/c^2)}} - m_0 c^2$$

or finally,

$$\boxed{\text{KE} = (m - m_0)c^2} \tag{6-17}$$

Although this was derived for the special case of the force being in the same direction as the displacement, this is a general expression that is applicable to any case.

We can easily reduce this expression for the kinetic energy to the classical form, $\text{KE} = \tfrac{1}{2} m_0 v^2$, when $v \ll c$. To show this, expand Equation (6-17) by the binomial expansion,

$$(1 + x)^m = 1 + mx + \frac{m(m-1)}{2!} x^2 + \cdots$$
$$+ \frac{m(m-1)\cdots(m-n+1)}{n!} x^n + \cdots$$

The kinetic energy then becomes

$$\text{KE} = m_0 c^2 \left[\left(1 - \frac{v^2}{c^2}\right)^{-1/2} - 1 \right]$$

or

$$\text{KE} = m_0 c^2 \left\{ \left[1 + \left(-\frac{1}{2}\right)\left(-\frac{v^2}{c^2}\right) + \frac{-\frac{1}{2}(-\frac{3}{2})}{2!}\left(-\frac{v^2}{c^2}\right)^2 + \cdots \right. \right.$$

$$\left. \left. + \frac{m(m-1)\cdots(m-n+1)}{n!}\left(-\frac{v^2}{c^2}\right)^n + \cdots \right] - 1 \right\}$$

or

$$\text{KE} = m_0 c^2 \left(1 + \frac{1}{2}\frac{v^2}{c^2} + \frac{3}{8}\frac{v^4}{c^4} + \cdots - 1 \right)$$

$$= m_0 c^2 \left(\frac{v^2}{2c^2} + \frac{3}{8}\frac{v^4}{c^4} \right)$$

As $v/c \to 0$, the higher powers of v/c can be neglected, and then

$$\text{KE} = m_0 c^2 \left(\frac{1}{2}\frac{v^2}{c^2} \right) = \tfrac{1}{2} m_0 v^2$$

which is a statement of the correspondence principle.

EXAMPLE 6–2: Although the computer program below is written for students with some background in computer programming, the BASIC language in which it is written is mathematical enough in form that it is not too difficult for a student of physics to follow. The comments following the single quotation mark in some statements are for description and play no part in the computation.

From the expanded expression for kinetic energy above, note that the nth term T_n compared to the $(n-1)$th term T_{n-1} is

$$\frac{T_n}{T_{n-1}} = \left(\frac{n - \frac{1}{2}}{n}\right)\left(\frac{v^2}{c^2}\right)$$

This relation is used in the following BASIC program to evaluate and compare the expansion of the relativistic kinetic energy with the classical kinetic energy.

In the program, $F = V/C$ is the ratio of a given speed with respect to the speed of light, and N is the number of terms in the expansion to be used. When a large number of terms in the expansion are calculated, an individual term may become so small that it adds an insignificant amount to the calculation. Statement 100 is a check that will just skip over terms that are too small. The program determines the kinetic energy for an electron, but statement 10 can be changed to include any mass desired.

CHAPTER 6: RELATIVISTIC MECHANICS · 79

```
10   LET MO = 9.1091E-31
20   LET C = 2.99793E+8
30   PRINT"V/C","REL KE-JOULES","CL KE-JOULES,"% "DIFF","NO. TERMS"
40   READ F,N              'F = V/C AND N = NUMBER OF TERMS IN
                            THE EXPANSION
50   LET V = F*C
60   LET S = 0             'INITIALIZE SERIES SUM
70   LET S1 = 1            'INITIALIZE THE EVALUATION OF EACH TERM
80   FOR I = 1 TO N        'BEGIN LOOP FOR CALCULATION
90   LET S1 = S1*((I-.5)/I)*F↑2   'CALCULATES EACH TERM
100  IF S1 < 1E-30 THEN 130 'TERM IS TOO SMALL—GO OUT OF THE
                            LOOP
110  LET S = S + S1        'SUMMATION OF SERIES
120  NEXT I
130  LET K1 = MO*C↑2*S     'RELATIVISTIC KE IN TERMS OF SERIES
                            EXPANSION
140  LET K2 = (MO*V↑2)/2   'CLASSICAL KE
150  PRINT F,K1,K2,100*(K1-K2)/K2,N
160  IF I > 100 THEN 190 ' A MEANS OF SKIPPING A SPACE BETWEEN DATA
170  GO TO 40
180  DATA .95,2,.95,5,.95,10,.95,20,.95,100,.95,1000
190  PRINT
200  PRINT
210  GO TO 40
220  DATA .999,100,.95,100,.80,100,.50,100,.10,100,.01,100
230  DATA .001,100,1E-6,100
240  END
```

(a) To see the effect of the expansion, let $F(\equiv V/C) = 0.95$ and make runs for $R = 2, 5, 10, 20, 100, 1000$ terms (see data statement 160).

V/C	REL KE-JOULES	CL KE-JOULES	% DIFF	NO. TERMS
0.95	6.19493 E-14	3.69433 E-14	67.6875	2
0.95	1.0767 E-13	3.69433 E-14	191.447	5
0.95	1.44292 E-13	3.69433 E-14	290.576	10
0.95	1.70079 E-13	3.69433 E-14	360.379	20
0.95	1.8032 E-13	3.69433 E-14	388.098	100
0.95	1.80321 E-13	3.69433 E-14	388.102	1000

(b) Next compare the kinetic energies as a function of velocity. Let $R = 100$ and make runs for $F = 0.999, 0.95, 0.80, 0.50, 0.10, 0.01, 0.0001$.

0.999	7.87098 E-13	4.08526 E-14	1826.68	100
0.95	1.8032 E-13	3.69433 E-14	388.098	100
0.8	5.45792 E-14	2.6198 E-14	108.333	100
0.5	1.26651 E-14	1.02336 E-14	23.7604	100
0.1	4.1244 E-16	4.09344 E-16	0.756303	100
0.01	4.09375 E-18	4.09344 E-18	7.50202 E-3	100
0.001	4.09344 E-20	4.09344 E-20	7.4002 E-5	100
0.000001	4.09344 E-26	4.09344 E-26	9.40984 E-7	100

OUT OF DATA IN 40

6-4 TOTAL ENERGY

From Equation (6–17), if a body moving with a speed v_1 increases its speed to v_2, the net work required to do this, or the change in kinetic energy, will be

$$\Delta KE = (m_2 - m_0)c^2 - (m_1 - m_0)c^2$$

or

$$\boxed{\Delta KE = (m_2 - m_1)c^2 = (\Delta m)c^2} \tag{6–18}$$

Thus, a change in speed (or kinetic energy) will produce a change in mass $\Delta m = m_2 - m_1$.

For a body moving in a conservative force field, the conservation of energy (valid both in relativistic and classical mechanics) shows that

$$\boxed{K_1 + V_1 = K_2 + V_2 = \text{constant}} \tag{6–19}$$

where K is the kinetic energy at a given point and V is the potential energy at that same point. From Equations (6–18) and (6–19), we conclude that

$$K_2 - K_1 = V_1 - V_2 = (\Delta m)c^2$$

or

$$\Delta m = \frac{K_2 - K_1}{c^2} = \frac{V_1 - V_2}{c^2} \tag{6–20}$$

Thus,

$$\boxed{\text{change in mass} = \frac{\text{change in KE}}{c^2} = \frac{\text{change in PE}}{c^2}}$$

Since the rest energy is defined as $E_0 = m_0 c^2$, the *total energy* will be defined as

$$\boxed{E = E_0 + K} \tag{6–21}$$

and since $E = m_0 c^2 + (m - m_0)c^2$,

$$\boxed{E = mc^2} \tag{6–22}$$

Notice that this definition of total energy in relativity does not include potential energy.

The equivalence between mass and energy [as expressed in Equation (6–22)] is one of the most important consequences of the special theory of relativity. It gives a different and more accurate insight into the conservation of energy. It now becomes the principle of mass-energy conservation, which for an isolated system can be stated as

$$\boxed{\sum (\text{rest energy} + \text{kinetic energy} + \text{potential energy}) = \text{constant}} \quad (6\text{-}23)$$

This was a consequence of the principle of the conservation of linear momentum from Equation (6–2) and the definition of force as found in Equation (6–14).

Another useful relationship that includes the total energy E can be obtained directly from the mass formula $m\sqrt{1 - (v^2/c^2)} = m_0$. Multiplying both sides of this equation by c^2 and then squaring and simplifying will give

$$m^2c^4 = m_0^2c^4 + m^2v^2c^2 \quad (6\text{-}24)$$

Since $p = mv$, this can also be written as

$$\boxed{E^2 = E_0^2 + p^2c^2} \quad (6\text{-}25)$$

If the body is moving at a very high speed, then E_0^2 is negligible compared to p^2c^2 and

$$E = pc \quad (6\text{-}26)$$

At high speeds E_0 is also small compared to K, and Equation (6–21) shows that

$$E \cong K$$

or, from Equation (6–26),

$$K \cong pc \quad (6\text{-}27)$$

Particles at very high speeds for which Equations (6–27) and (6–26) are useful are said to be in the *extreme relativistic* region.

Another interesting relationship involving the total energy is obtained by differentiating Equation (6–25). This is

$$\frac{dE}{dp} = \frac{pc^2}{E}$$

or

$$\boxed{\frac{dE}{dp} = \frac{pc^2}{mc^2} = v} \quad (6\text{-}28)$$

Now, if the body is moving at the speed of light, that is, $v = c$, then $dE = c\,dp$ or

$$E = pc + \text{constant}$$

For $p = 0$, $E = E_0$, and therefore

$$E - E_0 = pc \quad (6\text{-}29)$$

But Equation (6–25) shows that

$$E^2 - E_0^2 = p^2 c^2$$

and thus these two equations give

$$E + E_0 = pc \tag{6-30}$$

Comparison of Equations (6–30) and (6–29) implies that $E_0 = 0$ or $m_0 = 0$. In other words, if a body is moving at the speed of light, its rest mass and its rest energy must be zero. The reciprocal conclusion must also be true: If an entity has no rest mass nor rest energy, it must be traveling at the speed of light. Although it does not make sense for a body to have a mass of zero from a classical point of view, it is the correct relativistic description of a photon and of a neutrino.

R. V. Pound and G. A. Rebka, Jr., performed an experiment in 1960 using the Mossbauer effect and found that the mass of a photon moving at the speed of light (this is the only speed at which it can move) is given by $m = h\nu/c^2$, in agreement with the theoretical equation $E = h\nu = mc^2$.

EXAMPLE 6–3: Compute the mass of a proton, a neutron, and an electron in atomic mass units, and compute the equivalent rest mass energy of these particles.

SOLUTION: The *electron volt* (eV) is a convenient unit of energy defined as the kinetic energy gained by a body containing 1 electronic charge as it is accelerated through a potential difference of 1 V. Since the absolute charge of the electron is $q = 1.60 \times 10^{-19}$ coulombs (C), we have

$$qV = (1.60 \times 10^{-19} \text{ C})(1 \text{ V}) = 1.60 \times 10^{-19} \text{ J} = 1 \text{ eV}$$

where the accelerating potential is 1 V. Some convenient multiples of the electron volt are

$$1 \text{ MeV} = 10^6 \text{ eV}$$
$$1 \text{ BeV} = 10^9 \text{ eV}$$

In modern usage, the BeV is giving way to the European "GeV." The magnitudes of both are the same. Unless otherwise stated, the energy of a particle is given as the kinetic energy. Thus, a 1.0-MeV electron has a kinetic energy of 1.0 MeV, and not a total energy of 1.0 MeV.

The atomic mass unit (amu) is defined as one-twelfth of the mass of the neutral carbon C-12 atom (the most common isotope of carbon), and it is

$$1 \text{ amu} = 1.660 \times 10^{-27} \text{ kg}$$

The rest energy, which corresponds to 1 amu, is

$$E_0 = m_0 c^2 = (1.66 \times 10^{-27})(3.00 \times 10^8)^2 = 14.9 \times 10^{13} \text{ J}$$

$$\boxed{E_0 = 14.9 \times 10^{-13} \text{ J} = 931 \text{ MeV}}$$

The electron rest mass is $m_e = 9.11 \times 10^{-31}$ kg, and its rest energy is

$$E = m_e c^2 = (9.11 \times 10^{-31})(3.00 \times 10^8)^2 \left(\frac{1}{1.6 \times 10^{-13}} \frac{\text{MeV}}{\text{J}}\right)$$
$$= 0.511 \text{ MeV}$$

Since we have determined that there is an equivalence between mass and energy, it is often convenient to express the atomic mass unit and its energy equivalent in MeV interchangeably. Thus, although admittedly dimensionally inconsistent, we write

$$\boxed{1 \text{ amu} = 1.66 \times 10^{-27} \text{ kg} \,(= 931 \text{ MeV})}$$

For the electron, then,

$$m_e = \frac{0.511 \text{ MeV}}{931 \text{ MeV/amu}} = 0.00055 \text{ amu}$$

and the rest masses of the neutron and the proton are

$$m_n = 1.675 \times 10^{-27} \text{ kg} \quad \text{(neutron)}$$
$$m_p = 1.672 \times 10^{-27} \text{ kg} \quad \text{(proton)}$$

By a similar procedure we get

neutron rest energy = 939.6 MeV = 1.00867 amu
proton rest energy = 938.3 MeV = 1.00783 amu

A summary of these results is given below*:

	REST MASS (amu)	REST MASS (kg)	REST ENERGY (MeV)
atomic mass unit	1	1.660×10^{-27}	931
electron	0.00055	9.109×10^{-31}	0.511
neutron	1.00867	1.675×10^{-27}	939.6
proton	1.00729	1.673×10^{-27}	938.3

EXAMPLE 6–4: The speed of an electron in a uniform electrical field changes from $v_1 = 0.98c$ to $v_2 = 0.99c$.

(a) Compute the change in mass.
(b) Compute the work done on the electron to change its velocity.
(c) Calculate the accelerating potential in volts.

* For recent values, see B. N. Taylor, D. N. Langenberg, and W. H. Parker, "The Fundamental Constants," *Sci. Am.*, October 1970, pp. 62–73.

SOLUTION

(a) Evidently, the two masses will be

$$m_1 = \frac{m_0}{\sqrt{1 - 0.98^2}} = 5.0 m_0$$

and

$$m_2 = \frac{m_0}{\sqrt{1 - 0.99^2}} = 7.1 m_0$$

where $m_0 = 9.11 \times 10^{-31}$ kg is the rest mass of the electron. The change in mass will be

$$\Delta m = m_2 - m_1 = (7.1 - 5.0) m_0 = 19.1 \times 10^{-31} \text{ kg}$$

(b) Since the work done will be the change in kinetic energy,

$$\Delta K = K_2 - K_1 = (\Delta m) c^2 = 2.1 m_0 c^2$$
$$= 2.1 \times 0.511 = 1.07 \text{ MeV}$$

(c) $K = qV$ and

$$V = \frac{K}{q} = \frac{1.07 \times 1.6 \times 10^{-13}}{1.5 \times 10^{-19}}$$
$$= 1.07 \times 10^6 \text{ V}$$

6-5 SCHEMATIC REVIEW

When physicists understood the implications of the two postulates from Einstein's special theory of relativity,

1. the physical laws of nature are the same in all inertial frames of reference, and
2. the speed of light is the same in all inertial frames of reference,

the concepts of Newtonian mechanics, although they had been broadly successful, had to give way to relativistic mechanics.

Table 6-1 summarizes schematically the features of the theory of special relativity. It represents a logical scheme, but it is not necessarily in chronological order of development nor even in the only logical order. For example, the law of conservation of linear momentum is the most general law in physics, but Newton's laws, developing the ideas of force, were the first to be formulated. Also, the theoretical physicists can argue that the Michelson-Morley experiment should follow the principles of special relativity because it substantiates those ideas first presented in special relativity.

Table 6-1 Schematic summary of the special theory of relativity

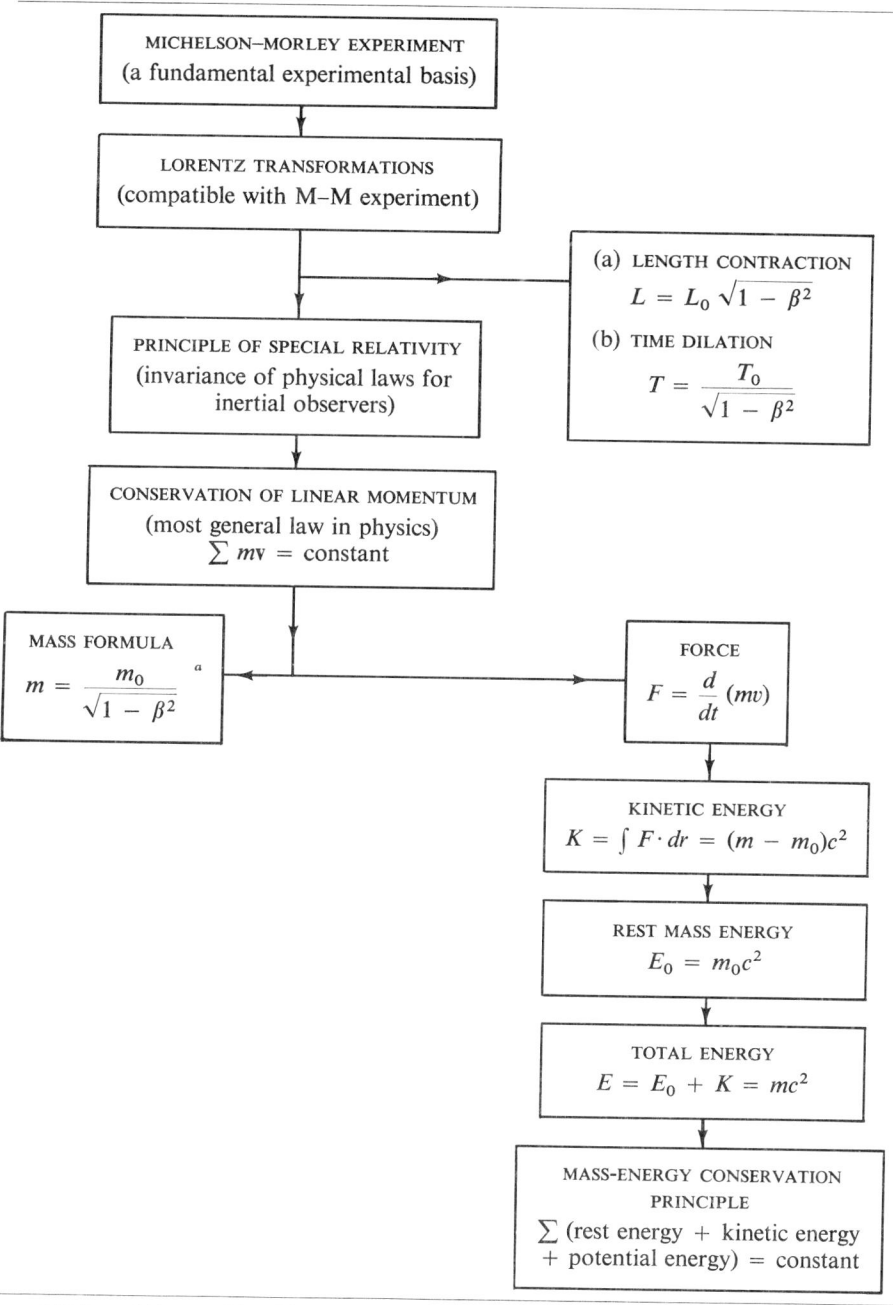

[a] This could be considered fundamental and the conservation of momentum derived from it—hence the double arrow. The necessary and sufficient condition for the conservation of linear momentum (assuming as a fact the Lorentz transformation) is that

$$m = \frac{m_0}{\sqrt{1 - \beta^2}}$$

PROBLEMS

6-1 Determine the total energy of a proton traveling at $0.800c$.

6-2 At what speed will an electron have a mass that is twice its rest mass? What is the total energy of the electron at this speed?

6-3 What is the momentum of an electron with a velocity of $0.980c$?

6-4 Show that the total energy and the rest mass energy can be related by

$$E = \frac{E_0}{\sqrt{1 - (v/c)^2}}$$

6-5 With reference to Problem 6-4, find v in terms of E_0 and E.

6-6 Find the mass and momentum of a 1.00-BeV proton.

6-7 (a) The accelerator at the Stanford Linear Accelerator Center produces highly relativistic 20.0-GeV electrons. Determine the speed, momentum, and wavelength of these electrons. (Hint: See Problem 6-4 and solve for v by the binomial expansion.)
(b) The electrons are accelerated over a distance of 3200 m (about 2 miles). What would be the length of path of the electrons as measured by an observer moving with the electron?

6-8 Determine the speed and momentum of a particle of rest mass m_0 when its kinetic energy is equal to twice its rest energy.

6-9 Calculate the work required to accelerate an electron (a) from rest to 4000 m/sec, (b) from rest to $0.800c$, and (c) from $0.980c$ to $0.999c$.

6-10 A laboratory observer sees that a proton moving at $0.500c$ makes a head-on collision with a second proton moving in the opposite direction at $0.600c$. (a) Determine the kinetic energy and momentum of the system as measured by the laboratory observer. (b) Determine the kinetic energy and momentum of the system as measured by an observer moving along with the proton of speed $0.500c$.

6-11 A unit of measure of momentum often used is 1.00 MeV/c. Find its numerical value in MKS units (kilogram-meters per second).

6-12 For a proton, determine the total energy when it has a momentum of (a) 2.00 BeV/c, and (b) 1.00 MeV/c.

6-13 Show that the ratio of the relativistic kinetic energy $K = (m - m_0)c^2$ to the approximate expression $K' = \frac{1}{2}m_0 v^2$ is given by $K/K' = 1 + \frac{3}{4}\beta^2$.

6-14 Proton A leaves a linear accelerator with a speed of $0.800c$ relative to the laboratory frame and collides with proton B at rest in the same laboratory frame. (a) Calculate the momentum and kinetic energy of the protons in the laboratory frame. (b) Calculate the speed of the center of mass (cm) of this system in the laboratory frame. (c) Do the same computations as in (a) for the cm frame.

6-15 A proton and an electron each have a kinetic energy of 10.0 MeV. (a) Compute their momenta and speeds following a classical approach. (b) Do the same computation using a relativistic approach. (c) What can you conclude from a comparison of the results of the two computations?

6-16 (a) What is the minimum speed that a particle must have in order that its total energy can be written as $E = pc$ with an error in its kinetic energy not greater than 1%?
(b) What are the values of the momentum and kinetic energy of a proton moving at that speed?

6-17 (a) Compute the maximum speed that a particle must have in order that its kinetic energy can be written as $K = \frac{1}{2}m_0v^2$ with an error not greater than 1%.
(b) Under these circumstances, compute the momentum and kinetic energy of an electron.

6-18 In a β^- decay process, the reaction

$$_0n^1 \rightarrow \,_1p^1 + \,_{-1}\beta^0 + \bar{\nu}$$

takes place, where n is a neutron at rest, p is a proton, and $\bar{\nu}$ is an antineutrino whose rest mass is zero. Compute the total kinetic energy of the decay products (proton + electron + antineutrino). (Hint: Use the mass-energy conservation principle.)

6-19 (a) Starting with the equation $E = \sqrt{E_0^2 + p^2c^2} = E_0 + K$, show that the linear momentum of a particle can be written as $p = (\sqrt{2E_0 K + K^2})/c$.
(b) Prove that this expression reduces to $p = m_0 v$ when $\beta = v/c \rightarrow 0$.
(c) Prove that the expression reduces to $p = E/c = K/c$ when $\beta \rightarrow 1$.

6-20 Prove that when a particle is moving perpendicular to a magnetic field B, it will describe a circle whose radius is given by $R = (\sqrt{2E_0 K + K^2})/qcB$, where q is the electrical charge of the particle.

6-21 The momentum of a proton moving in a circular path and perpendicular to a magnetic field of $1.00\,T$ has a constant magnitude of 2.40×10^{-22} kg-m/sec. Calculate (a) the radius of the circle, and (b) the kinetic energy of the proton.

6-22 An electron moves in a circular path whose radius is 0.600 m with constant speed and perpendicular to a magnetic field of $0.0300 T$ (Wb/m^2). In terms of its rest mass, find (a) its relativistic mass, (b) its kinetic energy, (c) its total energy, (d) its linear momentum, and (e) its angular momentum.

6-23 Show that the density of a cube moving with a velocity v in a direction parallel to one edge is

$$\rho = \frac{m_0}{V_0} \frac{1}{[1 - (v/c)^2]}$$

where V_0 is the rest volume and m_0 is the rest mass.

RECOMMENDED READING

BERTOZZI, W., "Speed and Kinetic Energy of Relativistic Electrons," *Am. J. Phys.* **32**, 551 (1964)

EINSTEIN, A., *The Meaning of Relativity*, Princeton University Press, Princeton, N.J., 1956.

FEYNMAN, R. P., LEIGHTON, R. B., and SANDS, L. M., *The Feynman Lectures on Physics*, Addison-Wesley, Reading, Mass., 1963, Vol. I, Chaps. 15, 16, and 17.

FRENCH, A. P., *Speed Relativity*, Norton, New York, 1966.
A well-written text with a good literary style.

GOOD, R. H., *Basic Concepts in Relativity*, Van Nostrand Reinhold, New York, 1968, Chaps. 1, 2, and 3.
Extremely good and clear presentation of the special theory of relativity. Excellent as collateral reading for any student of modern physics.

KACSER, C., *Introduction to the Special Theory of Relativity*, Prentice-Hall, Englewood Cliffs, N.J., 1967, pp. 148–56.

LEWIS, G. N., and TOLMAN, R. C., *Phil. Mag.* **18**, 510 (1909).
An original article on the theory of relativity.

MOLLER, C., *The Theory of Relativity*, Oxford University Press, London, 1952.
A very comprehensive study of the special and general theories of relativity.

SHANKLAND, R. S., "Conversations with Einstein," *Am. J. Phys.* **31**, 47 (1963).

Special Relativity Theory, Selected Reprints, American Institute of Physics, New York, 1963.

TWO Particles and Waves

> The Compton effect—a photoelectric game of billiards.
>
> MAX BORN
> *The Restless Universe*, 1936

This title to a section of Born's book succinctly describes the Compton effect. In Compton's experiment it was shown that photons of light not only have a wavelike character, but also behave much like particles when they are bounced off electrons. De Broglie reasoned that if light—known to be a wave phenomenon—could exhibit particle characteristics, why could not the reverse be true? He proposed that particles, such as electrons, could exhibit wavelike characteristics. The Davisson–Germer experiments did show that electrons scattered from crystals behaved as if they were diffracted—a characteristic of wave phenomena.

7
Photoelectric Effect

Robert Andrews Millikan
(1868–1953)

A native of Morrison, Illinois, Millikan studied at Oberlin College and received his doctorate from Columbia University in 1895. From 1921 to 1945 he was director of the Norman Bridge Laboratory of Physics at the California Institute of Technology. His early research was in the field of x rays and the free expansion of gases. In 1910 he isolated the electron and measured its electric charge while also working to verify Einstein's photoelectric equations. For his study of the elementary electrical charge and the photoelectric effect, he was awarded the 1923 Nobel Prize.

7–1 QUANTA OF ELECTRICITY
7–2 ELECTRON EMISSION
7–3 PHOTOELECTRIC EFFECT

7-1 QUANTA OF ELECTRICITY

"Science walks forward on two feet, namely theory and experiment."* R. A. MILLIKAN's work on the fundamental electrical unit and the photoelectric effect illustrates steps forward in an experimental direction. Weber, in 1871, and Stoney, in 1881, respectively, had theoretically developed the concept and numerical value for the ultimate electrical unit—the charge on the electron. In 1897 Thomson and Zeeman were able to determine the ratio of the charge to the mass of the electron, but Millikan's oil drop experiment was the first direct measurement of the charge alone.

By introducing a charged oil drop into an electrical field between two plates, Millikan was able to observe the effects of both electrical and gravitational forces on the drops. From many very careful and painstaking observations he was able to show that although the speeds for different drops in the electrical field were not always the same, they were always integral multiples of the same value. This experimental fact was attributed to the quantization of the charge picked up by individual oil drops. The value of charge on the electron, $e = 1.602 \times 10^{-19}$ C, can be determined from the physical constants and the equations of motions involved. Anderson's *Discovery of the Electron* (see references at the end of the chapter) gives a good presentation of the oil drop experiment, the physical quantities used, and an analysis of typical data.

7-2 ELECTRON EMISSION

It has been found that electrons can be removed from metals by the following mechanisms:

1. thermionic emission (the Edison effect)—electrons are emitted from a heated metal surface
2. secondary emission—energetic particles incident on some materials release still other electrons from the surface
3. field emission—a strong electrical field extracts electrons from a metal surface
4. photoelectric effect—light incident on a metal ejects electrons from the surface.

* From Millikan's Nobel Lecture, May 1924.

The *photoelectric effect* was accidentally discovered by Hertz in 1887 as he was investigating the electromagnetic waves predicted by Maxwell's theory of the electromagnetic field. After publication of this discovery, numerous other investigators began studies of this effect. Figure 7–1(a) is a schematic for a typical experimental photoelectric effect. Chief among the early experimenters was PHILIPP LENARD,* whose experiments showed the following:

1. When light of frequency $v \geqq 10^{15}$ Hz falls on a clean metal plate K of a metal such as tungsten or zinc, negatively charged particles are emitted from the metal and travel toward positive electrode P.
2. This emission occurs even when the tube is highly evacuated and so the charge carriers are not gaseous ions carriers.
3. A magnetic field applied in the region between K and P deflects the charged carriers as if they were negative.
4. The measured ratio of the charge to mass e/m for the charged carriers was found to be

$$\frac{1.60 \times 10^{-19} \text{ C}}{9.10 \times 10^{-31} \text{ kg}} = 1.76 \times 10^{-11} \text{ C/kg}$$

which was the same as the value found for electrons by Millikan and Thomson.

The above experimental evidence identified the charged carriers as *photoelectrons*.

Lenard, using monochromatic incident radiation at constant intensity, plotted the number of electrons emitted from the metal (the photoelectric current i_p) that reached P versus the accelerating potential between K and P to get a set of curves like those of Figure 7–1(b). The *saturation current* $i_{p(max)}$ for a given intensity I is reached when all electrons emitted from the metal surface reach P. Note that even when $V = 0$, there is still a photoelectric current i_{po}, which means that some of the electrons must have been emitted with finite initial velocity. The stopping potential is an indication of the kinetic energy of the emitted electrons. The potential between K and P may be made negative ($V = -V_0$) until only the most energetic electrons can reach P. At this point,

$$\boxed{K_{max} = eV_0} \tag{7-1}$$

with V_0 defined as the *stopping potential*, and K_{max} as the maximum kinetic energy of the emitted electrons.

When the intensity of the incident radiation is increased, the saturation current $i_{p(max)}$ increases (more electrons are emitted), but *none of the electrons are more energetic* since the stopping potential remains the same. The stopping potential is independent of the intensity of the incident light.

---

* Philipp Lenard (Germany), a student of Hertz, received the 1905 Nobel Prize for his studies of cathode rays.

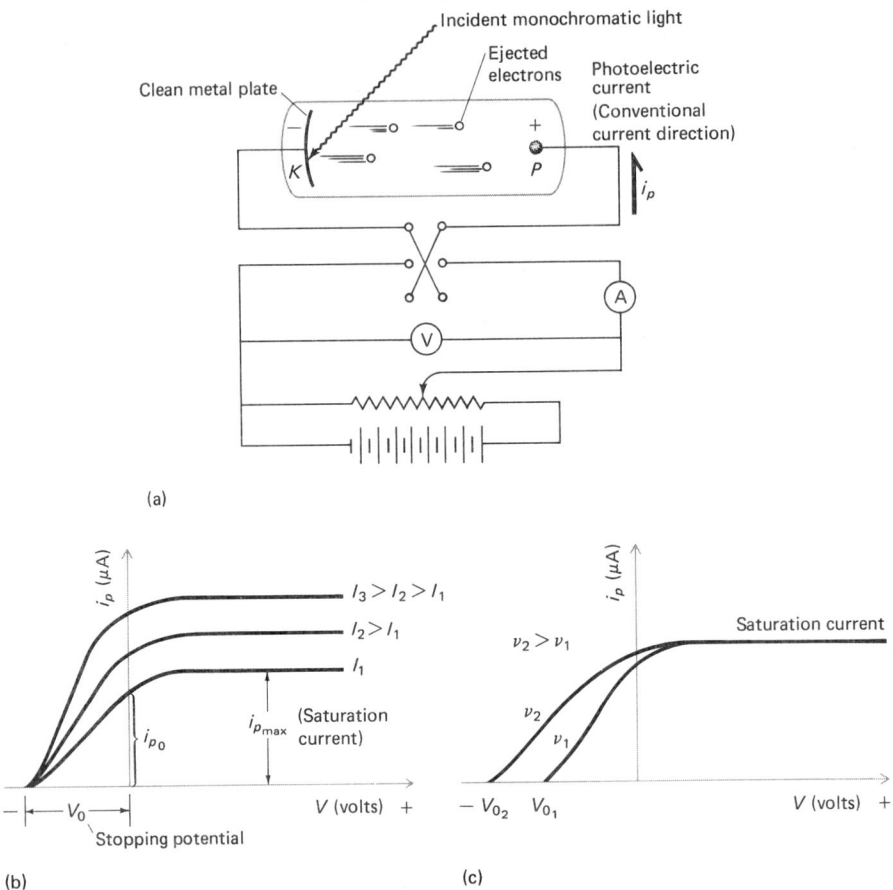

Figure 7–1 (a) Schematic for photoelectric experiment. (b) Photoelectric current versus the accelerating potential V for incident monochromatic light of wavelength λ. (c) Photoelectric current versus accelerating potential to show frequency dependence.

When incident light of different frequencies but with the same intensity is used, the number of electrons emitted in each case is the same, but the most energetic electrons are those emitted by the light of the highest frequency. This is illustrated in Figure 7–1(c), where the saturation current depends on the intensity and not on the frequency, but the stopping potential gets larger (more negative) as the incident frequency increases.

7-3 PHOTOELECTRIC EFFECT

It is expected that the greater the intensity of the incident light, the more electrons will be emitted from the surface of the metal. However, the classical theory is inadequate for the explanation of other aspects of the photoelectric phenomena.

Classical theory says that the more intense the incident radiation, the more energetic will be electrons emitted from the metal. Also, if the intensity of the incident radiation is very weak, classically it is expected that some time will be required to build up enough energy in the metal to eject electrons. However, experiments have shown that the kinetic energy of photoelectrons does not depend on the intensity but increases with the increase in the incident frequency, and that there is no appreciable time lag before electrons are ejected even for light of very weak intensity.

In 1905 Einstein, using the new concepts of quantum mechanics, assumed the incident radiation to be bundles of localized energy $E = hv$ that travel with the speed of light. He correctly developed the theory of the photoelectric effect. When photons fall on a metallic surface, the following can happen:

1. The photons can be reflected according to the laws of optics.
2. The photons can disappear, giving up their total energies in order to knock out electrons.

In Figure 7–1(c), the photoelectric current is again plotted versus the potential, but this is done for different sources of increasing frequency. Note that with increasing frequency it takes a still greater stopping potential to reduce the photoelectric current to zero. From a graph like Figure 7–1(c), data can be obtained to produce a plot of K_{max} ($= eV_0$) versus the incident frequency v as shown in Figure 7–2 for three different metals. Two interesting features are seen in the graph in

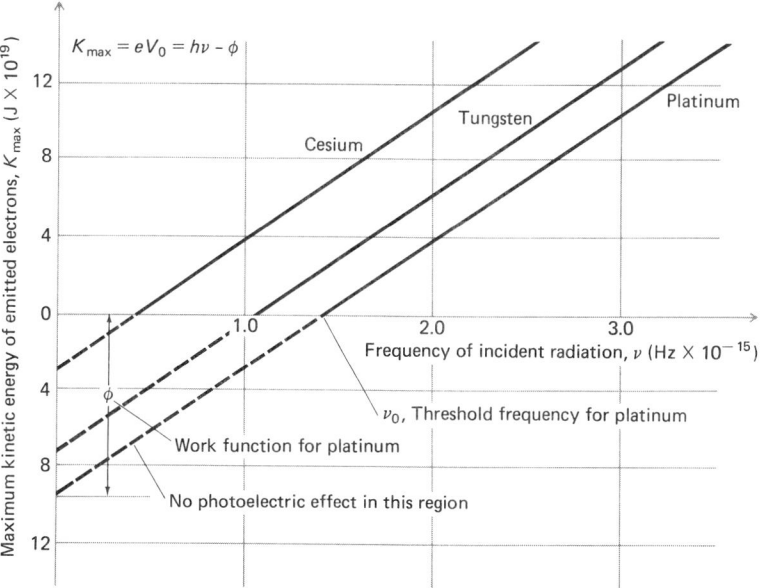

Figure 7–2 The maximum kinetic energy of photoelectrons K_{max} ($= eV_0$) versus the frequency of incident radiation.

Figure 7–2. First, the relation between K_{max} and v is a linear one, and second, the plots for different metals are not the same, although they are parallel.

The linear equation of the straight lines plotted in Figure 7–2 is known as Einstein's photoelectric equation:

$$\boxed{K_{max} = eV_0 = hv - \phi} \qquad (7\text{–}2)$$

The slope of the line h is *Planck's constant*, and the intercept ϕ is called the *work function*. The work function is the minimum amount of energy required to knock an electron from the metal surface and is dependent on the metal used. Equation (7–2) explains why each plot in Figure 7–2 is parallel (each has the same slope h) and why they are set apart (each has its own characteristic work function ϕ). When $K_{max} = 0$, $v = v_0$, is the *threshold frequency*. This is a minimum frequency of the incident light that will just knock electrons from the metal surface. From Equation (7–2) this now gives

$$hv_0 = \phi \qquad (7\text{–}3)$$

which is convenient for determining the work function. Equation (7–2) can now be written as

$$K_{max} = hv - hv_0 \qquad (7\text{–}4)$$

For a frequency $v < v_0$ or a wavelength $\lambda_0 > c/v_0$ (threshold wavelength), there is not enough incident energy to remove an electron from the metal surface and no photoelectric effect can be seen. An extrapolation of the linear graph to intercept the energy axis would give the work function ϕ directly from the graph.

In 1914 Millikan produced the first direct experimental proof of the equation developed by Einstein, Equation (7–2), and at the same time made the first direct photoelectric determination of Planck's constant h. The accepted value of Planck's constant is

$$h = 6.625 \times 10^{-34} \text{ J-sec}$$

The following is a brief summary of the photoelectric effect:

1. The number of electrons liberated is proportional to the intensity of the incident radiations.
2. The maximum kinetic energy of photoelectrons depends on the frequency, not the intensity of incident light.
3. K_{max} is linearly related to v through Equation (7–2).
4. The stopping potential V_0 depends on the work function ϕ.
5. There exists a threshold frequency v_0 below which no photoelectric effect occurs.
6. Emission starts with no observable time lag at $v \geq v_0$, even for very low intensity incident light.

PROBLEMS

7-1 If the work function for zinc is 4.3 eV, what is the maximum kinetic energy of electrons knocked out of a clean zinc surface by the 2537-Å ultraviolet line of mercury?

7-2 Nickel has a work function of 5.0 eV. (a) What is the maximum kinetic energy of photoelectrons knocked out of a nickel surface by a 1.0-mW ultraviolet source at 2000 Å? (b) What is the maximum kinetic energy of photoelectrons knocked out by a 15-W argon laser source at a wavelength of 4658 Å?

7-3 A maximum wavelength of 5450 Å is required to eject photoelectrons from sodium metal. (a) Determine the maximum velocity of electrons ejected by light of wavelength 2000 Å. (b) What is the stopping potential for photoelectrons ejected from sodium by light of wavelength 2000 Å?

7-4 The stopping potential for electrons ejected from a zinc surface is 2.42 V for the 1849-Å ultraviolet mercury line. What is the stopping potential for the 2537-Å mercury line?

7-5 The radiation from a 5.0-mW helium–cadmium laser ($\lambda = 3250$ Å) ejects electrons from a cesium surface that has a stopping potential of 1.91 V. (a) What is the work function in electron volts for cesium? (b) What will be the stopping potential when the incident radiation is 10.0 mW?

7-6 In an experiment the following data of stopping potentials of photoelectrons produced by prominent wavelengths of mercury spectrum were collected.

WAVELENGTH (Å)	5460	4920	4360	4050	3690	3130
STOPPING POTENTIAL (V)	0.40	0.60	0.90	1.20	1.50	2.10

Use this data to plot a graph, and from this graph determine values for Planck's constant and the work function of the metal used in this experiment.

RECOMMENDED READING

ANDERSON, David L., *Discovery of the Electron*, Van Nostrand Reinhold, New York, 1964.
   A very readable, interesting account of the development of the atomic concept of electricity.

MELISSINOS, Adrian, *Experiments in Modern Physics*, Academic, New York, 1966.
   Describes clearly and gives typical data for the Millikan oil drop experiment and the photoelectric effect.

MILLIKAN, R. A., in *Nobel Lectures*, Elsevier, New York, 1965.
   Brief biography and Nobel Lecture of Millikan on the electron and the photoelectric effect.

MORRISON, Philip, and MORRISON, Emily, "Heinrich Hertz," *Sci. Am.*, December 1957.

8
X Rays

Arthur Holly Compton
(1892–1962)

Compton was born in Wooster, Ohio, and earned his doctorate at Princeton University in 1916. He served as director of the Metallurgical Atomic Project at the University of Chicago (1942–1945) and as chancellor of Washington University in St. Louis (1945–1953). In 1923, by assuming that photons have momentum, Compton correctly explained change in the wavelength of x rays scattered by matter. For his discovery of the Compton effect, with C. T. R. Wilson, he received the 1927 Nobel Prize in physics.

8–1 ROENTGEN
8–2 X RAYS
8–3 DIFFRACTION OF X RAYS
8–4 DIFFRACTION OF X RAYS BY RULED GRATING
8–5 COMPTON EFFECT

8-1 ROENTGEN

In an experiment designed to study cathode rays, Professor WILHELM ROENTGEN carefully shielded a discharge tube with black cardboard. When the room was darkened and a discharge passed through the tube, Roentgen was somewhat startled to see a shimmer of light across the room in the vicinity of a work bench. Since he realized that cathode rays can travel only a few centimeters in air, he repeated the procedure and found the same shimmer of light. He lit a match and discovered that the source of the mysterious light was the fluorescence of a small barium platinocyanide screen on the bench.

Now he realized that he was witnessing a new radiation phenomenon. Upon further observation, he named this new radiation x rays and summarized their properties:

1. Many substances are quite transparent to x rays.
2. X rays cannot be reflected or refracted and do not show interference effects. (These phenomena were present, but they were too subtle for Roentgen to observe at the time.)
3. Photographic plates can be exposed by x rays.
4. X rays are not deflected by either electrical or magnetic fields.
5. Electrified bodies, positive or negative, are discharged by x rays.
6. X rays cause many substances to fluoresce.

The discoveries of x rays in November, 1895, and natural radioactivity shortly afterwards—radiations that readily penetrated matter—opened a new era of physics, a period of unusual growth and activity. The time for x rays was ripe: Physicists would not have been prepared much earlier for this discovery, and yet it seems unlikely that in this period of great scientific activity it would have been very long before someone else would have come upon the same event.

8-2 X RAYS

Roentgen's x rays were *electromagnetic radiations of very short wavelength* produced by the collision of high-speed electrons on the glass wall of the discharge tube. See Table 8-1 for a comparison of the wavelength of x rays with the rest of the electromagnetic spectrum. Figure 8-1(a) shows a typical modern x-ray tube, in

Table 8–1 Electromagnetic spectrum

NAME	WAVELENGTH RANGE (m)	ENERGY RANGE (J)	SOURCE	METHOD OF DETECTION
gamma rays	5×10^{-13} to 1×10^{-11}	4×10^{-13} to 2×10^{-14}	radioisotopes	Geiger counter, photographic emulsions, crystal counters
x rays	1×10^{-11} to 5×10^{-9}	2×10^{-14} to 4×10^{-17}	bremsstrahlung effect, electronic transitions for atomic shells nearest nucleus	photography, ionization chamber
ultraviolet	5×10^{-9} to 4×10^{-7}	4×10^{-17} to 5×10^{-19}	electronic transitions	photography, fluorescence
visible	4×10^{-7} to 7×10^{-7}	5×10^{-19} to 2.8×10^{-19}	electronic transitions in outer atomic shells	eye, photography, photocell
infrared	7×10^{-7} to 3.5×10^{-4}	2.8×10^{-19} to 5.7×10^{-22}	vibration and rotation of molecules and atoms	bolometer, thermopile, photoconductor
short Hertzian, television, radar	3.5×10^{-4} to 4×10^1	5.7×10^{-22} to 5×10^{-27}	oscillating electrical circuit	electrical resonance
radio	4×10^1 to 2×10^4 to 1×10^8	5×10^{-27} to 1×10^{-29} to 2×10^{-33}	oscillating electrical circuit	electrical resonance
micropulsations[a]	1×10^8 to 5×10^{10}	2×10^{-33} to 4×10^{-36}	unknown	magnetic receivers

[a] See James Heirtzler, "Longest Electromagnetic Waves," *Sci. Am.*, March 1962, p. 128.

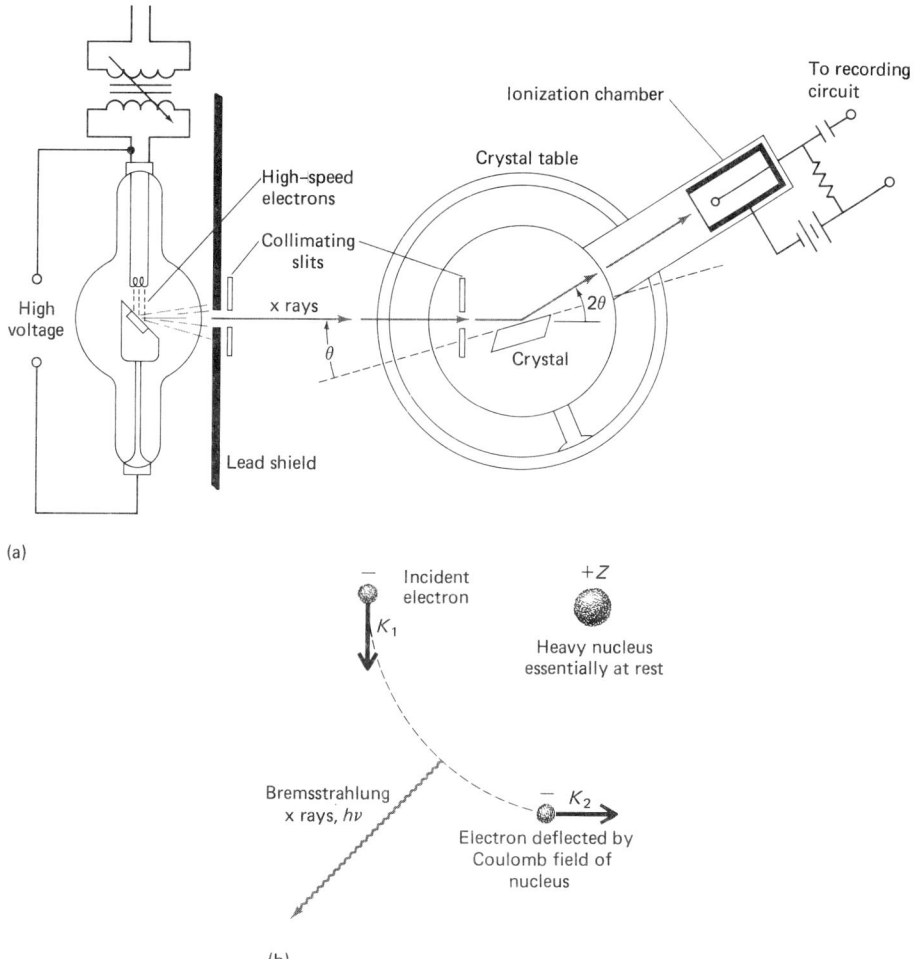

Figure 8-1 (a) Schematic of Bragg crystal spectrometer for investigation of x-ray spectra. (b) Bremsstrahlung produced by acceleration of an electron in a Coulomb field.

which thermionic electrons produced at the cathode are accelerated to high speeds through a potential difference and then stopped by impact with a metal target. When these electrons interact with the Coulomb field, as shown in Figure 8–1(b), they are decelerated and the radiation produced is that described by classical electromagnetic theory for an accelerated charged particle. As the incident electron slows down and loses kinetic energy, the energy lost is used to create a photon with energy

$$h\nu = K_1 - K_2 \tag{8-1}$$

where kinetic energy of the heavy recoiling nucleus has been neglected. Radiation produced by the acceleration of a charged particle is called *bremsstrahlung* (literally, German for "stopping radiation"). Bremsstrahlung is also produced in large accelerators, where charged particles are accelerated to high speeds. The difference between the approaches of the electromagnetic theory and quantum mechanics to the problem of *bremsstrahlung* is that the electromagnetic theory predicts a continuous radiation from each collision while the electron is being decelerated, but quantum theory predicts the creation of a single photon of energy $h\nu$ that is, in general, different for each collision. The continuous x-ray spectrum in Figure 8–2(a) is the bremsstrahlung. In addition to this continuous spectrum, Figure 8–2(a) shows a characteristic sharp line spectrum, which is dependent on the target material, superimposed on the continuous spectrum. Although the intensity of the continuous bremsstrahlung spectrum for a given potential depends on the physical characteristics of the target, the sharp wavelength cutoffs at points a, b, c, and d in the figure are independent of the target material. Figure 8–2(b) shows that when

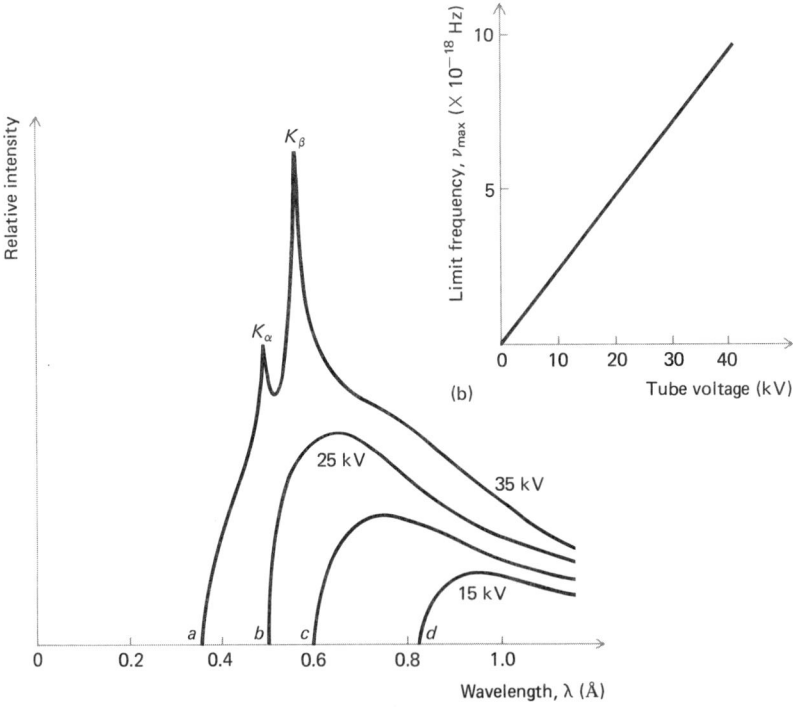

Figure 8–2 (a) X-ray spectrum of silver showing the bremsstrahlung and characteristic spectra and dependence of the shortest wavelength limit on the potential across the tube. (b) Simple linear relation between the maximum cutoff frequency and the tube voltage.

the maximum cutoff frequency v_{max} is plotted against the accelerating potential V, there is a simple linear relation:

$$\frac{v_{max}}{V} = \left(\frac{c}{\lambda_{min}}\right)\frac{1}{V} = \text{constant} \tag{8-2}$$

The classical theory makes no provision for this distinct cutoff, which experimental evidence shows so clearly. The path of the incident high-speed electron is changed when it is scattered by the Coulomb field of the atoms in the target material. According to Planck's quantum hypothesis, the radiation from the accelerated electrons must be given off in quanta, hv. However, the maximum energy of any one of the quanta cannot be greater than the energy of the most energetic incident electron,

$$hv_{max} = K_{max} = eV \tag{8-3}$$

where K_{max} is the maximum kinetic energy that an electron can reach, or

$$\frac{hc}{\lambda_{min}} = eV \tag{8-4}$$

and V is the accelerating potential across the x-ray tube. The accelerating potential is of the order of several thousand volts, but about 98% of the energy given off by the electrons when they collide with the target goes into heat energy, and the temperature of the target rises. Because the cutoff frequencies are distinct, and the wavelengths can be determined accurately, x-ray spectra provide a good experimental method for determining the value of Planck's constant h.

The characteristic sharp line spectrum is a function of the target material, although its intensity for a given target depends on the accelerating potential. The electrons of an atom are ordered according to their arrangement in shells about the nucleus. Those electrons nearest to the nucleus, those most tightly bound, are in the K shell. Those in the next most tightly bound position are in the L shell, and then the M shell, N shell, and so on. When highly energetic incident electrons knock an electron from the K shell, an electron in the L shell gives up energy in the form of an x ray as the electron fills the vacancy left in the K shell. This radiation, which is characteristic of the target material, is labeled the K_α line. The electron in the M shell that fills a vacancy in the K shell gives up energy as an x ray called the K_β line. These transitions from the shells L, M, N, and so on to the K shell give rise to a series of lines K_α, K_β, K_γ, and so on called the K series. When incident electrons dislodge electrons from the L shell, and these are filled by electrons from the remaining M, N, O, and remaining shells, these transitions give rise to the L series, the first line of which is L_α. The nomenclature of these transitions is illustrated in Figure 8-3.

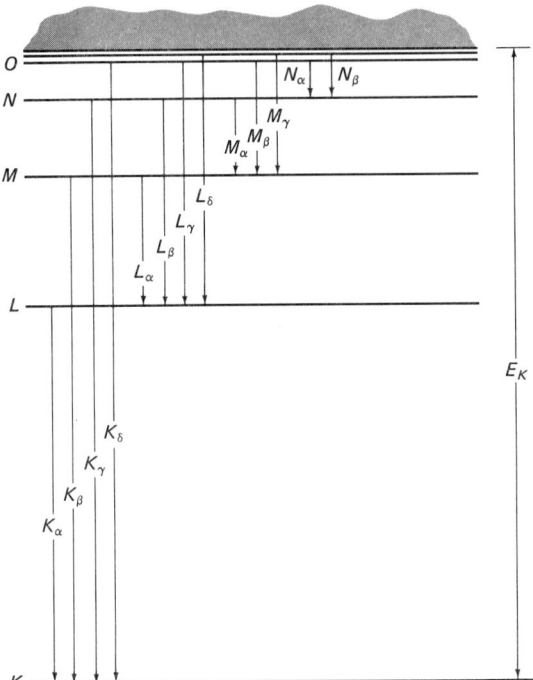

Figure 8–3 Electronic transitions for shells nearest the nucleus that give rise to characteristic x-ray spectra.

As the accelerating voltage in the x-ray tube is increased, incident electrons produce a continuous bremsstrahlung spectrum until at some "critical" voltage the electrons have enough energy to dislodge the electrons in the inner shells, that is, the K, L, or M shells. Only when the critical potential V_c is reached will incident electrons be energetic enough for the K transitions to occur. Thus,

$$\boxed{eV_c \geq E_k} \tag{8-5}$$

where E_k is the energy necessary to free a K electron from an atom. X rays associated with the more energetic K series are called "hard x rays," and those associated with the less energetic L, M, N series are "soft x rays." The hard x rays are more penetrating.

8-3 DIFFRACTION OF X RAYS

Roentgen's first attempts to verify experimentally the wave nature of x rays were unsuccessful. The unusually short wavelength of x rays made interference and diffraction effects difficult to observe. Early experiments indicated that the wavelength of x rays was of the order of 10^{-8} cm. By calculating for a crystal the number of molecules per unit volume, von Laue calculated the average distance between atoms in solids to be between 10^{-7} and 10^{-8} cm. Von Laue suggested that the regular array of atoms in crystals could be used as a kind of diffraction grating for x rays, and in 1914 he received the Nobel Prize for establishing the wave character of x rays by diffraction from crystals. W. H. Bragg and his son, Lawrence Bragg, were awarded the Nobel Prize the next year for perfecting von Laue's concepts of x ray diffraction from crystals.

As an example, consider a simple cubic crystal such as rock salt (NaCl), with a molecular weight of 58.46 and a density of 2.165 g/cm³. This means that the molecule of rock salt will have a mass of

$$m = \frac{M}{N_A} = \frac{58.45 \text{ g/g-mole}}{6.023 \times 10^{23} \text{ molecules/g-mole}} \quad (8\text{-}6)$$

$$= 9.705 \times 10^{-23} \text{ g/molecule}$$

where N_A is Avogadro's number. The number of molecules per unit volume becomes

$$N = \frac{\rho}{m} = \frac{2.165 \text{ g/cm}^3}{9.705 \times 10^{-23} \text{ g/molecule}} \quad (8\text{-}7)$$

$$= 2.24 \times 10^{22} \text{ molecules/cm}^3$$

Actually, since each molecule of rock salt contains two atoms (one Na and one Cl), the number of atoms per cubic centimeter is

$$2N = 4.48 \times 10^{22} \text{ atoms/cm}^3$$

If there are n atoms along a 1-cm edge of a cubic centimeter of rock salt, then

$$n^3 = 4.48 \times 10^{22} \text{ atoms/cm}^3$$

and

$$n = 3.55 \times 10^7 \text{ atoms/cm}$$

Thus, the distance between atoms is

$$d = \frac{1}{n} = 2.82 \times 10^{-8} \text{ cm} = 2.82 \text{ Å}$$

This calculation for the average distance between the atoms in the *simple cubic structure* of rock salt is given by

$$\boxed{d = \sqrt[3]{\frac{M}{2\rho N_A}}} \quad (8\text{-}8)$$

Von Laue recognized that the symmetry of a crystal comes from a unit of molecular or atomic size that is arranged in a regular repeating order. The layers of units are separated by successive uniform distances, which are of the order of the wavelength of x rays. These regularly spaced discontinuities form the basis of a diffraction grating similar to that used for visible light, with the exception that the crystal forms a three-dimensional grating. Also, unlike the two-dimensional grating, the three-dimensional grating will not diffract monochromatic light falling on it from any arbitrary angle. The diffracted beam of x rays will be reinforced for constructive interference only when a wavelength λ meets planes of atoms a distance d apart and meets these planes at a certain angle θ.

Bragg used this idea to analyze the diffraction patterns of x rays reflected from the crystal planes. Figure 8-4 shows an incident beam at an angle θ with

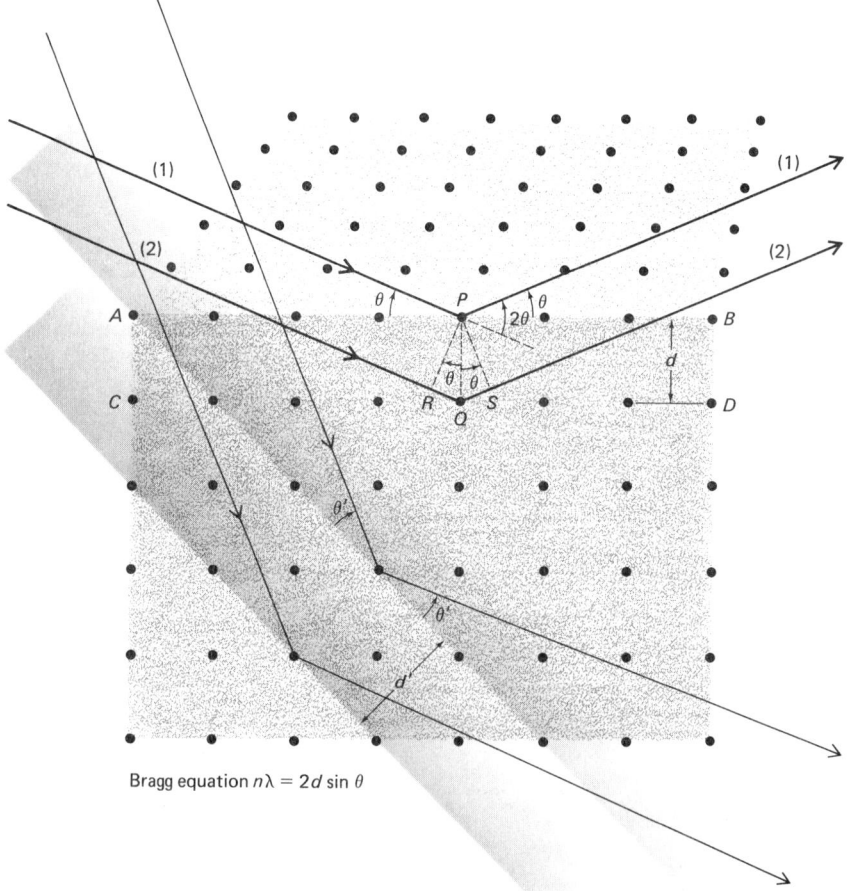

Bragg equation $n\lambda = 2d \sin \theta$

Figure 8-4 Exaggerated view of Bragg reflections from various planes in a crystal.

respect to a crystal plane that is rich in atoms. When rays 1 and 2 are reflected from points P and Q, respectively, the path difference between the rays is

$$\delta = RQ + QS = 2RQ$$

Since

$$RQ = PQ \sin \theta = d \sin \theta$$

this gives

$$\boxed{\delta = 2d \sin \theta}$$

Notice that in Bragg's reflection, the incident and reflection angles θ are the angles that the incident and reflected rays make with the crystal planes and not with the normal as is done in optics. This angle θ is also called the *grazing angle*. Thus, the conditions for reinforcement, known as *Bragg's laws of reflection*, are

> 1. The angle of incidence must equal the angle of reflection.
> 2. $\delta = 2d \sin \theta = n\lambda$ for $n = 1, 2, 3, \ldots$, where n is the order of reflection.

(8-9)

Figure 8-4 shows that there are many planes within the crystal from which rays from various angles of incidence could be reflected to interfere constructively. Note that, depending on the plane chosen, some planes will contain more atoms as reflecting agents than others. In general, rays reflected from these planes will be those of greatest intensity.

A diffraction pattern can also be obtained with x rays incident on a powdered crystal sample. Each crystal fragment within the powder contains atomic planes that will reflect x rays incident on them at a given angle. Since the fragments may be oriented at any azimuth angle about the incident x rays, there will be some that will reflect light of a given wavelength at one particular angle to form a circular pattern about the incident beam. Those properly oriented at still different angles will form other concentric circles. The resulting diffraction patterns are called *Laue patterns*. From Figure 8-5, the condition for maxima to occur is

$$n\lambda = 2d \sin \theta$$

where θ is the angle identified in the figure and is actually the angle of incidence for some of the powder fragments. Many of the physical properties of crystals can now be investigated with great accuracy by x ray techniques.

8-4 DIFFRACTION OF X RAYS BY RULED GRATING

Because the wavelength of x rays is so short and x rays are so penetrating, initial efforts to diffract x rays with a ruled diffraction grating did not meet with much success. It was not possible mechanically to build rule gratings with the grating

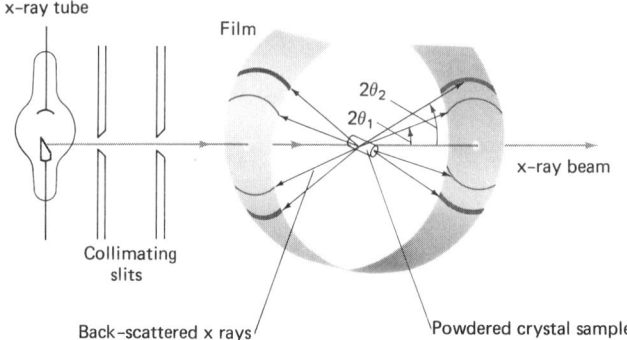

Figure 8-5 Laue x-ray pattern from fragments of powdered crystal.

spacings of the order of 1 Å, a typical wavelength of x rays. Also, most mirror surfaces do not reflect x rays well, since they are so penetrating. However, in 1925 Compton and Doan were able to measure the wavelength of x rays using an ordinary ruled diffraction grating at very small grazing angles (of the order of 0.001 rad).

When rays 1 and 2 in Figure 8-6 are diffracted from the ruled grating surface, the path difference between rays is

$$\delta = CB - AD = d\cos\theta - d\cos(\theta + \phi) \tag{8-10}$$

For these rays to interfere constructively, then,

$$d\cos\theta - d\cos(\theta + \phi) = n\lambda \tag{8-11}$$

where n is the order of diffraction. Written in the form of an approximation for small angles, $\cos x = 1 - x^2/2 + \cdots$, Equation (8-11) becomes

$$\boxed{n\lambda = d\phi\left(\theta + \frac{\phi}{2}\right)} \tag{8-12}$$

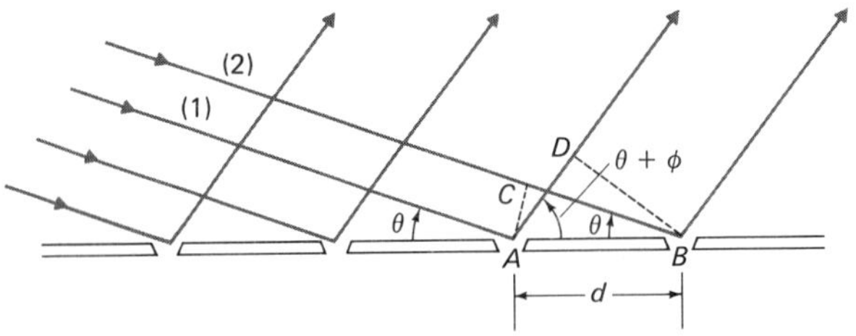

Figure 8-6 Diffraction of x rays incident at grazing angles on ruled diffraction grating.

Interestingly enough, when the wavelengths of x rays were first measured by the ruled grating, they were consistently found to be larger than the values found by crystal diffraction. After reexamination it was found that Avogadro's number in Equation (8-8) was not precise because of an error in the accepted value of the electronic charge that was used in the determination of N_A.

8-5 COMPTON EFFECT

Early experiments showed that when x rays were scattered, the secondary x rays involved were less penetrating than primary x rays. At first these secondary rays were thought to consist of fluorescent radiation characteristic of the radiating element. Although fluorescence is a characteristic of heavier elements, later experiments showed a difference in the penetration of secondary x rays from the lighter elements such as carbon from which no fluorescence radiation of the type observed can appear. This lead to the speculation by some physicists that a new radiation had been found, and it was labeled "J" radiation. However, under careful spectroscopic analysis, this idea of "J" radiation did not stand up.

Careful experimental evidence determined the following properties of the secondary x rays from the scattering process:

1. The scattered radiation consists of two wavelengths, the original λ_0 and an additional wavelength λ_s that has nearly the same value as λ_0.
2. λ_s is always greater than λ_0.
3. λ_s depends on the scattering angle θ but not on the scattering medium.

Following a mathematical analysis of this situation by G. E. M. JAUNCEY, A. H. COMPTON in 1923 boldly proposed that *the x-ray photons have a momentum* just as a particle has, and that the scattering process is an elastic collision between a photon and an electron. The shift in wavelength of x-ray photons due to elastic scattering from electrons is known as the *Compton effect*.

From Equation (6-34), the energy of a photon, a particle of rest mass zero, is $E = pc$, but the energy of a photon is also $E = h\nu$, where ν is the frequency. Thus, the momentum of a photon is

$$p = \frac{h\nu}{c} = \frac{h}{\lambda} \tag{8-13}$$

The x-ray photon in Figure 8-7 is scattered elastically from a stationary free electron. The conservation of momentum for this event can be written as

$$p_0 = p_s \cos \theta + p_e \cos \gamma \quad \text{(horizontal axis)} \tag{8-14}$$

and

$$p_s \sin \theta = p_e \sin \gamma \quad \text{(vertical axis)} \tag{8-15}$$

where p_s is the momentum of the scattered photon and p_e is the momentum of the scattered electron.

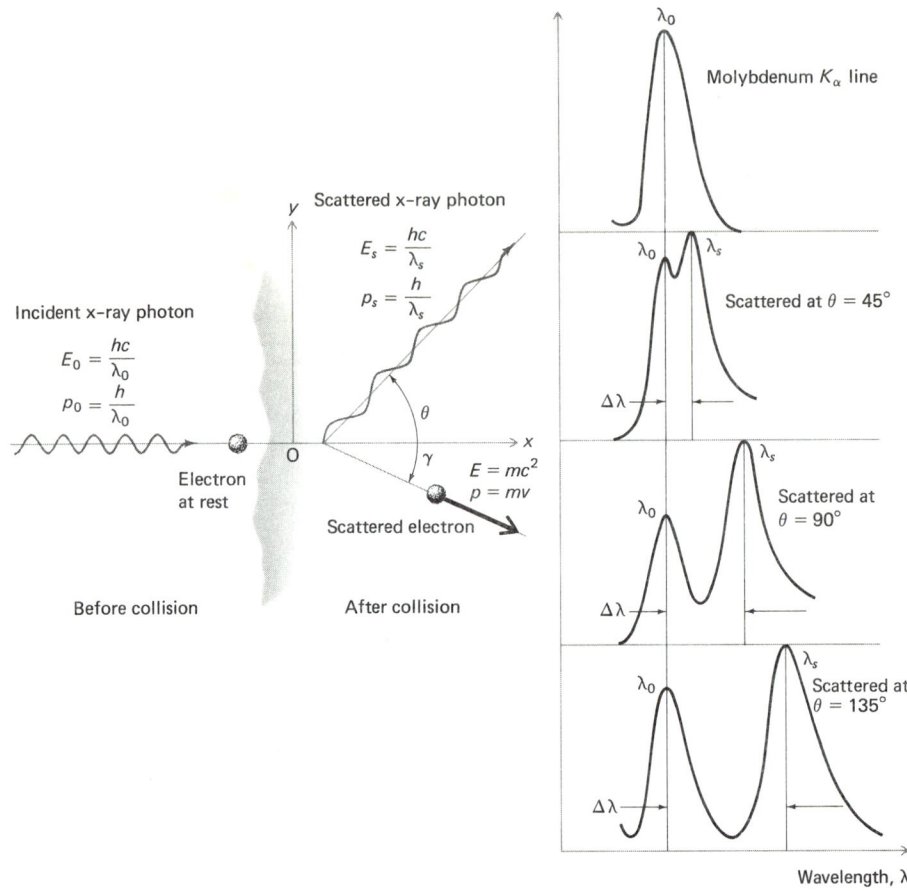

Figure 8–7   Compton scattering of a photon from an electron at rest. The graphs at the right show the shift in the K_α radiation from molybdenum scattered from carbon.

From the conservation of energy,

$$\underset{\substack{\text{incident photon}\\\text{energy, }h\nu_0}}{E_0} + \underset{\substack{\text{rest mass of}\\\text{electron}}}{m_0 c^2} = \underset{\substack{\text{scattered}\\\text{photon energy,}\\h\nu_s}}{E_s} + \underset{\substack{\text{total energy}\\\text{of scattered}\\\text{electron}}}{m_0 c^2 + K} \qquad (8\text{–}16)$$

or

$$K = E_0 - E_s = h(\nu_0 - \nu_s) \qquad (8\text{–}17)$$

Next put Equation (8–14) in the form

$$p_0 - p_s \cos\theta = p_e \cos\gamma$$

and then by squaring and adding this equation and Equation (8–15), we get

$$p_0^2 - 2p_0 p_s \cos \theta + p_s^2 = p_e^2 \tag{8-18}$$

Since $E_0 = h\nu_0 = p_0 c$ and $E_s = h\nu_s = p_s c$, Equation (8–17) becomes

$$K = (p_0 - p_s)c \tag{8-19}$$

For an electron,

$$E^2 = (m_0 c^2)^2 + p_e^2 c^2$$
$$E = K + m_0 c^2 \tag{8-20}$$

Eliminate E from Equations (8–20) to get

$$\frac{K^2}{c^2} + 2Km_0 = p_e^2 \tag{8-21}$$

Using K from Equation (8–19) and p_e from Equation (8–18) and multiplying through by h, Equation (8–21) becomes

$$\frac{h}{p_s} - \frac{h}{p_0} = \frac{h}{m_0 c}(1 - \cos \theta) \tag{8-22}$$

When h/p_s is replaced by the scattered wavelength λ_s and h/p_0 by λ_0, this equation takes the more useful form

$$\boxed{\begin{aligned} \lambda_s - \lambda_0 &= \frac{h}{m_0 c}(1 - \cos \theta) \\ \Delta\lambda &= \lambda_c(1 - \cos \theta) \end{aligned}} \tag{8-23}$$

where $\lambda_c = h/m_0 c = 0.024$ Å is defined as the *Compton wavelength*.

A summary of experimental results supporting Equation (8–23) as an explanation for the shift in scattered x rays is, briefly:

1. In 1923 Compton confirmed the results of Equation (8–23) experimentally.
2. Late in 1923 Bothe and Wilson observed recoil electrons.
3. In 1925 Bothe and Geiger showed that the scattered photon $E_s = h\nu_s$ and the recoil electron appear simultaneously.
4. In 1927 Bless checked the energy of the recoil electron experimentally.

Notice that Figure 8–7 shows that *both* the scattered and the incident wavelengths are detected at the angle θ. Some photons are scattered from electrons that are not free but that are bound to the atom. Thus, the mass m_0 in Equation (8–23) must be replaced by the mass of the entire atom that recoils. The large value of this mass, when compared to that of the electron, will make $\Delta\lambda$ very small, and the wavelengths of these scattered photons will be very nearly the same as those of the incident photons. A calculation for scattering from just a proton will show this to be the case.

PROBLEMS

8-1 To understand Roentgen's difficulty in observing interference effects with x rays, design a Young's double-slit arrangement that will produce fringes 1° apart on a distant screen. Assume that the incident x rays have a wavelength of $\lambda = 5.0$ Å. Discuss some of the difficulties in physically setting up such an experiment.

8-2 Compare the energy of the K_α line of tungsten (W-74) at 0.0210 Å to the lasing line of a CO_2 infrared laser at 10.6 μ.

8-3 The K_α line of thulium (Tm-69) has a wavelength of 0.246 Å. Compare the energy of this K_α photon with the rest mass energy of an electron.

8-4 From Figure 8-2(a), determine the accelerating potential for the curve that terminates at point c.

8-5 Determine the voltage applied to an x-ray tube that will give a short wavelength limit of 1.0 Å.

8-6 (a) What is the most energetic x ray emitted when a metal target is bombarded by 40-KeV electrons?
(b) What is the maximum frequency of x rays produced by electrons accelerated through a potential difference of 20,000 V?

8-7 From Figure 8-2(a), determine the maximum kinetic energy of the electrons that produce x-ray spectra terminating at points a, b, c, and d.

8-8 The graph in Figure 8-8 represents an x-ray spectrum from a hypothetical metal. If the energy required to remove an electron from the K shell is 20.0 KeV, determine from this graph (a) the energy required to remove an electron from the L shell, and (b) the maximum kinetic energy of the electrons incident on this target.

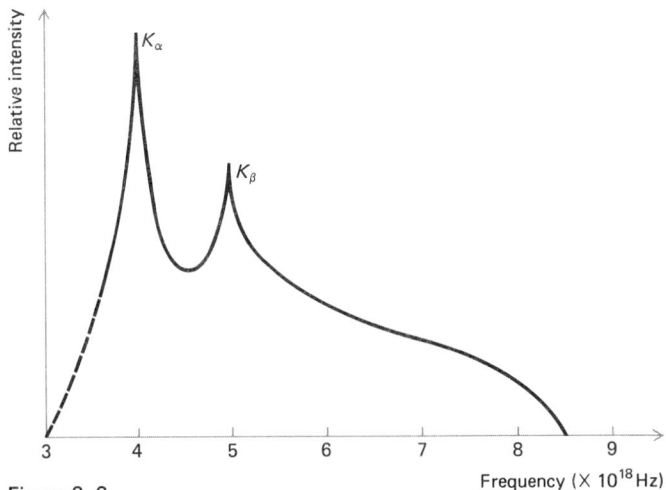

Figure 8-8

8-9 A first-order Bragg reflection occurs when a monochromatic x ray incident at an angle of 30° is reflected from a rock salt (NaCl) crystal. Determine the wavelength of the incident x ray.

8-10 Find the glancing angles on the face of a sylvine crystal (KCl), $d = 2.82$ Å, that correspond to the first- and second-order Bragg reflection maxima for x rays of wavelength $\lambda = 0.58$ Å.

8-11 NiO crystals with a molecular weight of $M = 74.69$ and a density of $\rho = 7.45$ g/cm^3 has a simple cubic structure like NaCl. Determine the angle at which a NiO crystal must be oriented relative to an incident x ray of wavelength $\lambda = 2.00$ Å to produce a first-order Bragg reflection.

8-12 Electrons accelerated through a potential of 35.0 kV produce x rays that are analyzed by a Bragg crystal spectrometer using a calcite (CaCO$_3$) crystal. If the spacing between crystal planes is 3.03 Å for calcite, what is the smallest angle between the incident beam and the crystal plane at which a strong reflected beam will be found?

8-13 X rays of wavelength 1.0 Å are incident on a ruled diffraction grating at an angle $\theta = 0.0010$ rad as shown in Figure 8-7, and a maximum is observed at an angle $\phi = 0.0020$ rad. How many lines per millimeter must this diffraction grating have? Why must the diffraction gratings be used at very small grazing angles of incidence?

8-14 A beam of potassium K_β radiation at $\lambda = 3.44$ Å is incident on a diffraction grating with 200 lines/mm at an angle of 20' to its surface. What is the angle between the first- and second-order beams?

8-15 Complete the algebraic manipulation to arrive at Equation (8-22).

8-16 X rays of wavelength 0.040 Å are scattered from a carbon block. Determine (a) the momentum of a photon scattered at an angle of 30°, and (b) the kinetic energy of the recoil electron.

8-17 When photons of wavelength 0.024 Å are incident on a target, scattered photons are detected at an angle of 60°. Calculate (a) the wavelength of the scattered photon, and (b) the angle at which the electron is scattered.

8-18 In a scattering experiment, 0.200-MeV incident photons produce scattered photons at an angle of 90° with respect to the incident beam. What is the energy in MeV of the scattered photons and of the recoil electrons? Show whether energy has been conserved.

8-19 What is the difference between scattering photons from electrons or protons? What would be the shift in wavelength of a 0.00200-Å γ ray scattered from a proton at an angle of 90°? What experimental difficulties would be involved in measuring the scattered photon wavelength?

8-20 Use the BASIC program (or your own) to show the difference in wavelength (*D*) between the incident photon and the photon that results from Compton scattering. In the program *MO* is the scattering agent, *L* is the incident wavelength in angstroms, and *A* is the scattering angle in degrees.

```
10   LET H = 6.6256E-34
20   LET C = 2.99793E + 8
30   PRINT "LAMBDA, ANGLE, MO–";
40   INPUT L,A,MO
50   LET A1 = 0.0174533*A
60   LET D = H*((1 – COS(A1))/(MO*C))*1E + 10
70   PRINT "DELTA L = "; D; "ANGSTROMS; L-PRIME = "; D + L
71   PRINT "DELTA L/L = "; D/L
72   PRINT
80   GO TO 30
99   END
```

(a) For an incident x ray of wavelength $L = 1.0$ Å scattered from an electron $MO = 9.1091\text{E-}31$ (kg), determine the shift in wavelength for scattering at angles of $A = 0°, 45°, 90°, 135°, 180°$.

(b) For a scattering angle $A = 90°$ and incident wavelength of $L = 1.0$ Å determine the shift in wavelength for protons scattered from an electron $[MO = 9.1091\text{E-}31 \text{ (kg)}]$, from a proton $[MO = 1.6725\text{E-}27 \text{ (kg)}]$, and from a carbon atom $[MO = 19.920\text{E-}27 \text{ (kg)}]$.

(c) For scattering from a free electron $MO = 9.1091\text{E-}31$ (kg) at an angle of $90°$, determine the shift in wavelength when the incident radiation is hard x rays ($L = 0.01$ Å), ultraviolet rays ($L = 1800$ Å), visible light ($L = 6000$ Å), and infrared radiation ($L = 100{,}000$ Å).

RECOMMENDED READING

ALONSO, M., and FINN, E. J., *Fundamental University Physics*, Addison-Wesley, Reading, Mass., 1968, pp. 176–178.
A very clear presentation of the x-ray spectra.

BEISER, A., *Perspectives in Modern Physics*, McGraw-Hill, New York, 1968, pp. 59–73.
A very lucid discussion on the production of x rays and the Compton effect.

BRAGG, W. L., *Proc. Cambridge Phil. Soc.* **17**, 43 (1912).

BRAGG, W. L., *Nature* **90**, 410 (1912).

COMPTON, A. H., "Wavelength Measurements of Scattered X Rays," *Phys. Rev.* **21**, 715 (1923).

COMPTON, A. H., "The Spectrum of Scattered X Rays," *Phys. Rev.* **22**, 409 (1923).

COMPTON, A. H., and DOAN, R. L., *Proc. Natl. Acad. Sci. U.S.* **11**, 598 (1925).

GUNST, S. B., and PAGE, L. A., *Phys. Rev.* **93**, 970 (1953).

MAGIE, W. F., *A Source Book in Physics*, Harvard University Press, Cambridge, Mass., 1963, p. 600.

MOSELEY, H. G. J., *Phil. Mag.* **26,** 1024 (1913).

RICHTMYER, F. K., *Phys. Rev.* **27,** 715 (1923).

THOMSON, J. J., and THOMSON, G. P., *Conduction of Electricity Through Gases*, Dover, New York, 1928, p. 321.

9
Pair Production

Carl David Anderson
(1905–)

A native of New York City, Anderson received his Ph.D. in 1930 from the California Institute of Technology, where he is now professor. In 1933 he discovered an intermediate mass particle that he assumed to be the predicted Yukawa particle. In fact, it was the daughter particle (mu meson, muon, lepton) of the Yukawa particle (pi meson). Very active in research on x rays, gamma rays, and cosmic rays, Anderson is a recipient of the Cresson medal and the Ericsson medal. In 1936 he, along with V. F. Hess, was awarded the Nobel Prize for his work.

9–1 RADIATION INTERACTION WITH MATTER
9–2 PAIR PRODUCTION
9–3 PAIR ANNIHILATION
9–4 ABSORPTION OF PHOTONS

9–1 RADIATION INTERACTION WITH MATTER

In the last several chapters, the interaction of radiation with matter was illustrated in bremsstrahlung, the Compton effect, and the photoelectric effect. Recall that bremsstrahlung is radiation produced when an energetic electron is slowed down as it interacts with matter to create photons whose energies are given by

$$h\nu = K_1 - K_2 \tag{9-1}$$

where $K_1 - K_2$ represents the loss of kinetic energy by the electron in each of many collisions. Another example of the interaction of radiation with matter is the Compton effect, in which a photon of energy $h\nu_0$ interacts with a free electron. The incident photon disappears, and a second photon of energy $h\nu_s < h\nu_0$ is created. The kinetic energy of the recoil electron is

$$K = h\nu_0 - h\nu_s \tag{9-2}$$

In the photoelectric effect, a photon of energy $h\nu$ interacts with a bound electron. The photon disappears as the electron is knocked out of the atom. The maximum kinetic energy of the liberated electrons is

$$K = h\nu - \phi \tag{9-3}$$

where ϕ is the energy that binds the electron to the atom.

These effects hint at the interesting question of whether the total energy of photons can be converted into rest mass.

9–2 PAIR PRODUCTION

In a paper published in 1928, P. A. M. DIRAC noted that there were twice as many solutions for the relativistic wave equation of the electron as expected. He stated that half of these must refer to electron states with negative values of energy. Because the quantum theory allows discontinuous transitions to take place, negative energy states cannot be ignored as being simply nonsense solutions. Dirac associated the "unwanted" solution with an electron of charge $+e$ and labeled it the *positron*. The enigmatic positron remained only a theoretical hypothesis until 1932, when CARL ANDERSON found particle traces similar to the one shown in Figure 9-1. Anderson had not set out to find Dirac's positron. He and Millikan had set up a cloud chamber within a strong magnetic field to study cosmic rays, and the positron tracks showed up in the chamber. Positrons had actually been "seen" by others, but it was Anderson who first identified them correctly.

In many cloud chamber pictures, Anderson observed tracks of charged particles typical of electrons, except that the tracks curved in the "wrong" direction in the presence of the magnetic field. The particles could be identified as "electron-like particles" by the kind of condensation trail they left in the cloud chamber. Alpha particles have two charges and are very massive compared to electrons, and for these reasons they leave a short, thick trail as compared to the long, thin trail produced by electrons. To determine the direction of travel of the particles, a lead plate was placed in the cloud chamber. Figure 9-1 shows a particle trail in a cloud chamber in a uniform magnetic field. Notice that the radius of curvature of the path is less in the upper portion of the cloud chamber. As the particle passes through the lead plate it loses some momentum, and so its radius of curvature in the magnetic field is reduced. This gives the direction in which the particle is traveling, and then a knowledge of the direction of the magnetic field yields the sign of the charge on the moving particle. The electron and the positron have charges of identical magnitudes, but the positron charge is positive while that of the electron is negative. The positron is an *antiparticle*, an antielectron.

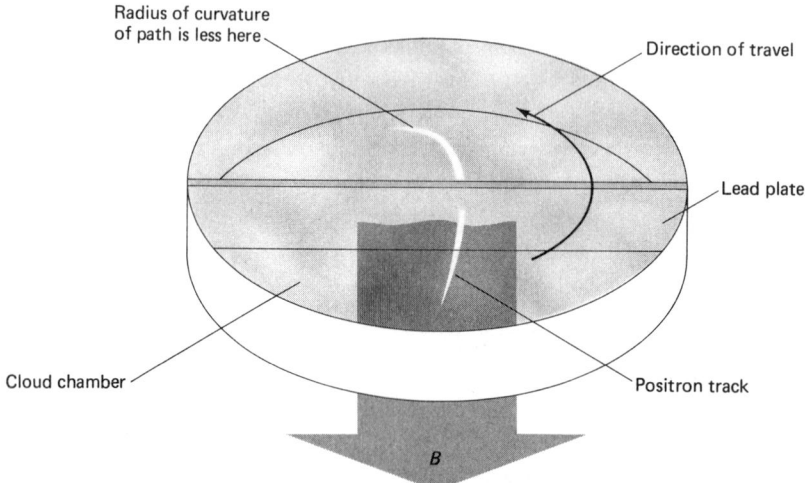

Figure 9-1 Schematic of positron track in the cloud chamber in a magnetic field of intensity B. From the direction of the magnetic field, the type of tracks made, and the decrease in the radius of curvature, Anderson was able to identify the positron.

Positrons are formed in electron pair production by γ rays. In this case, the energy of a photon is transformed into rest and kinetic energy. Pair production does not occur in free space, because the energy and momentum of a single photon cannot be simultaneously conserved by producing two electrons unless the photon passes close to a heavy nucleus. The nucleus takes some of the momentum and energy in the interaction. For this interaction [shown in Figure 9–2(a)] to occur, energy, momentum, and electrical charge must be conserved. As a photon of energy $E = h\nu$ interacts with a nucleus, an electron–positron pair is formed. By the conservation of energy, the minimum energy of the incident photon is

$$\boxed{h\nu_{\min} = m_0^- c^2 + m_0^+ c^2 = 2m_0 c^2} \qquad (9\text{–}4)$$

or

$$h\nu_{\min} = 1.02 \text{ MeV}$$

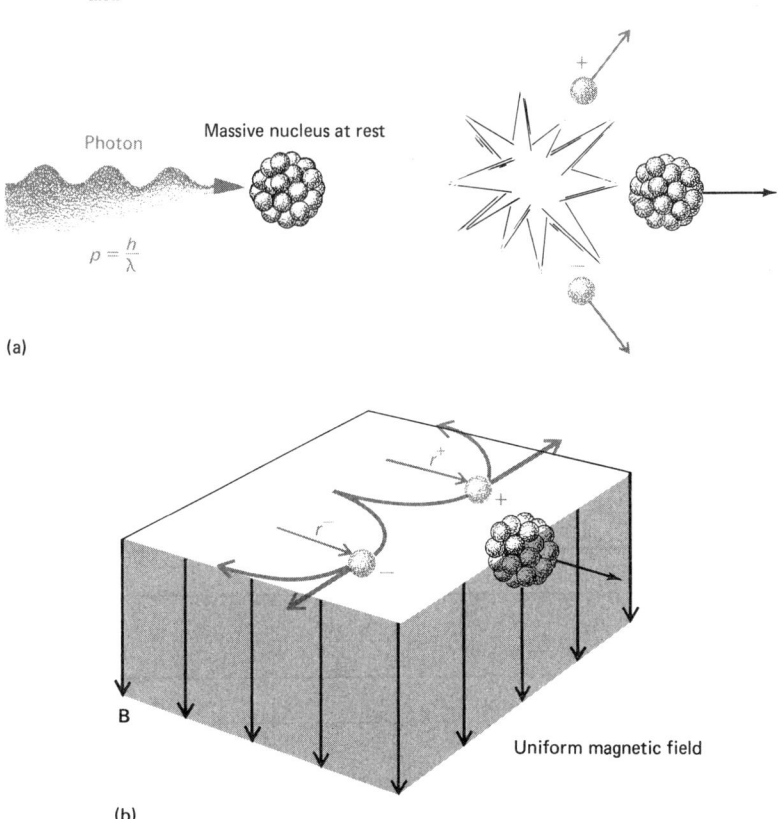

Figure 9–2 (a) Pair production occurs in the vicinity of a heavy nucleus; momentum and energy are conserved. (b) In a magnetic field, the electron and positron have opposite curvature of paths.

The wavelength of this photon is

$$h\nu_{min} = \frac{hc}{\lambda_{max}} = 1.02 \text{ MeV}$$

or

$$\lambda_{max} = 0.0122 \text{ Å}$$

This is a highly energetic γ-ray photon.

Figure 9-2(b) shows an electron–positron pair being formed in the presence of a uniform magnetic field. In this case, the particles move in opposite circular paths according to the sign of their charge. In general, for incident γ-ray photons of energy greater than 1.02 MeV,

$$\underset{\substack{\text{energy of} \\ \text{incident} \\ \text{photon}}}{h\nu} + \underset{\substack{\text{rest energy} \\ \text{of nucleus}}}{M_0 c^2} = \underset{\substack{\text{total energy} \\ \text{of electron}}}{(m_0^- c^2 + K^-)} + \underset{\substack{\text{total energy} \\ \text{of positron}}}{(m_0^+ c^2 + K^+)}$$

$$+ \underset{\substack{\text{total energy} \\ \text{of nucleus}}}{(M_0 c^2 + K_n)} \qquad (9\text{-}5)$$

with the excess energy now appearing as kinetic energies K^- and K^+ of the electron and the positron, respectively, and K_n, the kinetic energy of the nucleus. Since the nucleus is very massive and has a velocity much less than the lighter electrons and positrons, $K_n \cong 0$. Equation (9-5) is then written as

$$h\nu = (m_0^- c^2 + K^-) + (m_0^+ c^2 + K^+) \qquad (9\text{-}6)$$

From Figure 9-2(b), the magnetic field intensity B is known and the radius r of the circular paths can be measured. The momentum of a particle from the electron–positron pair is then determined from

$$p = mv = eBr \qquad (9\text{-}7)$$

where e is the charge of the positron or electron. Therefore, the total energy of a particle, from Equation (6-33), is

$$E = \sqrt{E_0^2 + p^2 c^2} \qquad (9\text{-}8)$$

The energy of one of the particles, either the electron or the positron, is given by

$$E = \sqrt{(m_0 c^2)^2 + (eBr)^2 c^2} \qquad (9\text{-}9)$$

The positron is an antiparticle of the electron. In 1955 Chamberlain, Segré, Wiegand, and Ypsilantis at the University of California observed pair production for the proton–antiproton combination, and in the same year neutron–antineutron pairs were also experimentally verified. Antiparticles will be discussed at length in Chapters 37, 38, and 39.

9-3 PAIR ANNIHILATION

The inverse effect of pair production also occurs; it is known as *pair annihilation*. In pair annihilation, a particle and one of its antiparticles come together and are converted completely into radiation energy. When a positron is formed, it is very shortlived. After losing most of its kinetic energy in collisions, it forms a kind of an atom with an electron. This atom is called *positronium* [Figure 9–3(a)], and it exists until the positron and electron are mutually annihilated.

The total energy, including the rest mass of the electron–positron pair, is changed into photon energy. Again, just as in pair production, *the momentum must be conserved*. It sometimes happens that the two particles are essentially at rest and hence have zero momentum before annihilation. Then momentum is conserved when two photons of equal but oppositely directed momentum are given off [Figure 9–3(b)] in the annihilation of the two particles. For an electron and a positron essentially at rest before the interaction,

$$m_0^- c^2 + m_0^+ c^2 = 2h\nu_{min}$$

or

$$2m_0 c^2 = 2h\nu_{min} = \frac{2hc}{\lambda_{max}} \tag{9-10}$$

from which $\lambda_{max} = 0.0244$ Å, the maximum wavelength of a created photon. For particles that have an initial kinetic energy before the collision, Equation (9–10) becomes

$$\boxed{(m_0^- c^2 + K^-) + (m_0^+ c^2 + K^+) = 2h\nu} \tag{9-11}$$

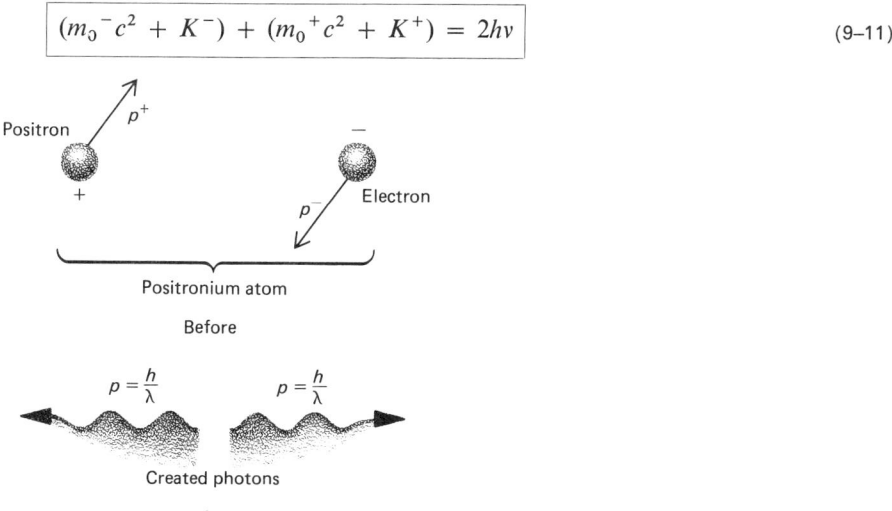

Figure 9–3 Photons are created from the annihilation of an electron–positron pair in collision.

9-4 ABSORPTION OF PHOTONS

The intensity of electromagnetic radiation in terms of photons is defined as

$$\text{intensity} = \left(\frac{\text{number of photons}}{\text{area} \perp \text{to propagation} \times \text{time}}\right) \times \frac{\text{energy}}{\text{photon}}$$

where the *photon flux* is

$$N = \frac{\text{number of photons}}{\text{area} \perp \text{to propagation} \times \text{time}}$$

From these definitions, the intensity is

$$\boxed{I = Nh\nu} \tag{9-12}$$

The intensity of radiation in a beam is reduced both by scattering of the beam and by absorption. An absorption coefficient μ is used as a measure of the ability of the medium to attenuate a beam of radiation.

Figure 9-4(a) shows an incoming flux of photons incident on an absorbing material of absorption coefficient μ and thickness dx. The change of flux in a beam in passing through a material of thickness dx is negative and is proportional to the thickness of the absorber and the original flux. The change in flux is thus

$$dN = -\mu N\, dx$$

where μ is a constant of proportionality. This integrates to give

$$N = N_0 e^{-\mu x} \tag{9-13}$$

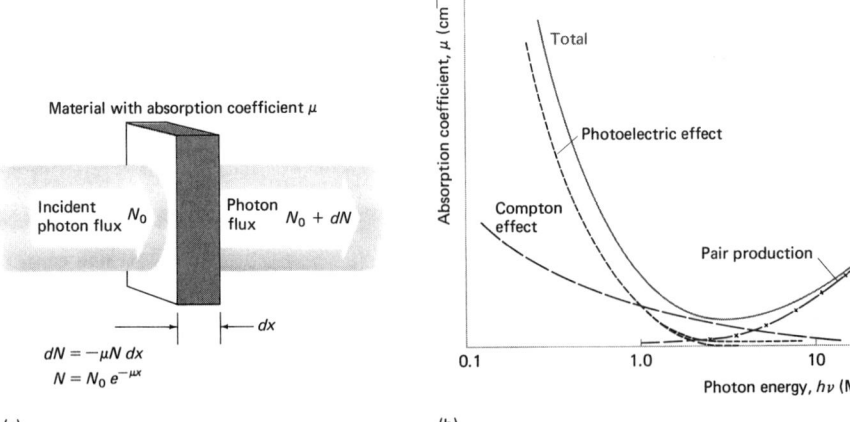

Figure 9-4 (a) Photon flux absorbed in passing through a material of absorption coefficient μ. (b) Absorption coefficients for γ-ray photons in lead.

where N_0 is the incident flux and μ appears as the absorption coefficient. Since, for a given frequency of radiation, the flux is directly proportional to intensity, the intensity of the transmitted radiation becomes

$$I = I_0 e^{-\mu x} \tag{9-14}$$

where $I_0 = N_0 h\nu$.

The absorption coefficient μ depends on both the absorbing material and the frequency ν of the incident radiation. The processes already mentioned—photoelectric effect, Compton effect, and pair production—are principally effective in absorbing the radiation. The total effective absorption coefficient is a result of all three effects:

$$\mu_{\text{total}} = \mu_{\text{Compton}} + \mu_{\text{photoelectric}} + \mu_{\text{pair production}} \tag{9-15}$$

The graph in Figure 9–4(b) shows the effect of each individual process as a function of incident energy. For photons of energy less than about 1.0 MeV, the absorption process is mainly from the photoelectric effect, although there is some contribution from the Compton effect. Pair production, according to Equation (9–4), starts at 1.02 MeV, and it predominates more and more at higher energies.

PROBLEMS

9–1 Show that the radius of curvature of a positron moving perpendicular to a uniform magnetic field is reduced when the positron has passed through a lead plate.

9–2 What is the energy and wavelength of the photon that will just create a proton–antiproton pair?

9–3 Determine the total kinetic energy of the electron and positron formed by pair production from a γ ray of wavelength 0.00247 Å.

9–4 What is the momentum of the photons created in the annihilation of a proton and antiproton, each with an original kinetic energy of 1.00 MeV?

9–5 A photon creates an electron–positron pair, each with a kinetic energy of 0.500 MeV. Compare the wavelength of the incident photon with the de Broglie wavelength of one of the particles produced.*

---

* The de Broglie wavelength associated with a particle is given by $\lambda = h/p$, where p is the momentum of the particle. See Equation (8–13) and Chapter 10.

9-6 A photon enters a cloud chamber located in a magnetic field of 0.0300 tesla to produce an electron–positron pair. The radii of the curved paths perpendicular to the magnetic field of the electron and positron are 2.00 cm and 1.50 cm, respectively. What was the energy in MeV of the incident photon?

9-7 An electron and a positron, each traveling at $0.80c$ in opposite directions, collide and are annihilated in the form of radiation. Calculate (a) the de Broglie wavelength of the electron, (b) the wavelength of the photons formed, and (c) the momentum of each of the photons.

9-8 The radiation from CO_2 lasers ($\lambda = 10.6\ \mu$) reaches typical values of 100 W/cm² normal to a surface. (a) What is the photon flux, that is, the number of photons incident on a unit area in unit time? (b) What would be the photon flux of γ rays of wavelengths 5.00×10^{-3} Å that would produce the same intensity?

9-9 The absorption coefficient of low-energy γ rays in lead is 1.50 cm^{-1}. What thickness of lead is required to reduce the intensity of the γ rays (a) to half the original intensity? (b) to 0.0100 of the original intensity?

9-10 For γ rays of a given wavelength, the coefficient of absorption for lead is 0.900 cm^{-1} and for aluminum 0.280 cm^{-1}. For the absorption of these γ rays, what thickness of aluminum would have the same absorption as 1.00 cm of lead?

9-11 By what mechanism would a 12-MeV photon be absorbed in a piece of metal? Explain.

9-12 Data is recorded for the intensity of "hard" x rays as the x rays penetrate various thicknesses of copper. From the data given, plot a graph and determine (a) the absorption coefficient of copper, and (b) the thickness of copper that will transmit 37% of the incident x rays.

THICKNESS (cm)	0.10	0.20	0.30	0.35	0.38	0.40	0.45	0.50	0.62
RELATIVE I/I_0	0.90	0.80	0.74	0.70	0.68	0.67	0.64	0.57	0.50

9-13 A ruby laser ($\lambda = 6983$ Å) produces a 50.0-J pulse at the rate of 96 pulses/min. How many photons are in a single pulse?

9-14 Assume that pair production can occur *without* the presence of a heavy nucleus and show that

$$\left(\frac{h\nu}{c}\right)^2 = p_-^2 + p_+^2 + 2p_-p_+ \cos(\theta + \phi)$$

(see Figure 9-5). Then show that this equation leads to $h\nu < E_+ + E_-$, which is a contradiction of the conservation of energy.

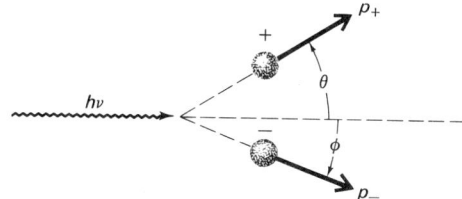

Figure 9-5

RECOMMENDED READING

ALONSO, M., and FINN, E. J., *Fundamentals of University Physics*, Addison-Wesley, Reading, Mass., 1968, Vol. III, pp. 32, 383.

ANDERSON, C. D., *Science* **76,** 238 (1932).

ANDERSON, Carl D., in *Nobel Lectures*, Elsevier, New York, 1965.
Brief biography and Nobel Lecture of Anderson on production and properties of positrons.

ANDERSON, David L., *Discovery of the Electron*, Van Nostrand Reinhold, New York, 1964.

COWAN, C. L., "The Absorption of Gamma Radiation," *Phys. Rev.* **74,** 12, 1841–1845 (1948).

FRISCH, David H., and THORNDIKE, Alan M., *Elementary Particles*, Van Nostrand Reinhold, New York, 1964.
Well-written, interestingly presented discussion on particles and antiparticles.

HOUSTON, W. C., *Frontiers of Nuclear Physics*, American Education Pub., Columbus, Ohio, 1964.
A very elementary presentation of the elementary particles. The photograph on p. 28 is a very good illustration of the problem of pair annihilation.

MEYERHOF, W. E., *Elements of Nuclear Physics*, McGraw-Hill, New York, 1967, pp. 100–104.
A more advanced description of the pair production and annihilation on the basis of the Dirac theory.

10
Wave Nature of Particles

Prince Louis Victor de Broglie (1892–)

Born in Dieppe, France, de Broglie earned his doctor of science degree from the Sorbonne in 1924. There he was an instructor and later served as the director of the Henry Poincaré Institute. In 1924 he proposed that matter and radiation have both wave and particle properties. In 1926 his hypothesis was incorporated into a mathematical formalism called wave mechanics. De Broglie has since made great efforts to find a causal (as opposed to probabilistic) interpretation of wave mechanics. For his discovery of the wave character of electrons, he was awarded the 1929 Nobel Prize in physics.

10–1 WAVE-PARTICLE DILEMMA
10–2 DE BROGLIE WAVES
10–3 EXPERIMENTAL CONFIRMATION OF PARTICLE-WAVES
10–4 WAVE PACKETS
10–5 THE UNCERTAINTY PRINCIPLE
10–6 ANOTHER FORM OF THE UNCERTAINTY PRINCIPLE

10-1 WAVE-PARTICLE DILEMMA

From the time of the philosophers of antiquity to Isaac Newton, scientists considered light to be rapidly moving corpuscles. In 1801 Thomas Young's interference experiments established the wave theory of light on a firm experimental basis, and at that time it seemed that physicists could reject the idea of an atomic structure of light. However, atomic theories began to answer more and more questions about the physical properties of solids, liquids, and gases. By the close of the nineteenth century, experiments by J. J. Thomson and H. A. Lorentz yielded the concept of the electricity corpuscle—the electron. Finally, by 1900 the cycle was completed when a law of blackbody radiation was established after Max Planck postulated that a light source does not emit radiation continuously, but in equal and finite quantities called *quanta*. By 1923 the wave-particle dilemma was again emphasized when Arthur Compton discovered that x-ray quanta have momentum as well as energy. Thus, light seemed to have two distinct sets of properties that were in direct conflict with each other.

Finally, in 1924 LOUIS DE BROGLIE delivered his thesis for his doctorate degree, *Recherches sur la Théorie des Quanta*, in which he suggested that since light had much of the character of particles, that particles, in particular electrons, could by the symmetry of nature be expected to exhibit wavelike properties. In 1927 his ideas were verified independently by G. P. THOMSON, and by C. J. DAVISSON and L. G. GERMER when they showed that electrons, like light, could be diffracted.

10-2 DE BROGLIE WAVES

Planck had related the energy of light corpuscles (photons) and the frequency of light by

$$E = h\nu \tag{10-1}$$

where ν is the frequency of the photon and the energy E is the same expression found in the special theory of relativity, that is $E = E_0 + K$ where, for the case of the photon, $E_0 = m_0 c^2 = 0$ and the total energy $E = K$ is entirely kinetic. However, frequency had no real place in a purely corpuscular or particle theory. De Broglie reasoned that there must be waves of some type associated with these photons to explain such purely wave phenomena as interference. Compton had shown that light waves have a momentum, usually associated only with particles. The energy expression for a photon could also be expressed as

$$E = mc^2 = pc \tag{10-2}$$

where p is the momentum associated with the photon. Combining Equations (10–1) and (10–2) to get

$$\boxed{p = \frac{h\nu}{c} = \frac{h}{\lambda}} \tag{10-3}$$

relates the wave character of photons, λ the wavelength and ν the frequency, to the particle character, the momentum p.

Then de Broglie put forth an astounding idea: If light waves can have a particle nature, then particles, such as electrons, can possess wavelike characteristics. Thus, an electron can have a momentum,

$$\boxed{p = mv = \frac{h}{\lambda}} \tag{10-4}$$

where m is the relativistic mass. This expression has now related to the electron, a particle, the wavelike character of a frequency ν. By applying photon Equation (10–3) to particle Equation (10–4) the wavelength of the electron becomes

$$\boxed{\lambda = \frac{h}{mv}} \tag{10-5}$$

The wavelength can be written in terms of the momentum as

$$\lambda = \frac{h}{p}$$

where $p = (1/c)\sqrt{E^2 - E_0^2}$ from Equation (6–33). Now

$$\lambda = \frac{h}{(1/c)\sqrt{E^2 - E_0^2}} = \frac{h}{(1/c)\sqrt{(mc^2)^2 - (m_0 c^2)^2}}$$

which becomes

$$\lambda = \frac{h}{m_0 c \sqrt{1/[1 - (v^2/c^2)] - 1}}$$

and simplifies to

$$\lambda = \frac{h\sqrt{1 - (v^2/c^2)}}{m_0 v} = \frac{h}{mv}$$

Although de Broglie's initial work considered the problem of an electron particle, the equations is valid for all material bodies. An official American League baseball of mass 0.14 kg (5 oz) thrown as a fast ball has a velocity of about 40 m/sec. The de Broglie wavelength associated with this baseball is

$$\lambda = \frac{h}{m_0 v} = \frac{6.63 \times 10^{-34}}{0.14 \times 40} \times \left(10^{10} \frac{\text{Å}}{m}\right) = 1.2 \times 10^{-24} \text{ Å}$$

which is an extremely small wavelength that could not be detected. An electron traveling with the same velocity will have a wavelength

$$\lambda = \frac{h}{m_0 v} = \frac{6.63 \times 10^{-34}}{9.1 \times 10^{-31} \times 40} \times \left(10^{10} \frac{\text{Å}}{m}\right) = 1.8 \times 10^{5} \text{ Å}$$

which can be readily measured in the laboratory.

10–3 EXPERIMENTAL CONFIRMATION OF PARTICLE-WAVES

Louis de Broglie's idea of giving wavelike characteristics to particles was an ingenious and novel solution to a problem. This idea was novel enough that, had it not been for the experimental confirmation that quickly followed, it might have been some time before it could be fully accepted by fellow physicists. In 1925 a series of experiments was begun on secondary electron emission at the Bell Telephone Laboratories in the United States under C. J. Davisson and C. H. Kunsman and later Davisson and L. H. Germer. In these experiments, a beam of electrons incident on a nickel crystal caused electrons (secondary electrons) to be reemitted from the crystal. When one of the nickel targets accidentally received an oxide coating, it was heated to remove this oxide. Experiments performed on this crystal target gave quite different experimental results. The prolonged heating had changed the polycrystalline target into a single large crystal. Electrons were still emitted at all angles, but at certain given angles a larger number of electrons were detected than at any previous time.

A beam of 54-eV electrons was found to give an increase in the number of electrons emitted at an angle of $\theta = 50°$ (Figure 10–1). For the electrons, $K = 54 \times 1.60 \times 10^{-19}$ J and

$$p = \sqrt{2 m_0 K}$$

from which the de Broglie hypothesis predicts a wavelength for free electrons of

$$\lambda = \frac{h}{p} = \frac{6.63 \times 10^{-34}}{\sqrt{2 \times 9.11 \times 10^{-31} \times 54 \times 1.60 \times 10^{-19}}} \times \left(10^{10} \frac{\text{Å}}{m}\right)$$

$$= 1.67 \text{ Å}$$

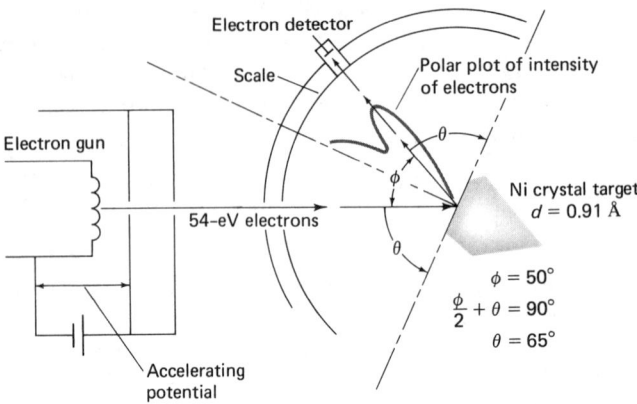

Figure 10-1   Diffraction of electrons from a single nickel crystal.

which is the *associated de Broglie wavelength*. Now, on the basis of a wave diffracted from the Bragg planes within the crystal, a first-order diffraction ($n = 1$) from the nickel crystal with spacing $d = 0.91$ Å will give a wavelength

$$\lambda = 2d \sin \theta = 2 \times 0.91 \times \sin 65°$$
$$= 1.65 \text{ Å}$$

which is the *wavelength of a diffracted wave*. This comparison is good and gives strong evidence that, indeed, the electrons do have a wave character as well as a particle character.

At about the same time the above experiments were being conducted, G. P. Thomson and some of his students were conducting a different experiment that was to arrive at the same conclusion. In the same manner as von Laue's x-ray diffraction experiments, Thomson scattered high-energy electrons from very thin metal foils made of a polycrystalline material with a random orientation of crystal axes. The experimental setup is shown in Figure 10–2. In his experiment he sent a collimated beam of electrons (energy about 10^4 eV) through the foil about 10^{-5} cm thick. He was able to obtain a series of diffraction rings in which the scattering angles θ_1, θ_2, θ_3, and so on verified the well-known equation for diffraction by transmission, $n\lambda = d \sin \theta$ ($n = 1, 2, 3, \ldots$), where θ is the angle of the diffracted beam with the direction of the incident beam.

Several other interesting experimental verifications of the wave nature of particles include the following:

1. In Germany, Rupp measured the wavelength of electrons directly by the diffraction of an electron beam incident on an optical diffraction grating at a grazing angle of incidence.
2. In 1931 Johnson demonstrated the diffraction effects of hydrogen beams scattered from crystals.

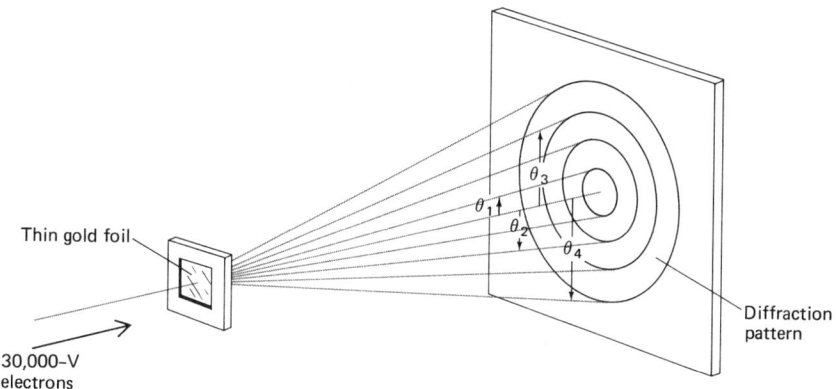

Figure 10-2 Diffraction pattern produced by passing high-speed electrons through a thin gold foil.

3. In a striking experiment (1938), Eastermann, Frisch, and Stern produced diffraction effects with a beam of helium molecules incident on LiF crystals.

The above experiments present conclusive evidence of the wave nature of particles, but other questions quickly arise. For example, since photons and electrons both exhibit particle and wave characteristics, it would seem that there is really no distinction to be made between particles and waves. However, the wave nature of photons still does not give a satisfactory description of such experiments as Compton scattering or the photoelectric effect, and particles such as electrons can never travel at the speed of light.

Niels Bohr suggested the *principle of complementarity* to resolve the apparent wave-particle contradictions. His principle of complementarity states that neither waves nor particles will ever exhibit *both* their wave and particle characters in any single experiment. A complete description of either radiation or electrons requires in either case that both the wave and particle features be used—but each in its own sphere of application.

10-4 WAVE PACKETS

The electrical field associated with a charged pith ball is not localized in the ball alone, but spreads throughout space. This electrical field and the charged particle cannot be separated as a single entity; they are two aspects of the same quantity. In the same manner, an entity of matter—a particle—will be represented by waves. By the very fact that it is represented by a wave, the particle has lost some aspects of being localized. Such a system of waves will form an envelope that propagates with a velocity different from the velocity of either of the component waves. This envelope of waves, which represents a material point, is called a *wave packet* (or wave group).

A combination of many individual waves is required for a wave packet that is to represent a material point. As a simple approach to wave packets, consider a wave packet made up of two waves,

$$\psi_1 = A \cos(\omega_1 t - k_1 x)$$
$$\psi_2 = A \cos(\omega_2 t - k_2 x)$$

where $\omega_1 \cong \omega_2 = \omega$ and $k_1 \cong k_2 = k$. By the principle of superposition, these two waves can be added to get a resultant wave,

$$\psi = \psi_1 + \psi_2 = A_R \cos\left[\left(\frac{\omega_1 + \omega_2}{2}\right)t - \left(\frac{k_1 + k_2}{2}\right)x\right]$$

with a resultant amplitude,

$$A_R = 2A \cos\left[\left(\frac{\omega_1 - \omega_2}{2}\right)t - \left(\frac{k_1 - k_2}{2}\right)x\right]$$

Figure 10–3(a) shows two waves that have been added together to produce a resultant wave. Note that the addition of these two waves has produced *beats* or wave groups. Each wave will propagate with its own velocity (its *phase velocity*), but the wave groups that are formed move with a still different velocity (the *group velocity*). Since the ω's and k's are approximately equal, the phase velocities of ψ_1 and ψ_2 are approximately equal. From Equations (10–1) and (10–3), the phase velocity becomes

$$v_{ph} = \nu\lambda = \frac{E}{h}\frac{h}{p} = \frac{mc^2}{mv} = \frac{c^2}{v} \tag{10–6}$$

or

$$v_{ph} = \nu\lambda = \frac{\omega \times 2\pi}{2\pi \times k} = \frac{\omega}{k} \tag{10–7}$$

where $\omega = 2\pi\nu$ is the angular frequency and the propagation constant is $k = 2\pi/\lambda$. In Equation (10–6), ν and λ are, respectively, the associated de Broglie frequency and the wavelength.

Next, the meaning of the de Broglie wave comes into still sharper focus when it is shown that the velocity of a particle is the same as the group velocity of the de Broglie waves. The particle speed can be determined from

$$v = \frac{dE}{dp} \tag{10–8}$$

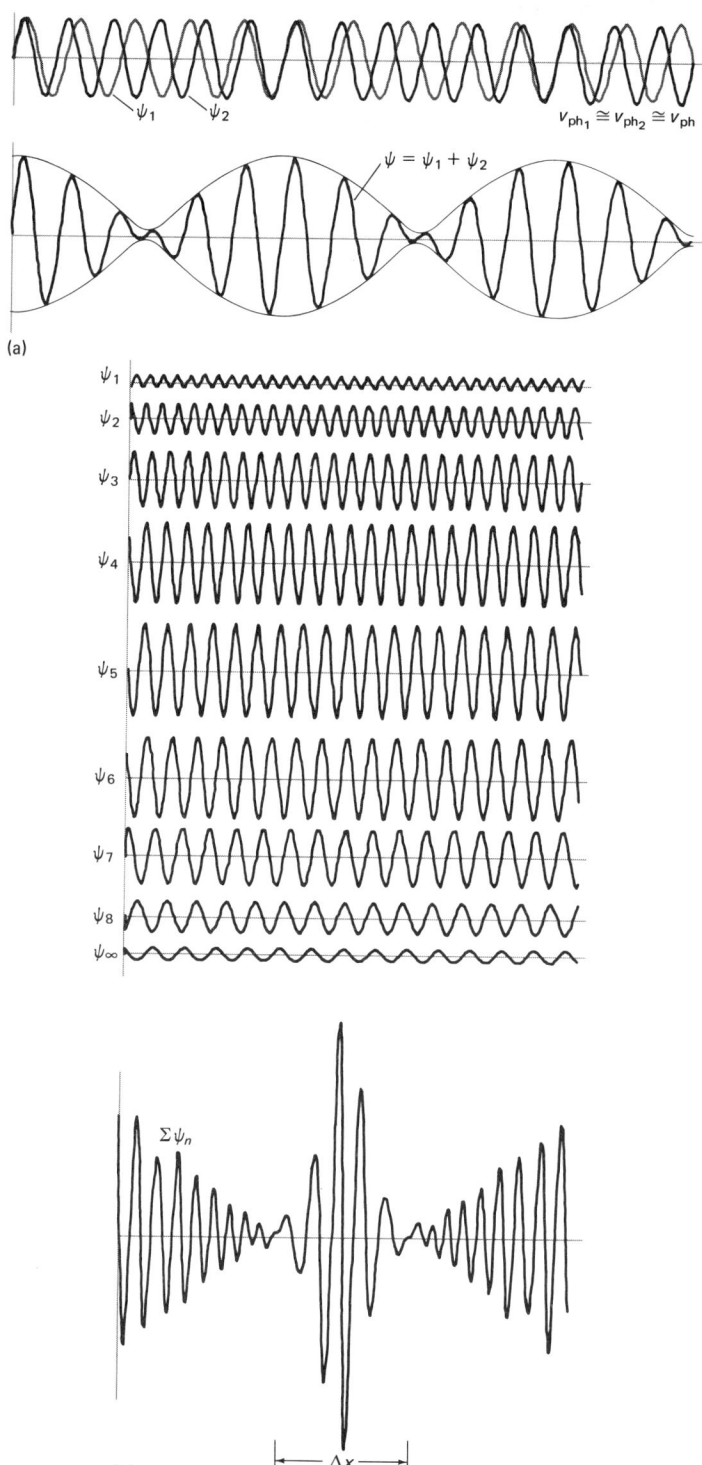

Figure 10–3 (a) The addition of two waves of nearly the same frequency to produce wave groups (or beats). (b) A rather sharply defined wave group resulting from the superposition of many different waves.

and group velocity is defined from wave theory as

$$v_g = \frac{d\omega}{dk} \tag{10-9}$$

which can also be written as

$$v_g = \frac{dv}{d(1/\lambda)} = -\lambda^2 \frac{dv}{d\lambda} \tag{10-10}$$

From $\lambda = h/p$,

$$d\lambda = \frac{-h\,dp}{p^2} \tag{10-11}$$

from $E^2 = E_0^2 + p^2 c^2$,

$$dE = \frac{pc^2}{E}\,dp \tag{10-12}$$

and from $v = E/h$,

$$dv = \frac{dE}{h} = \frac{pc^2}{Eh}\,dp \tag{10-13}$$

Substituting Equations (10-11) and (10-13) for $d\lambda$ and dv into Equation (10-10) will show that

$$v_g = -\lambda^2 \frac{dv}{d\lambda} = \frac{pc^2}{E} \tag{10-14}$$

or

$$v_g = \frac{mvc^2}{mc^2}$$

and

$$\boxed{v_g = v}$$

The group velocity thus equals the velocity of the particle, or the de Broglie waves move along with the particle. When Equation (10-10) is rewritten, replacing v by v_{ph}/λ, the group velocity becomes

$$v_g = \frac{d(v_{ph}/\lambda)}{d(1/\lambda)}$$

which then becomes

$$\boxed{v_g = v_{ph} - \lambda \frac{dv_{ph}}{d\lambda}} \tag{10-15}$$

where the group velocity has been related to the phase velocity. Note that if the waves are not in a dispersive medium, that is, v_{ph} = constant, then $v_g = v_{ph}$.

What about the group velocity of photons? Does this compare with the group velocity of particles? For a photon, $pc = E$, and Equation (10–14) becomes

$$v_g = \frac{pc^2}{E} = \frac{Ec}{E}$$

or

$$\boxed{v_g = c}$$

This is by now an expected result—*the group velocity for photons is equal to the velocity of the photon.*

Figure 10–3(b) shows a more ideal case, in which a large number of waves $\psi_1, \psi_2, \ldots, \psi_n$ have been added together. The wave packet is sharply defined, and its size Δx has been reduced considerably.

10–5 THE UNCERTAINTY PRINCIPLE

In 1927 the German physicist WERNER HEISENBERG provided an interesting addition to the meaning of the wave-particle concepts. This is the *uncertainty principle* or the *principle of indeterminacy*, and it expresses a fundamental limit to the simultaneous determination of certain pairs of variables, such as position and momentum. Figure 10–4 represents a particle moving with a velocity v, which is located within a wave packet that moves with velocity $v_g = v$. From Equations (10–10) and (10–11), the group velocity can be written in still a different form,

$$v_g = h \frac{dv}{dp} \tag{10–16}$$

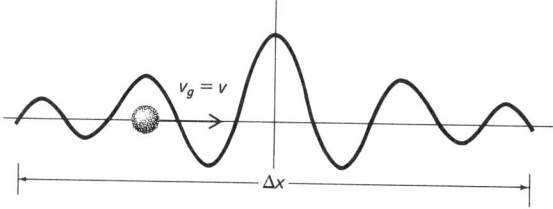

Figure 10–4   A moving particle is represented in a wave packet moving with a velocity v_g.

From Figure 10–4 and Equation (10–16),

$$v_g = \frac{\Delta x}{\Delta t} = h \frac{\Delta \nu}{\Delta p}$$

or

$$\Delta p \, \Delta x = h \, \Delta \nu \, \Delta t \qquad (10\text{–}17)$$

If the frequency of the wave is to be measured, the least time of measurement will be the time required for one complete wavelength to pass a reference point. This time, related to the frequency, is $\Delta t \geq 1/\Delta \nu$, the time of passage of one complete cycle.

Now, from

$$\Delta \nu \, \Delta t \geq 1 \qquad (10\text{–}18)$$

Equation 10–18 takes the form

$$\boxed{\Delta p \, \Delta x \geq h} \qquad (10\text{–}19)$$

which is one expression of the Heisenberg uncertainty principle. This makes a considerable impact on our ideas of measurement. This fundamental limit ($\geq h$) implies not only that there is a limit to the precision of a measurement, but that the more closely a particle is located in position, the greater the uncertainty in the measurement of its momentum and vice versa.

The momentum in Equation (10–19) should actually be the component of the momentum in the x direction, and it is more properly written

$$\boxed{\Delta p_x \, \Delta x \geq h} \qquad (10\text{–}20)$$

Pairs of variables such as p_x and x, in which the uncertainty in one places a limit on the accuracy of measurement of the other, are called *conjugate variables*. There is no limitation in the precision of measurement of the components of the momentum along the y and z axes (Δp_y and Δp_z) and the position along the x direction.

Due to the small value of h, this uncertainty is not relevant in the macroscopic world. This uncertainty has nothing to do with the natural uncertainty that appears in the macroscopic world, namely, the interaction that appears between the measuring instrument and the magnitude we are trying to measure. For example, if we try to measure the temperature of a body, we need a thermometer; but when this is placed in contact with the object, we modify the temperature we want to measure, and so on.

The principle of uncertainty, which is a natural law, does not destroy the law of causality that was taken for granted in the macroscopic world. The special relativity, which we have discussed, is based solidly on the absolute distinction between cause and effect. A better name for the principle is the *indeterminacy principle.* In this respect, the classical mechanics of Newton and Galileo is *deterministic.* (If we know the force acting on a particle and the initial conditions—position and momentum—then we can predict the subsequent motion of the particle with absolute precision.) The microscopic world is essentially *nondeterministic.*

The problem of uncertainty is a logical consequence of the dual behavior of matter. A deeper insight into the nature of the uncertainty principle can be accomplished by an idealized experiment called "the Heisenberg microscope," which is shown in Figure 10–5.

In Figure 10–5, an imaginary microscope with a very high resolving power is being used to try to measure simultaneously the position and the linear momentum of an electron. Geometrical optics shows that the resolving power of the microscope permits an uncertainty in measurement of the position of

$$\Delta x \cong \frac{\lambda}{2 \sin \theta} \tag{10-21}$$

where θ is the angle indicated in Figure 10–5. When the distance is less than Δx, the two points will be seen as just one point, and hence this Δx represents the smallest uncertainty in the position of the traveling electron that we are trying to measure with the microscope. From Equation (10–21), the smaller the wavelength λ of the incident light with which the object is illuminated, the smaller the Δx and consequently the more precisely the position of the electron can be fixed.

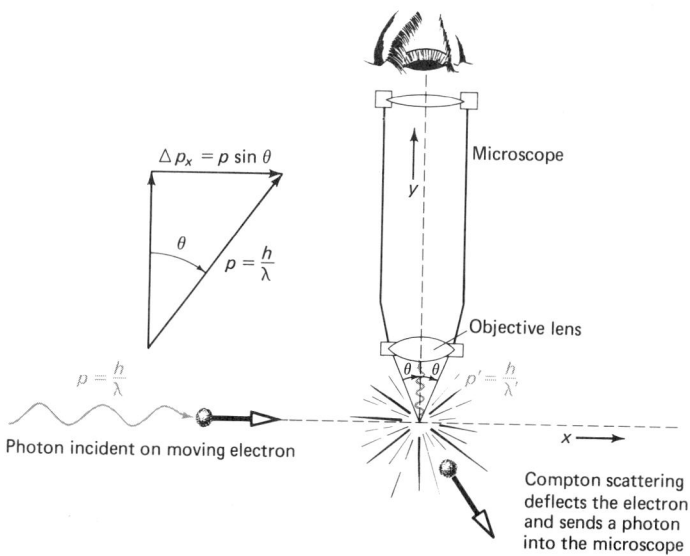

Figure 10–5 Heisenberg microscope. An attempt is being made to measure simultaneously the position along the x axis and the linear momentum in the x direction.

However, the incoming photon will interact with the electron through the Compton effect. To be able to see the electron, the scattered photon should enter the microscope within the angle 2θ. The momentum (Figure 10–5) of the photon thus has an uncertainty in the x direction of

$$\Delta p_x \cong p \sin \theta \tag{10-22}$$

where $p = h/\lambda$ is the momentum of the incoming photon. According to the conservation of linear momentum, Equation (10–22) must also give the minimum uncertainty in the momentum of the recoil electron. Therefore, for the recoil electron, Equations (10–21) and (10–22) show that

$$\Delta p_x \, \Delta x \geq (p \sin \theta) \left(\frac{\lambda}{2 \sin \theta} \right) = \frac{h}{\lambda} \frac{\lambda}{2} = \frac{h}{2}$$

$$\geq \frac{h}{2}$$

which is the uncertainty principle. A more sophisticated approach will show that

$$\boxed{\Delta p_x \, \Delta x \geq \hbar} \tag{10-23}$$

where $\hbar = h/2\pi$ and is called "$\hbar$-bar."

EXAMPLE 10–1: Assume the uncertainty in position of a hydrogen molecule of mass about 2×10^{-27} kg is of the order of the diameter, about 10^{-10} m. The uncertainty in momentum is then

$$\Delta p_x \geq \frac{h}{\Delta x} = \frac{6.6 \times 10^{-34}}{10^{-10}}$$

$$\geq 6.6 \times 10^{-24} \text{ kg-m/sec}$$

The momentum of this molecule moving at 2000 m/sec (thermal velocity at room temperature),

$$p_x = mv = 2 \times 10^{-27} \times 2 \times 10^3 = 4 \times 10^{-24} \text{ kg-m/sec}$$

will give a fractional uncertainty

$$\frac{\Delta p_x}{p_x} = \frac{6.6 \times 10^{-24}}{4 \times 10^{-24}} = 1.7$$

Thus, the momentum of this molecule cannot be specified with any more precision than 170% of the original value. However, a 50-g bullet fired at a speed of 1000 m/sec, whose position is known within 1.0 mm, has an uncertainty in momentum of

$$\Delta p_x \geq \frac{6.6 \times 10^{-34}}{10^{-3}} = 6.6 \times 10^{-31} \text{ kg-m/sec}$$

With a momentum of

$$p_x = 0.050 \times 1000 = 50 \text{ kg-m/sec}$$

the fractional uncertainty is

$$\frac{\Delta p_x}{p_x} = \frac{6.6 \times 10^{-31}}{50} = 1.3 \times 10^{-32}$$

This number is so small that no real laboratory apparatus will be affected by it.

10–6 ANOTHER FORM OF THE UNCERTAINTY PRINCIPLE

Another uncertainty relation between energy and time comes again from Equation (10–18),

$$\Delta v \, \Delta t \geq 1$$

or

$$\Delta \left(\frac{E}{h}\right) \Delta t \geq 1$$

which becomes

$$\boxed{\Delta E \, \Delta t \geq h} \tag{10-24}$$

Just as momentum and displacement cannot be determined simultaneously with infinite precision, so the energy E and the time t (which are another set of conjugate variables) cannot be determined simultaneously with infinite precision. The more precise we are in our measurement of time (i.e., the smaller Δt), the less precise we will be in the determination of energy. Again, a more sophisticated derivation of Equation (10–24) would give

$$\boxed{\Delta E \, \Delta t \geq \hbar} \tag{10-25}$$

EXAMPLE 10–2: The lifetime of an excited state of an atom is about 10^{-8} sec. (An excited state of an atom is one in which the energy is greater than that of the usual state of minimum energy or the ground state.) The minimum uncertainty in the determination of the energy of the excited state is, from Equation (10–25),

$$\Delta E \geq \frac{\hbar}{\Delta t} = \frac{6.6 \times 10^{-34}}{2\pi \times 10^{-8}}$$

$$\geq 1.0 \times 10^{-26} \text{ J} = 6.5 \times 10^{-8} \text{ eV}$$

This is known as the *energy width* of an excited state.

PROBLEMS

10-1 What is the wavelength associated with (a) a 100-eV electron? (b) a golf ball (1.65 oz) with a velocity of 60.0 m/sec?

10-2 An electron with an associated wavelength of 2.00 Å will have what speed?

10-3 Determine the momentum and energy for (a) an x-ray photon, and (b) an electron, each with a wavelength of 1.00 Å.

10-4 A Van de Graaff generator accelerates the bare nuclei of lithium atoms through a potential of 5.00×10^6 V. What are the speed and wavelength of these nuclei?

10-5 (a) What is the relativistic mass of an electron with a wavelength of 0.0420 Å?
(b) From $E = hc/\lambda = mc^2$, an effective mass of a photon $m^* = h/\lambda c$ can be calculated. What is the effective mass of a photon with a wavelength of 0.0420 Å?

10-6 If an accelerator gives an electron a kinetic energy of 0.511 MeV, what is its de Broglie wavelength?

10-7 In Figure 10-1, if the accelerating potential is 100 V, at what angle will the peak occur for the scattered electrons?

10-8 A beam of neutrons produced by a nuclear reaction is incident on a crystal with a spacing of 1.50 Å. Determine the speed of these neutrons if first-order Bragg reflection occurs at 30°.

10-9 (a) An optical diffraction grating (Figure 10-6) was used by Rupp to show the diffraction of electrons. For grazing angles of incidence, that is, θ is very small, show that

$$n\lambda = d\left(\frac{\alpha^2}{2} + \alpha\theta\right) \quad n = 1, 2, 3, \ldots$$

where d is the grating spacing and α is the angle of diffraction (see Section 10-4).
(b) At what angle α will 100-eV electrons incident at an angle of $\theta = 10^{-3}$ rad produce a diffraction maximum from a grating with a spacing of 5.00×10^{-6} m?

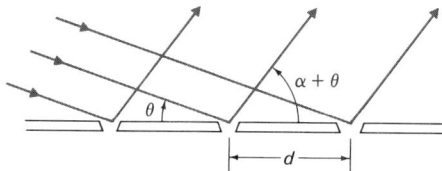

Figure 10–6

10–10 (a) For a free particle moving at a relativistic speed v, show that $v_{ph} = c^2/v$.
(b) For a free particle moving at a nonrelativistic speed v', show that $v_g = v'/2$.

10–11 Show that the group velocity of a particle can be expressed as

$$v_g = \frac{1}{\hbar}\frac{dE}{dk}$$

where E is the total energy and k is the propagation constant.

10–12 Start with the definition of group velocity, $v_g = dE/dp$, and show that

$$v_g = v + k\frac{dv}{dk}$$

where $k = 2\pi/\lambda$ is the propagation constant and v is the phase velocity.

10–13 From Equation (10–10), prove that

$$v_g = -v_{ph}\frac{d(\ln \lambda)}{d(\ln p)}$$

10–14 (a) Compute the minimum uncertainty in the determination of the speed of a truck whose mass is 2000 kg if you want to determine the position of its center of mass to within 2.00 Å.
(b) Compute the percentage of uncertainty in the momentum for the same case.

10–15 The speed of a nuclear particle (proton or neutron) moving in the x direction is measured to an accuracy of 10^{-6} m/sec. Determine the limit of accuracy to which its position can be located. (a) along the x axis, and (b) along the y axis. Solve the same problem if the particle is a positron.

10–16 A particle moving along the x axis has an uncertainty in its position equal to its de Broglie wavelength. Find the percentage uncertainty in its velocity.

10–17 The uncertainty in the position of an electron moving in a straight line is 10 Å. Calculate the uncertainty in (a) its momentum, (b) its velocity, and (c) its kinetic energy.

10–18 (a) The lifetime of an excited state in an atom is about 10^{-8} sec. Compute the spread in energy of the emitted photon (energy width).

(b) If the emitted photons are in the visible spectrum ($\lambda \cong 4000$ Å), calculate the line width in angstroms.

RECOMMENDED READING

Воhм, D., *Causality and Chance in Modern Physics*, Harper Torchbooks, Harper & Row, New York, 1961.
A penetrating philosophical approach to the basic principles of quantum mechanics.

Christy, R. W., and Pytte, A., *The Structure of Matter*, W. A. Benjamin, Menlo Park, Calif., 1965, pp. 314–320.
Contains a very clear explanation of the uncertainty principle at an elementary level.

Darrow, K. K., "The Quantum Theory," *Sci. Am.*, March 1952.

Gamow, G., *Mr. Tompkins in Wonderland*, Cambridge University Press, London, 1965.
A delightful book for anyone with some interest in science.

Heisenberg, W., *Physics and Philosophy*, Harper & Row, New York, 1961.
Excellent book, should be read by anyone interested in the philosophy of physics.

Planck, Max, *The Philosophy of Physics*, Norton, New York, 1963.
Chapter 2 is particularly interesting for those who want to know the relation between the uncertainty principle and the law of causality.

II
The Rutherford Experiment

Sir Ernest Rutherford
(1871–1937)

A native of New Zealand, Rutherford worked under J. J. Thomson at Trinity College. He was research professor at McGill University (1898–1907), director of the physics laboratory at Victoria University (1907–1919), and director of the Cavendish Laboratory (1919–1937). Arguing that the atom consists of a small positively charged central nucleus balanced by negative electrons that revolve around the nucleus, he established the nuclear model for the atom. For his work, Rutherford received the 1908 Nobel Prize in chemistry.

11–1 THE NUCLEAR MODEL OF THE ATOM
11–2 EXPERIMENTAL SETUP
11–3 IMPACT PARAMETER AND SCATTERING ANGLE
11–4 RUTHERFORD SCATTERING FORMULA

11–1 THE NUCLEAR MODEL OF THE ATOM

By 1898 Sir J. J. THOMSON had discovered the electron and he then proposed a physical model of the atom known as the "plum pudding" atom. The atom, as he pictured it, was a *positive* plum pudding in which were embedded *negative electron raisins* distributed in such a manner as to make the entire thing neutral.

In 1911 Professor ERNEST RUTHERFORD (1871–1937) (who incidentally had been a student of Thomson's) and two of his students, HANS GEIGER and ERNEST MARSDEN, performed a number of experiments on the scattering of α particles by a very thin gold foil. As a result of these famous experiments, the idea of the "plum pudding" model was ruled out in favor of the now generally accepted model. In this model, the atom is said to consist of a very small nucleus (dimensions of the order of 10^{-14} m) in which all of the positive charge and most of the mass is concentrated, and a cloud of negatively charged electrons surrounds this nucleus. Since the dimensions of the atom are of the order of 10^{-10} m, most of the space within the atom is empty; and for neutral atoms, the charge on the electrons about the nucleus is equal to the positive charge on the nucleus.

Let us review Rutherford's experiment, in order to make a detailed study of the atom. Rutherford proposed that a thin foil of gold ($Z = 79$) be bombarded with high-speed α particles from a Po-214 source. A study of the scattering or deflection angles of α particles passing through the foil would then give details of the target atoms that are doing the scattering. An α particle is just a helium nucleus and consists of two protons and two neutrons. At that time the existence of the neutron was not known, but Rutherford and Thomas Royds had previously (in 1909) determined the charge of the α particle to be $2e$.

A theoretical study of the scattering angle θ of the Thomson model and the model proposed by Rutherford was made by Rutherford, and a comparison with the experimental results was carried out. Figure 11–1 compares the Rutherford and Thomson models and shows the expected electrical field associated with each. An α particle penetrating an atom like the Thomson model [Figures 11–1(a) and 11–1(c)] would experience only a small deflection, since the electrical field within such an atom would be weak, especially when compared to the Rutherford model. In the Rutherford model, the electrical field for the same distance from the nucleus is much stronger, because all of the positive charge of the atom, $+Ze$, is concentrated in the small volume of the nucleus, and hence the scattering angle θ will be greater than in the Thomson models [Figures 11–1(b) and 11–1(d)].

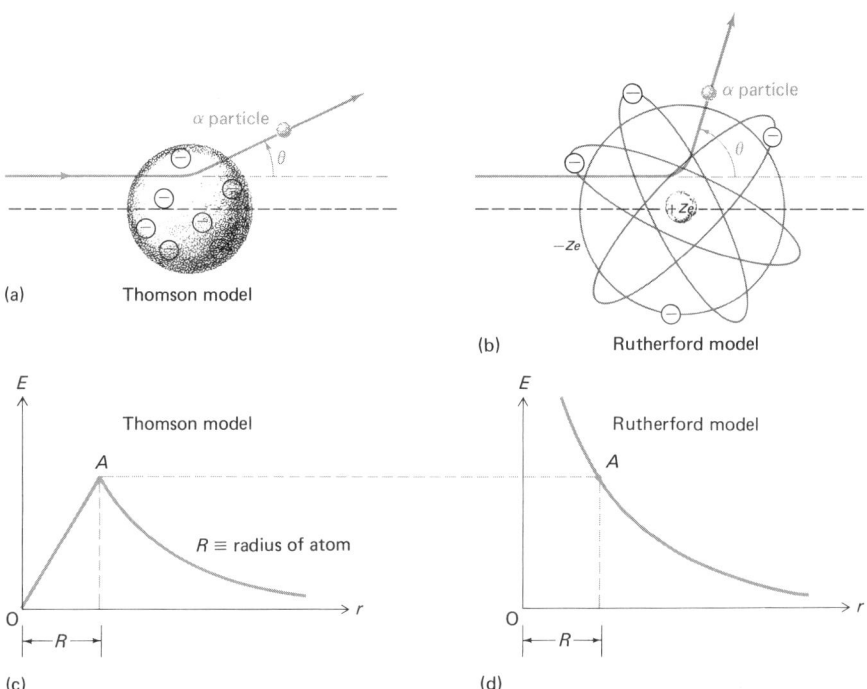

Figure 11–1 (a) The expected deflection of the α particle is small because the electrical field within the atom is small. (b) The positive charge is concentrated in the small volume of the nucleus, and the deflection of the α particle is greater. (c) The electrical field increases linearly to a surface where it is a maximum. For $r > R$, it decreases according to $E = k(Ze/r^2)$. (d) The electrical field decreases with distance to the nucleus according to $E = k(Ze/r^2)$. At $r = R$, it is the same as the Thomson model, but for $r < R$, it becomes larger.

11–2 EXPERIMENTAL SETUP

Geiger had performed many times the experiment of sending a beam of α particles through a thin foil of metal and noting the scattering of the particles. However, it was almost as an afterthought that Rutherford and Geiger suggested to Marsden that he look for scattering at large angles, even as great as 90°. When it was found that α particles were being scattered backwards, it was such a surprise that Rutherford stated: "It is just as surprising as if a gunner fired a shell at a single piece of paper and for some reason or other the projectile bounded back again."

Figure 11–2 shows the Rutherford α-particle scattering experiment. Polonium-214 is a monoenergetic source of 7.68-MeV α particles. The thin gold foil ($t = 6 \times 10^{-7}$ m) allows most of the α particles to pass through it without experiencing any deviation at all. Some, however, are scattered through various angles θ to produce

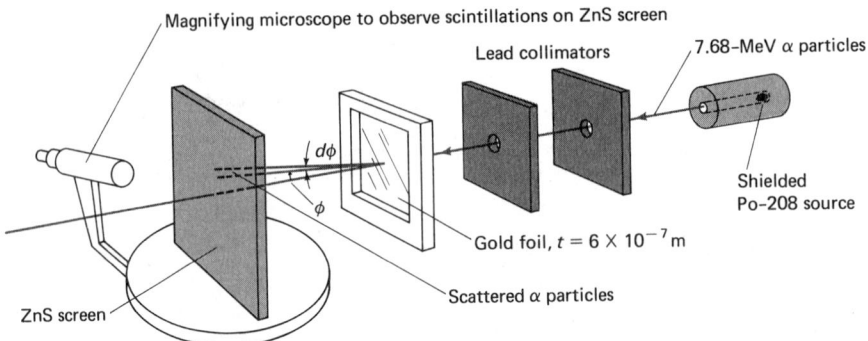

Figure 11–2 Schematic diagram of α particles scattered from atoms within thin gold foil.

scintillation that can be observed and counted by a magnifying microscope. The experiment is to count the number of particles per unit time that are deflected with scattering angles between ϕ and $\phi + \Delta\phi$ and compare these results according to the expected values in the Rutherford and Thomson models. The average deflection angle predicted by both models was about 1°, but the big difference in the two models was in the predicted deflection for very large scattering angles. For example according to the Thomson model only one out of every 10^{3500} α particles would experience a deflection of $\phi \geq 90°$, but experimental results show that one of about 8000 particles were deflected through $\phi \geq 90°$. This figure was in close agreement with the Rutherford model, and it brought acceptance of the nuclear model of the atom proposed by Rutherford.

11–3 IMPACT PARAMETER AND SCATTERING ANGLE

Figures 11–3(a) and 11–3(b) illustrate an α particle scattered from a nucleus. The *impact parameter b* in each figure is the minimum distance that the α particle would approach the nucleus if there were no forces between them. The Coulomb electrostatic repulsion between the α particle and the gold nucleus located at N will cause the α particle to follow the path ACB. The repulsive Coulomb force,

$$F = \frac{1}{4\pi\varepsilon_0} \frac{2Ze^2}{r^2} \qquad (11\text{–}1)$$

follows an inverse square law, and the trajectory must be the hyperbola ACB with the nucleus N at the focus of the hyperbola. For a head-on collision, it is evident that the impact parameter $b = 0$. The axis of the hyperbola will be Nz, and Nx and Ny are asymptotic directions passing through N and parallel to the direction of travel when the α particle is very far from the nucleus, before and after interaction.

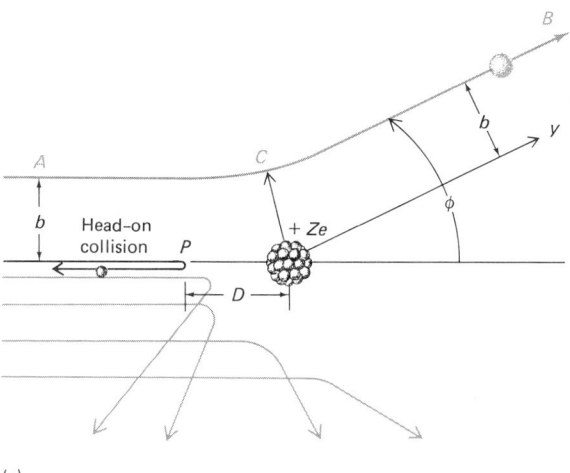

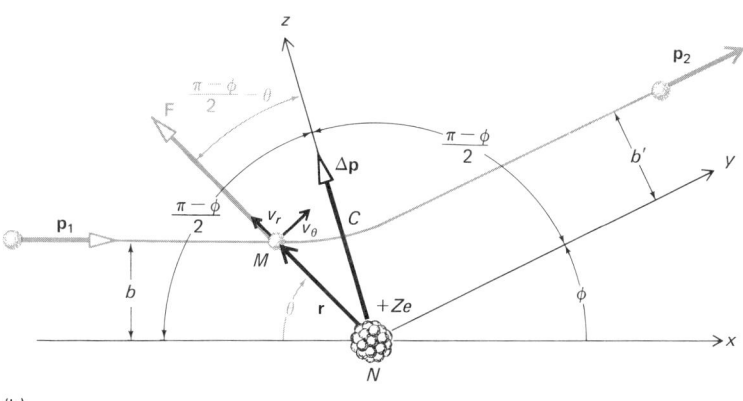

Figure 11–3 (a) The impact parameter b is the distance by which the α particle would miss the nucleus if there were no forces involved. The scattering angle ϕ depends on the impact parameter. (b) Scattering of α particle by nucleus with charge $+Ze$. The polar coordinates locating the α particle at M are r and θ.

The impact parameter b should not be confused with the distance of closest approach D. To determine the distance of closest approach, consider an α particle at a great distance from the nucleus but approaching a head-on collision with a kinetic energy K_α. At the point P in Figure 11–3(a), the repulsive force of the nucleus stops the approaching α particle momentarily, and all of its kinetic energy is transformed into potential energy. So we can write

$$K_\alpha = \frac{1}{4\pi\varepsilon_0} \frac{2Ze^2}{D}$$

or the distance of closest approach is

$$D = \frac{1}{4\pi\varepsilon_0} \frac{2Ze^2}{K_\alpha} \qquad (11\text{-}2)$$

If the collision is not "head-on," the distance of closest approach will be NC as seen in Figure 11-3(b). Also note from Figure 11-3(b) that, approximately,

$$b \cong NC \sin\left(\frac{\pi - \phi}{2}\right) = NC \cos\frac{\phi}{2} \qquad (11\text{-}3)$$

For a head-on collision, $b = 0$ and from Equation (11-3), $\phi = 180°$, which is an expected result.

In deriving the relationship between b and ϕ, we shall assume the following:

1. The α particle and nucleus are point charges.
2. The scattering is due to Coulomb electrostatic repulsive forces between the α particle and the positive charge of the nucleus (Ze).
3. The gold nucleus (mass ≅ 197 amu) is massive enough compared to the α particle (mass ≅ 4 amu) that its recoil energy can be ignored.
4. The α particles do not penetrate the nuclear region and the strong interaction nuclear forces are not involved.

If $\mathbf{p}_1$ and $\mathbf{p}_2$ in Figure 11-3(b) are the linear momenta of the α particle when it is far from the nucleus before and after the interaction, respectively, we can write

$$\frac{p_1^2}{2m} = \frac{p_2^2}{2m} \qquad (11\text{-}4)$$

and then $p_1 = p_2$. The α particle moves under the action of a central force $F = (1/4\pi\varepsilon_0)(2Ze^2/r^2)$ directed along the radius vector. Hence, angular momentum is conserved because according to Newton's laws $\mathbf{r} \times \mathbf{F} = d\mathbf{L}/dt = 0$, since $\mathbf{r}$ and $\mathbf{F}$ are aligned in the same direction. Therefore, $L = bp_1 = b'p_2$ and $b = b'$. Also, from Newton's second law,

$$\Delta\mathbf{p} = \mathbf{p}_2 - \mathbf{p}_1 = \int_0^\infty F\, dt \qquad (11\text{-}5)$$

From Figure 11-4, then,

$$|\Delta p| = 2p_1 \sin\frac{\phi}{2} \qquad (11\text{-}6)$$

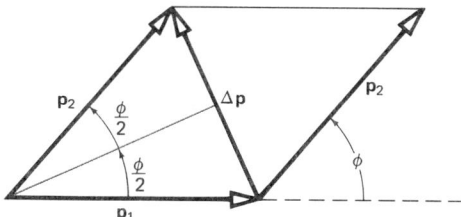

Figure 11-4 From Equation (11-4), the linear momenta $\mathbf{p}_1$ and $\mathbf{p}_2$ have equal magnitudes, but $\Delta \mathbf{p} \neq 0$ since $\mathbf{p}_1$ and $\mathbf{p}_2$ have different directions.

and the vector $\Delta \mathbf{p}$ is directed along the N axis, as can be seen by comparing Figures 11-3(b) and 11-4.

Since the total impulse $\int_0^\infty F\, dt$ is directed along the Nz axis, a scalar equation combining Equations (11-5) and (11-6) is

$$2p_1 \sin \frac{\phi}{2} = \int_{t=0}^{t=\infty} F \cos\left(\frac{\pi - \phi}{2} - \theta\right) dt \tag{11-7}$$

or, by changing variables,

$$2p_1 \sin \frac{\phi}{2} = \int_{t=0}^{t=\infty} F \sin\left(\theta + \frac{\phi}{2}\right) \frac{dt}{d\theta}\, d\theta$$

$$= \int_{\theta=0}^{\theta=\pi-\phi} \frac{F}{\omega} \sin\left(\theta + \frac{\phi}{2}\right) d\theta \tag{11-8}$$

where $d\theta/dt = \omega$ is the angular velocity of the α particle. Again, from the conservation of angular momentum,

$$mv_1 b = mv_\theta r \tag{11-9}$$

where $v_\theta = \omega r$ is the transverse component of velocity, and then

$$v_1 b = \omega r^2 \tag{11-10}$$

When F and ω from Equations (11-1) and (11-10) are replaced in Equation (11-8), we get

$$2p_1 \sin \frac{\phi}{2} = \int_{\theta=0}^{\theta=\pi-\phi} \frac{1}{4\pi\varepsilon_0} \frac{2Ze^2}{r^2} \frac{r^2}{v_1 b} \sin\left(\theta + \frac{\phi}{2}\right) d\theta$$

which upon integration gives

$$2p_1 \sin \frac{\phi}{2} = \frac{2Ze^2}{4\pi\varepsilon_0 v_1 b}\left[-\cos\left(\theta + \frac{\phi}{2}\right)\right]_{\theta=0}^{\theta=\pi-\phi} = \frac{4Ze^2}{4\pi\varepsilon_0 v_1 b} \cos \frac{\phi}{2}$$

Using the fact that $K_\alpha = mv_1^2/2$, we get from this equation

$$b = \frac{Ze^2}{K_\alpha 4\pi\varepsilon_0} \cot \frac{\phi}{2} \qquad (11\text{–}11)$$

which is the relation between the impact parameter b and the scattering angle ϕ.

Figure 11–5 shows in a more realistic way how the α particles are scattered by the nuclei in a gold foil. Most of the particles pass undeflected; only those passing near a nucleus are deflected.

The scattering angle ϕ is dependent on b; and, as seen from Equation (11–11) and Figure 11–3(a), the smaller b is, the larger the scattering angle ϕ will be. Equation (11–11) can also be written in terms of the distance of closest approach D from Equation (11–2),

$$b = \frac{D}{2} \cot \frac{\phi}{2} \qquad (11\text{–}12)$$

11–4 RUTHERFORD SCATTERING FORMULA

There is no way to check Equation (11–12) against experimental results, because it is impossible to measure directly the impact parameter b. In an actual experiment, it is the number of particles dN with scattering angles between ϕ and $\phi + d\phi$ or within a solid angle $d\Omega = dS/r^2$ that is measured. Figure 11–6 shows that $dS = 2\pi r^2 \sin \phi \, d\phi$ is the area on the screen impacted by scattered α particles. Hence, the solid angle in which these dN particles are enclosed is $d\Omega = 2\pi \sin \phi \, d\phi$.

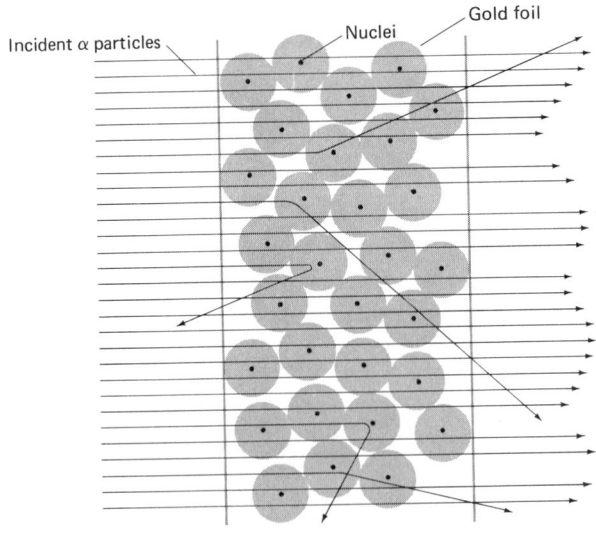

Figure 11–5 Alpha particles scattered by nuclei in gold foil.

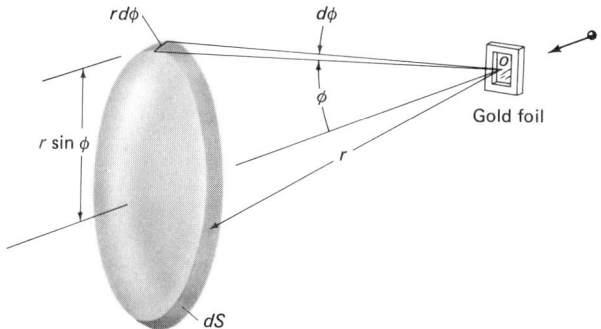

Figure 11–6 The area of the screen being hit by those α particles with scattering angles between ϕ and $\phi + d\phi$. The particles are within a solid angle $d\Omega = dS/r^2 = 2\pi \sin\phi \, d\phi$ as seen from the point O in the gold foil.

All α particles that approach the nucleus with an impact parameter $\leq b$ will be scattered by an angle $\geq \phi$. The area around each nucleus in Figure 11–7(a) with a radius equal to an impact parameter b is called the *integral cross section*. This area is

$$\sigma = \pi b^2 \tag{11-13}$$

Figure 11–7(b) shows an enlarged portion of gold foil with a surface area A and a thickness t so thin ($t = 6.0 \times 10^{-7}$ m) that the individual cross sections σ of

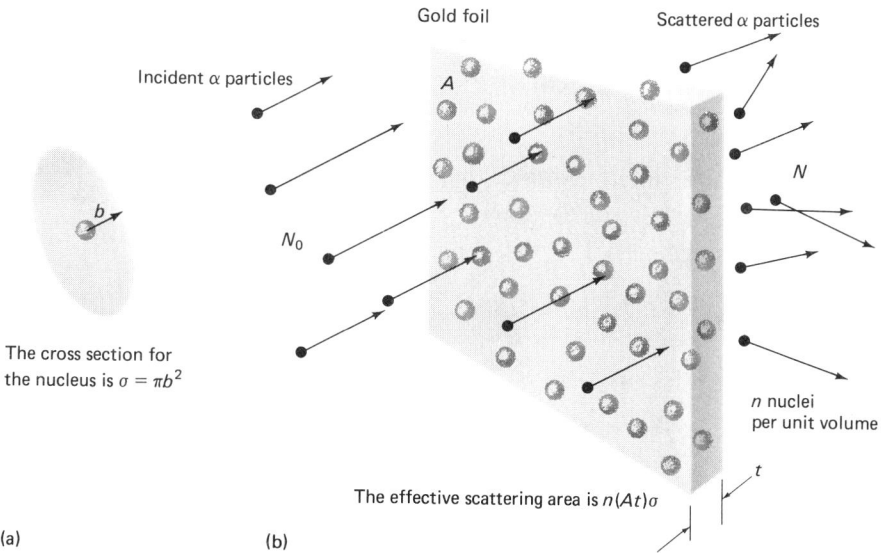

Figure 11–7 The fraction of α particles scattered is $f = (N_0/N) = n(At)\sigma/A$ and $f = n\sigma t$.

the different nuclei do not overlap. When a number of α particles N_0 is directed perpendicular to A, those interacting with different nuclei (all with the same cross section $\sigma = \pi b^2$) will experience only "single scattering." If n is the number of nuclei per unit volume, the gold foil will contain $n(At)$ nuclei, and therefore the target area T offered by those nuclei in order to have scattering angles greater than ϕ and impact parameter less than b will be

$$T = n\sigma(At) \tag{11-14}$$

Since A is the total target area, the fraction of α particles experiencing scattering angles greater than ϕ will be

$$f = \frac{\text{target area offered by nuclei in the foil}}{\text{total target area}} = \frac{n\sigma(At)}{A}$$

or

$$\boxed{f = n\sigma t = n\pi b^2 t} \tag{11-15}$$

This ratio is also given by $f = N/N_0$, where N is the number of α particles experiencing scattering and N_0 is the total number directed against the target. Combining Equations (11-12) and (11-15) now give

$$f = \frac{N}{N_0} = \frac{n\pi t D^2}{4} \cot^2 \frac{\phi}{2} \tag{11-16}$$

This equation is still not a practical formula to be checked experimentally. To compute the number of particles per unit solid angle, $d\Omega = 2\pi \sin\phi\, d\phi$, differentiate Equation (11-16) with respect to ϕ and divide by $d\Omega$ to get

$$\frac{df}{d\Omega} = \frac{dN}{N_0\, d\Omega} = -\frac{n\pi t D^2 \cot(\phi/2) \csc^2(\phi/2)\, d\phi}{8\pi \sin\phi\, d\phi}$$

The minus sign simply indicates that df and $d\phi$ have opposite signs. Taking just the absolute value of $df/d\Omega$ and, since $\sin\phi = 2\sin(\phi/2)\cos(\phi/2)$, we have

$$\boxed{\left|\frac{df}{d\Omega}\right| = \left|\frac{dN}{N_0\, d\Omega}\right| = \frac{nD^2 t}{16 \sin^4 \phi/2}} \tag{11-17}$$

Since $d\Omega = dS/r^2$, we can write

$$\frac{dN}{N_0\, d\Omega} = \frac{dN}{N_0\, dS/r^2}$$

and finally,

$$\boxed{\left|\frac{dN}{dS}\right| = \frac{N_0 n D^2 t}{16 r^2 \sin^4 \phi/2}} \tag{11-18}$$

which is the number of α particles per unit area striking the screen into a ring of area $dS = 2\pi r^2 \sin\phi\, d\phi$. Equation (11–18), known as the *Rutherford scattering formula*, was verified experimentally, and thus Rutherford should be credited with the discovery of the nucleus. Experimental verification of the Rutherford scattering formula shows that if 10^6 of the incident particles have a scattering angle $\geq 10°$, then only about 230 are deflected by $\phi \geq 90°$.

The number of nuclei per unit volume can be computed from

$$n = \frac{\rho N_A}{M} \tag{11-19}$$

where ρ (kg/m^3) is the density of the material of the foil, Avogadro's number $N_A = 6.02 \times 10^{26}$ atoms/kg-mole, and M (kg/kg-mole) is the atomic weight. In their experimental verification, Geiger and Marsden used different materials from carbon to gold.*

EXAMPLE 11–1

(a) The kinetic energy of the α particles in the Geiger–Marsden experiment was $K_\alpha = 7.68$ MeV. Calculate the distance D of closest approach to a gold nucleus ($Z = 79$).

At the distance of closest approach, the K_α of the α particle is transformed into the potential energy of the system. Hence, from Equation (11–2),

$$D = \frac{2Ze^2}{4\pi\varepsilon_0\, K_\alpha}$$

$$= \frac{2 \times 79 \times (1.60 \times 10^{-19})^2}{4 \times 3.14 \times 8.85 \times 10^{-12} \times 7.68 \text{ MeV}(1.60 \times 10^{-13} \text{ J/MeV})}$$

$$= 2.96 \times 10^{-14}\text{ m}$$

The radius of the gold nucleus should have an order of magnitude of about 10^{-14} m or smaller, which is about 10^4 smaller than the radius of the atom as a whole.

(b) Again continuing with the Geiger–Marsden experiment and hence considering gold as the material of the foil, calculate the impact parameter b in order to produce scattering angles $\phi \geq 90°$.

When the value of D above and $\phi = 90°$ are used, Equation (11–12) gives

$$b = \frac{D}{2}\cot\frac{\phi}{2} = \frac{2.96 \times 10^{-14}}{2}\cot\frac{90°}{2}$$

---

* The interested student should read a complete description of this experiment in H. Geiger and E. Marsden, "Deflection of Alpha Particles Through Large Angles," *Phil. Mag.* **25**, 605 (1913).

which yields

$$b = 1.48 \times 10^{-14} \text{ m}$$

(c) What is the corresponding cross section in this case?

From Equation (11–13), the cross section is

$$\sigma = \pi b^2 = \pi (1.48 \times 10^{-14})^2 = 6.87 \times 10^{-28} \text{ m}^2$$

The cross section is often measured in terms of a unit called the *barn*, where 1 barn = 1×10^{-28} m². Thus, in this calculation $\sigma = 6.87$ barns.

(d) What fraction of the α particles are deflected by an angle of 90° or more if the thickness of the gold foil is $t = 6.00 \times 10^{-7}$ m?

For gold ($\rho = 1.93 \times 10^4$ kg/m³, $M = 197$ kg/kg-mole), the number of nuclei per unit volume from Equation (11–19) is

$$n = \rho \frac{N_A}{M} = \frac{1.93 \times 10^4 \times 6.02 \times 10^{26}}{197} = 5.91 \times 10^{28} \text{ atoms/m}^3$$

Now, from Equation (11–15),

$$f = \frac{N}{N_0} = n\sigma t$$

or, upon substituting the calculated values,

$$f = \frac{N}{N_0} = 5.91 \times 10^{28} \times 6.87 \times 10^{-28} \times 6.00 \times 10^{-7} = 2.44 \times 10^{-5}$$

which means that only approximately two out of 100,000 α particles experience deflections $\geq 90°$. The conclusion is that a gold foil of this thickness is relatively transparent to α particles.

PROBLEMS

$$\rho_{\text{gold}} = 1.93 \times 10^4 \text{ kg/m}^3 \quad Z = 79 \quad M = 197 \text{ kg/kg-mole}$$

11-1 What is the velocity of an α particle with a kinetic energy of 7.68 MeV?

11-2 (a) If the distance of closest approach of an α particle directed against a gold nucleus is 3.00×10^{-13} m, calculate the kinetic energy in MeV of the α particle.
(b) What is the potential energy of the system at this distance of closest approach?

11-3 Use the data from Problem 11-2 and (a) determine the impact parameter in order to produce a scattering angle $\phi \geq 60°$, (b) compute the corresponding cross section in barns, and (c) find the fraction of α particles that will be scattered by 60° or more from a gold foil with a thickness of 3.00 × 10⁻⁷ m.

11-4 (a) For 7.68-MeV α particles directed against a gold foil 3.00 × 10⁻⁷ m thick find the fraction of α particles whose scattering angles are between $\phi_1 = 10°$ and $\phi_2 = 12°$.
(b) If the total number of α particles directed toward the gold foil is 1.00×10^6, how many will be scattered between angles of 10° and 12°? [Hint: For the data given in (a), compute n and D and then make use of Equation (11-16).]

11-5 If a sodium target ($Z = 11$, $A = 23$) scatters 1.00×10^4 α particles in a given direction, (a) how many will be scattered by the same angle if the sodium target is replaced by a gold foil ($Z = 79$, $A = 197$) of the same thickness? (b) How many will be scattered in the same direction if the thickness of the sodium target is half the original value? The density of sodium is 0.93×10^4 kg/m³.

11-6 For a given foil, find the ratio of α particles scattered between 60° and 90° to those scattered by 90° or more. Assume the kinetic energy of the α particles to be the same in both cases.

11-7 The number of α particles scattered by an angle of 10° or more by a given foil is 1.00×10^6 particles/sec. (a) How many particles per second will be scattered by an angle of 30° or more? (b) How many particles per second will have scattering angles between 10° and 30°? (c) How many particles per second will have scattering angles between 10° and 30° if the thickness of the foil is half that in (b)?

11-8 What fraction of 7.68-MeV α particles directed at a gold foil 5.00 μ thick will be scattered by an angle of less than 10°?

11-9 For equal numbers of incident particles, determine the ratio of protons to α particles of the same energy that will be scattered by angles greater than 90° from a gold foil.

11-10 It has been found experimentally that the radius of a nucleus is given satisfactorily by $R = R_0 A^{1/3}$, where $R_0 = 1.3 \times 10^{-15}$ m. (a) Compute (in MeV) the height of the electrostatic potential barrier at the surface of a gold nucleus for an approaching α particle. (b) Do the same computation for an approaching proton.

11-11 Using the data from Problem 11-10, plot the Coulomb potential energy versus r, the distance of the approaching particle to the gold nucleus, from the surface of the nucleus to infinity (a) when the approaching particle is an α particle, and (b) when it is a proton.

11-12 The width of the potential energy barrier of a nucleus for an approaching charged particle is given by $D - R$, where D is the distance of closest approach and $R = (1.3 \times 10^{-15} \text{ m})A^{1/3}$. (a) In the case of the gold nucleus, compute the width of the potential barrier for an approaching α particle whose energy is 7.68 MeV when it is very far from the nucleus. (b) What will be the kinetic energy in MeV of the α particle when its distance to the center of the nucleus is 3.20×10^{-14} m? (Assume a head-on collision.)

11-13 An 8.00-MeV α particle is scattered by a gold nucleus at an angle of 45°. (a) Compute the impact parameter b. (b) If the gold foil is 0.400 μ thick, what fraction of the α particles is scattered by an angle greater than 45°? (c) What fraction is scattered by an angle less than 45°?

11-14 What fraction of 5.00-MeV deuterons will be scattered between $\phi_1 = 10°$ and $\phi_1 + d\phi = 12°$ when they are incident on gold foil 6.00×10^{-7} m thick?

11-15 (a) In the case of α particles with $K = 7.7$ MeV approaching a gold foil with a thickness $t = 4.0 \times 10^{-7}$, calculate the fraction of α particles per unit solid angle (steradian) when the scattering angle is $\phi = 45°$.
(b) What will be the fraction corresponding to a solid angle $d\Omega = 4\pi \times 10^{-2}$ steradian if $\phi = 45°$?
(c) If a screen is placed at a distance $R = 2.0$ cm, how many α particles per unit area will hit the screen if $\phi = 45°$ and the total number of particles directed at the screen is $N_0 = 1.0 \times 10^6$? [Hint: Make use of the relation $r = R \tan \phi$ and then apply Equation (11-18).]
(d) How many particles will hit the screen inside a ring of internal radius r and external radius $r + dr$, where $dr = 1.0$ mm?

RECOMMENDED READING

BADASH, L., *Rutherford and Boltwood, Letters on Radioactivity*, Yale University Press, New Haven and London, 1969.

GEIGER, H., and MARSDEN, E., *Proc. Roy. Soc.* A **82**, 495 (1909).

GEIGER, H., and MARSDEN, E., *Phil. Mag.* **25**, 604 (1913).

HOFSTADER, R., "The Atomic Nucleus," *Sci. Am.*, July 1956.

KAPITZA, P. L., "Recollections of Lord Rutherford," *Selected Lectures of the Royal Society*, Academic, London, 1967, p. 119.

MELISSINOS, A., *Experiments in Modern Physics*, Academic, New York, 1966.
Chapter 6 describes a modern experimental setup to study Rutherford scattering.

RUTHERFORD, E., *Phil. Mag.* **5**, 576 (1911).

RUTHERFORD, E., *Phil. Mag.* **21**, 669 (1911).

RUTHERFORD, E., CHADWICK, J., and ELLIS, C. D., *Radiations from Radioactive Substance*, Cambridge University Press, London, 1930.

12
The Bohr Model I

Niels Henrik David Bohr
(1885–1962)

Born in Copenhagen, Bohr earned his doctorate from the University of Copenhagen in 1911. At Cavendish Laboratory in Cambridge he worked under J. J. Thomson, and at Victoria University he worked with Ernest Rutherford. In 1920 he became the head of the Institute of Theoretical Physics in Copenhagen. For his studies in the structure of atoms and their radiation, he received the Nobel Prize in 1922. Concerned with the peaceful applications of atomic energy, Bohr organized the first Atoms for Peace Conference in Geneva in 1955.

12–1 PLANETARY MODEL
12–2 ATOMIC SPECTRA
12–3 THE BOHR MODEL—POSTULATES
12–4 THE BOHR MODEL—ENERGY STATES
12–5 THE RYDBERG CONSTANT AND SPECTRAL SERIES
12–6 THE BOHR MODEL AND THE PRINCIPLE OF CORRESPONDENCE

12–1 PLANETARY MODEL

According to the Rutherford model, the atom consists of a very small but massive nucleus (size of the order of 10^{-14} m) carrying a charge $+Ze$. Around this central region are located the Z electrons of the neutral atom. The diameter of an atom is about 10^{-10} m, or 10,000 times greater than the size of the nucleus. We consider this model to be a dynamic model. If a static model is assumed, all the electrons surrounding the nucleus would be attracted toward it because of the Coulomb force between the nucleus and the electrons, and the atom would soon collapse. In the dynamic planetary model the massive nucleus is essentially at rest, with the electrons revolving about it in circular and elliptical paths.

Let us consider the simplest atomic structure, the hydrogen atom, with this model in mind. Figure 12–1 shows the simplest approximation of a proton (with a charge $+e$) at the center and an electron (of mass m and charge $-e$) revolving around the nucleus with uniform circular motion. In this first approximation the

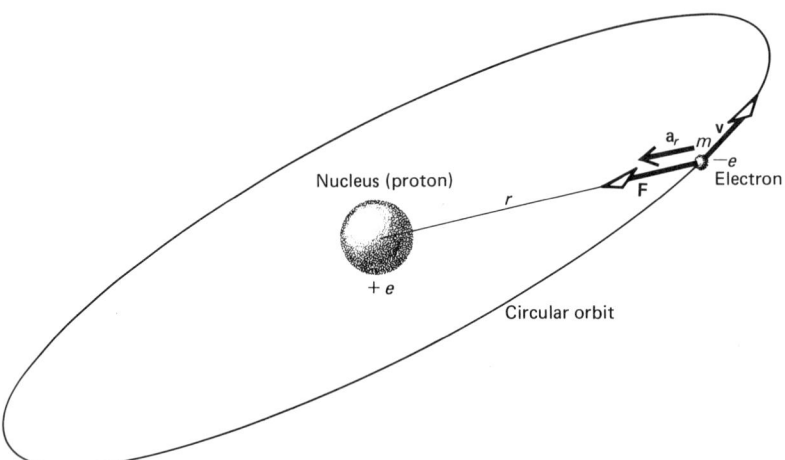

Figure 12–1 The planetary model of the hydrogen atom. An electron of mass m revolves around the nucleus with a uniform circular motion. The driving force is the Coulomb electrostatic attraction F between the nucleus (proton) and the electron.

motion of the proton, with a mass about 1836 times larger than the mass of the electron, will be neglected.

The driving force F is provided by the electrostatic attraction between the proton and the electron. This is a central force whose magnitude is given by

$$F = \frac{1}{4\pi\varepsilon_0} \frac{e^2}{r^2} \tag{12-1}$$

where r is the radius of the circular path of the electron. According to Newton's second law,

$$\frac{1}{4\pi\varepsilon_0} \frac{e^2}{r^2} = m \frac{v^2}{r} \tag{12-2}$$

where $a_r = v^2/r$ is the centripetal acceleration (Figure 12-2). From Equation (12-2), the kinetic energy of the electron can be obtained from

$$K = \tfrac{1}{2}mv^2 = \frac{1}{8\pi\varepsilon_0} \frac{e^2}{r} \tag{12-3}$$

assuming a classical approach. The potential energy of the system is

$$V = -\frac{1}{4\pi\varepsilon_0} \frac{e^2}{r} \tag{12-4}$$

The minus sign is an indication that the system is one of attraction rather than

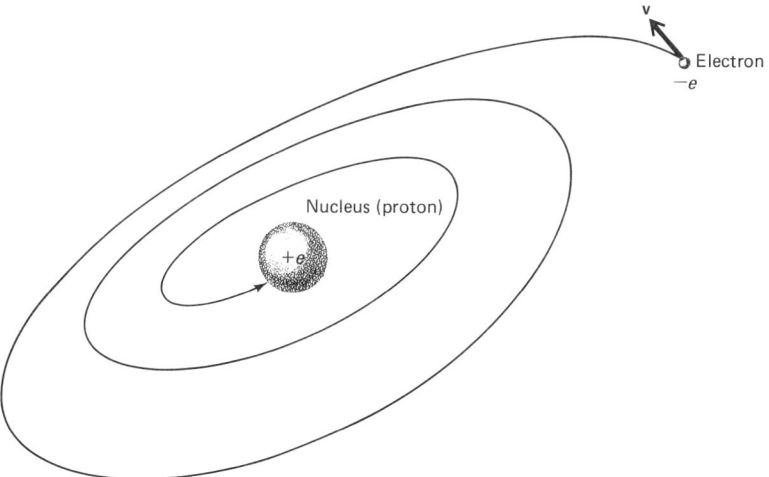

Figure 12-2 In the planetary model of the atom, the electron would spiral around the nucleus (proton) until final collapse occurred.

repulsion, since the electron is attracted to the positive nucleus. The total energy of this system is

$$E = K + V = -\frac{1}{8\pi\varepsilon_0}\frac{e^2}{r} \tag{12-5}$$

where the minus sign indicates that it is a *bound system*.

The binding energy of an electron is defined as the minimum energy required to remove the electron completely from the atom or, in other words, to ionize the atom. From experimental work, the binding energy of the hydrogen atom has been found to be 13.6 eV. When this value is substituted into Equation (12-5) for E, the radius r can be found:

$$r_1 = 0.53 \times 10^{-10} \text{ m} = 0.53 \text{ Å}$$

This value for r_1 is called the *Bohr radius* and is in good agreement with values obtained by other experimental techniques.

The linear velocity v is related to the frequency f of revolution of the electron in its orbit by

$$v = \omega r = 2\pi f r \tag{12-6}$$

Replacing this value in Equation (12-3), we have

$$m(2\pi f r)^2 = \frac{1}{4\pi\varepsilon_0}\frac{e^2}{r}$$

from which we obtain

$$\boxed{f = \frac{1}{2\pi}\sqrt{\frac{e^2}{4\pi\varepsilon_0 m r^3}}} \tag{12-7}$$

for the number of revolutions per second made by an electron in an orbit. Using the value of r given above and the known values of e and m for the electron yields a value $f = 7 \times 10^{15} \text{ sec}^{-1}$. This also agrees with the determination of the frequency of revolution obtained by other methods.

However, in spite of these initial achievements, physicists found that this planetary model had to be abandoned because, according to classical electrodynamics,

1. an accelerated charge must continuously radiate electromagnetic energy, and
2. the frequency of the emitted radiation must be equal to the frequency of revolution.

According to this model, therefore, the total energy of the atom should decrease (become more negative), while the frequency of rotation [Equation (12-7)] must continuously increase. A simple calculation shows that only about 10^{-6} sec is required for complete collapse. According to this model, the optical spectrum of hydrogen (as well as the spectra of other elements) is *continuous*, and all atoms must collapse in a very short time. Both conclusions, of course, contradict experimental evidence. Atoms have proved themselves to be most stable against vanishing; moreover, the optical spectra of gases show just discrete frequencies ("lines") and not a continuous distribution in frequency at all. The planetary model was soon abandoned.

12-2 ATOMIC SPECTRA

The light from an electrical discharge through a tube containing a monatomic gas at low pressure exhibits a series of very distinct lines when analyzed by a prism spectrometer as shown in Figure 12-3. These lines, characteristic of the gas used in the tube, are called the *line spectrum* of the gas. The visible spectrum of hydrogen shown in Figure 12-3 is called the *Balmer series* after J. J. BALMER, who discovered it in 1885. If nitrogen gas is used in the discharge tube, the spectrum is a regular arrangement of very closely spaced lines known as a *band spectrum*. The band spectrum is characteristic of the diatomic molecule N_2 and is evidently of a

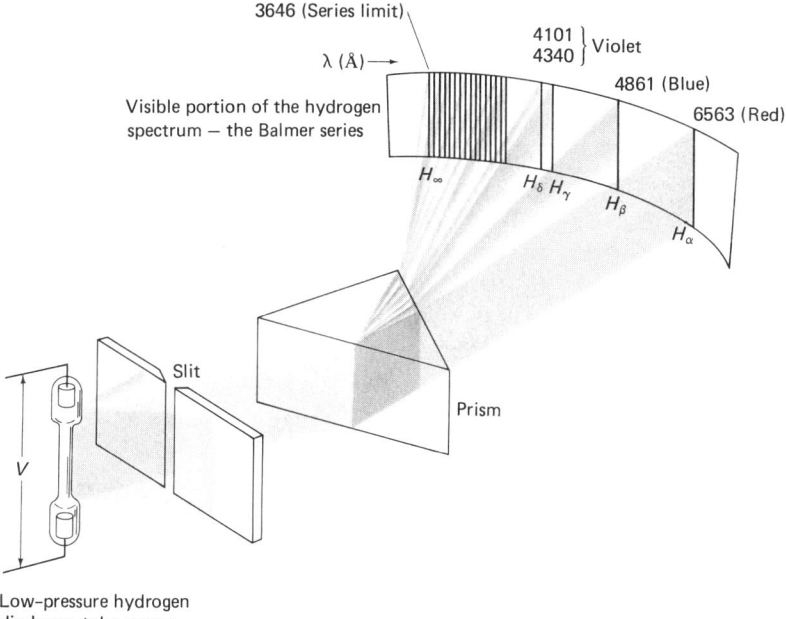

Figure 12-3 Schematic of a prism spectrograph. Light from a low-pressure hydrogen discharge tube is refracted through a prism to produce a line spectrum.

different origin from that of the line spectra. The white light from an incandescent source such as a light bulb is a continuous spectrum and contains a continuum of wavelengths. When the light from a continuous spectrum is passed through a monatomic gas such as hydrogen, an *absorption spectrum* is produced. A spectrograph then shows dark lines against a bright background. The positions of these dark lines correspond to the wavelengths of the bright spectral lines from hydrogen. The gas in this case is absorbing the incident radiation of these wavelengths.

The Swedish spectroscopist J. R. RYDBERG (1854–1919) found an empirical formula,

$$\frac{1}{\lambda} = R\left(\frac{1}{2^2} - \frac{1}{n^2}\right) \quad \text{for } n = 3, 4, 5, \ldots \tag{12-8}$$

from which the wavelengths of the Balmer series could be calculated. The Rydberg constant R has a value of $R = 1.0973731 \times 10^{-3}$ Å for $n = 3$, $\lambda = 6563$ Å, which is identified as the red H_α line; for $n = 4$, $\lambda = 4861$ Å, the blue H_β line; and for increasing values of n, the wavelengths get closer and closer together, and the intensities become weaker and weaker until the series limit at $n = \infty$, $\lambda = 3645$ Å is reached. In addition to the Balmer series, Table 12–1 shows other series found in the ultraviolet and infrared regions.

Table 12–1 Spectral series for hydrogen

SERIES	SPECTRAL REGION	SERIES EQUATION	SERIES LIMIT ($n = \infty$)
Lyman	ultraviolet	$\frac{1}{\lambda} = R\left(\frac{1}{1^2} - \frac{1}{n^2}\right)$ $n = 2, 3, 4, \ldots$	911.27 Å
Balmer	visible	$\frac{1}{\lambda} = R\left(\frac{1}{2^2} - \frac{1}{n^2}\right)$ $n = 3, 4, 5, \ldots$	3645.1 Å
Paschen	infrared	$\frac{1}{\lambda} = R\left(\frac{1}{3^2} - \frac{1}{n^2}\right)$ $n = 4, 5, 6, \ldots$	8201.4 Å
Brackett	infrared	$\frac{1}{\lambda} = R\left(\frac{1}{4^2} - \frac{1}{n^2}\right)$ $n = 5, 6, 7, \ldots$	14,580 Å
Pfund	infrared	$\frac{1}{\lambda} = R\left(\frac{1}{5^2} - \frac{1}{n^2}\right)$ $n = 6, 7, 8, \ldots$	22,782 Å

12-3 THE BOHR MODEL—POSTULATES

NIELS BOHR received his doctorate in 1911 in Copenhagen, and the same year he traveled to England to study under both J. J. Thomson and Ernest Rutherford. From Rutherford's description of the atom, it was evident to Bohr that the atom had to consist of a heavy nucleus about which, and some distance away, electrons would revolve. Bohr then proposed a remarkable set of postulates as the basis for a new model of the atom. Leon Cooper expresses it well when he states: "There was a certain presumption in asserting what was contrary to Maxwell's electrodynamics and Newton's mechanics, but Bohr was young."*

The Bohr model of the atom, although now replaced by the more powerful quantum model of Heisenberg, Schrödinger, Dirac, and others, remains an effective pictorial way to introduce the concept of stationary states. Bohr's model was the first satisfactory explanation of atomic structure, and it was improved during the next ten years by Sommerfeld, Wilson, and others. Because of the difficulty they found in making it compatible with new experimental discoveries in spectroscopy, it was replaced between 1924 and 1926 by the quantum mechanical model.

To correct the failures of the planetary model of the atom, Bohr based his model of the hydrogen atom on the following postulates:

1. The electron revolves around the proton in the hydrogen atom with a uniform circular motion due to the Coulomb force and in agreement with Newton's laws.
2. The only orbits that are allowed are those in which the angular momentum of the revolving electron is an integral multiple of $h/2\pi = \hbar$. The angular momenta of the only permissible orbits are given by

$$L = mvr = n\frac{h}{2\pi} = n\hbar, \quad n = 1, 2, 3, \ldots \quad (12\text{-}9)$$

where h is Planck's constant and $\hbar = 1.05 \times 10^{-34}$ J-sec.

3. When an electron is in an allowed orbit, the atom does not radiate energy. (Classical electromagnetic theory predicts that any accelerated charge will radiate electromagnetic energy.)
4. If the electron jumps from an initial orbit of energy E_i to a final orbit of energy E_f ($E_i > E_f$), a photon of frequency

$$v = \frac{E_i - E_f}{h} \quad (12\text{-}10)$$

is emitted.

---

* From Leon Cooper, *An Introduction to the Meaning and Structure of Physics,* Harper & Row, New York, 1968, p. 457.

In Figure 12–4, if an electron jumps from the orbit $n = 5$ to the orbit $n = 4$, a photon of frequency $v = (E_5 - E_4)/h$ is emitted. (This explains the discrete frequencies obtained in an emission spectrum.) On the other hand, if a photon of energy $hv = E_5 - E_4$ is incident on the atom, it can be absorbed, and an electron will jump from C in orbit $n = 4$ to D in orbit $n = 5$. This is the mechanism responsible for the absorption spectrum.

12–4 THE BOHR MODEL—ENERGY STATES

The starting point of the Bohr model (Figure 12–5) is the same as that of the planetary model. The first postulate of the Bohr model, the application of Coulomb's law and Newton's second law, gives the total energy of the system as it was in Equation (12–5) for the planetary model,

$$E = -\frac{1}{8\pi\varepsilon_0}\frac{e^2}{r}$$

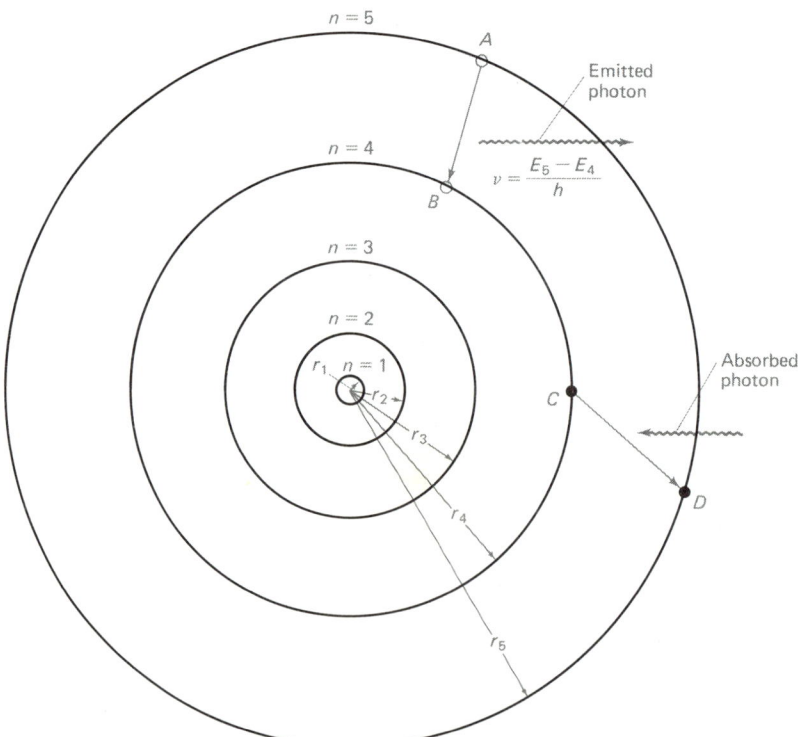

Figure 12–4 When an electron is in one of the allowed orbits it will not radiate energy in spite of its centripetal acceleration, in contradiction with classical electrodynamics.

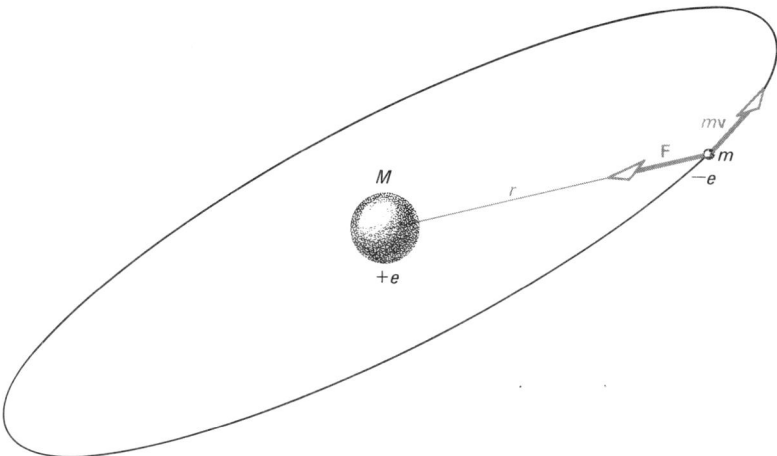

Figure 12-5 The hydrogen atom consists of an electron of mass m and charge $-e$ revolving about a very massive proton of mass $M \gg m$ and charge $+e$.

Now a remarkable departure is made from classical physics with the application of the second postulate and Equation (12-9),

$$L = mvr = n\hbar$$

In classical physics, the spectrum of values of the angular momentum L is continuous, that is, all values of L are possible; but Equation (12-9) means that the values of the angular momentum L must now be chosen from a *discrete spectrum* of values. Thus, the angular momentum is *quantized* and the "allowed" values are $1\hbar, 2\hbar, 3\hbar, \ldots$, and $\hbar$ should be considered as a natural unit of angular momentum. This situation is similar to the quantization of electrical charge in classical physics.

According to the third postulate, when the atom is in any one of the quantized states designated by the angular momentum in Equation (12-9), it will not radiate energy as would be expected from classical electromagnetic theory. These states, or "nonradiating" orbits, are called *stationary states*. The state of least energy is the one defined for $n = 1$, and it is called the *ground* or *normal state* because this is the lowest energy state and the one in which the atom is found to be most of the time. The states where $n = 2, 3, 4, \ldots$ are the *excited states*, because the atom then has more energy than it has in the normal state. In the Bohr model, there is no "explanation" of the assertion that, when in stationary states, the atom does not radiate energy. *This is simply taken as a postulate.* It is impossible to show experimentally that the electron moves in a circular orbit around the nucleus. These difficulties are removed, however, when the hydrogen atom is analyzed in terms of quantum or wave mechanics. As was previously stated, the Bohr model has its limitations, but it is a good mechanical model to introduce energy states and other physical concepts.

Now, from Equation (12–9),

$$v = \frac{n\hbar}{mr}$$

and the kinetic energy from Equation (12–3) becomes

$$\frac{1}{2} m \left(\frac{n\hbar}{mr}\right)^2 = \frac{1}{8\pi\varepsilon_0} \frac{e^2}{r}$$

or finally,

$$\boxed{r = r_n = \frac{4\pi\varepsilon_0 n^2 \hbar^2}{me^2} \qquad n = 1, 2, 3, \ldots} \qquad (12\text{–}11)$$

which gives the radii of the "nonradiating" orbits. For the ground state, $n = 1$ and

$$\boxed{r_1 = \frac{4\pi\varepsilon_0 \hbar^2}{me^2} = 0.53 \text{ Å}} \qquad (12\text{–}12)$$

which is called the *Bohr radius*. This result agrees with the radius of the atom obtained previously from Equation (12–5) using the planetary model.

From Equation (12–11),

$$r_n = n^2 r_1 \qquad (12\text{–}13)$$

which shows that the radii of the orbits for stationary states are also quantized and are given by r_1, $4r_1$, $9r_1$, and so on. These radii are proportional to the square of the integer number n, called the *principal quantum number*. Now, if r in Equation (12–5) is replaced by Equation (12–11), we obtain

$$\boxed{E = E_n = -\frac{me^4}{32\pi^2 \varepsilon_0^2 \hbar^2} \left(\frac{1}{n^2}\right)} \qquad (12\text{–}14)$$

where the negative sign again indicates a bound system. Thus, a second consequence is that the *energy is quantized*. The only allowed values of the energy are those given by Equation (12–14) when n takes the values of $n = 1, 2, 3, \ldots$. By using the values of $m = 9.11 \times 10^{-31}$ kg and $e = 1.60 \times 10^{-19}$ C for the mass and charge of the electron, we can evaluate Equation (12–14) to give

$$E_n = -\frac{13.6}{n^2} \text{ eV} \qquad \text{for } n = 1, 2, 3, \ldots \qquad (12\text{–}15)$$

The state of lowest energy or ground state corresponds to $n = 1$, and its energy is $E_1 = -13.6$ eV.

Figure 12–6 is an energy level diagram representing the allowed energies for the hydrogen atom. Note that all the states from $n = 1$ to $n = \infty$ are bound states, since they have negative energies. When n increases and approaches $n = \infty$, the energy states get closer and closer together until the energy difference between two consecutive states becomes so small that the distribution gives a practically continuous spectrum in agreement with the classical planetary model and the correspondence principle. Above the line given by $n = \infty$, the energy states have positive energy, $E > 0$, and the spectrum of the states is continuous. The system is then *unbound*, meaning that the electron is free.

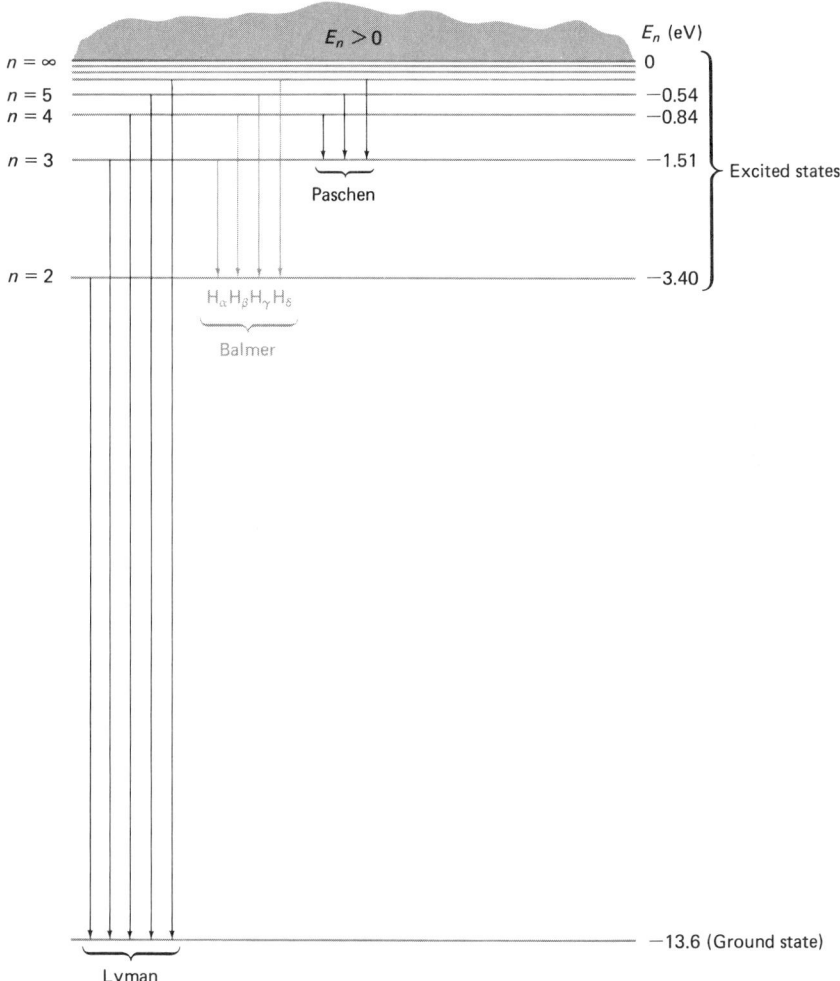

Figure 12–6 Energy level diagram for the hydrogen atom.

From Equation (12–14), it can easily be seen that if the atom is in its ground state, 13.6 eV will be necessary to liberate the electron from the atom. Therefore, the binding energy (BE) or ionization energy for the hydrogen atom in its ground state is

$$BE = E_i = 13.6 \text{ eV}$$

This result (which has been checked experimentally) was used in Equation (12–5) for the planetary model of the atom to obtain the Bohr radius $r_1 = 0.53$ Å.

In relation to the energy level diagram (Figure 12–6), a few definitions are important.

The *excitation energy* E_e is the energy that must be supplied to the atom to raise the electron from the ground state to an excited state. For example, $E_e = -3.40 - (-13.6) = 10.2$ eV is the excitation energy for the state $n = 2$ (first excited state).

The *ionization energy* E_i is the energy that we must supply to liberate the electron from the atom when the electron is in the ground state. In Figure 12–6, evidently, $E_i = 13.6$ eV.

The *binding energy* BE for a given state is the energy that must be supplied to the atom to remove an electron when the electron is in a given excited state. For example, the BE for the state $n = 2$ is 3.40 eV. If the atom is in the ground state, the BE for that state is equal to the ionization energy (13.6 eV). When we speak about BE without mentioning the state, it is understood that the BE and the ionization energy have the same numerical value. It is in this respect that we have said that the BE of the hydrogen atom is 13.6 eV.

12–5 THE RYDBERG CONSTANT AND SPECTRAL SERIES

Now, according to Bohr's postulate, if one electron jumps from an initial state n_i (energy E_i) to another state of lower energy n_f (energy E_f), the frequency of the emitted photon is, from Bohr's formula (Equation 12–10),

$$\nu = \frac{E_i - E_f}{h} = \frac{E_i - E_f}{2\pi\hbar}$$

When the expressions for the energy from Equation (12–14) are introduced, the frequency of the emitted photon becomes

$$\nu = \frac{c}{\lambda} = \frac{me^4}{64\pi^3\hbar^3\varepsilon_0^2}\left(\frac{1}{n_f^2} - \frac{1}{n_i^2}\right) \qquad (12\text{–}16)$$

or finally, the wavelength of the emitted photon is

$$\boxed{\frac{1}{\lambda} = \frac{me^4}{64\pi^3\hbar^3\varepsilon_0^2 c}\left(\frac{1}{n_f^2} - \frac{1}{n_i^2}\right)} \qquad (12\text{–}17)$$

Notice that this equation is similar to the ones for the spectral series given in Table 12-1. If n_i is replaced by n and $n_f = 1$, this equation has the same form as the empirical equation for the Lyman series; or if $n_f = 2$, it becomes the Balmer series, and so on. Therefore, from this comparison we find that

$$R = \frac{me^4}{64\pi^3 \hbar^3 \varepsilon_0^2 c} \qquad (12\text{-}18)$$

is the theoretical value of the Rydberg constant. Equation (12–17) can now be written as

$$\frac{1}{\lambda} = R\left(\frac{1}{n_f^2} - \frac{1}{n_i^2}\right) \qquad (12\text{-}19)$$

If we replace the correct numerical values in Equation (12–18), the computation of the Rydberg constant $R = 1.0974 \times 10^7$ m^{-1} is in agreement with the experimental value given in Section 12–2.

A very practical equation for the photon energy released in a transition between the stationary states n_i and n_f can be obtained from Equations (12–10) and (12–15) when the numerical values of the constants involved are replaced:

$$E_i - E_f = 13.6\left(\frac{1}{n_f^2} - \frac{1}{n_i^2}\right) \text{eV} \qquad (12\text{-}20)$$

The energy level diagram of Figure 12–6 represents the possible transitions from the states $n = 2, 3, 4, \ldots$ to the ground state $n = 1$ (Lyman series), the Balmer series for transitions to $n = 2$ from $n = 3, 4, 5, \ldots$, the Paschen series for transitions to $n = 3$ from $n = 4, 5, 6, \ldots$, and so on. Transitions between states with negative energy will give rise to line spectra, whereas transitions between states with positive energy $E > 0$ and states $E < 0$ will result in a continuous spectrum.

12–6 THE BOHR MODEL AND THE PRINCIPLE OF CORRESPONDENCE

A beautiful application of the principle of correspondence (recall Section 4–2) can be made by a comparison of the frequency of emitted photons when we apply the Bohr model to the macroscopic world (large quantum numbers) and the frequency of the revolution in the classical planetary model. According to classical electromagnetic theory, the latter must be equal to the frequency of the radiated electromagnetic waves.

According to classical theory, the orbital frequency [Equation (12–7)] is

$$f = \frac{1}{2\pi}\sqrt{\frac{e^2}{4\pi\varepsilon_0 m r^3}}$$

but the radii of the stationary orbits, according to the Bohr's model [Equation (12–11)], are given by

$$r_n = \frac{4\pi\varepsilon_0 n^2 \hbar^2}{me^2}$$

Replacing this in the expression for the frequency then gives

$$f = \frac{me^4}{64\pi^3 \varepsilon_0^2 \hbar^3}\frac{2}{n^3} \qquad (12\text{–}21)$$

Now, according to the Bohr model, the frequency of the photon emitted in a transition from n_i to n_f is

$$\nu = \frac{me^4}{64\pi^3 \hbar^3 \varepsilon_0^2}\left(\frac{1}{n_f^2} - \frac{1}{n_i^2}\right)$$

which can be rearranged as

$$\nu = \frac{me^4}{64\pi^3 \hbar^3 \varepsilon_0^2}\frac{(n_i - n_f)(n_i + n_f)}{n_f^2 n_i^2}$$

When n_i and n_f are very large numbers and are at the same time close to each other, the following substitutions can be made:

$$n_i - n_f = \Delta n$$
$$n_i + n_f \cong 2n_i = 2n$$
$$n_f^2 n_i^2 \cong n^4$$

If these are replaced in the above equation, the frequency is

$$\nu = \frac{me^4}{64\pi^3 \hbar^3 e_0^2}\frac{2\,\Delta n}{n^3}$$

which, if we make $\Delta n = 1$, is identical to the classical frequency in Equation (12–21). For $\Delta n = 2, 3, 4, \ldots$, we obtain harmonics of the fundamental frequency.

The conclusion is that when we apply the Bohr model (which is especially designed for the microscopic world) to problems in the macroscopic world, we find results identical to those obtained with classical methods. This is the basic philosophy of the principle of correspondence.

PROBLEMS

12-1 Assume that the planetary model describes the motion of the electron in the hydrogen atom. If the radius of the electron orbit is 0.53 Å, calculate (a) the angular frequency of the electron, (b) its linear speed, (c) its kinetic energy in electron volts, (d) the potential energy of the atom in electron volts, and (e) its total energy in electron volts. What is the minimum energy in electron volts needed to ionize the atom (binding energy)?

12-2 By finding the ratio of the gravitational force of attraction between an electron and a proton and the Coulomb force of attraction between the same two particles, show that the gravitational force can be neglected in the study of the hydrogen atom. (If only gravitational forces were involved, the radius of the first Bohr orbit would be $r_1 = 1 \times 10^{26}$ miles!)

12-3 In the planetary model of the atom, the radius of the orbit is 0.53 Å and the linear velocity is approximately 2.2×10^6 m/sec. Find (a) the centripetal acceleration, (b) the centripetal force, and (c) the electrostatic force of attraction between the proton and the electron. (d) Compare the forces calculated in parts (b) and (c). What do you conclude?

12-4 (a) Find the wavelength in angstroms of the first three lines of the Lyman series of hydrogen.
(b) From Figure 12-6, determine the wavelength in angstroms of the H_α line.

12-5 Spectroscopists often identify spectra according to wave numbers defined by $\bar{v} = 1/\lambda$ (not to be confused with the frequency v). (a) What is the physical meaning of a wave number? (b) Compute the wave numbers in reciprocal centimeters (cm^{-1}) for parts (a) and (b) in Problem 12-4.

12-6 Show that the energy levels of the hydrogen atom can be described by $E_n = (-2\pi\hbar c/n^2)R$, where R is the Rydberg constant.

12-7 A spectroscope uses light from a hydrogen discharge tube incident normally on a diffraction grating of 15000 lines/in. If the first-order spectrum of the Balmer series shows the H_α line diffracted at an angle $\theta = 23°$, compute (a) the wavelength of the H_α line (red line in the Balmer series), and (b) the Rydberg constant in reciprocal meters (m^{-1}).

12-8 In the Bohr model of the hydrogen atom, the orbits $n = 1, 2, 3, \ldots$ are represented symbolically by the letters $K, L, M, \ldots$, and so on. For the electrons in each of the orbits K, L, and M, compute (a) the radii, (b) the frequencies of revolution, (c) the linear speeds, (d) the angular momenta, and (e) the total energy of the system. (f) For each orbit, compute the ratio v/c and decide if the classical treatment can be justified.

12-9 The ratio $\alpha = v_1/c$, where v_1 is the linear speed of the electron in the K ($n = 1$) orbit of the Bohr hydrogen atom, is called the fine structure constant. (a) Show that $\alpha = e^2/4\pi\varepsilon_0\hbar c$. (b) By substituting the numerical values, show that $\alpha = 1/137$. (c) Show that the energy levels can be written as $E_n = -\alpha^2 mc^2/2n^2$.

12-10 An electron in a hydrogen atom makes a transition from the $n = 5$ to the $n = 1$ ground state. (a) Find the energy and momentum of the emitted photon. (b) Find the speed and momentum of the recoil atom.

12-11 The lifetime of an excited state is about 10^{-8} sec. Compute how many revolutions an electron in the excited state $n = 4$ will make before jumping to the ground state.

12-12 Compute the first three wavelengths for the Paschen series of hydrogen. In what region of the spectrum do the lines of the Paschen series lie?

12-13 (a) For an electron revolving in the first ($n = 1$) orbit about a proton, determine the frequency of revolution.
(b) What is the value in amperes of the equivalent current?
(c) Calculate the magnetic flux density B (in teslas = Wb/m^2) at the center of this circular path. How is the flux density aligned with respect to the orbital angular momentum?

12-14 (a) Illustrate graphically, by means of an energy level diagram, the excitation energy E_e, binding energy BE, and ionization energy E_i for a given state.
(b) For any given n, show that $E_i = E_e + BE$.
(c) Find the excitation energy for $n = 4$ in the hydrogen atom.
(d) Find the BE for the electron in the same state $n = 4$, and check part (b) numerically.

12-15 In the hydrogen atom, an electron experiences a transition from a state whose binding energy is 0.54 eV to another state whose excitation energy is 10.2 eV. (a) What are the quantum numbers for these states? (b) Compute the wavelength of the emitted photon. (c) To what series does this line belong?

12-16 Calculate the minimum energy that must be given to a hydrogen atom for it to be able to emit the H_γ line of the Balmer series. How many possible spectral lines can be expected if the electron finally goes to the ground state?

12-17 Find the de Broglie wavelength of an electron in the $n = 3$ orbit of the hydrogen atom. In what region of the spectrum would a photon of the same wavelength be classified?

12-18 In the inelastic collision of an electron of mass m with a stationary hydrogen atom of mass M, the atom is excited to a level whose energy is E above the ground state. (a) Prove that the minimum kinetic energy K of the electron must be $K = [(m + M)/M]E$. (b) Find the minimum kinetic energy of an electron that makes an inelastic collision with a hydrogen atom at rest and raises the hydrogen atom from the ground state ($n = 1$) to the second excited state ($n = 3$). (c) Solve the same problem if the incoming particle is a proton.

12-19 A photon of energy 12.1 eV absorbed by a hydrogen atom, originally in the ground state, raises the atom to an excited state. What is the quantum number of this state?

RECOMMENDED READING

BALMER, Johan J., "The Hydrogen Spectral Series," in W. F. Magie, *A Source Book on Physics*, Harvard University Press, Cambridge, Mass., 1963, pp. 360–365.

BANET, L., "The Evolution of the Balmer's Series," *Am. J. Phys.* **34,** 496 (1966).

BOHR, A., et al., "Papers Given at the Niels Bohr Memorial Session," *Phys. Today*, October 1963.

BOHR, Niels, *The Theory of Spectra and Atomic Constitution*, 2nd ed., Cambridge University Press, London, 1924.
Excellent for those interested in the cultural aspects of physics. It contains Bohr's interpretation of the development of his theory of atomic structure.

GAMOW, G., *Thirty Years That Shook Physics*, Doubleday, Garden City, New York, 1963.

HERZBERG, G., *Atomic Spectra and Atomic Structure*, Dover, New York, 1946.

WHITE, H. E., *Introduction to Atomic Spectra*, McGraw-Hill, New York, 1934.

YOUNG, H. D., *Fundamentals of Optics and Modern Physics*, McGraw-Hill, New York, 1967, Vol. I, pp. 155–173.

13
The Bohr Model II

James Franck
(1882–1964)

Born in Hamburg, Germany, Franck studied at the universities of Berlin, California, and Haifa. He was head of the Kaiser Wilhelm Institute, director of the physics institute at the University of Göttingen, and professor at Johns Hopkins University and the University of Chicago. He established the principle of constancy of atom distances by electron jumps and discovered the energy transmission in atom systems in fluorescence. For their discovery of laws governing the impact of an electron on an atom, Franck and G. Hertz received the 1925 Nobel Prize.

Gustav Ludwig Hertz
(1887–)

A native of Hamburg, Germany, Hertz received his Ph.D. in physics from the University of Berlin. He has worked in the physics laboratory of the Phillips Incandescent Lamp Factory, was director of the Research Laboratory of Siemens Company, and has served as director of the physics institute in Leipzig, Germany. In 1931 he and J. Franck, by bombarding atoms of mercury vapor with electrons, gave early independent evidence of discrete energy levels in atoms, confirming Bohr's theory of atomic spectra. With Franck, he was awarded the 1925 Nobel Prize in physics.

13–1   HYDROGENLIKE ATOMS
13–2   CORRECTION FOR NUCLEAR MOTION
13–3   THE FRANCK–HERTZ EXPERIMENT
13–4   THE FRANCK–HERTZ EXPERIMENT—INTERPRETATION

13–1 HYDROGENLIKE ATOMS

As we have seen the Bohr theory is limited, and so far we have applied it only to the hydrogen atom. We shall see that the usefulness of the Bohr theory can be extended, however, by considering *hydrogenlike* atoms. These are atoms in which the nuclear charges are Ze, but where only one electron revolves around the nucleus of each. These include such atoms as singly ionized helium He^+ (in which $Z = 2$), doubly ionized lithium Li^{2+} ($Z = 3$), and so on. The fundamental Newton's second-law equation in this case is

$$F = \frac{1}{4\pi\varepsilon_0} \frac{Ze^2}{r^2} = \frac{mv^2}{r} \tag{13-1}$$

The second basic equation is the same angular momentum equation that was used when the Bohr theory was applied to the hydrogen atom,

$$L = mvr = n\hbar \tag{13-2}$$

Table 13-1 lists useful equations for the hydrogen and hydrogenlike atoms for comparison. Notice that everywhere e^2 appears for the hydrogen atom, it is simply replaced by Ze^2 for the hydrogenlike atom.

For the same value of the quantum number n, the radius of the electron orbit in a hydrogenlike atom is smaller than the one in the hydrogen atom by a

Table 13–1 Comparison of hydrogen and hydrogenlike atoms according to the Bohr theory

HYDROGEN	HYDROGENLIKE
$r_n = \dfrac{4\pi\varepsilon_0 n^2 \hbar^2}{me^2}$	$r_n = \dfrac{4\pi\varepsilon_0 n^2 \hbar^2}{mZe}$
$E_n = -\dfrac{me^4}{32\pi^2\varepsilon_0^2\hbar^2}\dfrac{1}{n^2} = -\dfrac{1}{n^2}\,13.6\text{ eV}$	$E_n = -\dfrac{mZ^2 e^4}{32\pi^2\varepsilon_0^2\hbar^2}\dfrac{1}{n^2} = -\dfrac{Z^2}{n^2}\,13.6\text{ eV}$
$R = \dfrac{me^4}{64\pi^3\hbar^3\varepsilon_0^2 c}$	$R' = \dfrac{mZ^2 e^4}{64\pi^3\hbar^3\varepsilon_0^2 c} = RZ^2$
$\dfrac{1}{\lambda} = R\left(\dfrac{1}{n_f^2} - \dfrac{1}{n_i^2}\right)$	$\dfrac{1}{\lambda} = RZ^2\left(\dfrac{1}{n_f^2} - \dfrac{1}{n_i^2}\right)$

factor $1/Z$. The energy levels for the same n are made more negative by a value of $1/Z^2$. In particular, for singly ionized helium He^+ ($Z = 2$), the energy of the ground state is $E_1 = -(13.6/1^2)2^2 = -54.4$ eV. For $n = 2$, the energy level for He^+ is $E_2 = -(13.6/2^2)2^2 = -13.6$ eV, which coincides with the $E_1 = -13.6$ eV for the ground state of hydrogen. Also, for He^+, $E_4 = (-13.6/4^2)2^2 = -3.40$ eV, which coincides with the $n = 2$ state of hydrogen, $E_2 = -3.40$ eV. Hence, a transition from $n = 2$ to $n = 1$ in hydrogen releases a photon of the same wavelength as a transition from $n = 4$ to $n = 2$ in He^+. These transitions are illustrated in Figure 13–1. Many lines in the Lyman series in hydrogen (transitions to $n = 1$)

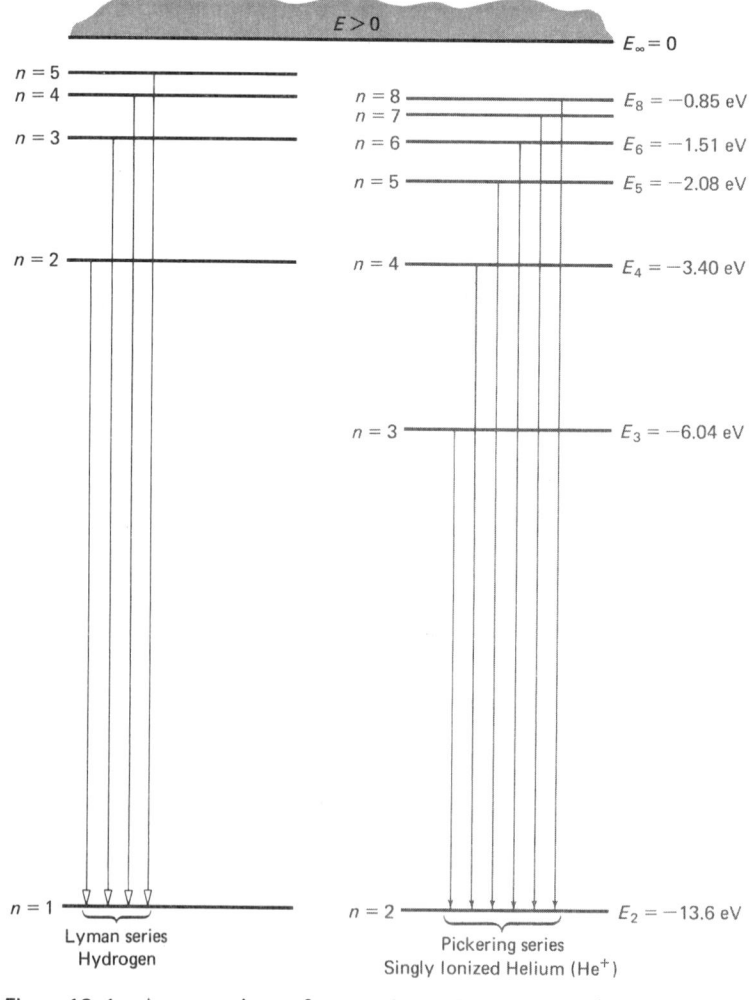

Figure 13–1 A comparison of energy levels for H and He^+.

coincide with some of the lines in the *Pickering series* (transitions to $n = 2$) in He^+; this was a source of confusion for early spectroscopists. The Rydberg constant R, and hence the wave number $k = 1/\lambda$, are Z^2 times greater in He^+ than in H for any given transition ($n_i \rightarrow n_f$).

13–2 CORRECTION FOR NUCLEAR MOTION

Thus far in the development of the Bohr theory, it has been assumed that the massive nucleus is essentially at rest while the electron revolves about it. A more realistic picture of the hydrogen atom, shown in Figure 13–2, has the electron of mass m and the proton of mass M each revolving about their common center of mass c. If r_e and r_n are the respective distances of the electron and the nucleus to the center of mass, Figure 13–2 shows that

$$r = r_e + r_n \tag{13-3}$$

From the definition of the center of mass,

$$Mr_n = mr_e \tag{13-4}$$

and these two equations give

$$r_e = \left(\frac{M}{M + m}\right) r \tag{13-5}$$

$$r_n = \left(\frac{m}{M + m}\right) r \tag{13-6}$$

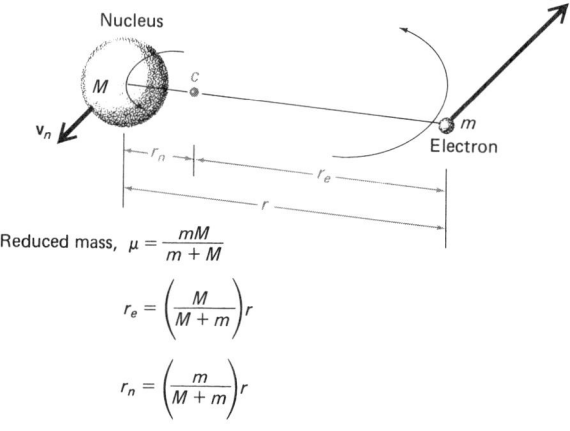

Figure 13–2 The electron and the nucleus revolve around the common center of mass c.

Application of Bohr's second postulate now gives the total angular momentum relative to the center of mass as

$$L = Mv_n r_n + m v_e r_e = n\hbar \tag{13-7}$$

where $v_n = \omega r_n$ and $v_e = \omega r_e$ are the respective linear speeds of the nucleus and the electron. Equation (13-7) can thus be written as

$$L = M\omega r_n^2 + m\omega r_e^2 = n\hbar \tag{13-8}$$

The expressions for r_n and r_e from Equations (13-5) and (13-6) substituted into Equation (13-8) give

$$\boxed{\mu \omega r^2 = n\hbar} \tag{13-9}$$

where

$$\boxed{\mu = \frac{mM}{m+M}} \tag{13-10}$$

is called the *reduced mass*.

Equation (13-9) is similar to Equation (12-9), $L = mvr = n\hbar$, which was developed ignoring the motion of the nucleus. This similarity is more evident if we assume a stationary state and write

$$L = m\omega r^2 = n\hbar \tag{13-11}$$

since $v = \omega r$.

The basic Equation (13-9) is now identical to Equation (12-9), except that the mass of the electron has been replaced by the reduced mass μ. It is easily seen that Equation (13-11) is just an approximation, since $M \gg m$ and $\mu = mM/(m+M) \cong m$.

The potential energy of the system is

$$V = -\frac{1}{4\pi\varepsilon_0} \frac{e^2}{r}$$

and the kinetic energy is

$$K = \tfrac{1}{2} m v_e^2 + \tfrac{1}{2} M v_n^2$$
$$= \frac{\omega^2}{2}(m r_e^2 + M r_n^2)$$

which can be written after some simplification as

$$\boxed{K = \tfrac{1}{2}\mu \omega^2 r^2} \tag{13-12}$$

Now, applying Newton's second law to the motion of the electron, we can write

$$\frac{1}{4\pi\varepsilon_0}\frac{e^2}{r^2} = m\frac{v_e^2}{r_e} = m\omega^2 r_e$$

and by using Equation (13–5),

$$\frac{1}{4\pi\varepsilon_0}\frac{e^2}{r^2} = \frac{mM}{(m+M)}\omega^2 r$$

or

$$\boxed{\frac{1}{4\pi\varepsilon_0}\frac{e^2}{r^2} = \mu\omega^2 r} \tag{13-13}$$

From Equation (13–11), we can conclude that

$$\frac{1}{4\pi\varepsilon_0}\frac{e^2}{r^2} = \mu\left(\frac{n\hbar}{\mu r^2}\right)^2 r$$

and the radii of the stationary orbits are

$$\boxed{r_n = r = \frac{4\pi\varepsilon_0 n^2 \hbar^2}{\mu e^2}} \tag{13-14}$$

which is identified as Equation (12–11), where m is replaced by μ.

Combining Equations (13–12) and (13–13) gives the kinetic energy as

$$K = \frac{1}{2}\mu\omega^2 r = \frac{1}{8\pi\varepsilon_0}\frac{e^2}{r}$$

and the total energy, the kinetic energy plus the potential energy, becomes

$$E = -\frac{1}{8\pi\varepsilon_0}\frac{e^2}{r}$$

or, when $r = r_n$ from Equation (13–14) is substituted,

$$\boxed{E_n = -\frac{\mu e^2}{32\pi^2 \varepsilon_0^2 \hbar^2}\left(\frac{1}{n^2}\right)} \tag{13-15}$$

Applying Bohr's formula for a transition between an initial energy state E_i and a final state E_f (where $E_i > E_f$) gives the frequency of the photon emitted as

$$\nu = \frac{c}{\lambda} = \frac{E_i - E_f}{h}$$

and when the expression for energy from Equation (13–15) is used,

$$v = \frac{\mu e^4}{64\pi^3 \hbar^3 \varepsilon_0^2} \left(\frac{1}{n_f^2} - \frac{1}{n_i^2} \right) \qquad (13\text{–}16)$$

Thus, the wavelength of the photon is

$$\frac{1}{\lambda} = \frac{\mu e^4}{64\pi^3 \hbar^3 \varepsilon_0^2 c} \left(\frac{1}{n_f^2} - \frac{1}{n_i^2} \right) \qquad (13\text{–}17)$$

with the Rydberg constant now evidently given by

$$R_\mu = \frac{\mu e^4}{64\pi^3 \hbar^3 \varepsilon_0^2 c} \qquad (13\text{–}18)$$

The wavelength is now more correctly written as

$$\frac{1}{\lambda} = R_\mu \left(\frac{1}{n_f^2} - \frac{1}{n_i^2} \right) \qquad (13\text{–}19)$$

The ratio of R_μ to the Rydberg constant R (recall Equation 12–18), with correction for nuclear motion, is

$$\frac{R_\mu}{R} = \frac{\mu}{m} = \frac{1}{1 + m/M} < 1 \qquad (13\text{–}20)$$

A comparison of energy levels with and without corrections for nuclear motion from Equations (13–15) and (12–14), respectively, shows that for the same value of n, the energy levels calculated with the corrections are less negative than the corresponding levels without the corrections; that is,

E_n (with correction) $>$ E_n (without correction)

Consequently, the energy levels with the corrections are slightly displaced in the positive direction, as shown in Figure 13–3.

A comparison of Equations (13–19) and (12–19) also shows now that

$$\left[\frac{1}{\lambda} \text{(with correction)} \right] < \left[\frac{1}{\lambda} \text{(without correction)} \right]$$

This means that when the above corrections are taken into account, the calculated wavelengths of the emitted photons are slightly larger. A recalculation of the Rydberg constant now gives

$R = 1.0973731 \times 10^7$ m^{-1} (without correction)

$R_\mu = 1.0967758 \times 10^7$ m^{-1} (with correction)

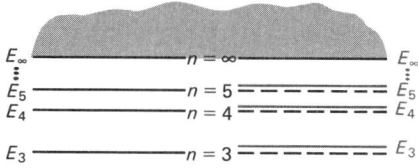

(a) Without correction   (b) With correction

Figure 13-3 Energy levels of the hydrogen atom with and without corrections due to the motion of the nucleus. The energy levels in (a) are to scale, but the differences shown in (b) are exaggerated so that they can be seen.

The deuterium atom ^2_1D, an isotope of hydrogen, has a nucleus composed of a proton and a neutron. Since the mass of the neutron is only slightly different from that of the proton, the reduced mass for deuterium is

$$\mu_D = \frac{m}{1 + m/2M} \tag{13-21}$$

which makes $\mu_D > \mu$. Since the Rydberg constant is directly proportional to the reduced mass, it is evident that the Rydberg constant for deuterium is slightly larger than that for hydrogen; that is, $R_{\mu_D} > R_\mu$. This small discrepancy played an important role in the discovery of deuterium (heavy hydrogen) by the American physicist H. C. UREY. For this discovery, Urey was awarded the 1934 Nobel Prize in chemistry.

13-3 THE FRANCK–HERTZ EXPERIMENT

A direct and striking demonstration of the existence of the discrete stationary states postulated by the Bohr theory of the atom was first provided by an experiment designed by JAMES FRANCK (1882–1964) and GUSTAV HERTZ (1887–). For a better understanding of the conclusions from this experiment, let us review briefly the excitation and ionization of atoms in the "optical" levels.

In a heavy atom such as mercury $^{202}_{80}$Hg, the electrons in the innermost shells of the atom are difficult to remove because of the strong electrostatic attraction of the nucleus. They have binding energies typically in the range of a few KeV. The outermost (valence) electrons are somewhat shielded from the nucleus by the screening effect of the electrons in the inner shells. Thus, the binding energy of these electrons is only a few eV. In the Franck–Hertz experiment, only the outer valence electrons are involved, and the corresponding energy level for this electron is shown in Figure 13-4. These energy levels are usually called *optical levels*, because any transitions among these levels involve photons with wavelengths in the visible or near visible region of the spectrum.

In Figure 13-4, the energy of the valence electron in the *ground state* (G) is $E_G = -10.42$ eV. The other energy levels, H, I, and so on, are excited states. The first excited state (H) has an energy $E_H = -5.44$ eV. I is the *second* excited state,

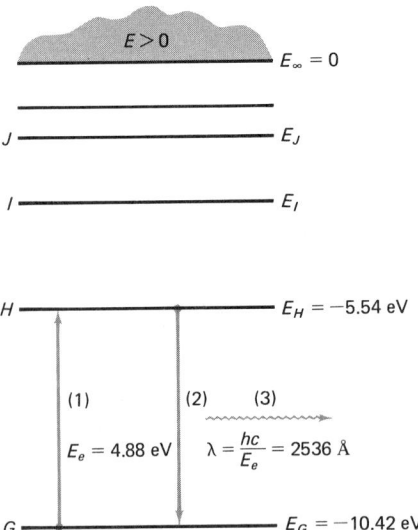

Figure 13-4 Optical energy levels for the valence electron for $^{202}_{80}$Hg.

J is the *third* excited state, and so on. The energy required to raise the electron from the ground state to the first excited state H (line 1 in Figure 13-4) is

$$E_e = E_H - E_G = -5.44 - (-10.42) \text{ eV} = 4.88 \text{ eV}$$

and is called the *first excitation potential* of mercury. If for some reason the mercury atom is excited to the first excited state, the electron will return in a very short time (about 10^{-8} sec) to the ground state (line 2). In this transition, a photon (3) of energy $E_e = 4.88$ eV and wavelength $\lambda = hc/E_e = 2536$ Å will be emitted. From the same Figure 13-4, the ionization energy is 10.42 eV.

Consider the case of a beam of slow electrons traveling through mercury vapor at low pressure. If the kinetic energy of the electrons is less than 4.88 eV, the collision will be elastic; that is, the *translational kinetic energy will be conserved*. The electrons will lose some kinetic energy according to the expression*

$$\Delta K = \frac{4mM}{(m+M)^2} K = \frac{4m}{M} K \qquad (13\text{-}22)$$

where m is the mass of the electron, M is the mass of the mercury atom, and $K = \frac{1}{2}mv^2$ is the kinetic energy of the incident electron. This loss in kinetic energy ΔK is quite small, since $m \ll M$. The energy ΔK is transferred to the mercury atom and appears as its recoil energy, represented schematically by

β	+	A	$\rightarrow$	A'	+	β
slow electron $K_1 < 4.88$ eV		atom at rest		atom with some recoil energy, ΔK		slower electron $K_2 = K_1 - \Delta K$

Since ΔK is so small, the electron will experience many collisions along a zigzag trajectory before it comes to rest (Figure 13-5).

However, if the kinetic energy of the electron is larger than $E_H - E_G = 4.88$ eV, an inelastic collision may occur in which part of the kinetic energy, is transferred to the atom in the form of internal energy, raising the electron from the ground state to the first excited state, E_H. The kinetic energy of the electron after the inelastic collision is

$$K_2 = K_1 - (E_H - E_G) = K_1 - 4.88 \text{ eV}$$

This situation is represented schematically by

β	+	A	$\rightarrow$	A^*	+	β
$K_1 > E_H - E_G$ or $K_1 > 4.88$ eV		atom at rest and in the ground state		excited atom $\quad\downarrow$ A + $h\nu$ ground state emitted photon $\lambda = 2536$ Å		$K_2 = K_1 - (E_H - E_G)$

*See D. Halliday and R. Resnick, *Physics for Students of Science and Engineering*, Wiley, New York, 1960, Chap. 10.

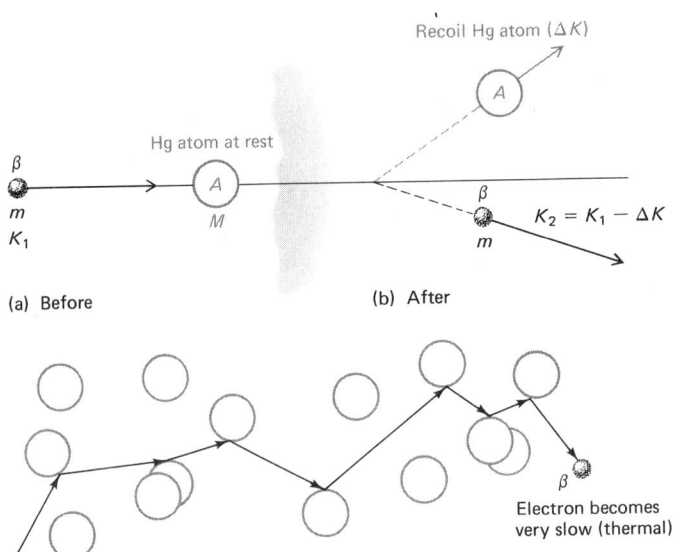

Figure 13–5 Elastic collisions of electrons of energy $K_1 < 4.88$ eV with a mercury atom at rest. The recoil energy of an atom is $\Delta K = K_1 - K_2$, where $K_2 < K_1$ is the kinetic energy of an electron after the collision. The electron makes a zigzag trajectory in mercury vapor.

A second process takes place almost immediately after the collision (the life of an excited state is about 10^{-8} sec). The excited atom A^* will return to the ground state by the emission of a photon of energy $E_H - E_G = 4.88$ eV (see Figure 13–4) and wavelength $\lambda = 2536$ Å. If K_1, the energy of the incident electron, is just slightly greater than 4.88 eV, then $K_2 < 4.88$ eV, and no more inelastic collisions can take place. All other collisions will be elastic. If $K_1 \gg 4.88$ eV, then $K_2 > 4.88$ eV and other inelastic collisions can take place.

13–4 THE FRANCK–HERTZ EXPERIMENT—INTERPRETATION

The collision mechanisms discussed above were checked experimentally by Franck and Hertz in 1914 using an experimental arrangement shown schematically in Figure 13–6(a). Tube T in the figure contains mercury vapor at low pressure and at a temperature of about 150°. The tube contains a filament F, supplied by battery C, a grid G, and a plate P. Between the filament and the grid there is an accelerating potential V_a that can be varied between 0 and 60 V. Between the plate P and the grid G is a small retarding potential V_n (around 0.5 V). Finally, a very sensitive electrometer D in series with the plate measures the plate current of about 10^{-9} A.

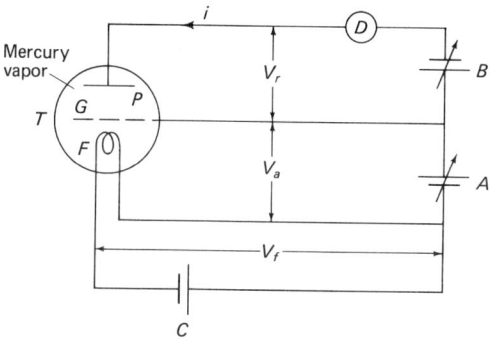

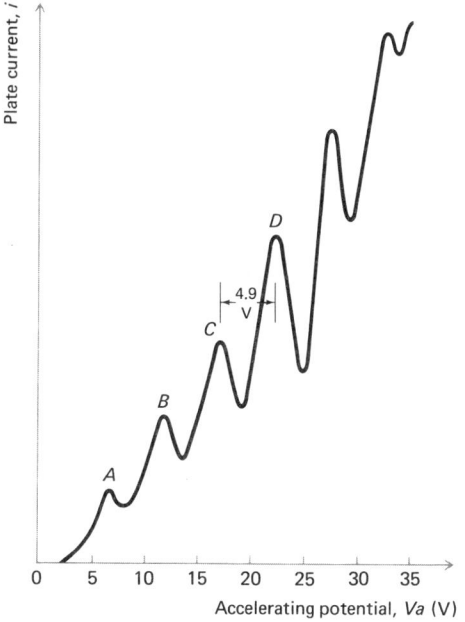

Figure 13-6 (a) Experimental setup for the Franck–Hertz experiment. (b) Plate current versus accelerating voltage in the Franck–Hertz experiment. The separation between any two consecutive peaks is about 4.9 V.

When the accelerating potential V_a is increased, the plate current increases as shown in Figure 13-6(b).

As V_a is increased, the plate current increases as in any electronic tube, except that a significant decrease in the plate current occurs each time the accelerating potential is increased by approximately 5 V. Some of those electrons with energies slightly greater than 4.88 eV will experience inelastic collisions and will be left with such small energy that they will not be able to reach the plate, because of the presence of the retarding voltage. If V_a is increased by an additional 5 V, some of the electrons that were left with almost no kinetic energy will experience another inelastic collision and will not reach the plate. This explains the second dip at a potential of about 5 V greater than the first dip. Hence, this second dip corresponds to those electrons that have experienced two inelastic collisions; the third dip corresponds to three inelastic collisions, and so on. Each time there is an inelastic collision, the mercury atoms will be excited and return to the ground state by the emission of photons. By using spectroscopic techniques, the wavelength of the radiation coming from the tube was found to be 2536 Å, corresponding to transitions from the first excited state of mercury to the ground. This result, together with the fact that the energy difference between two consecutive dips is about 4.9 V, shows in a very convincing way the existence of discrete energy levels in the mercury atom. It is also possible, by using different voltages and improved resolution, to measure the excitation of other atomic energy levels. It is little wonder that Franck and Hertz were awarded the Nobel Prize in physics (1925) for this work!

PROBLEMS

13-1 Without taking into account the corrections due to the motion of the nucleus, (a) calculate for singly ionized helium He^+ the value of the Rydberg constant, and (b) the energy levels for $n = 1, 2, 3,$ and so on. (c) Draw an energy level diagram for He^+ along with one for hydrogen. What can you conclude from these energy level diagrams?

13-2 Repeat the computations of Problem 13-1 using doubly ionized lithium Li^{2+}.

13-3 Apply the Bohr theory to He^+ and calculate for each orbit, $n = 1, 2, 3,$ (a) the radius, (b) the frequency of revolution, (c) the linear speed of the electron, (d) the total energy of the electron, (e) the angular momentum, and (f) the ratio v/c, and decide whether the classical treatment can be used or not.

13-4 Repeat Problem 13-3 using Li^{2+}.

13-5 (a) Compute the first and second excitation potentials for singly ionized helium He^+.
(b) What wavelengths are emitted when the He^+ returns to the ground state from these excited states?

13-6 Tritium, ^3_1H, a hydrogen isotope with a nucleus of one proton and two neutrons, is mixed with ordinary hydrogen. What is the resolution of the spectroscopic instrument that will just separate the H_α lines observed?

13-7 Determine the wavelength of the first two lines of singly ionized helium that corresponds to the first two lines of the Balmer series.

13-8 A tube such as that in Figure 13-6(a) contains hydrogen gas instead of mercury vapor. Assume that only the first excitation potentials are involved and determine (a) the accelerating potential of electrons before the first dip in plate current can be observed, and (b) the wavelength of the light given off from the tube.

13-9 For the positronium atom (see Chapter 9), calculate (a) the reduced mass, (b) the Rydberg constant, and (c) a few lines of the Balmer series and the series limit.

13-10 When hydrogen gas is used in the Franck–Hertz experiment [Figure 13-6(a)], the first and second Lyman lines appear when the energy of the incident electrons exceeds the quantum energy of the second Lyman line but is less than that of the third. What is the accelerating potential of the electrons that will produce the first three lines of the Lyman series?

13-11 How much energy is required to remove an electron entirely free from the nucleus of singly ionized helium if the electron is originally in the ground state? If the electron is in the $n = 3$ state?

RECOMMENDED READING

BORN, M., *Problems of Atomic Dynamics*, M.I.T. Press, Cambridge, Mass., 1926.
DIRAC, P. A. M., "The Quantum Theory of the Emission and Absorption of Radiation," *Proc. Roy. Soc.* **114**, 243 (1927).
DIRAC, P. A. M., "The Quantum Theory of the Electrons," *Proc. Roy. Soc.* **117**, 610 (1928).
FOOTE, P. D., MEGGERS, W. F., and MOHLER, F. L., *Astrophys. J.* **55**, 145 (1922).
    A detailed study of the ionization potentials for sodium and potassium.
FRANCK, B. J., and HERTZ, G., *Verhandl. Deut. Physik Ges.* **16**, 512 (1914).
    In German, no English translation available.
MELISSINOS, A., *Experiments in Modern Physics*, Academic, New York, 1966, pp. 8–18.
PAULING, L., and GOUDSMIT, S., *The Structure of Line Spectra*, McGraw-Hill, New York, 1934.
PAULING, L., and WILSON, E. B., *Introduction to Quantum Mechanics*, McGraw-Hill, New York, 1935.
SOMMERFELD, A., *Ann. Physik* **51**, 1 (1916).
    Written in German, no English translation available.
SOMMERFELD, A., *Wave Mechanics*, Methuen, London, 1930.
WILSON, W., *Phil. Mag.* **29**, 795 (1915).

THREE
The Atom

> It was quite the most incredible thing that has ever happened to me in my life. It was almost as incredible as if you fired a 15-inch shell at a piece of tissue paper and it came back and hit you.
>
> ERNEST RUTHERFORD
> *Background to Modern Science*, 1940

The quotation is Rutherford's comments on the results of Marsden's experiments of scattering α particles from gold nuclei. Rutherford's analysis of what was causing the unusual scattering led to his "discovery" of the nucleus of the atom. It was not long after this that Bohr postulated the model of the "modern" atom. The Schrödinger equation and quantum mechanics further refined the model of the atom to the concepts that we use today.

14
The Schrödinger Equation I

Erwin Schrödinger
(1887–1961)

Born in Vienna and educated at the University of Vienna, Schrödinger succeeded Max Planck as professor at the University of Berlin (1927–1933). From 1940 to 1955 he was professor at the Institute of Advanced Study in Dublin. In the early 1920s, he showed that wave mechanics and the matrix mechanics of W. Heisenberg are equivalent. His findings placed the quantum theory on a new basis and constituted the groundwork on which atomic theory was built. In 1933 he received the Nobel Prize with P. A. M. Dirac for his work on wave mechanics and atomic structure.

14–1 BLACKBODY RADIATION
14–2 WAVE FUNCTIONS
14–3 THE SCHRÖDINGER EQUATION
14–4 THE TIME-INDEPENDENT SCHRÖDINGER EQUATION

14–1 BLACKBODY RADIATION

We have seen that the description of the motion of a body given by the classical mechanics has proved to be inadequate when the speed of the body approaches the speed of light. For this case, the limitations of the classical mechanics force us to adopt the relativistic mechanics. Another limitation on the classical mechanics occurs when we study matter of very small dimensions—within the microscopic world of atomic and nuclear structure and of the elementary particles. It is here that the remarkably successful quantum mechanics, and its modern version called quantum field theory, takes over the story from classical approaches.

We may trace the beginning of quantum concepts to about 1900, when an unexplained puzzle existed concerning the wavelength spectrum of light emitted by heated solid bodies. In spite of attempts by noted physicists at the time, a single classical theory could not explain adequately the shape of the curve of the power radiated from a blackbody as a function of wavelength. The idealized concept of a *blackbody*, a theoretical object that absorbs all light of whatever wavelength that falls on it, was conceived to simplify the problem. This concept removed the parameters that depend on the particular kind of solid emitting the light. An experimental approach to such an object is to use a small hole in the wall of a cavity that is heated (Figure 14–1). The light emitted through this hole is close to that which would be emitted by an ideal, heated blackbody.

Since we conceive of light as having a wavelike character, it is reasonable to assume that the light in the blackbody is emitted by *harmonic oscillators*. We may carefully construct a model that allows these "oscillators" to have any frequency and that looks on the light within the blackbody cavity as standing waves extending from wall to wall. This model leads to the Rayleigh-Jeans distribution law.

According to the Rayleigh-Jeans law, the spectrum of the light from the blackbody radiation has an energy distribution of

$$\rho_\lambda \, d\lambda = 8\pi k T \frac{d\lambda}{\lambda^4} \qquad (14\text{–}1)$$

where ρ_λ is the energy density per unit wavelength of radiation in a small wavelength range $d\lambda$ centered at the particular wavelength λ, T is the absolute temperature, and k is the Boltzmann constant. This law was found to describe the spectrum

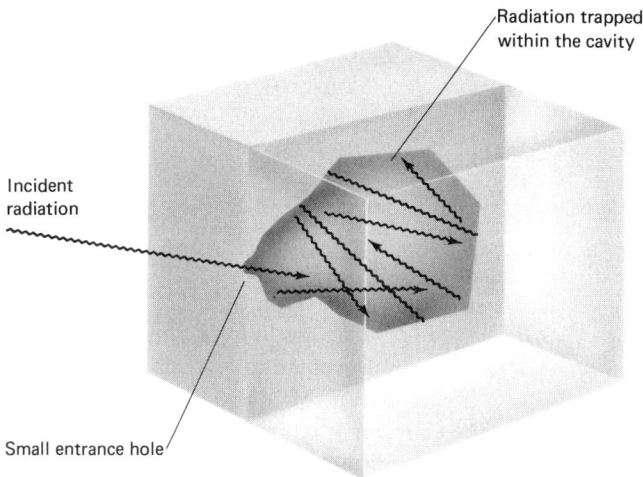

Figure 14-1   The small entrance hole to an irregularly shaped cavity acts like a blackbody because it absorbs most of the radiation incident on it. The hole itself is the blackbody.

of blackbody radiation quite well for large values of λ; but it can be seen that if very short wavelengths are present, and in particular if the wavelengths are arbitrarily short, the energy density ρ_λ becomes very high and approaches infinity. This obviously does not happen, since only a fixed and finite amount of energy is radiated per unit time from a blackbody, and only a fixed and finite amount of energy is contained in one.

At this juncture in the development of the theory, MAX PLANCK made the following radical assumptions:

1. The oscillators in the blackbody do not emit light continuously but only in the process of changing their amplitudes—a transition to a smaller from a larger amplitude results in the emission of light, while a transition to a larger amplitude constitutes the absorption of light by the oscillator.
2. An oscillator may emit energy to the radiation field or absorb energy from it only in units of energy called *quanta* having the magnitude $h\nu$, where h is a constant (now called Planck's constant) and ν is the frequency of the oscillator.

These assumptions led Planck to the energy distribution law*:

$$\rho_\lambda \, d\lambda = \frac{c_1}{\lambda^5} \frac{d\lambda}{e^{c_2/\lambda T} - 1} \qquad (14\text{-}2)$$

---

* Planck's energy distribution law was presented on October 19, 1900, before the Berlin Physical Society.

where $c_1 = 2\pi c^2 h$ and $c_2 = hc/k$ are constants (see Figure 14–2). Here the exponential term in the denominator forces the energy density to zero at extremely short wavelengths. This is the distribution that is actually observed in laboratory measurements of blackbody spectra. Planck's assumption that radiation interacts with matter units or quanta of energy $h\nu$ rather than by continuous absorption was used by Einstein in 1905 to explain successfully the mechanism of the photoelectric effect.

Light, which we have already noted displays many wavelike properties such as diffraction and interference, has been shown to have particle-like properties as well, in that it can be shown that its energy is carried in discrete, small bundles of energy $h\nu$. The special theory of relativity now forces us to associate an effective mass with each photon of $h\nu/c^2$ and a momentum of $p = h/\lambda = h\nu/c$. Further experimentation with very weak light beams in attempts to work with single photons, and with very narrow beams to investigate the question of the possible spreading out sideways of a photon as it travels in a beam, were found to support the photon model. It was found that a photon does not spread out but remains small in lateral dimensions at all times. These experiments were capped by the discovery by A. H. Compton (see Chapter 8) that photons of x rays are scattered by electrons as if they were small elastic particles of effective mass $h\nu/c^2$ and momentum $h\nu/c$. The dual nature of light was verified by experimental observations and was thus incorporated into modern physical theory as quantum electrodynamics.

We may well ask, if a packet of radiant energy (a photon) has some properties typical of a material particle, then when a particle (such as an electron) moves, will it also have properties that are associated with a "frequency" and, therefore, a "wavelength"? The answer to this question is affirmative, as was proved in 1927

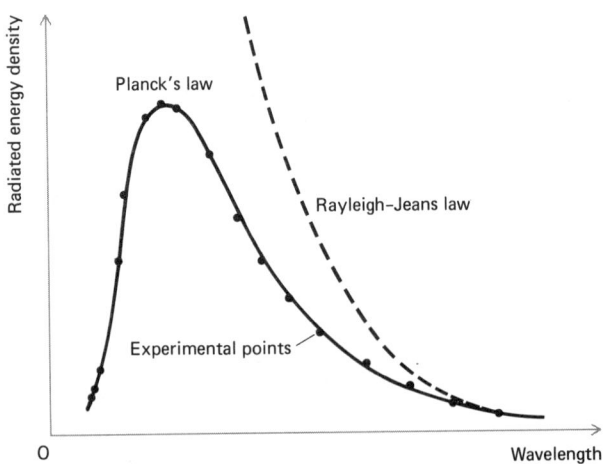

Figure 14–2 The graph indicates how Planck's energy distribution law, which is based on the idea of quantized oscillators, agreed with experiment.

by the Davisson–Germer experiment (Chapter 10), in which a diffraction pattern was found when a stream of rapidly moving electrons fell on the surface of a crystal. This concept grew among physicists in the early decades of the twentieth century, especially following the successes of the Bohr theory, which applied Planck's assumptions to the energy levels of atoms and led to even deeper puzzles. The energy levels of neutral helium were not given by Bohr's theory, for example, while those of hydrogen fitted beautifully.

De Broglie's concept of a wavelength $\lambda = h/p$ associated with an electron of momentum p proved of immediate value in discussing the energy and momentum states of electrons in atoms. It provided a formulation of the problem in which the fixed and definite energy states in an atom are associated with the de Broglie wavelengths of the electrons in the atom. The states are those in which the electron wave is "stationary" or "standing" in its distribution around the nucleus. Later, experiments verified de Broglie's wavelengths for protons, neutrons, and atoms. We now associate a de Broglie wave with every particle and material object.

The wave-particle duality shown by particles and radiation does not represent a conflict as was originally assumed; now we conceive of both modes of behavior as simply the manifestations of matter. The quanta described by Planck represent discrete units of energy, given by Equation (10–1) as

$$E = h\nu$$

Then, implausible as it was to many at the time, Bohr proposed that energy levels of *all* matter are of this same form. In the particular case of a photon, there is an associated electromagnetic wave for which the amplitude of the electromagnetic field is given by a function $\Psi_E(x, t)$. The electromagnetic field is the source of information about such concrete quantities as the linear momentum and energy of a photon. In general, for any given particle, whether a photon or electron, there is an associated *matter-field* whose amplitude is given by a function $\Psi(x, t)$, known as a *wave function*. The matter-field is also the source of information of quantities such as the linear momentum and energy of particles such as electrons or α particles. The frequency and wavelength associated with the matter-field wave are determined from $\nu = E/h$ and $\lambda = h/p$, respectively.

14–2 WAVE FUNCTIONS

The intensity of a wave is proportional to the square of the amplitude of the wave. Hence, the intensity of the matter-field associated with a particle is proportional to the square of the matter-field amplitude $\Psi(x, t)$. Since the wave function can be a complex function (it can contain complex numbers of the form $a + ib$, where $i = \sqrt{-1}$), the intensity is proportional to

$$|\Psi(x, t)|^2 = \Psi^*\Psi \tag{14-3}$$

in which Ψ^* is the complex conjugate of Ψ.

What is the physical significance of the wave function? What feature of the particle character of matter can be measured by the wave function? The wave function will be found to be as much a physical quantity as the electrical field or the magnetic field. The wave function must describe something about the location of the particle in the space-time universe, as the particle is more likely to be located in those places where intensity is large. MAX BORN has given the following meaning to it: The wave function has a probability interpretation and $|\Psi|^2$ is proportional to the probability per unit length of finding the particle at a given point in space and a given time. The probability of finding the particle within an element of length dx is

$$\Psi^*\Psi \, dx \qquad (14\text{-}4)$$

More precisely, this expression is *normalized* as

$$\boxed{\int_{-\infty}^{\infty} \Psi^*\Psi \, dx = 1} \qquad (14\text{-}5)$$

since the probability of finding the particle somewhere must be 1 (this represents certainty). In the more general case, $\Psi = \Psi(x, y, z, t)$ and $\Psi^*\Psi \, dx \, dy \, dz$ is the probability of finding a particle in an element of volume $dv = dx \, dy \, dz$ and

$$\int_{-\infty}^{\infty}\int_{-\infty}^{\infty}\int_{-\infty}^{\infty} \Psi^*\Psi \, dv = 1$$

Because of the uncertainty relation, the deterministic principles of classical mechanics must be abandoned. That is, we cannot predict *exactly* the subsequent motion of the particle, because position and velocity cannot be measured simultaneously with absolute precision. Hence, in quantum mechanics we cannot talk about the trajectory of a particle. The only thing we can do is to evaluate the probability per unit volume $\Psi^*\Psi$ of finding the particle at a given position and at a given time.

14-3 THE SCHRÖDINGER EQUATION

ERWIN SCHRÖDINGER approached the wave-particle duality of nature by adopting the de Broglie and Planck relationships, $\lambda = h/p$ and $\nu = E/h$, and by defining the total energy of a particle by

$$\boxed{E = \frac{p^2}{2m_0} + V} \qquad (14\text{-}6)$$

In this equation, m_0 is the rest mass, $k = p^2/2m_0$ is the classical kinetic energy, and p is the linear momentum of the particle. Note that this is a nonrelativistic form of the energy, and the rest energy $E_0 = m_0 c^2$ has not been included.

The group velocity of the wave packet, from Equation (10–9), is $v_g = d\omega/dk$, where $\omega = 2\pi\nu$ is the angular frequency and $k = 2\pi/\lambda$ is the propagation constant. In Equations (10–8) and (10–9), it was also shown that

$$v_g = \frac{d\omega}{dk} = \frac{dE}{dp}$$

When the energy E is expressed by Equation (14–6), the group velocity becomes

$$v_g = \frac{dE}{dp} = \frac{d}{dp}\left(\frac{p^2}{2m_0} + V\right) = \frac{p}{m_0} = v$$

Thus, in the Schrödinger approach, the group velocity is still equal to the particle velocity. Incidentally, we should recall that in Chapter 10 it was also shown that for a photon the group velocity is equal to c, the speed of light.

The *phase velocity* of a free particle will be found to be different when Schrödinger's approach is compared with that of de Broglie. From the de Broglie theory,

$$\boxed{\begin{array}{c}v_{ph} \\ \text{(de Broglie)}\end{array} = \lambda\nu = \frac{h}{p}\frac{E}{h} = \frac{E}{p} = \frac{mc^2}{mv} = \frac{c^2}{v}}$$

but from the Schrödinger expression for the energy,

$$\begin{array}{c}v_{ph} \\ \text{(Schrödinger)}\end{array} = \lambda\nu = \frac{hE}{ph} = \frac{E}{p}$$

and

$$\boxed{\begin{array}{c}v_{ph} \\ \text{(Schrödinger)}\end{array} = \frac{E}{p} = \frac{p^2/2m_0}{p} = \frac{p}{2m_0} = \frac{v}{2}}$$

where the potential energy function V has been set equal to zero, which is the case for a free particle.

The "matter-wave" equation in one dimension relating the de Broglie theory and the wave function,

$$\boxed{-\frac{\hbar^2}{2m}\frac{\partial^2 \Psi(x,t)}{\partial x^2} + V\Psi(x,t) = i\hbar\frac{\partial \Psi(x,t)}{\partial t}} \quad (14\text{–}7)$$

is known as the *Schrödinger equation*. Although Schrödinger developed this equation from an insight into the wave character of matter, it is not derivable from first principles. This equation, like Newton's second law, is itself a first principle. This particular form of the equation is known as the *time-dependent Schrödinger equation*, because the potential energy is general enough to be a function of both position and time.

Finally, there are certain requirements that must be set for the wave function to be a useful tool in describing the physical world. The wave function as a solution of the Schrödinger equation must meet the following requirements.

1. It must be consistent with the following relationships:

$$\lambda = \frac{h}{p}$$

$$\nu = \frac{E}{h}$$

$$E = \frac{p^2}{2m} + V$$

2. It must be linear in $\Psi(x, t)$; that is, if $\Psi_1(x, t), \Psi_2(x, t), \ldots, \Psi_n(x, t)$ are solutions of the Schrödinger equation, then

$$\Psi(x, t) = a_1\Psi_1 + a_2\Psi_2 + \cdots + a_n\Psi_n = \sum_{i=1}^{n} a_i\Psi_i$$

where $a_1, a_2, \ldots, a_n$ are constants, must also be a solution.
3. The function $\partial\Psi(x, t)/\partial x$ must also be linear.
4. The function $\Psi(x, t)$, as well as its derivative $\partial\Psi(x, t)/\partial x$, must be *well behaved*; that is, they must be single valued, finite, and continuous.
5. When $x \to \pm\infty$, then $\Psi(x, t)$ must approach zero:

$$\lim_{x \to \pm\infty} \Psi(x, t) \to 0$$

In 1925 WERNER HEISENBERG developed a matrix algebra mathematical model for treating the same quantum mechanics problems as the Schrödinger equation. The approach was new and difficult, and it was a while before physicists realized that the two approaches were equivalent although expressed in different mathematical languages.

14-4 THE TIME-INDEPENDENT SCHRÖDINGER EQUATION

We shall begin by showing that the matter-field wave function of the form

$$\Psi(x, t) = A \exp[-(i/\hbar)(Et - px)] \quad (14\text{–}8)$$

is a solution to the time-independent Schrödinger Equation (14–7) and that it represents the quantum mechanical description of a *free* particle with a total energy E and a linear momentum p. Since the particle is a free particle, both E and p are constant; they are related by

$$E = \frac{p^2}{2m} + V = \text{constant}$$

in agreement with the Schrödinger approach of Equation (14-6). In the nonrelativistic case, the rest mass m of the free particle is a constant and the potential energy is a constant. For a particle that is not free moving in a conservative field, $V = V(x)$ is independent of time and p is a variable, but the total energy E is a constant.

The second derivative of Equation (14-8) with respect to position is

$$\frac{\partial^2 \Psi(x, t)}{\partial x^2} = -\frac{p^2}{\hbar^2} \Psi(x, t) \tag{14-9}$$

and differentiation with respect to time gives

$$\frac{\partial \Psi(x, t)}{\partial t} = -i \frac{E}{\hbar} \Psi(x, t) \tag{14-10}$$

Replacing Equations (14-9) and (14-10) in the time-dependent Equation (14-7),

$$-\frac{\hbar^2}{2m} \frac{\partial^2 \Psi(x, t)}{\partial x^2} + V\Psi(x, t) = i\hbar \frac{\partial \Psi(x, t)}{\partial t}$$

we get

$$-\frac{\hbar^2}{2m} \left[-\frac{p^2}{\hbar^2} \Psi(x, t) \right] + V\Psi(x, t) = i\hbar \left[-i \frac{E}{\hbar} \Psi(x, t) \right]$$

and by cancelling out the common factor Ψ and simplifying, we obtain

$$E = \frac{p^2}{2m} + V$$

which proves that Equation (14-8) is a solution of the time-dependent Schrödinger equation and represents the mathematical description of a free particle.

We shall now consider again Equation (14-8), and write it in the following way,

$$\Psi = (Ae^{ipx/\hbar})(e^{-iEt/\hbar}) \tag{14-11}$$

in which the variables x and t have been separated. If the spatial part is

$$\psi(x) = Ae^{ipx/\hbar} \tag{14-12}$$

then Equation (14-11) can be written

$$\Psi(x, t) = \psi(x) e^{-iEt/\hbar} \tag{14-13}$$

Now, by differentiating twice with respect to position, we get

$$\frac{\partial^2 \Psi(x, t)}{dx^2} = \frac{d^2 \psi(x)}{dx^2} e^{-iEt/\hbar} \tag{14-14}$$

and differentiation with respect to time gives

$$\frac{\partial \Psi(x, t)}{\partial t} = -\frac{i}{\hbar} E\psi(x) e^{-iEt/\hbar} \tag{14-15}$$

Substitution of Equations (14-13), (14-14), and (14-15) into Equation (14-7) gives

$$-\frac{\hbar^2}{2m} \frac{d^2 \psi(x)}{dx^2} e^{-iEt/\hbar} + V\psi(x) e^{-iEt/\hbar} = i\hbar \left(-\frac{i}{\hbar} E\psi(x) e^{-iEt/\hbar} \right)$$

Finally, cancellation of the common factor $e^{-iEt/\hbar}$ and some simplification yields

$$-\frac{\hbar^2}{2m} \frac{d^2 \psi(x)}{dx^2} + V\psi(x) = E\psi(x) \tag{14-16}$$

which is the "steady state" or *time-independent Schrödinger equation*.
Sometimes it is more convenient to write Equation (14-16) as

$$\frac{d^2 \psi(x)}{dx^2} + \frac{2m}{\hbar^2}(E - V)\psi = 0$$

In this equation, $\psi(x)$ is also called the wave function. $V(x)$, the potential function, does not contain the time explicitly, and E, the total energy of the particle, is a constant.

PROBLEMS

14-1 A wave function $\psi(x) = A_n \sin(2n\pi x/L)$ is defined only within a region $0 \leq x \leq L$. Use this normalization condition to evaluate the constant A_n.

14-2 Determine the constant A_n for a wave function

$$\Psi(x, t) = A_n \sin \frac{n\pi x}{L} e^{-iE_0 t/\hbar}$$

which is defined within a region $0 \leq x \leq L$.

14-3 For the wave function in Problem 14-2, the probability of finding the particle within a range $(a, b)(0 \leq a < b \leq L)$ is $\int_a^b \Psi^* \Psi \, dx$. (a) Determine the probability of finding the particle within the dimensions $x = 0$ to $x = L/4$. (b) What is the average probability per unit length?

14-4 Prove that $\Psi(x, t) = Ae^{(i/\hbar)(p_x x - Et)}$ is a solution to the Schrödinger equation. Is $\Psi + \Psi^*$ a solution?

14-5 Show whether $\Psi(x, t) = A \sin(kx - \omega t)$ is a solution to the Schrödinger equation.

14-6 Are $\Psi_1(x, t) = A_1 e^{-i\omega_1 t} \cos k_1 x$ and $\Psi_2(x, t) = A_2 e^{-i\omega_2 t} \sin k_2 x$ each solutions to the Schrödinger equation?

14-7 Show that for a free particle

$$\Psi(x, t) = A \cos\left(kx - \frac{E}{\hbar}t\right) + iA \sin\left(kx - \frac{E}{\hbar}t\right)$$

where $k = \sqrt{2mE}/\hbar$ is a solution to the Schrödinger equation.

14-8 For an electron with a de Broglie wavelength of 1.0 Å, determine (a) the group velocity, (b) the phase velocity (de Broglie), and (c) the phase velocity (Schrödinger).

RECOMMENDED READING

BEISER, Arthur, *The World of Physics*, McGraw-Hill, New York, 1960, pp. 195-208.

BORN, Max, *Physics in My Generation*, Springer-Verlag, New York, 1969.
    A collection of readable essays that give some insight into the nature of physics as well as of the author, Max Born.

DIRAC, P. A. M., "The Evolution of the Physicists' Picture of Nature," *Sci. Am.*, May 1963.

HEISENBERG, Werner, *The Physical Principles of the Quantum Theory*, Dover, New York, 1930.

SPOSITO, Garrison, *An Introduction to Quantum Mechanics*, Wiley, New York, 1970.
    A very clear description of the subject of quantum mechanics at an intermediate level.

15
The Schrödinger Equation II

Max Born
(1882–1970)
A native of Breslau, Prussia, Born taught at the universities of Göttingen, Cambridge, and Edinburgh. His research helped to develop quantum mechanics and also contributed to the study of crystallography, atomic structure, and the kinetic theory of fluids; he shared in the formulation of the Born–Oppenheimer theory of molecules. In discussing the political responsibilities of scientists, Born has condemned warlike uses of scientific knowledge. With Walter Bothe, he received the 1954 Nobel Prize for his statistical interpretation of wave functions.

15–1   THE HAMILTONIAN
15–2   OPERATORS
15–3   THE POTENTIAL WELL

15–1 THE HAMILTONIAN

To recapitulate somewhat, let us remind ourselves that we arrived at a form of Schrödinger's Equation (14–16) that did not contain a time dependence explicitly,

$$-\frac{\hbar^2}{2m}\frac{d^2\psi}{dx^2} + V\psi = E\psi \tag{15-1}$$

where $\psi = \psi(x)$ is the wave function, $V = V(x)$ is the potential energy, and E is the total energy. The equation is applicable only to conservative fields, that is, those in which the total energy

$$E = \tfrac{1}{2}mv^2 + V = \text{constant} \tag{15-2}$$

is a constant of the motion. When the kinetic energy is expressed in terms of the momentum p, rather than in terms of the speed v, Equation (15–2) is written

$$H = \frac{p^2}{2m} + V = E = \text{constant} \tag{15-3}$$

In classical mechanics, the function $H(p, x)$ is called the *Hamiltonian* of the system. Since only functions of position appear in Equation (15–1), this is commonly called the "steady state" or *time-independent* form of Schrödinger's wave equation. It calls to mind the picture of a standing wave of some sort. If, for a particular case, this equation has a solution $\psi = \psi(x)$, then we may suspect that the wave is reflected back and forth, somehow, to set up a standing wave. There must be two places along the x axis that act as mirrors for the ψ wave. The particle (which, after all, is being described by the function) must be bouncing elastically back and forth between two walls.

It is possible, of course, to have a standing wave system where the amplitude is dying out with time. In such cases it is better to use the general Equation (14–7),

$$-\frac{\hbar^2}{2m}\frac{\partial^2\Psi(x, t)}{\partial x^2} + V\Psi(x, t) = i\hbar\frac{\partial\Psi(x, t)}{\partial t} \tag{15-4}$$

but with the wave function separated as in Equation (14–13),

$$\Psi(x, t) = \psi(x)e^{-iEt/\hbar} \tag{15–5}$$

If the particle is moving in a conservative field of force, then the potential function $V = V(x)$ is time-independent, the total energy $H = (p^2/2m) + V = E$ is a constant, and Equation (15–1) must be used. In the particular case of a free particle, $V(x)$ is a constant that may be arbitrarily set equal to zero. The frequency and wavelength of the associated wave are given by the equations $\nu = E/h$ and $\lambda = h/p$.

15–2 OPERATORS

The forms of the wave equation that we have so far described may be very conveniently cast into operator forms. An *operator*, in general, is any expression that acts on a function within some domain to produce new values occupying a given range. For example, the number 2, acting as a multiplier, transforms every value in the domain of a function into twice that value in the range. It is then correct to say that 2 is an *arithmetic operator*. The *differential operator* d/dx applied to a function $f(x)$ in the usual sense transforms every value of $f(x)$ into the value of its derivative $(d/dx)f(x) = f'(x)$. Now, inspection of the time-independent Equation (15–1) shows that it may be written as

$$\left(-\frac{\hbar^2}{2m}\frac{d^2}{dx^2} + V\right)\psi = E\psi \tag{15–6}$$

The full expression in parentheses on the left-hand side can now be defined as an operator,

$$H = -\frac{\hbar^2}{2m}\frac{d^2}{dx^2} + V \tag{15–7}$$

called the *Hamiltonian operator* because of its similarity with the *Hamiltonian function* of classical mechanics [see Equation (15–3)]. Then the time-independent Equation (15–6) is written

$$H\psi = E\psi \tag{15–8}$$

The function ψ cannot be cancelled out of this equation, because H is not a simple scalar multiplier, while E is the value of an energy. Equation (15–8) must be interpreted as

(operator H) acting on function ψ
$\qquad\qquad\qquad\qquad\qquad$ = (total energy) multiplying function ψ

In the three-dimensional case, Equation (15-1) must be written as

$$-\frac{\hbar^2}{2m}\left(\frac{\partial^2 \psi}{\partial x^2} + \frac{\partial^2 \psi}{\partial y^2} + \frac{\partial^2 \psi}{\partial z^2}\right) + V\psi = E\psi \qquad (15-9)$$

where $\psi = \psi(x, y, z)$, $V = V(x, y, z)$, and the three-dimensional *Hamiltonian operator* is

$$\boxed{H = -\frac{\hbar^2}{2m}\left(\frac{\partial^2}{\partial x^2} + \frac{\partial^2}{\partial y^2} + \frac{\partial^2}{\partial z^2}\right) + V} \qquad (15-10)$$

Let us now compare Equations (15-3) and (15-6). They are the same, provided that both sides of Equation (15-3) are interpreted as operators and each is operating on the wave function ψ—and if the momentum p, as it appears in Equation (15-3), is defined as the operator

$$\boxed{p = \frac{\hbar}{i}\frac{d}{dx}} \qquad (15-11)$$

In view of the more general operator H in three dimensions, Equation (15-10), we must use a partial derivative operator to represent that particular component of the momentum **p**, and write

$$\boxed{p_x = \frac{\hbar}{i}\frac{\partial}{\partial x}} \qquad (15-12)$$

The other two components of the momentum operator are

$$p_y = \frac{\hbar}{i}\frac{\partial}{\partial y} \quad \text{and} \quad p_z = \frac{\hbar}{i}\frac{\partial}{\partial z}$$

Returning for a moment to the general time-dependent, one-dimensional Schrödinger Equation (15-4), we can, by a simple algebraic manipulation, write

$$-\frac{\hbar^2}{2m}\frac{\partial^2 \Psi}{\partial x^2} + V\Psi = -\frac{\hbar}{i}\frac{\partial \Psi}{\partial t} \qquad (15-13)$$

or

$$\left(-\frac{\hbar^2}{2m} + V\right)\Psi = -\frac{\hbar}{i}\frac{\partial \Psi}{\partial t} \qquad (15-14)$$

But $H = -\hbar^2/2m + V$ is the Hamiltonian operator, hence in a more abbreviated form the time-dependent Schrödinger equation may be written

$$\boxed{H\Psi = -\frac{\hbar}{i}\frac{\partial \Psi}{\partial t}} \qquad (15-15)$$

A comparison of Equations (15–8) and (15–15) suggests that we can now define an energy operator by the expression

$$E = -\frac{\hbar}{i}\frac{\partial}{\partial t} \quad (15\text{–}16)$$

So, in operator form the time-dependent Schrödinger equation can be written

$$H\Psi = E\Psi \quad (15\text{–}17)$$

which is very similar to the time-independent form, Equation (15–8).

Again, the function $\Psi(x, t)$ cannot be cancelled out in both terms of Equation (15–17), because the meaning of the equation is

(operator H) acting on function Ψ = (operator E) acting on function Ψ

and both H and E are not simple scalar multipliers.

If the general Equation (15–15) describes a free particle that experiences no forces anywhere, then the total energy E may have any value. This results in an infinite number of possible solutions $\Psi_i(x, y, z, t)$ of the wave equation. If the free particle is contained in some finite enclosure of rigid walls, this represents a physical situation that is time-independent, and Equation (15–8), $H\psi = E\psi$, where $\psi(x, y, z)$ is a function depending only on position, should then be used. In this case, only particular values of the energy E_i are allowed, and these values correspond to particular solutions $\psi_i(x, y, z)$ of the wave equation. They are typified by systems of standing waves, where each wavelength and frequency corresponds to a different solution ψ_i. The allowed solutions ψ_i are often called *eigenfunctions*, and the corresponding energies E_i are called *eigenvalues*.

15–3 THE POTENTIAL WELL

As a simple application of the steady-state Schrödinger equation, let us consider the case of a particle trapped in an infinitely deep *potential well*. Imagine this well as having zero potential along some finite region of the x axis and as having an infinitely large potential everywhere else on that axis (see Figure 15-1). We may visualize this situation as describing a particle moving along the x axis inside a box that has infinitely hard, perfectly rigid walls from which the particle rebounds elastically.

In terms of the boundary conditions imposed by the problem, the potential function is

$$\begin{aligned} V &= 0 \quad \text{for } 0 < x < L \\ V &= \infty \quad \text{for } x \leq 0 \\ V &= \infty \quad \text{for } x \geq L \end{aligned} \quad (15\text{–}18)$$

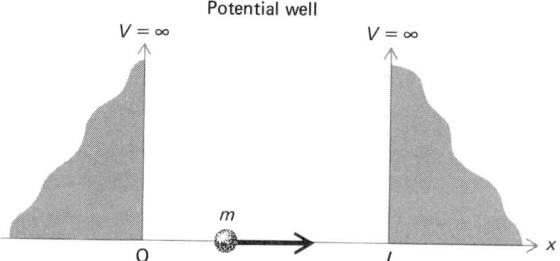

Figure 15-1 A particle of mass m is constrained to move in one dimension in a potential well. The particle rebounds perfectly elastically from the walls of the infinite potential.

As the problem has been set up, there is certainty that the particle is inside the box, and of course there is no possibility that the particle is anywhere outside the box. This sets the conditions on the wave function such that

$$\psi(x) = 0 \quad \text{for} \quad \begin{cases} x \leq 0 \\ x \geq L \end{cases} \tag{15-19}$$

and

$$\int_0^L \psi^*\psi \, dx = 1$$

It is not known exactly where the particle is at any moment inside the box, so no time-dependent data will be used. Recalling the time-independent Schrödinger Equation (15-1), and making $V = 0$ according to conditions (15-18), we obtain

$$-\frac{\hbar^2}{2m} \frac{d^2}{dx^2} \psi(x) = E\psi(x)$$

which we may write as

$$\frac{d^2\psi(x)}{dx^2} + \alpha^2 \psi = 0 \tag{15-20}$$

where

$$\alpha^2 = \frac{2mE}{\hbar^2} \tag{15-21}$$

Equation (15-20) describes the situation of the particle in the box. The equation has the solution

$$\psi(x) = Ae^{+i\alpha x} + Be^{-i\alpha x} \tag{15-22}$$

which represents the superposition of two waves in the box, each traveling in different directions along the x axis. This is just the condition necessary for standing waves if taken along with proper boundary conditions (reflecting walls). It is helpful to verify that Equation (15–22) is a solution of the Schrödinger Equation (15–20).

The boundary conditions, Equation (15–19), can now be used to evaluate the constants A and B in Equation (15–22). For $\psi(x) = 0$ at $x = 0$, the equation becomes

$$0 = A + B$$

and

$$A = -B$$

Then

$$\psi(x) = A(e^{i\alpha x} - e^{-i\alpha x})$$

or, by Euler's relation,

$$\psi(x) = 2iA \sin \alpha x$$

The second boundary condition, $\psi(x) = 0$ for $x = L$, gives

$$0 = 2iA \sin \alpha L$$

or, since $A \neq 0$,

$$\sin \alpha L = 0$$

or

$$\alpha L = n\pi$$

and

$$\alpha = \frac{n\pi}{L} \quad \text{for } n = 1, 2, 3, \ldots$$

The energy from Equation (15–21) is then, for each value of n,

$$\boxed{E_n = \frac{\hbar^2 \alpha^2}{2m} = \frac{n^2 \pi^2 \hbar^2}{2mL^2} \quad n = 1, 2, 3, \ldots} \tag{15-23}$$

The particle can have only those values of energy given by Equation (15–23). We express this by saying that *the energy is quantized into discrete values* or *levels* and that the particle is in any one of a number of discrete states available to it. It can, of course, have only one value of these possible values at any time. To take on another value of energy, it must receive or lose some of its energy. In either case, the amount received or lost must be just sufficient to put the particle in another one of the possible states.

Note also that the particle cannot have zero energy. The minimum possible value given by Equation (15–23) is obtained when $n = 1$, or

$$\boxed{E_1 = \frac{\pi^2 \hbar^2}{2mL^2}} \tag{15-24}$$

and the other values of energy are $4E_1$, $9E_1$, $16E_1$, ... corresponding to $n = 2, 3, 4, \ldots$. For this minimum value E_1 to be appreciably different from zero, however, the product mL^2 must be small and of the order of $\hbar^2$. Since $h = 6.625 \times 10^{-34}$ J-sec, the magnitude of the denominator must be very small. The value E_1 given by Equation (15–24) is called the *zero point energy*. In other words, the particle cannot have a zero energy. This conclusion contradicts classical mechanics because it is a result of the uncertainty principle. It is possible to see clearly the reason: Since the particle is bounded by an infinite potential, its position is known within an uncertainty $\Delta x \cong L$, hence, according to Heisenberg's principle, the uncertainty in its momentum must be $\Delta p \geq \hbar/L$. Therefore, according to this principle the energy can never be zero, because this would imply that $\Delta p = 0$.

The linear momentum conjugate of any one of the allowed values E_n is obtained by writing

$$E_n = \frac{n^2 \pi^2 \hbar^2}{2mL^2} = \frac{p_n^2}{2m}$$

or

$$\boxed{p_n = \frac{n\pi\hbar}{L} \quad n = 1, 2, 3, \ldots} \tag{15-25}$$

and so the momentum is also quantized into discrete allowed values. Again, it is seen that the dimension of the box must be very small. The student should satisfy himself that the dimensions of the right-hand side of Equation (15–23) are those of an energy and that the dimensions of Equation (15–25) are those of a linear momentum.

The expression for the wave function ψ is now

$$\psi = 2iA \sin\left(\frac{n\pi}{L} x\right)$$

so

$$\psi^* = -2iA \sin\left(\frac{n\pi}{L} x\right)$$

and the probability density is

$$\boxed{\psi^*\psi = 4A^2 \sin^2\left(\frac{n\pi}{L} x\right)} \tag{15-26}$$

Using the normalizing condition Equation (14–5), which expresses certainty that the particle is somewhere inside the box, we write

$$\int_0^L \psi^*(x)\psi(x)\, dx = \int_0^L 4A^2 \sin^2\left(\frac{n\pi}{L} x\right) dx = 1$$

Integration then gives

$$4A^2 \int_0^L \sin^2\left(\frac{n\pi}{L} x\right) dx = 2A^2 \left[x - \frac{L}{2n\pi} \sin\left(\frac{2n\pi}{L} x\right) \right]_0^L = 2A^2 L$$

Since this must equal 1, the constant is evaluated to be $A = 1/\sqrt{2L}$, and the normalized eigenfunctions are, then,

$$\psi_n(x) = i \frac{2}{\sqrt{2L}} \sin\left(\frac{n\pi x}{L}\right)$$

or

$$\boxed{\psi_n(x) = i \sqrt{\frac{2}{L}} \sin\left(\frac{n\pi x}{L}\right)} \qquad (15\text{--}27)$$

Thus, for the case of our particle in an infinitely deep potential well (in a box with perfectly rigid and perfectly reflecting walls), the probability for finding it inside a small interval given by $x_1 = a$ and $x_2 = b$, where the interval lies completely inside the box, is

$$\int_a^b \psi^* \psi \, dx = \int_a^b \frac{2}{L} \sin^2\left(\frac{n\pi}{L} x\right) dx \qquad (15\text{--}28)$$

The results of the particle-in-the-box problem are summarized in Table 15–1.

Table 15–1

n	EIGENFUNCTION, $\psi(x)$	PROBABILITY DENSITY, $\psi^*\psi$	ENERGY EIGENVALUE, E_n
1	$i\sqrt{\dfrac{2}{L}} \sin \dfrac{\pi x}{L}$	$\dfrac{2}{L} \sin^2 \dfrac{\pi x}{L}$	$\dfrac{\pi^2 \hbar^2}{2mL^2}$
2	$i\sqrt{\dfrac{2}{L}} \sin \dfrac{2\pi x}{L}$	$\dfrac{2}{L} \sin^2 \dfrac{2\pi x}{L}$	$\dfrac{4\pi^2 \hbar^2}{2mL^2}$
3	$i\sqrt{\dfrac{2}{L}} \sin \dfrac{3\pi x}{L}$	$\dfrac{2}{L} \sin^2 \dfrac{3\pi x}{L}$	$\dfrac{9\pi^2 \hbar^2}{2mL^2}$
4	$i\sqrt{\dfrac{2}{L}} \sin \dfrac{4\pi x}{L}$	$\dfrac{2}{L} \sin^2 \dfrac{4\pi x}{L}$	$\dfrac{16\pi^2 \hbar^2}{2mL^2}$
⋮	⋮	⋮	⋮
n	$i\sqrt{\dfrac{2}{L}} \sin \dfrac{n\pi x}{L}$	$\dfrac{2}{L} \sin^2 \dfrac{n\pi x}{L}$	$\dfrac{n^2 \pi^2 \hbar^2}{2mL^2}$

Figure 15–2 shows graphs of the probability density (the probability per unit length), for $n = 1, 2, 3, \ldots$. For the $n = 1$ eigenfunction, the probability of finding the particle at $x = L/2$ is greater than at any other position. Note, however, that

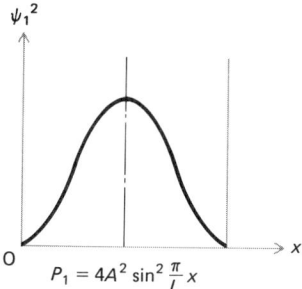

$P_1 = 4A^2 \sin^2 \frac{\pi}{L} x$

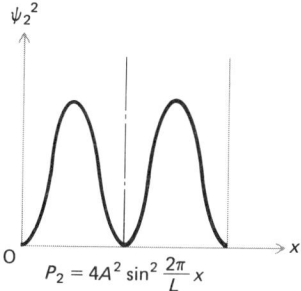

$P_2 = 4A^2 \sin^2 \frac{2\pi}{L} x$

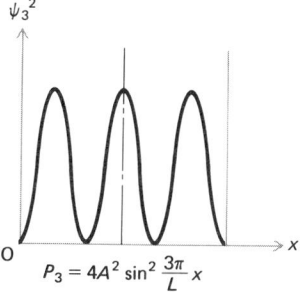

$P_3 = 4A^2 \sin^2 \frac{3\pi}{L} x$

Figure 15-2 Probability densities for the first three wave functions for a particle in a rigid box.

for the $n = 2$ eigenfunction, the probability of finding the particle at $x = L/2$ is zero. For the energy E_2, it is impossible for the particle to be at $x = L/2$.

EXAMPLE 15-1: The following computer program, written in the BASIC language, is designed to evaluate by Simpson's rule* the integral approximated by

$$\int_a^b f(x)\,dx = \frac{h}{3}\{f(a) + 4f(a + h) + 2f(a + 2h) + \cdots + 4f[a + (2n - 1)h] + f(b)\}$$

* See, for example, Ivan S. Sokolnikoff and R. M. Redheffer, *Mathematics of Physics and Modern Engineering*, McGraw-Hill, New York, 1958, pp. 715-720.

This program will be used to evaluate the integral of Equation (15–28), where $f(x)$ is identified in the program as FNF(X). The inputs include the width of the well (which is actually normalized so that any value will do), the quantum number associated with the energy of the particle, and the limits of integration. In statement 30, below A and B, are the fractions of the total width of the well. For example, for integration from $X_A = 0.49L$ to $0.51L$, A and B are input as 0.49 and 0.51, respectively. The program computes the probability of finding the particle within the limits and the probability per unit length for various quantum numbers.

The BASIC program for the computation of probabilities of finding the particle in a given interval and the corresponding probabilities per unit length for any quantum number is given below:

```
10   PRINT"INFINITE PØTENTIAL WELL ØF WIDTH X = 0 TØ X = ";
15   INPUT L
20   PRINT"QUANTUM NUMBER,N = ";
25   INPUT N
30   PRINT"LIMITS ØF INTEGRATIØN, A TØ B –";
35   INPUT A,B
40   LET A = A*L
45   LET B = B*L
50   LET H = (B–A)/(20*N)
55   DEF FNF(X) = (2/L)*(SIN(N*3.14159*X/L))↑2
60   LET R = 0
65   FØR K = 1 TØ (20*N – 1)
70   LET R = R + ((3 – (–1)↑K))*FNF(A + K*H)
75   NEXT K
80   LET S = H*(R + FNF(A) + FNF(B))/3
85   PRINT"PRØBABILITY ØF FINDING THE PARTICLE WITHIN"
90   PRINT"X(A) = ";(A/L);"L AND X(B) = "(B/L);"L IS";S
95   PRINT"PRØB/UNIT LENGTH = "; S*L/(B – A)
100  PRINT
105  GØ TØ 20
999  END
```

PRINTØUTS FØR X(A) = 0.49L, X(B) = 0.51L FØR
QUANTUM NUMBERS N = 1 AND N = 2

INFINITE PØTENTIAL WELL OF WIDTH X = 0 TØ X= ?4E–10
QUANTUM NUMBER, N = ?1
LIMITS ØF INTEGRATIØN, A TØ B – ?.49, .51
PRØBABILITY ØF FINDING THE PARTICLE WITHIN
X(A) = 0.49 L AND X(B) = 0.51 L IS 3.99868E–2
PRØB/UNIT LENGTH = 1.99934

QUANTUM NUMBER,N = ?2
LIMITS ØF INTEGRATIØN, A TØ B –? .49, .51
PRØBABILITY ØF FINDING THE PARTICLE WITHIN
X(A) = 0.49 L AND X(B) = 0.51 L IS 5.25963E–5
PRØB/UNIT LENGTH = 2.62981E–3

Table 15–2 shows the probabilities per unit length for the interval $X_A = 0.49L$ to $X_B = 0.51L$ as functions of the quantum numbers. Compare these values with Figure 15–2.

Table 15–2

QUANTUM NUMBER, N	PROBABILITY PER UNIT LENGTH
1	1.9993
2	0.0026
3	1.9941
4	0.0105
5	1.9836
25	1.6366
26	0.3891
99	0.9899
100	1.0000

PROBLEMS

$$\hbar = 1.054 \times 10^{-34} \text{ J-sec}$$

15–1 For the potential well problem in Section 15–3, assume that the particle is an electron confined within a dimension $L = 2.0$ Å. Determine for this particle (a) the smallest possible energy E_1 that it may have in electron volts, (b) the difference in energy between the smallest energy E_1 and the next highest energy E_2, $\Delta E = E_2 - E_1$, and (c) the wavelength of a photon with energy ΔE.

15–2 For an electron confined within a dimension $L = 2.0$ Å, calculate (a) the smallest value of the linear momentum, and (b) the percent uncertainty in momentum of an electron within the box.

15–3 If the particle in the potential well is a grain of sand with a mass of 1.0×10^{-7} kg confined within a dimension $L = 1.0$ mm, determine (a) the smallest energy E_1 in electron volts, and (b) the difference in energy between the smallest energy E_1 and the next highest energy E_2 ($\Delta E = E_2 - E_1$). Compare this value with that in Problem 15–1.

15–4 Find an approximate value of n for (a) an electron confined to a box of length $L = 5.0$ Å and moving with a velocity of 7.3×10^6 m/sec, (b) an oxygen molecule ($m = 5.3 \times 10^{-26}$ kg) confined to a box of length 10,000 Å and moving with a speed of 460 m/sec, and (c) a particle of mass 1.0×10^{-6} kg confined to a box of length $L = 1.0$ mm and moving with a velocity of 0.0010 m/sec.

15-5 Use the BASIC program in Example 15-1 to evaluate the probability and probability per unit length of finding a particle of energy E_3 in intervals of $0.10L$ from $X = 0$ to $X = L$. Now compare the results of the probability and the probability per unit length for the intervals $X_A = 0.30L$ to $X_B = 0.36L$ and $X_A = 0.499L$ to $X_B = 0.501L$. Why is the probability per unit length greater in the latter case?

15-6 (a) For an infinite potential well, use Equation (15-27) to determine the probability of finding an electron in the situations given below.

QUANTUM NUMBER, N	INTERVAL
1	$0 \rightarrow \frac{1}{2}L$
1	$\frac{1}{4}L \rightarrow \frac{3}{4}L$
2	$0 \rightarrow \frac{1}{2}L$
2	$\frac{1}{4}L \rightarrow \frac{3}{4}L$

(b) Find the probability per unit length corresponding to the midpoints of the intervals for the quantum numbers given.

RECOMMENDED READING

BORN, Max, *Physics in My Generation*, Springer-Verlag, New York, 1956.
See particularly p. 140, "The Interpretation of Quantum Mechanics."

FONG, Peter, *Elementary Quantum Mechanics*, Addison-Wesley, Reading, Mass., 1962.
Quantum mechanics has been developed close to classical mechanics to show that it is a natural extension and to emphasize the physical picture of quantum mechanics rather than the mathematical abstraction.

PLANCK, M., *Scientific Autobiography*, Philosophical Library, New York, 1949, pp. 43-46.

SCHRÖDINGER, E., *Collected Papers on Wave Mechanics*, Blackie, Glasgow, 1928.
An interesting account of Schrödinger's early work on quantum mechanics.

SCHRÖDINGER, E., *What is Life? and Other Scientific Essays*, Doubleday, Garden City, N.Y., 1956.
Every physics student should read these thought-provoking essays.

SCHRÖDINGER, E., et al., *Letters on Wave Mechanics*, M. L. Klein (Ed.), Philosophical Library, New York, 1967.
A very well selected set of personal letters written by these outstanding physicists concerning the birth of quantum mechanics.

SHERWIN, Chalmers W., *Introduction to Quantum Mechanics*, Holt, Rinehart & Winston, New York, 1959.
A popular, readable text aimed at the undergraduate level. Many problems are solved by numerical methods to reach a basic understanding without advanced mathematics.

16
Some Applications of the Schrödinger Equation

Werner Karl Heisenberg
(1901–)

Born in Würzburg, Germany, Heisenberg studied at the University of Munich under Arnold Sommerfeld. He worked three years at Copenhagen with Niels Bohr, and was later director of the Max Planck Institute for Physics at Göttingen. He is considered the founder of quantum mechanics (1925). In 1927 Heisenberg developed the uncertainty principle, which implies that the less space occupied by an electron, the larger its energy must be. For his prediction of two allotropic forms of hydrogen and the matrix formulation of quantum mechanics, he was awarded the Nobel Prize in 1932

16–1 THE CLASSICAL HARMONIC OSCILLATOR
16–2 THE QUANTUM MECHANICAL HARMONIC OSCILLATOR
16–3 THE TUNNEL EFFECT

16-1 THE CLASSICAL HARMONIC OSCILLATOR

Classical mechanics is a special case of the more general quantum mechanics. A simple yet striking example of the contrast between the two "mechanics" is furnished by the treatment of the motion of the harmonic oscillator. The idealized harmonic oscillator problem is one of the few cases that can be treated completely using Schrödinger's equation, and it supplies a valuable first approximation method for more complex problems, such as the treatment of vibrational energy of molecules. (Strictly speaking, the only problem that can be treated exactly by quantum mechanics is the free particle problem.)

As a brief review of the classical treatment of a simple oscillator, consider the particle of mass m in Figure 16–1. It will execute simple harmonic motion when it is displaced a distance x from 0 and is acted upon by the force

$$F = -kx \tag{16-1}$$

where k is a constant and F is the magnitude of a vector always directed toward the fixed point at 0. When Newton's second law is applied this becomes

$$m \frac{d^2x}{dt^2} = -kx \tag{16-2}$$

which can also be written

$$m \frac{dx}{dt} \frac{d^2x}{dt^2} dt = -kx \, dx$$

Integration then gives

$$\boxed{\tfrac{1}{2}mv^2 + \tfrac{1}{2}kx^2 = \text{constant} = E} \tag{16-3}$$

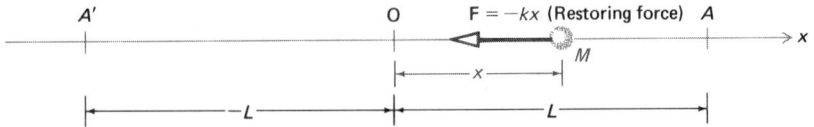

Figure 16–1 Linear harmonic oscillator.

CHAPTER 16: SOME APPLICATIONS OF THE SCHRÖDINGER EQUATION · 217

The first term is the kinetic energy of the particle,

$$K = \tfrac{1}{2}mv^2 \tag{16-4}$$

And the second is the potential energy,

$$V = \tfrac{1}{2}kx^2 \tag{16-5}$$

so that the total mechanical energy of the system is a constant:

$$\boxed{K + V = E = \text{constant}} \tag{16-6}$$

For any finite value, the particle will oscillate between two points, say A at $x = L$ and A' at $x = -L$. Since E can have any value depending on x and v, the spectrum of allowed values for E is continuous.

If we define

$$\omega^2 = \frac{k}{m} \tag{16-7}$$

then Equation (16-2) can be written as

$$\frac{d^2x}{dt^2} + \omega^2 x = 0 \tag{16-8}$$

This is strikingly similar in form to the wave Equation (15-14) and, like that expression, has a solution

$$x = Ae^{i\omega t} + Be^{-i\omega t} \tag{16-9}$$

The values of the constants A and B can be determined from some initial values of position and velocity. Euler's relationship allows the solution to be written as

$$x = C \cos \omega t + D \sin \omega t \tag{16-10}$$

This is an equation of motion giving the position of the particle as a function of time. The speed of the particle at any instant is

$$v = \frac{dx}{dt} = -C\omega \sin \omega t + D\omega \cos \omega t \tag{16-11}$$

Let the particle be at $x = L$ at time $t = 0$ and have a velocity at that instant $v = 0$. With these initial conditions, Equations (16-10) and (16-11) yield $C = L$ and $D = 0$. These two equations in final form are

$$x(t) = L \cos \omega t \tag{16-12}$$

and

$$v(t) = -L\omega \sin \omega t \tag{16-13}$$

and the total energy is

$$E = \tfrac{1}{2}mL^2\omega^2 \sin^2 \omega t + \tfrac{1}{2}kL^2 \cos^2 \omega t \tag{16-14}$$

The maximum speed occurs as the particle passes through the origin at $x = 0$ and is $v_{max} = \omega L$. At the origin the potential energy is zero, and the total energy is

$$\boxed{E = \tfrac{1}{2}mv_{max}^2 = \tfrac{1}{2}m\omega^2 L^2} \tag{16-15}$$

When the particle is at either bound A or A', however, the kinetic energy is zero when the speed $v = 0$, so that the total energy is then contained in the potential energy term; hence,

$$\boxed{E = \tfrac{1}{2}kx_{max}^2 = \tfrac{1}{2}kL^2} \tag{16-16}$$

16–2 THE QUANTUM MECHANICAL HARMONIC OSCILLATOR

Treatment of the same mechanical system by the methods of wave mechanics involves the solution of Schrödinger's equation as applied to the system. We must, therefore, set up the equation to describe the same harmonic oscillator as we have treated classically. We should realize that the wave function, however, is not localized at some point along the x axis, and so no certain location for the particle at any given time can be defined. The product $\psi^*\psi$ yields the *probability density* for finding the particle in any small region dx along x. Thus, a force equation that is a function of the position cannot be used to set up the problem as was done for the classical oscillator by using Equation (16–1). In fact, the concept of force loses its relevance in the quantum mechanical model, yet the concepts of momentum and energy still prevail. For the same reasons, we should not expect to find results giving the position of the particle as a function of time as given for the classical oscillator by Equation (16–12), nor for the velocity as a function of time as given by Equation (16–13).

However, the energy of the system may be considered, since this quantity appears in terms of potential energy as a function of x in both treatments. In the classical case, the potential energy given by Equation (16–5) results from applying Newton's laws to the force equation. In the quantum mechanical treatment, however, the potential energy function

$$V = \tfrac{1}{2}kx^2 \tag{16-17}$$

is an initial and primary condition imposed on the mechanical system. This condition, then, sets up the problem by defining $V(x)$.

Let us recall that in the classical case, Equation (16–16) sets a maximum displacement L ($=x_{max}$) on the particle and equates the *total* energy to the potential energy of the particle at this maximum displacement. This cannot be used for

CHAPTER 16: SOME APPLICATIONS OF THE SCHRÖDINGER EQUATION

defining x_{max} in terms of the total energy, and so unless new conditions are imposed on $V(x)$, it must extend in its defined form in Equation (16–17) to both $x = -\infty$ and $x = +\infty$. However, it would not do to have a wave function ψ that remains finite at infinity, for then it could not be normalized to yield a probability for finding the particle in finite regions of space. So the condition is imposed that the wave function ψ must vanish at infinite distances from the origin.

The picture now is no longer that of a particle tied to a point by an elastic force proportional to its displacement from the point. The picture is, rather, that of a wave system contained in a sort of potential energy bottle or "well" whose shape is given by Equation (16–17). We may discuss the probabilities for finding the particle in various regions in and around the well, and ask for its energy in all possible circumstances as contained in the well. As for any wave system that finds itself bounded by some means on all sides, it is not surprising to find that this also takes the form of standing waves within the well.

Thus, the problem is to find solutions ψ_n that are *eigenfunctions* representing the various possible standing wave systems, and to find the corresponding energies E_n that are the *eigenvalues*. The procedure outlined here for the harmonic oscillator in one dimension is easily generalized to any number of dimensions and is representative of potential well problems, which occur so often in quantum mechanics.

When the shape of our potential well is the potential energy associated with the classical harmonic oscillator, $V = \frac{1}{2}kx^2$, the Schrödinger equation then takes the form

$$-\frac{\hbar^2}{2m}\frac{d^2\psi}{dx^2} + \frac{kx^2}{2}\psi = E\psi \qquad (16\text{–}18)$$

It is interesting to note that this wave equation and the problem of its solution were well known to mathematicians before physicists applied it to real physical systems. The detailed technique for its solution is a bit tedious and somewhat beyond the scope of this text, so we shall pass immediately to the resulting eigenvalues and eigenfunctions.*

The eigenvalues (permitted values of the total energy) are given by

$$E_n = (n + \tfrac{1}{2})\hbar\omega = (n + \tfrac{1}{2})h\nu \qquad (16\text{–}19)$$

where $\omega = 2\pi\nu$ and $n = 0, 1, 2, 3, \ldots$. This spectrum of energy values is a discrete one, as distinct from the continuous spectrum permitted by the classical mechanics. The difference between energy levels in this spectrum is $h\nu$.

In what sense, then, can the classical mechanics be considered as a particular example of a more general quantum mechanics? The answer lies in considering

---

* For a detailed solution to the harmonic oscillator problem, see, for instance, L. Pauling and E. B. Wilson, *Introduction to Quantum Mechanics*, McGraw-Hill, New York, 1935, Chap. 3.

the particular application. Suppose, for instance, that we are treating a mechanical gadget such as a cymbal, or the column of air in an organ pipe, or a tuning fork. Then the frequency lies somewhere in the region of 1000–10,000 Hz, and the energy of the vibrating system may be several joules. The separation between permitted energy levels in building such gadgets is still $h\nu$, and since h is about 6.63×10^{-34} J-sec, the separation between levels is in the range of 10^{-30} J. Compared to the total energy involved, the separation between levels is so small as to appear effectively zero, so the spectrum of permitted tones appears to be continuous.

At atomic and nuclear dimensions, however, the frequencies may be greatly in excess of 10^{12} Hz, and the energy of the system may be 10^{-24} J or less. In these cases, the separation between levels ($h\nu = 6.63 \times 10^{-34} \times 10^{12} = 6.63 \times 10^{-22}$ J) becomes very pronounced, and the spectrum of allowed energy levels is noticeably discrete. It should be remembered that such discrete energy spectra are obtained only when the quantum mechanical system is somehow bounded. A "free" particle—one that lies in no field of force and is under the influence of no changing potential energy functions—may take on any value of energy whatsoever and so has a truly continuous energy spectrum.

Another surprising result is obtained from the quantum mechanical oscillator. It can *not* have zero energy. Equation (16-19) does not permit E to take the value of 0 as its lowest one. This sets the *zero point energy* as just $\frac{1}{2}h\nu$. [A similar situation was discussed in Chapter 15: See the explanation given in connection with Equation (15-23).]

Table 16-1 lists some of the energy eigenvalues E_n and the corresponding eigenfunctions ψ_n for different values of n.

The polynomials

$$H_0 = 1$$

$$H_1 = 2\alpha x$$

$$H_2 = 4\alpha^2 x^2 - 2$$

$$H_3 = 8\alpha^3 x^3 - 12\alpha x$$

$$\vdots$$

$$H_n = (-1)^n e^{\xi^2} \frac{d^n}{d\xi^n} (e^{-\xi^2})$$

where $\xi = \alpha x$, that appear in the corresponding eigenfunctions are called *Hermitian* or *Hermite polynomials* and are discussed in a number of mathematical textbooks.*

The probability per unit length for finding the particle in a given region along the x axis is given by $\psi^*\psi$ or, in usual notation, $|\psi|^2$. The values of this probability

---

*See, for example, M. R. Spiegel, *Mathematical Handbook of Formulas and Tables*, McGraw-Hill, New York, 1968, p. 151.

CHAPTER 16: SOME APPLICATIONS OF THE SCHRÖDINGER EQUATION · 221

Table 16–1 Eigenvalues and eigenfunctions for the harmonic oscillator

n	ENERGY EIGENVALUES, E_n	NORMALIZED EIGENFUNCTIONS, $\psi_n(x)$
0	$E_0 = \frac{1}{2}h\nu$	$\psi_0 = \left(\dfrac{\alpha}{\sqrt{\pi}}\right)^{1/2} e^{-\alpha^2 x^2/2}$
1	$E_1 = \frac{3}{2}h\nu$	$\psi_1 = \left(\dfrac{\alpha}{2\sqrt{\pi}}\right)^{1/2} 2\alpha x e^{-\alpha^2 x^2/2}$
2	$E_2 = \frac{5}{2}h\nu$	$\psi_2 = \left(\dfrac{\alpha}{8\sqrt{\pi}}\right)^{1/2} (4\alpha^2 x^2 - 2)e^{-\alpha^2 x^2/2}$
3	$E_3 = \frac{7}{2}h\nu$	$\psi_3 = \left(\dfrac{\alpha}{48\sqrt{\pi}}\right)^{1/2} (8\alpha^3 x^3 - 12\alpha x)e^{-\alpha^2 x^2/2}$
⋮	⋮	⋮
n	$E_n = (n + \frac{1}{2})h\nu$	$\psi_n = \left(\dfrac{\alpha}{\sqrt{\pi}2^n n!}\right)^{1/2} H_n(\alpha x)e^{-\alpha^2 x^2/2}$

where
$$\alpha^2 = \frac{4\pi^2 m\nu}{h}$$
and
$$H_{n+1} = 2(\alpha x)H_n - 2nH_{n-1}$$

density for a few values of permitted energies are plotted in Figure 16–2 along the potential energy function $V(x)$, which sets the potential well for the oscillator. The points A and A', B and B', and so on, represent those points where the potential energy equals the permitted total energy for that value of the quantum number n. A classical oscillator would not be found "outside" these points according to Equation (16–16). In the quantum mechanical case, the probability density has finite values beyond these limits, and so there is a small but finite probability of finding the particle in regions outside of the potential well.

16–3 THE TUNNEL EFFECT

It was illustrated in Figure 16–2 that the wave function penetrates a short distance "into" the potential well in each case, giving a finite probability for finding the particle beyond the classical limits set by the wall. The probability density function inside the potential well may be considered as the result of a standing wave system in the function ψ corresponding to each permitted energy level. Since a standing wave is the result of two wave trains traveling in opposite directions between reflecting boundaries, we may consider the wave function at any one of the walls as consisting of an incident wave and a reflected one. In this case, the wave penetrates

222 · PART THREE: THE ATOM

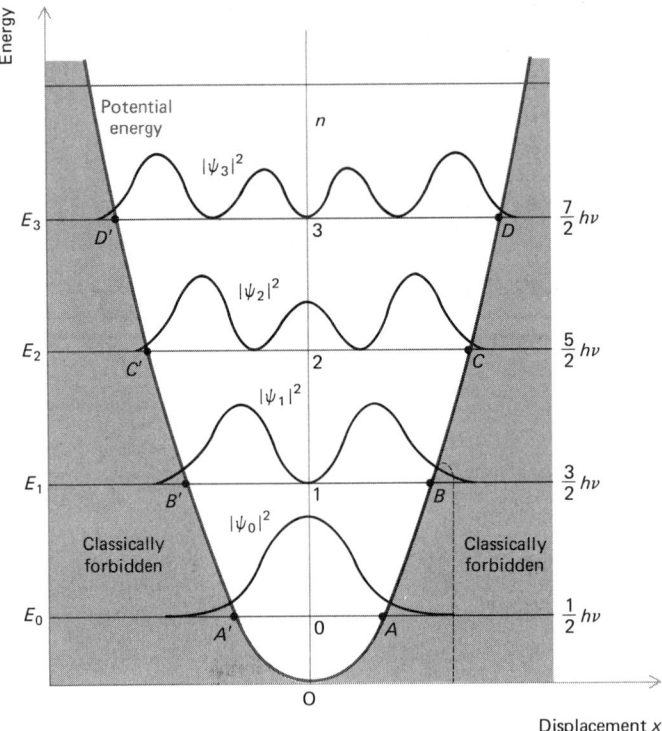

Figure 16-2 Energy level diagram and potential energy curve for the harmonic oscillator. The probability densities associated with each energy E_0, E_1, and so on, are sketched for the first four states.

a bit into the wall, and so the reflection takes place along this finite depth as well as at the surface of the wall itself.

Suppose, now, that the wall in the region of the penetrating wave function is very thin; in other words, the potential energy function turns down sharply toward zero just beyond the point A or B or C, as does the dashed line near point B in Figure 16-2. Then the wave function may have finite amplitude at this point. What happens to it a bit further on?

This situation may be treated in a simplified manner by using a thin potential wall—a *potential barrier*. Let this be a small region on the x axis that is bounded by sharp potential jumps: one from zero to a finite value V, and the other from V back to zero. In Figure 16-3, this situation is represented by putting the first jump at $x = 0$, the origin, and the second one at $x = A$. This divides the axis into three regions:

region I, $x < 0$, where the potential energy $= 0$,
region II, $0 < x < t$, where the potential energy $= V$, and
region III, $x > 0$, where the potential energy $= 0$.

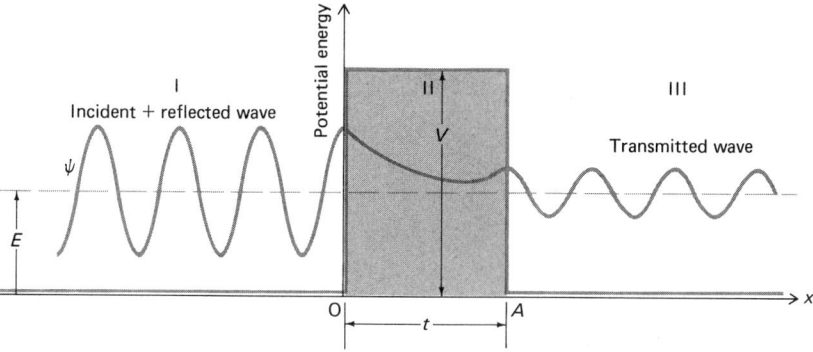

Figure 16-3 A beam of particles of kinetic energy E is incident on a potential barrier $V > E$ with a width $OA = t$.

Now let the wave train be incident on the barrier from the left. The barrier has been constructed such that it is thin compared to the depth of penetration of the wave into it, and there must be some finite wave amplitude in region III to the right.

We have developed this situation from the case of one of the energy levels indicated for the harmonic oscillator in Figure 16-2. Notice that there the total energy of the level (say, E_1) is less than the height of the barrier, which we turn down at x just a bit greater than at point B and indicated by the dashed line there. The potential energy at the barrier maximum is greater than the total energy of the particle in that level—yet we say that the wave function has a finite amplitude beyond the barrier. This implies that the probability of finding the particle outside of the barrier is finite, even though its total energy is less than the barrier height. We are forced to draw a finite wave function amplitude in region III as shown in Figure 16-3. Assigning the notation ψ_1, ψ_2, and ψ_3 to the respective wave functions in regions I, II, and III as indicated in the figure, the corresponding Schrödinger equations are

region I $\quad \dfrac{-\hbar^2}{2m} \dfrac{d^2\psi_1}{dx^2} = E\psi_1 \quad$ since $\quad V_I = 0$

region II $\quad \dfrac{-\hbar^2}{2m} \dfrac{d^2\psi_2}{dx^2} + V\psi_2 = E\psi_2 \quad$ since $\quad V_{II} = V \qquad (16\text{-}20)$

region III $\quad \dfrac{-\hbar^2}{2m} \dfrac{d^2\psi_3}{dx^2} = E\psi_3 \quad$ since $\quad V_{III} = 0$

Rearranging these equations and defining the quantities

$$\alpha^2 = \dfrac{2m}{\hbar^2} E \quad \text{and} \quad \beta^2 = \dfrac{2m(V - E)}{\hbar^2}$$

the equations become

region I $\quad \dfrac{d^2\psi_1}{dx^2} + \alpha^2\psi_1 = 0$

region II $\quad \dfrac{d^2\psi_2}{dx^2} - \beta^2\psi_2 = 0 \quad$ (16–21)

region III $\quad \dfrac{d^2\psi_3}{dx^2} + \alpha^2\psi_3 = 0$

The solutions to these equations are

region I $\quad \psi_1 = Ae^{i\alpha x} + Be^{-i\alpha x}$

region II $\quad \psi_2 = Fe^{-\beta x} + Ge^{\beta x} \quad$ (16–22)

region III $\quad \psi_3 = Ce^{i\alpha x} + De^{-i\alpha x}$

where the constants A, B, and so on are the amplitudes of the corresponding components of each wave. They may be recognized as follows:

A is the amplitude of the wave incident on the barrier from the left,
B is the amplitude of the reflected wave in region I,
F is the amplitude of the wave penetrating the barrier in region II,
G is the amplitude of the reflected wave (from the surface at A) in region II,
C is the amplitude of the transmitted wave in region III, and
D is the amplitude of a (nonexistent) reflected wave in region III.

It should be noted that we have drawn the wave function through the three regions in Figure 16–2 so that it is continuous and singly valued everywhere along the x axis. These are reasonable conditions to impose, and they make it possible to solve for the various amplitudes explicitly in terms of the energy of the particle, the barrier height, and its thickness.

Since the probability density associated with a wave function is proportional to the square of the amplitude of that function, we can define the barrier *transmission coefficient* as

$$\boxed{T = \dfrac{|C|^2}{|A|^2}} \quad (16\text{–}23)$$

and a *reflection coefficient* for the barrier surface at $x = 0$ as

$$\boxed{R = \dfrac{|B|^2}{|A|^2}} \quad (16\text{–}24)$$

CHAPTER 16: SOME APPLICATIONS OF THE SCHRÖDINGER EQUATION · 225

If the barrier is high compared to the total energy of the particle or is thick compared to the wavelength of the wave function, then the transmission coefficient becomes

$$T \approx 16 \frac{E}{V}\left(1 - \frac{E}{V}\right) e^{-(2t/\hbar)\sqrt{2m(V-E)}} \qquad (16\text{-}25)$$

where t is the physical thickness of the barrier.

We have reached the remarkable conclusion that if a particle with energy E is incident on a thin energy barrier of height greater than E, there is a finite probability of the particle penetrating the barrier. This phenomenon, called the *tunnel effect*, is a result in the quantum mechanics that is not allowed in the classical treatment.

Among the earliest successes of the quantum theory in nuclear physics was the application of the tunnel effect to radioactive α decay by Gamow in 1928 and by Condon and Gurney in 1929. The nucleons in the nucleus of, say, uranium, consist of neutrons and protons. These particles form shortlived clusters of two protons with two neutrons (α particles) within the nucleus. It is calculated on the basis of the tunnel effect that such an α particle, if it impinges on the inside of the nuclear force barrier that holds the nucleus together, stands about one chance in 10^{38} of penetrating the barrier and escaping from the nucleus. This escape, however, constitutes radioactive α decay.

The nucleus has a diameter of about 10^{-14} m, and the α particle moves within it with a velocity of about 10^7 m/sec, so that it makes about 10^{21} collisions/sec with the barrier. Thus, there are

$$\frac{10^{38} \text{ collisions}}{10^{21} \text{ collisions/sec}} = 10^{17} \text{ sec/penetration}$$

This is about 3×10^9 yr required for an α particle to make a probable escape, and it provides some understanding of the long radioactive half-life of uranium, which is over a billion years.

The height of the nuclear barrier of polonium is somewhat less than that of uranium, and an α particle has a chance of about one in 10^{17} of escaping from the nucleus per collision. Taking the collision rate as 10^{21}/sec, the probable escape time for an α particle from this isotope of polonium is of the order of 10^{-4} sec. The quantum mechanical tunnel effect in α decay exhibits extreme difference in radioactive lifetimes (ranging from billions of years to milliseconds) for rather small variations in the height of the potential barrier.

EXAMPLE 16-1: The potential barrier problem is a good approximation to the problem of an electron trapped inside but near the surface of a metal. Calculate the probability of transmission that a 1.0-eV electron will penetrate a potential barrier of 4.0 eV when the barrier width is 2.0 Å.

SOLUTION: From Equation (16–25), the transmission coefficient is*

$$T \approx 16 \left(\frac{1.0 \text{ eV}}{4.0 \text{ eV}}\right)\left(1 - \frac{1.0 \text{ eV}}{4.0 \text{ eV}}\right)$$

$$\times \exp\left[-\frac{2 \times 2 \times 10^{-10} \text{ m}}{1.05 \times 10^{-34} \text{ J-sec}} \sqrt{2(9.1 \times 10^{-31} \text{ kg})(4-1)(1.6 \times 10^{-19} \text{ J})}\right]$$

$$\approx 0.084$$

Thus, only about eight 1.0-eV electrons out of every hundred penetrate the barrier.

PROBLEMS

16–1 A pendulum in the first approximation is a harmonic oscillator. Determine the quantum zero point energy for a pendulum of length 10 m in the earth's gravitational field.

16–2 Use a table of integrals and show that the ψ_0 eigenfunction in Table 16–1 is normalized.

16–3 Use Table 16–1 and find the expression for the harmonic oscillator eigenfunction $\psi_4(x)$.

16–4 What is the frequency of vibration of an electron with a zero point energy of 15 eV? What is the next allowed value of energy for this electron?

16–5 When 1.0-eV electrons are incident on a potential barrier of 8.0 eV (such as the work function of a metal), what fraction of the electrons penetrate their barrier if it is 5.0 Å wide?

16–6 A particle of kinetic energy E is incident on a potential step $V > E$, as shown in Figure 16–4.
(a) Set up the Schrödinger equations for region I and region II and find the expression for the wave function for each region. (b) Use the boundary con-

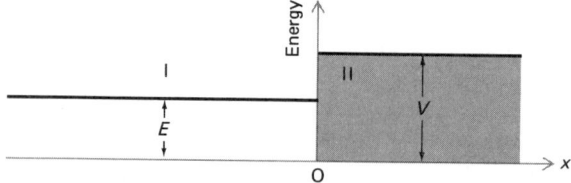

Figure 16–4

---
* Remember that exp(x) is a shorthand form of e^x.

ditions and the definition of the wave function to determine the constants of the wave functions. (If $\alpha^2 = (2m/\hbar^2)E$ and $\beta^2 = 2m(V - E)/\hbar^2$, then the constant associated with $e^{\beta x}$ must be zero.) (c) If A is the amplitude of the incident wave function and B is the amplitude of the reflected wave function, show that the reflection coefficient is equal to one, that is,

$$R = \frac{|B|^2}{|A|^2} = 1$$

What does this mean physically?

16-7 Electrons are trapped 3.0 Å within the surface of a metal plate. What is the probability that the electrons will get out of the plate if the potential barrier is 8.0 eV and the energy of the electrons is (a) 1.0 eV, (b) 4.0 eV, and (c) 7.0 eV?

16-8 Equation (16-25) is valid only when the barrier is either high or wide. The exact equation for the transmission coefficient is

$$T = \left\{1 + \frac{\sinh^2\sqrt{(2mVt^2/\hbar^2)[1 - (E/V)]}}{4(E/V)[1 - (E/V)]}\right\}^{-1}$$

(a) Show that when t or V are large, this equation reduces to Equation (16-25). (b) Solve Problem 16-6 using the exact equation for T and compare the results.

16-9 An α particle is trapped in a nucleus whose radius is $r_0 = 1.4 \times 10^{-15}$ m. What is the probability that an α particle will escape from the nucleus if its energy is (a) 2.0 MeV, or (b) 1.0 MeV? The potential barrier at the surface of the nucleus is 4.0 MeV.

16-10 For the classical harmonic oscillator, the probability P of finding the particle in a length dx is proportional to the time spent in dx, that is, dx/v. (a) Show that P is proportional to $dx/\sqrt{2m(E - \frac{1}{2}kx^2)}$. (b) Show that the constant of proportionality A in the integral

$$P = \int_a^b \frac{A \, dx}{a\sqrt{2m(E - \frac{1}{2}kx^2)}} = 1$$

is equal to k^2m^2/π. For the classical harmonic oscillator, what are the limits a and b?

RECOMMENDED READING

EISBERG, R. M., *Fundamentals of Modern Physics*, Wiley, New York, 1967.
   A well-written book appropriate for an intermediate approach to the subject of modern physics.

HARRIS, Louis, and LOEB, Arthur L., *Introduction to Wave Mechanics*, McGraw-Hill, New York, 1963.
A somewhat supplementary, but well done text. There are several practical problems discussed in good detail.

PAULING, Linus, and WILSON, E. Bright, *Introduction to Quantum Mechanics*, McGraw-Hill, New York, 1935.
Good, clearly written reference on the fundamentals of quantum mechanics.

PLANCK, M., *The Philosophy of Physics*, Norton, New York, 1936.
In Chapter 2 ("Causality in Nature"), Dr. Planck makes an exhaustive analysis of Heisenberg's principle of uncertainty and the principle of causality.

SHERWIN, Chalmers W., *Introduction to Quantum Mechanics*, Holt, Rinehart & Winston, New York, 1959.
Chapter 3 gives a solution for the amplitude equation of the harmonic oscillator using numerical methods.

SILVA, Andrade E., and LOCHAK, G., *Quanta*, P. Moore (Trans.), McGraw-Hill, New York, 1969.
A brilliant, clear, and precise study of the basic principles of quantum mechanics. Preface written by de Broglie.

17
Different Models of Mechanics

Paul Adrien Maurice Dirac
(1902–)

Born in Bristol, England, Dirac received his Ph.D. in physics from Cambridge University in 1926, where he was appointed Lucasian Professor of Mathematics in 1932. He has been visiting professor at the University of Wisconsin and Princeton University. In 1928 Dirac extended wave mechanics to the study of high-speed particles by combining it with the relativity theory. Consequently, he predicted the existence of the positron. For "discovery of new and fruitful forms of atomic theory," he and Erwin Schrödinger were awarded the 1933 Nobel Prize in physics.

17–1 MODELS OF MECHANICS
17–2 CLASSICAL MECHANICS
17–3 RELATIVISTIC MECHANICS
17–4 QUANTUM MECHANICS
17–5 WAVE-PARTICLE DUALITY
17–6 UNCERTAINTY PRINCIPLE

17-1 MODELS OF MECHANICS

Let us pause at this point and review the various basic models of the physical world that we refer to as "mechanics". In this chapter, we shall summarize each model briefly, leaving the detailed proofs and examples to the respective chapters where each was developed. Here only the principal results of each are presented, so that we may compare the different approaches. It is most important for us to realize that these models are not competing viewpoints, nor are they different and exclusive pictures of nature that are to be tested against one another so that all but one might some day be rejected. The various models that physicists have developed are in fact different *approximations* to the reality of nature, which apply in different circumstances. There is not yet a sufficient number of approximations to describe all that we observe in nature, nor is there a single unified theory that can be used to describe every situation. The progress of physics consists of finding new approximations to cover new observations and in developing generalizations that bring together these approximations into single theories. These various mathematical approximations, together with the set of concepts that tie them to a part of nature, are called *models*.

The models that we study in this text may be called (1) the Newtonian or classical mechanics, (2) the special relativistic mechanics, and (3) the quantum or wave mechanics.

17-2 CLASSICAL MECHANICS

The classical or Newtonian mechanics was, historically, the first system of mechanics to be developed in what we now call physics. Based on observations of the motion of ordinary objects in the everyday world, classical mechanics succeeded in developing a general description of the motion of these objects and their interactions. These objects were neither exceedingly large, such as whole galaxies, nor exceedingly small, such as individual atoms. When in motion they traveled with speeds that were not exceedingly high; the speeds were small compared with the speed of light, for instance. In general, classical mechanics successfully describes the motion of these objects.

The classical mechanics, in its most elementary form, may be considered as based on Newton's three laws of motion:

The law of inertia states that a free body is either at rest or moves with constant velocity.

The law of force states that the force **F** acting on a particle of mass m equals the time rate of change of momentum $\mathbf{p} = m\mathbf{v}$,

$$\mathbf{F} = \frac{d}{dt}(m\mathbf{v}) \tag{17-1}$$

The law of action and reaction states that when a body A exerts a force $\mathbf{F}_A$ on a body B by any means, then B in turn exerts an equal and opposite force $\mathbf{F}_B$ on A, so that

$$\mathbf{F}_A = -\mathbf{F}_B \tag{17-2}$$

The law of inertia defines the equilibrium condition for a body and implies the conservation of momentum. Both this law and the law of action and reaction can be derived from the law of force, Equation (17-1), and so we might take the second law to be the fundamental law of classical mechanics. An alternative approach is to consider the conservation of momentum as more basic.* Consider a cluster of n particles moving freely under no external forces. Let there be arbitrarily many forces acting *between* the particles, however, and let their masses and velocities be $m_1, m_2, \ldots, m_n$ and $\mathbf{v}_1, \mathbf{v}_2, \ldots, \mathbf{v}_n$. The law of conservation of momentum states that the total momentum of the cluster is composed of the vector sum of all of the momenta of the particles, and that this sum remains constant even though individual particle momenta may change. That is,

$$\boxed{\sum_{i=1}^{n} m_i \mathbf{v}_i = m_1 \mathbf{v}_1 + m_2 \mathbf{v}_2 + \cdots + m_n \mathbf{v}_n = \text{constant}} \tag{17-3}$$

Thus, if the momentum of one of the particles changes, then the momentum of at least one other particle must change also to preserve the constant sum in Equation (17-3). The interaction between these particles that causes this cooperative action is called a "force," and Newton's second and third laws follow as a logical consequence. The conservation principle in Equation (17-3), taken with its consequence, Equation (17-1), serves to define the notion of force. The concept of force loses much of its relevance, however, in both the microscopic

---

* The reason for this is that in more sophisticated models, like the quantum mechanical model, the concept of force loses its relevance, while the conservation of linear momentum is like a dogma in physics and prevails in all accepted models.

world of quantum mechanics and in the macroscopic world of the general theory of relativity (which we shall not study in this text).

To make a study of the classical motion of a particle, its mass m is taken as constant and Equation (17-1) is expanded as

$$\mathbf{F} = \frac{d}{dt}(m\mathbf{v}) = m\frac{d\mathbf{v}}{dt} = m\frac{d^2\mathbf{r}}{dt^2} \qquad (17\text{-}4)$$

The vector $\mathbf{r}$ is the position vector of the particle relative to the origin of an arbitrary inertial system of coordinates:

$$\mathbf{r} = \hat{\mathbf{i}}x + \hat{\mathbf{j}}y + \hat{\mathbf{k}}z \qquad (17\text{-}5)$$

for instance. The system need not be Cartesian—it may be a set of spherical or cylindrical or any other three orthogonal space coordinates.

The equation of motion giving the position as a function of time t is obtained by integrating Equation (17-4) to yield

$$\mathbf{r} = \mathbf{r}(t, c, \ldots, c_6) \qquad (17\text{-}6)$$

where the six c's are constants of integration. Evaluation of these constants is then based on the fundamental assumption that at some initial time, when $t = t_0$, both the position of the particle,

$$\mathbf{r}_0 = \hat{\mathbf{i}}x_0 + \hat{\mathbf{j}}y_0 + \hat{\mathbf{k}}z_0 \qquad (17\text{-}7)$$

and its velocity,

$$\mathbf{v}_0 = \hat{\mathbf{i}}\frac{d}{dt}x_0 + \hat{\mathbf{j}}\frac{d}{dt}y_0 + \hat{\mathbf{k}}\frac{d}{dt}z_0 \qquad (17\text{-}8)$$

are known *simultaneously* and with *absolute precision*. The theoretical possibility of having this knowledge is unquestioned; its attainment is based only on our skill in devising the measurement.

Our equation of motion, Equation (17-6), has been obtained in this case relative to some particular inertial coordinate system which we have chosen to begin with. If the corresponding equation is to be evaluated relative to some other system of coordinates, then we may either begin all over in the derivation or use a "coordinate transformation" to obtain it. The transformation is just a set of relations between the coordinates of the first reference system and those of the new one. In the classical mechanics, this is the Galilean transformation, named for Galileo.

Refer, for instance, to Figure 17-1, where $S_1(x_1, y_1, z_1, t_1)$ indicates some arbitrary inertial coordinate system and $S_2(x_2, y_2, z_2, t_2)$ is a second inertial system moving relative to S_1 with the constant velocity $\mathbf{v}$. Then the coordinates in S_1 of an event $E(x_1, y_1, z_1, t_1)$ taking place at a point P are related to the

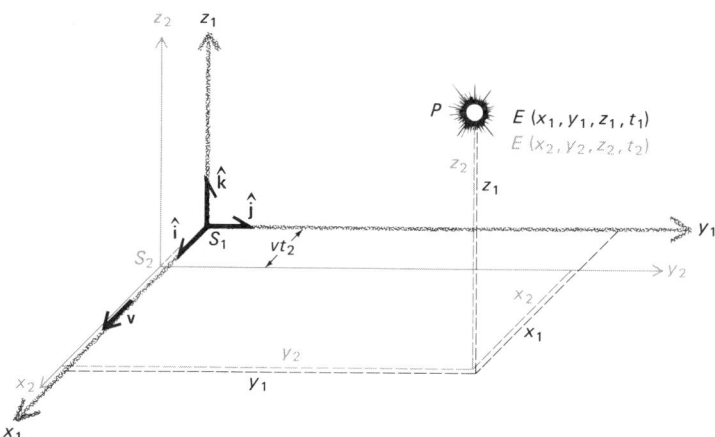

Figure 17–1 The coordinates of event $E(x_1, y_1, z_1, t_1)$ occurring at point P are related to the coordinates in S_2 of the same event $E(x_2, y_2, z_2, t_2)$ through the Galilean transformations. The unit vectors $\hat{\imath}$, $\hat{\jmath}$, and $\hat{k}$ are the same in both systems because the axes (x, y, z) in S_1 and S_2 are parallel.

coordinates in S_2 of the *same* event $E(x_2, y_2, z_2, t_2)$ at the *same* point P by

$$\begin{aligned} x_1 &= x_2 + vt_2 \\ y_1 &= y_2 \\ z_1 &= z_2 \\ t_1 &= t_2 \end{aligned} \tag{17-9}$$

This set of equations defines the *Galilean transformation of Cartesian axes*. It has counterparts when applied to other systems of coordinates, such as spherical or cylindrical coordinates.

The situation has been simplified by taking the relative velocity **v** as being directed along the x axis of both S_1 and S_2. This has the effect of setting the numerical values of the y and z coordinates in both systems equal, as indicated in Equation (17–9). The equality of t_1 and t_2—the value of the time at any time and at any place as read from clocks in either S_1 or S_2—is independent of the relative motion and is taken as a fundamental assumption.

From Equation (17–9), the corresponding Galilean transformation of velocities is found to be

$$\begin{aligned} v_{1x} &= v_{2x} + v \\ v_{1y} &= v_{2y} \\ v_{1z} &= v_{2z} \end{aligned} \tag{17-10}$$

where the velocity of a particle at point P as measured in S_1 is

$$\mathbf{v}_1 = \hat{\mathbf{i}} v_{1x} + \hat{\mathbf{j}} v_{1y} + \hat{\mathbf{k}} v_{1z} \tag{17-11}$$

and the velocity of the *same* particle at the *same* point P and the *same* time $t_1 = t_2$ as measured in S_2 is

$$\mathbf{v}_2 = \hat{\mathbf{i}} v_{2x} + \hat{\mathbf{j}} v_{2y} + \hat{\mathbf{k}} v_{2z} \tag{17-12}$$

The classical or Galilean composition of velocities is then given by

$$\boxed{\mathbf{v}_1 = \mathbf{v}_2 + \mathbf{v}} \tag{17-13}$$

and, finally, the accelerations of the particle as measured from the two coordinate systems are just

$$\boxed{\mathbf{a}_1 = \mathbf{a}_2} \tag{17-14}$$

Since in classical mechanics the mass m is a *universal constant*, we obtain from Equation (17-14)

$$m\mathbf{a}_1 = m\mathbf{a}_2$$

and Newton's laws are *invariant* in both systems S_1 and S_2. Hence, system S_2 is also inertial.

It is important to remember that the transformation is an *operation* that has various representations, one of which is given by the set of Equations (17-9). An immediate consequence of the Galilean transformation is the *classical principle of relativity*, which states that the *laws of mechanics are invariant in form for all inertial frames moving relative to one another with relative constant velocity small compared to the velocity of light in a vacuum.*

The classical mechanics has been described as being based on the concept of forces acting on masses (Newton's laws) or, as an alternative, being based on the principle of the conservation of momentum. There are other starting points in use in various approaches to the classical mechanics, and of course they must all yield the same results to any given problem. Each provides a somewhat different concept of the basic nature of the physical universe, and each has its own particular advantages in applications to actual problems. A very important approach to the classical mechanics—one that we shall not treat at length in this text but that is widely used in advanced classical dynamics and can be adapted to both relativistic and to quantum mechanics—is based on *Hamilton's principle*. This principle considers a dynamic situation in which, for instance, (Figure 17-2) a particle travels between two points A and B in a time $t = t_2 - t_1$, under the influence of forces. Kinetic energy K and potential energy V are defined as functions of position and time along the path. Hamilton's principle states that *the integrals $\int_{t_1}^{t_2} (K - V)\, dt$ of the different $(K - V)$ functions over the time $t = t_2 - t_1$ are the same when taken along the real path ACB or any slightly different path* (e.g., ADB).

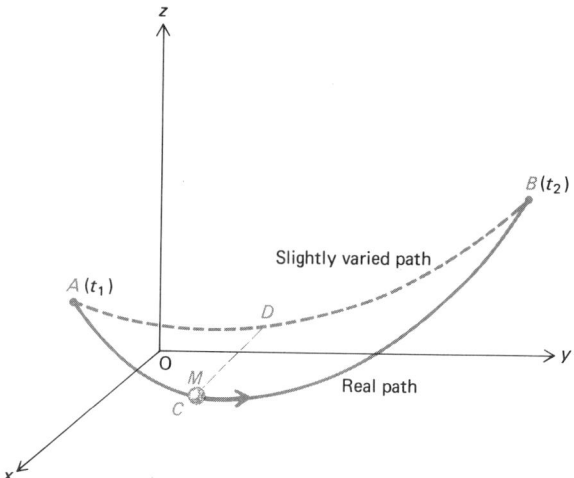

Figure 17-2 Hamilton's principle states that if *ACB* is the real path followed by a particle traveling between points *A* and *B*, and *ADB* is any "slightly different path" connecting the same points, the integral $\int_{t_1}^{t_2} (K - V)/dt$ has the same value for both paths. Or, in other words, $\delta I = 0$, which means that $\int_{t_1}^{t_2} (K - V)/dt$ has a "stationary value"; it can be a *minimum* or a *maximum*.

The quantity $L = K - V$ is called the *Lagrangian function* or the *kinetic potential*. The time integral between two points along a dynamical path is said to have a *stationary value* with respect to the same integral taken over any other permitted (or "varied") paths. The value of the integral along a given path is a minimum compared to its value along *any* other path in many cases of interest. All the laws of classical dynamics may be derived from Hamilton's principle, and it provides a system of mechanics based on *energies* rather than on vector quantities such as forces and momenta. Since energy (in its many forms) appears to be the primary "stuff" out of which all the physical universe is formed, perhaps a Hamiltonian approach to mechanics is the most fundamental one.

17-3 RELATIVISTIC MECHANICS

When a dynamic situation involves bodies moving with velocities that approach the speed of light, the approximation that must be used is called *relativistic mechanics*. If there are large accelerations involved, or if there are extremely large masses such as are found in neutron stars, then we must work in one of the systems of mechanics related to the *general relativistic theory*. This complication is avoided here by requiring that the masses be of ordinary size and that any velocities involved may be very large but are either constant or are changing in a very uniform way. These are the bounds of the special theory of relativity.

A primary experimental result, the *Michelson–Morley experiment*, provided much of the motivation for the development of this theory. A basic assumption stemming from consideration of this result is that the speed of a given packet of light in a vacuum (c) is the same for all inertial observers,* even though they may be moving relative to one another with arbitrary constant velocities. Because all observations of natural events are accomplished ultimately through the use of electromagnetic fields in some form, a coordinate transformation fundamentally different from the Galilean one, Equation (17–9), must be employed. We must now use the *Lorentz transformation*. When applied to the systems represented in Figure 17–1, the Lorentz transformation equations are

$$\boxed{\begin{aligned} x_1 &= \gamma(x_2 + vt_2) \\ y_1 &= y_2 \\ z_1 &= z_2 \\ t_1 &= \gamma\left(t_2 + v\frac{x_2}{c^2}\right) \end{aligned}} \qquad (17\text{–}15)$$

where γ is the *Lorentz* factor,

$$\gamma = \frac{1}{\sqrt{1 - (v^2/c^2)}} \qquad (17\text{–}16)$$

The ratio of the relative speed v to the speed of light c is often given the symbol

$$\beta = \frac{v}{c} \qquad (17\text{–}17)$$

The student should show that as v becomes very small relative to c, Equation (17–15) reduces to that of the Galilean transformation Equation (17–9). The relativistic or Lorentz transformation for velocities corresponding to Equation (17–15) is given by

$$\boxed{\begin{aligned} v_{1x} &= \frac{v_{2x} + v}{1 + \beta(v_{2x}/c)} \\ v_{1y} &= \frac{v_{2y}\sqrt{1 - \beta^2}}{1 + \beta(v_{2x}/c)} \\ v_{1z} &= \frac{v_{2z}\sqrt{1 - \beta^2}}{1 + \beta(v_{2x}/c)} \end{aligned}} \qquad (17\text{–}18)$$

* An inertial observer is an observer at rest relative to an inertial frame.

These, also, reduce to their Galilean approximations Equation (17–10) when $\beta = v/c \to 0$.

A consequence of the Lorentz transformation is that the difference between two coordinate values (say, a length along the x axis) is dependent on the relative velocity between that particular axis and the observer measuring the length. This is seen in the first of Equations (17–15), where γ is a common factor for any two values of x_1 and so multiplies any length along that axis. It is easily found that the transformation law for spatial lengths is

$$L_1 = \frac{1}{\gamma} L_2 \qquad (17\text{–}19)$$

Since the value of γ is always greater than unity, L_1 is always less than L_2, and we speak of length *contraction*.

A similar consequence arises from the last of Equations (17–15). The time interval between two values of t_1 is affected by the relative velocity as well. The transformation law that results for time intervals is

$$T_1 = \gamma T_2 \qquad (17\text{–}20)$$

and so they are *lengthened* or *dilated*.

We may now expand the definition of the principle of *classical relativity* by stating that *the laws of nature are invariant in form for all inertial frames moving relative to each other with constant but otherwise arbitrary velocity.*

The law of conservation of momentum given by Equation (17–3) is valid as it stands in relativistic mechanics as well as in the classical mechanics. This fact recommends it as a starting point in a model of nature. This law, taken together with the Lorentz transformation, results in a definition of mass that is also dependent on relative velocity,

$$m = \gamma m_0 \qquad (17\text{–}21)$$

where m_0, *the rest mass*, is taken as the mass measured by an observer relatively at rest. It is the smallest possible value of m for any object.

Force is defined in relativistic mechanics as it is in classical mechanics,

$$\mathbf{F} = \frac{d}{dt}(m\mathbf{v}) \qquad (17\text{–}22)$$

which is the same as Newton's second law except that m is now a function of v according to Equation (17–21).

Equation (17–22) may now be used to define the kinetic energy relativistically as

$$K = \int_0^s \mathbf{F}\cdot d\mathbf{s} = \int_0^s \frac{d}{dt}(mv)\,ds = \int_0^v v\,d(mv) \qquad (17\text{–}23)$$

where the kinetic energy is the work done by **F** acting on the body to change its speed from 0 to v. In Equation (17–23), we have assumed that **F** is acting parallel to **v**, but the result we are going to get is also valid for curvilinear motion.

The relativistic form for the kinetic energy from this equation is then

$$\boxed{K = (m - m_0)c^2} \qquad (17\text{–}24)$$

When $v/c \to 0$, this reduces to the classical expression

$$K = \tfrac{1}{2}m_0 v^2 = \frac{p^2}{2m} \qquad (17\text{–}25)$$

where $p = m_0 v$ is the classical momentum. The total energy is then defined by the equation

$$\boxed{E = E_0 + K = \sqrt{p^2 c^2 + E_0^{\,2}} = mc^2} \qquad (17\text{–}26)$$

where

$$E_0 = m_0 c^2 \qquad (17\text{–}27)$$

is the *rest energy* and

$$\boxed{\mathbf{p} = m\mathbf{v} = \gamma m_0 \mathbf{v}} \qquad (17\text{–}28)$$

is the relativistic momentum. The principle of conservation of momentum and the above definition of energy lead to the *principle of conservation of mass energy*, which states that, for an isolated system,

$$\boxed{\sum \text{rest energy} + \sum \text{kinetic energy} + \sum \text{potential energy} = \text{constant}}$$
$$(17\text{–}29)$$

17–4 QUANTUM MECHANICS

Quantum mechanics, an entirely different approach to mechanics, is especially applicable to systems of atomic dimensions and smaller. Our example here is from the nonrelativistic Schrödinger quantum theory. An equally successful matrix algebra formulation, different in form but basically equivalent to the Schrödinger treatment, was developed by Heisenberg. It yields the same results, of course. A relativistic treatment of the quantum mechanics was pioneered by P. A. M. Dirac and led to a new and broader picture of nature, including the concept of antimatter. It will not be treated in depth since it is beyond the scope of this text.

A complex wave function $\psi(x)$, springing from the concepts of the wave nature of matter and due to de Broglie, is defined to describe the entire physical state of a particle or of a system of particles such as an atom. At present no deeper physical significance is given to $\psi(x)$, but the product

$$\psi^*(x)\psi(x) = |\psi(x)|^2 \qquad (17\text{-}30)$$

is real and represents the probability per unit length (or per unit volume if defined in three dimensions) of finding the particle at a given point. The certainty of finding the particle *somewhere* by the so-called normalization condition is expressed by

$$\int_{-\infty}^{\infty} \psi^*(x)\psi(x)\, dx = 1 \qquad (17\text{-}31)$$

If the wave function $\Psi(x, t)$ depends on position and time, it verifies the *time-dependent Schrödinger equation*,

$$-\frac{\hbar^2}{2m}\frac{\partial^2 \Psi(x, t)}{\partial x^2} + V\Psi(x, t) = i\hbar \frac{\partial \Psi(x, t)}{\partial t} \qquad (17\text{-}32)$$

and the product $\Psi^*\Psi = |\Psi|^2$ represents the *probability per unit length* of finding the particle at a given point at a given time. When the equation describes a stationary state in which Ψ and V are not functions of time, Equation (17–32) reduces to the *time-independent* form,

$$-\frac{\hbar^2}{2m}\frac{d^2 \psi(x)}{dx^2} + V(x)\psi(x) = E\psi(x) \qquad (17\text{-}33)$$

By defining the total energy as

$$E = K + V = \frac{p^2}{2m} + V \qquad (17\text{-}34)$$

and a *Hamiltonian operator* as

$$H = -\frac{\hbar^2}{2m}\frac{d^2}{dx^2} + V(x) \qquad (17\text{-}35)$$

the time-independent Schrödinger Equation (17–33) may be written

$$H\psi(x) = E\psi(x) \qquad (17\text{-}36)$$

To arrive at results consistent with physical observations, several additional requirements are imposed on the wave function $\psi(x)$:

1. It must be well behaved, that is, single-valued and continuous everywhere.
2. If $\psi_1(x), \psi_2(x), \ldots, \psi_n(x)$ are solutions of Equation (17-36), then the linear combination $\psi(x) = a_1\psi_1(x) + a_2\psi_2(x) + \cdots + a_n\psi_n(x)$ must be a solution.
3. The wave function $\psi(x)$ must approach zero as $x \to \pm\infty$.

For a brief outline of a typical approach to quantum mechanical problems, we consider a simple *steady-state* case in which the particle (or system of particles) is contained in some sort of limits that set "boundary conditions" on the solution of the equation. The potential energy function $V(x)$ is determined explicitly and inserted in the Hamiltonian operator H. Solution of the equation then yields a set of functions $\psi_1, \psi_2, \ldots, \psi_n$ that are discrete solutions (eigenfunctions), each defining a different state of the system. Each eigenfunction satisfies the normalization condition of Equation (17-31) separately,

$$\int_{-\infty}^{+\infty} \psi_i^*(x)\psi_i(x)\,dx = 1 \tag{17-37}$$

as well as an *orthogonality condition* that guarantees the noninterference of the wave trains representing different states,

$$\int_{-\infty}^{+\infty} \psi_i^*(x)\psi_j(x)\,dx = 0 \quad (i \neq j) \tag{17-38}$$

In the simplest case, there is a single, unique value of the energy E_i associated with each solution ψ_i. These are the corresponding *eigenvalues*, and the spectrum constitutes the *quantized energy states* permitted in the system.

17-5 WAVE-PARTICLE DUALITY

In quantum mechanics, a particle of mass m, momentum $p = mv$, and total energy $E = (p^2/2m) + V$ is represented by a *wave packet*. Any wave packet may be considered to be just the superposition (or the result of the interference of) an infinite number of traveling waves whose amplitudes might be represented by

$$\Psi_i = \Psi_i \cos(\omega_i t - k_i x) \tag{17-39}$$

where $\omega_i = 2\pi\nu_i$ is the angular frequency and $k_i = 2\pi/\lambda_i$ is the propagation constant of the ith component wave. The particle is considered as being "located" in the region of maximum interference of the many waves. The frequency and wavelength at the center of the packet are $\nu = E/h$ and $\lambda = h/p$.

The group velocity of the packet as a whole is given by

$$v_g = \frac{dE}{dp} \qquad (17\text{--}40)$$

and equals the velocity of the particle represented by the packet, which is found by differentiating Equation (17–34),

$$\frac{dE}{dp} = \frac{d}{dp}\left(\frac{p^2}{2m} + V\right) = v \qquad (17\text{--}41)$$

because $V =$ constant for a "free particle."

The question naturally arises: "Are we dealing, fundamentally, with a particle described by a mass, an energy, and a momentum, or with a wave described by an amplitude, a frequency, and a wavelength?" The dual nature of the particle-wave is only apparent. The two characters do not appear simultaneously in any observation of nature. Whether a given system appears *wavelike* or *particlelike* depends on the means used for its observation and the questions asked in the measurement. The duality question is raised, principally, by the simplistic nature of our mathematical treatment and becomes much less apparent in the more sophisticated quantum field theory in use today.

17–6 UNCERTAINTY PRINCIPLE

Another feature of the microscopic world closely related to the problem of duality is *Heisenberg's uncertainty principle*. This principle states that there are pairs of quantities referring to a microscopic system that cannot be known *simultaneously* with infinite precision. Consider an electron, for example. Its position x and its momentum p are known only to a certain preciseness. If Δx is the uncertainty in position and Δp is the uncertainty in momentum,

$$\boxed{\Delta x \, \Delta p \geq h} \qquad (17\text{--}42)$$

where h is Planck's constant. If it happens that the position x is perfectly known, then it follows that we know nothing about the magnitude of p, and vice versa. The same relation holds for the energy and time related to any given event or states as a consequence of Equation (17–42),

$$\boxed{\Delta E \, \Delta t \geq h} \qquad (17\text{--}43)$$

The uncertainty principle stems from the fact that we are forced to represent a particle by a wave packet (see Figure 17–3) in which the infinite number of monochromatic waves forming the packet have an effective spread in frequency of

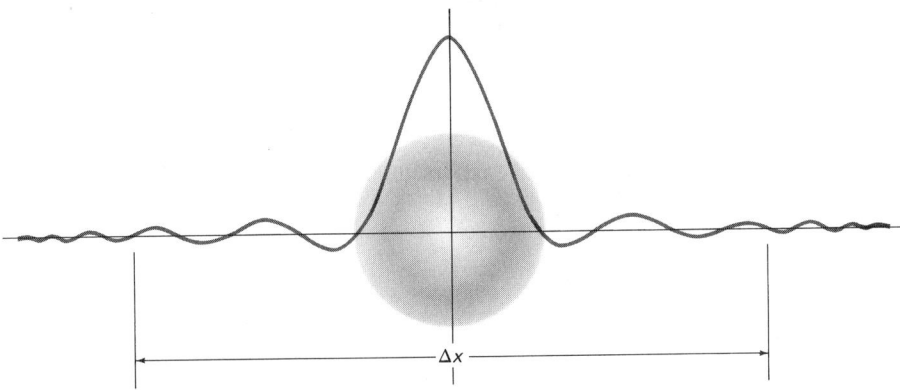

Figure 17-3 The region of maximum interference of a traveling wave packet represents a moving particle.

$\Delta v = \Delta E/h$. The particle is somewhere within the region Δx of the packet, and the uncertainty in momentum is

$$\Delta p = \Delta \left(\frac{h}{\lambda}\right) = -h \frac{\Delta \lambda}{\lambda^2} \qquad (17\text{-}44)$$

This uncertainty is intrinsic to the very nature of the systems we are discussing and represents an ultimate limit on what is knowable about them. It has nothing to do with any technical difficulties that may be encountered in actually building more precise measuring instruments.

PROBLEMS

17-1 In an inertial system S_1, a 2.0-kg mass moves with a velocity $v_i = 5.0\hat{\imath} + 3.0\hat{\jmath}$ m/sec and collides head on with a mass of 3.0 kg moving with a velocity $v_2 = -10\hat{\imath} - 6.0\hat{\jmath}$. (a) Determine the momentum of each mass after the collision. (b) Determine the momentum of each mass as measured by an observer moving with a velocity $v_0 = 6\hat{\imath}$ relative to system S_1.

17-2 A radar monitor sees two space ships, one with a velocity $v_1 = (0.54c)\hat{\imath} + (0.72c)\hat{\jmath}$ and the second with a velocity $v_2 = (0.54c)\hat{\imath}$. What is the velocity of the first ship as measured by the second ship?

17-3 Nobel laureate Ernest Lawrence proposed plans for a cyclotron with a 4000-ton magnet in which ions could be accelerated through a potential of 100 MeV. (a) What would be the relativistic mass of a proton accelerated through this potential? (b) What would be the relativistic mass of the magnet as measured by an observer on the proton?

17–4 Determine the wavelength of a quantum of light whose "effective mass" is equal to the rest mass of (a) an electron, and (b) a proton.

17–5 A particle of rest mass m_0 traveling with a velocity $(0.90c)\hat{\imath}$ makes a completely inelastic collision with an identical particle. (a) Determine the velocity of the combined masses as they move off together. (b) What is the change in kinetic energy?

17–6 Show that the wave function

$$\psi(x) = Ax \exp[-(\sqrt{mk}/2\hbar)x^2]$$

could be a solution of the Schrödinger equation for a harmonic oscillator of mass m with a spring constant k.

17–7 A pulsed ruby laser with an output of 2.0 GW (gigawatts) produces a pulse of 10 psec duration. What is the relative uncertainty in the measurement of the laser energy?

17–8 What is the speed of an electron with a relativistic mass equal to $1.1\ m_0$? How many electron volts of energy are required for the electron to reach this mass?

17–9 Determine the wavelength associated with an electron moving at (a) $0.80c$, and (b) $0.90c$.

RECOMMENDED READING

BEISER, Arthur, *The World of Physics*, McGraw-Hill, New York, 1960.
A small book with several interesting essays on some of the fundamental concepts of physics.

BORN, Max, *Problems of Atomic Dynamics*, M.I.T. Press, Cambridge, Mass., 1929.
Every serious student of physics should at least glance through the pages of this excellent book, because it contains the complete series of lectures given by Prof. Born at M.I.T. in 1925–1926.

GURNEY, R. W., *Elementary Quantum Mechanics*, University of Nebraska Press, Lincoln, Neb., 1934.
A very well-written book in which stress has been placed on the treatment of quantum mechanics by graphical methods.

KOMPANEYETS, A. S., *Basic Concepts in Quantum Mechanics*, Van Nostrand Reinhold, New York, 1966.
Translated from the Russian by Prof. L. F. Landavitz, Yeshiva University, and written at a very elementary level, this book contains a very precise approach to the subject of wave mechanics, stressing the physical meaning of its basic principles.

YOURGRAU, W., and MANDELSTON, W. B., *Variational Principles in Dynamics and Quantum Theory*, Saunders, Philadelphia, 1968.
A very clear presentation of the variational principle and its numerous applications to the fields of classical, relativistic, and quantum mechanics.

18
The Schrödinger Theory of the Hydrogen Atom

Arnold J. W. Sommerfeld
(1868–1951)

Born in Königsberg, Germany, Sommerfeld studied at Königsberg University. From 1906 to 1940 he was professor of theoretical physics at the University of Munich. Author and coauthor of nearly 300 scientific papers and a dozen books, Sommerfeld's research included the structure of spectral lines, the quantum theory of spectra, and applications of quantum theory to the Bohr atom model and spectroscopy. Among his many honors are the Planck medal (1931), the Lorentz medal (1939), and the Oersted medal (1948).

18–1 THE WAVE EQUATION: SEPARATION OF VARIABLES
18–2 THE AZIMUTHAL EQUATION
18–3 THE POLAR EQUATION
18–4 THE RADIAL EQUATION
18–5 THE COMPLETE WAVE FUNCTION

18-1 THE WAVE EQUATION: SEPARATION OF VARIABLES

Bohr's bold, ingenious theory of the atom provided answers to many problems plaguing experimental physicists at the time. However, in spite of the initial success of this theory, it was found to be inadequate in the explanation of such problems as

1. Why do transitions occur between certain energy levels but not others?
2. Why do electrons not radiate electromagnetically and spiral into the nucleus?
3. What is the origin of the complex spectra from more complex atoms like He and Li?

Erwin Schrödinger's theory, using the concepts of quantum mechanics and wave functions, successfully completed the picture of the atom beyond those concepts first introduced by Bohr. The hydrogen atom, the simplest atomic structure in nature, provides an excellent subject for a first application of the Schrödinger theory. The successful study of the details of hydrogenic spectra is one of the first great achievements of quantum mechanics. In our brief treatment here, we shall attempt to maintain contact with the physical meaning of the results rather than study the mathematical technique in detail.*

Figure 18-1 is a schematic representation of the hydrogen atom as it is visualized classically. A heavy proton (mass = $1836m_e$) is assumed at rest at the origin of a rectangular coordinate system, and an electron (mass = m_e) is pictured as orbiting about it at a radius r under the influence of the attractive Coulomb field of the system. Our picture is centered on the proton, whereas in a better approximation the two particles rotate about a common center of mass displaced a bit from the center of the proton. We shall neglect this effect here, however.

The potential energy function due to the Coulomb field is

$$V(r) = -\frac{e^2}{4\pi\varepsilon_0} \frac{1}{r} \qquad (18\text{-}1)$$

where e is the electronic charge and ε_0 is the permittivity of free space (8.85×10^{-12} C^2/N-m^2). From the viewpoint of quantum mechanics, however, the

---

*For a complete treatment, see L. Pauling and E. B. Wilson, *Introduction to Quantum Mechanics*, McGraw-Hill, New York, 1935.

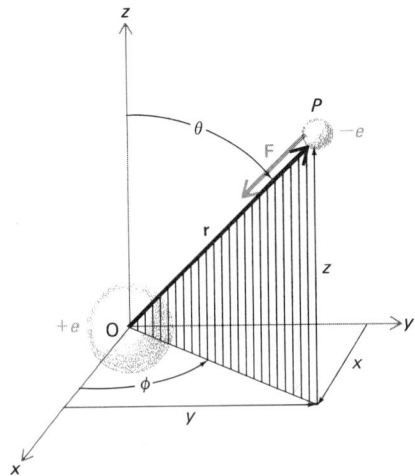

Figure 18–1 Schematic representation of the hydrogen atom. The electron of charge $-e$ at P revolves about the massive proton nucleus of charge $+e$ at the origin O. The force **F** is the Coulomb attraction of the electron to the proton.

electron is represented by a wave system bounded by the potential well of the Coulomb field. This circumstance results in a set of permitted standing wave systems, each corresponding to a particular possible value of the total energy.

The wave equation must now be expressed in three dimensions. Since the hydrogen atom is in its steady state, the time-independent Schrödinger equation is the appropriate form to use. Recall that the corresponding expression for the total energy is

$$E = \frac{p^2}{2m} + V \tag{18-2}$$

and the Hamiltonian operator expressed in rectangular coordinates for three dimensions is

$$\boxed{H = -\frac{\hbar^2}{2m}\left(\frac{\partial^2}{\partial x^2} + \frac{\partial^2}{\partial y^2} + \frac{\partial^2}{\partial z^2}\right) + V} \tag{18-3}$$

The geometrical symmetry of the problem suggests that a more convenient coordinate system would be a spherical one such as the system in Figure 18–1, where the spherical polar coordinates are

r, the radius vector,
θ, the polar angle, and
ϕ, the azimuth angle.

CHAPTER 18: THE SCHRÖDINGER THEORY OF THE HYDROGEN ATOM · 247

The rectangular coordinates are related to the spherical ones by

$$
\begin{aligned}
x &= r \sin \theta \cos \phi \\
y &= r \sin \theta \sin \phi \\
z &= r \cos \theta
\end{aligned}
\tag{18-4}
$$

By standard mathematical techniques, the Hamiltonian operator can be transformed from rectangular coordinates to spherical ones. When this is done, the resulting steady-state Schrödinger equation becomes

$$-\frac{\hbar^2}{2m}\left[\frac{1}{r^2}\frac{\partial}{\partial r}\left(r^2\frac{\partial \psi}{\partial r}\right) + \frac{1}{r^2 \sin \theta}\frac{\partial}{\partial \theta}\left(\sin \theta \frac{\partial \psi}{\partial \theta}\right) \right.$$
$$\left. + \frac{1}{r^2 \sin^2 \theta}\frac{\partial^2 \psi}{\partial \phi^2}\right] + \left(\frac{-e^2}{4\pi\varepsilon_0 r}\psi\right) = E\psi \tag{18-5}$$

and the wave function is now taken to depend upon r, θ, and ϕ,

$$\psi = \psi(r, \theta, \phi) \tag{18-6}$$

We shall now assume that this equation has solutions that may be expressed as products of functions of the three coordinates, that is,

$$\psi(r, \theta, \phi) = R(r)\Theta(\theta)\Phi(\phi) \tag{18-7}$$

Using this expression for the wave function in the Schrödinger Equation (18-5) and multiplying through by $2mr^2/\hbar^2$, we obtain

$$\Theta\Phi \frac{d}{dr}\left(r^2 \frac{dR}{dr}\right) + \frac{R\Theta}{\sin^2 \theta}\frac{d^2\Phi}{d\phi^2} + \frac{\Phi R}{\sin \theta}\frac{d}{d\theta}\left(\sin \theta \frac{d\Theta}{d\theta}\right)$$
$$+ \frac{2mr^2}{\hbar^2}\left(E + \frac{e^2}{4\pi\varepsilon_0 r}\right)R\Theta\Phi = 0 \tag{18-8}$$

Dividing this equation through by $\psi = R\Theta\Phi$ creates a form in which those terms that depend on the coordinate r do not contain any dependence on θ or ϕ. This circumstance permits us to collect the two portions, the *radial* one dependent on r only and the *angular* one dependent on θ and ϕ only, and set them each equal to a constant—the same constant. This can be carried out because the orthogonality of the three coordinates prevents either portion from being a function of the other. We define a constant $l(l + 1)$, obviously with later results in mind, and set the two portions equal to it:

$$\frac{d}{dr}\left(r^2 \frac{dR}{dr}\right) + \frac{2mr^2}{\hbar^2}\left(E + \frac{e^2}{4\pi\varepsilon_0 r}\right)R = l(l+1)R \qquad (18\text{--}9)$$

and

$$\frac{1}{\Phi \sin^2 \theta}\frac{d^2\Phi}{d\phi^2} + \frac{1}{\Theta \sin \theta}\frac{d}{d\theta}\left(\sin \theta \frac{d\Theta}{d\theta}\right) = l(l+1) \qquad (18\text{--}10)$$

Now Equation (18–10) can be separated in a similar way into two portions, one dependent only on the polar angle θ and the other only on the azimuth angle ϕ. This is accomplished by multiplying Equation (18–10) by $\sin^2 \theta$ and collecting terms:

$$\frac{1}{\Phi}\frac{d^2\Phi}{d\phi^2} + \frac{\sin \theta}{\Theta}\frac{d}{d\theta}\left(\sin \theta \frac{d\Theta}{d\theta}\right) - l(l+1)\sin^2 \theta = 0 \qquad (18\text{--}11)$$

As before, each portion will be set equal to a constant. Since any given solution of these two equations will carry a corresponding value of l as a parameter, a convenient new constant $-m_l^2$ is chosen. Setting the two portions equal to this constant and rearranging a bit, we obtain the two equations

$$\frac{m_l^2}{\sin^2 \theta} - \frac{1}{\Theta \sin \theta}\frac{d}{d\theta}\left(\sin \theta \frac{d\Theta}{d\theta}\right) = l(l+1) \qquad (18\text{--}12)$$

and

$$\frac{d^2\Phi}{d\phi^2} + m_l^2 \Phi = 0 \qquad (18\text{--}13)$$

The Schrödinger wave equation for the idealized hydrogen atom has now been separated into three equations, each dependent on only one of the coordinates, and two constants have been explicitly introduced into the problem in the process. It will be seen shortly that the set of solutions is dependent on not two but three constants as parameters.

18–2 THE AZIMUTHAL EQUATION

The most readily recognizable of the three equations is, perhaps, Equation (18–13), the azimuthal wave equation, which describes the behavior of the wave function as the system rotates about the z axis. Its form is the familiar one of the harmonic oscillator equation, which has among its solutions the two real and one complex periodic functions:

$$\Phi = A \sin m_l \phi$$
$$\Phi = A \cos m_l \phi \tag{18-14}$$
$$\Phi = A e^{i m_l \phi}$$

On obviously physical grounds, if the atom is rotated through a full turn about the z axis, the same value for the solution Φ should be reached when the angle ϕ is reached again. This is satisfied by the functions of Equation (18-14) whenever $m_l \phi$ is an integral multiple of 2π. Because ϕ is measured in radians, m_l must have an integral value. Since this includes both zero and rotations in the opposite direction, the only permitted values of m_l are

$$m_l = 0, \pm 1, \pm 2, \pm 3, \ldots \tag{18-15}$$

From the previously established nomenclature for this situation, the values of m_l in Equation (18-15), when squared, are the *eigenvalues* proper to the various *eigenfunctions* from Equation (18-14), which are, in turn, the permitted solutions to the azimuthal Equation (18-13). The constant m_l is the first example of a *quantum number* in this treatment of the hydrogen atom. For purposes that will become clear later, it is called the *magnetic quantum number*.

18-3 THE POLAR EQUATION

The other equation describing the behavior of the wave function in angle about the origin is the polar Equation (18-12). This is another well-known equation in the study of spherical harmonics, with a class of solutions known as *associated Legendre polynomials** represented by

$$\boxed{P_{lm_l}(\cos \theta)} \tag{18-16}$$

The details of the mathematics of these polynomials will not be discussed, except to note that they are functions of $\cos \theta$ and depend on two parametric constants, m_l and l. Since m_l can take on only positive or negative integral values and zero, from the known properties of the Legendre polynomials as applied to the system pictured in Figure 18-1, where the polar angle may take on values from 0 to π, the solutions given by Equation (18-16) exist only when l is integral and when it is equal to or greater than the absolute value of m_l. This imposes the conditions for a

---

* See, for example, M. R. Spiegel, *Mathematical Handbook of Formulas and Tables*, McGraw-Hill, New York, 1968, p. 149.

second quantum number—the *orbital quantum number*. A summary of the situation up to this point is that

$$l = 0, 1, 2, 3, \ldots$$
$$m_l = 0, \pm 1, \pm 2, \ldots, \pm l \tag{18-17}$$

Because l must be greater than or equal to the absolute value of m_l, if $l = 0$, then $m_l = 0$; or if $l = 1$, then m_l can equal 0 or $+1$ or -1; and so on. In general, for any given value of l there are $(2l + 1)$ possible solutions as m_l takes on its permitted values. This situation is expressed by saying that a state corresponding to a given value of l shows a $(2l + 1)$-*fold degeneracy* in m_l. If the $(2l + 1)$ *energy eigenvalues* corresponding to a given l are all equal, then the state is *degenerate*. If, on the other hand, some additional physical effects are present, which separate these eigenvalues from one another, then the degeneracy is removed and the resulting state is said to be *nondegenerate*. A magnetic field applied to the hydrogen atom can remove the degeneracy in m_l, hence the name *magnetic quantum number*.

18-4 THE RADIAL EQUATION

The remaining wave equation, Equation (18-9), is the radial equation, which specifies the behaviour of the wave function as it depends on distance from the proton. Again the solution is found in terms of a well-known form as the *Laguerre polynomials* $L_{n,l}(r)$.* As before, the details of the mathematics of this class of polynomial are left to independent study. Suffice it to say that the resulting set of solutions to the radial wave equation, the *radial eigenfunctions*, are given by

$$\boxed{R_{nl} = e^{-nr}r^l L_{n,l}(r)} \tag{18-18}$$

where n is any positive nonzero integer and l is the orbital quantum number of the other equations. The name *principal* or *total quantum number* is assigned to the number n. From the properties of the Laguerre polynomials, solutions to Equation (18-18) exist only when $n \geq (l + 1)$. We have now arrived at a system of three quantum numbers for the simple model of the hydrogen atom, which are related in the following way:

$$\begin{aligned}
\text{total quantum number} \quad & n = 1, 2, 3, \ldots \\
\text{orbital quantum number} \quad & l = 0, 1, 2, \ldots, (n-1) \\
\text{magnetic quantum number} \quad & m_l = 0, \pm 1, \pm 2, \ldots, \pm l
\end{aligned} \tag{18-19}$$

In this idealized case only, a pure Coulomb potential function has been assumed with no consideration of intrinsic angular momenta for the electron or

---

* See, for example, M. R. Spiegel, *Mathematical Handbook of Formulas and Tables*, McGraw-Hill, New York, 1968, p. 153.

proton or external perturbing physical influences such as magnetic fields. The total number of independent (factorable) solutions to the complete Schrödinger equation for a given value of n is

$$\boxed{\sum_{l=0}^{n-1}(2l+1) = n^2} \qquad (18\text{-}20)$$

18-5 THE COMPLETE WAVE FUNCTION

The complete wave equation for the model of the hydrogen atom being considered may now be obtained by first normalizing each of the three component equations over the relevant range for each coordinate and then multiplying them together in the same manner in which they are factored in Equation (18-7). When the resulting complete wave equation is then solved for its eigenfunctions, it is found that each is characterized by an oscillating amplitude. The space around the origin is divided by nodal surfaces into cells, in each of which the phase of the oscillation is opposite to that of its neighbors. The number of nodal surfaces is just $(n-1)$.

When the energy eigenvalues are calculated for a specific set of quantum numbers, it is found that only the total quantum number n remains. This is an expression of the degeneracy of the system in this highly simplified model, as mentioned before. The total energy of the electron must be less than the potential energy function at any particular value of the radius in order for quantized solutions to the problem to yield energy eigenvalues. *If the total energy is greater than $V(r)$, then a continuum of solutions exists, and the electron may have any energy above this value. The electron is unbounded.*

For bounded states, the energy eigenvalues are given by

$$\boxed{\begin{aligned} E_n &= \left(\frac{-me^4}{32\pi^2\varepsilon_0^2\hbar^2}\right)\frac{1}{n^2} \\ &= (-13.6)\frac{1}{n^2}\ eV \qquad n = 1, 2, 3, \ldots \end{aligned}} \qquad (18\text{-}21)$$

in complete agreement with the predictions of the Bohr theory. Many other results that are not available from the Bohr theory are obtainable from a quantum mechanical treatment. These are arrived at by removing the degeneracies that we have noted and by refining the model to conform better to the actual atomic system found in nature.

Figure 18-2 presents a schematic representation of the Coulomb potential well and the corresponding energy levels resulting from a bounded electronic system

252 · PART THREE: THE ATOM

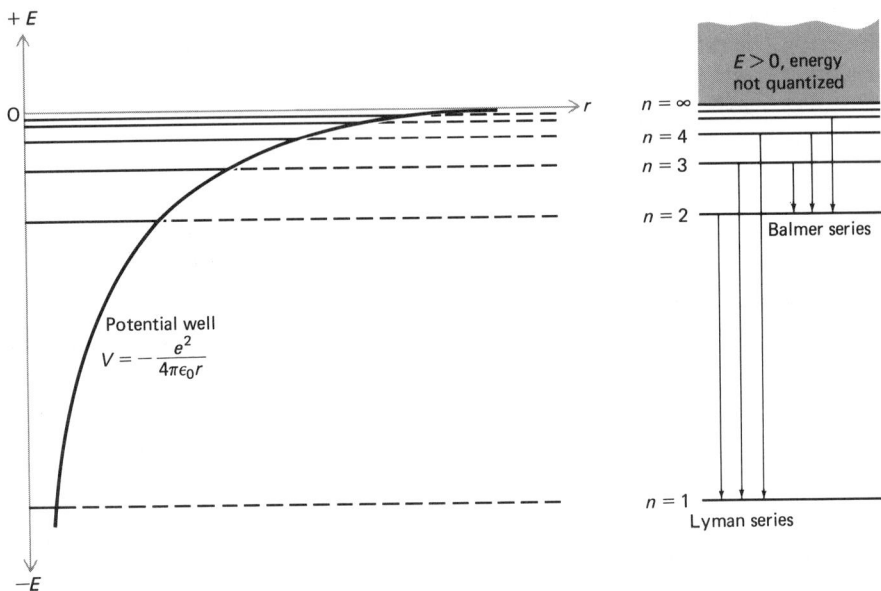

Figure 18-2 Electrostatic potential energy of the electron in the hydrogen atom $V = -(e^2/4\pi\epsilon_0 r)$. When $E < 0$, the only possible values of energy are quantized as shown in the energy level diagram. For $E > 0$ the electron is unbounded, it may have any positive value, and it is not quantized.

contained in it. Several optical transitions between these levels that yield the Balmer series for hydrogen are also indicated.

PROBLEMS

18-1 The normalized wave functions ψ_{nlm_l} [Equation (18-7)] for some of the states of lowest energy are

$$\psi_{100} = \frac{1}{\sqrt{\pi}} \left(\frac{1}{r_1}\right)^{3/2} e^{-(r/r_1)} \qquad n = 1, l = 0, m_l = 0$$

$$\psi_{200} = \frac{1}{2\sqrt{2\pi}} \left(\frac{1}{r_1}\right)^{3/2} \left(1 - \frac{r}{2r_1}\right) e^{-(r/2r_1)} \qquad n = 2, l = 0, m_l = 0$$

$$\psi_{210} = \frac{1}{2\sqrt{2\pi}} \left(\frac{1}{r_1}\right)^{3/2} \frac{r}{2r_1} e^{-(r/2r_1)} \cos\theta \qquad n = 2, l = 1, m_l = 0$$

where $r_1 = 4\pi\epsilon_0 \hbar^2/me^2$ is the Bohr radius. Solve for the corresponding eigenvalues E_{nlm_l} by substituting into Equation (18-5).

CHAPTER 18: THE SCHRÖDINGER THEORY OF THE HYDROGEN ATOM · 253

18-2 Show that if $x = \cos\theta$, $\Theta(\theta) = y(x)$, and $\sin\theta \, d\Theta/d\theta = -(1-x^2)\,dy/dx$, Equation (18-12) transforms to

$$(1-x^2)\frac{d^2y}{dx^2} - 2x\frac{dy}{dx} + \left[l(l+1) - \frac{m_l^2}{1-x^2}\right]y = 0$$

(When l is an integer and equal to or greater than m_l, this equation has well-behaved solutions known as associated Legendre polynomials.)

18-3 For the total quantum number $n = 4$, list the possible values of l and m_l.

18-4 Show that (a) $\Theta_{lm_l} = \sqrt{\frac{5}{8}}(3\cos^2\theta - 1)$, where $l = 2$ and $m_l = 0$; and (b) $\Theta_{lm_l} = \sqrt{\frac{15}{4}}\sin\theta\cos\theta$, where $l = 2$, and $m_l = \pm 1$ are solutions of Equations (18-12).

18-5 Show that the following equations are solutions of the radial Equation (18-9), assuming that the eigenvalues of the energy are given by Equation (18-21):

(a) $R_{nl} = \dfrac{1}{2\sqrt{2}}\left(\dfrac{1}{r_1}\right)^{3/2}\left(2 - \dfrac{r}{r_1}\right)e^{-(r/2r_1)}$ $n = 2, l = 0$

(b) $R_{nl} = \dfrac{1}{2\sqrt{6}}\left(\dfrac{1}{r_1}\right)^{3/2}\left(\dfrac{r}{r_1}\right)e^{-(r/2r_1)}$ $n = 2, l = 1$

(c) $R_{nl} = \dfrac{1}{9\sqrt{3}}\left(\dfrac{1}{r_1}\right)^{3/2}\left(6 - \dfrac{4r}{r_1} + \dfrac{4r^2}{9r_1^2}\right)e^{-(r/3r_1)}$ $n = 3, l = 0$

where r_1 is the Bohr radius. See problem 18-1 above.

18-6 Prove that the solutions of Problem 18-4 obey the normalization condition

$$\int_{\theta=0}^{\theta=\pi} \Theta^*\Theta \sin\theta \, d\theta = 1$$

18-7 Prove that the solutions of Problem 18-5 obey the normalization condition

$$\frac{1}{4\pi}\int_{r=0}^{\infty} R_{nl}^* R_{nl} \, dV = 1$$

where $dV = 4\pi r^2 \, dr$ is the volume of any shell of radii r and $r + dr$, and $R_{nl}^* R_{nl}/4\pi$ is the probability per unit volume and per unit solid angle taken along r.

18-8 By using the normalization condition

$$\int_{\phi=0}^{\phi=2\pi} \Phi_{m_l}^* \Phi_{m_l} \, d\phi = 1$$

prove that in the solutions

$$\Phi_{m_l} = Ae^{im_l\phi}$$

of the azimuthal Equation (18-3), the normalization constant is $A = 1/2\pi$.

RECOMMENDED READING

ALONSO, M., and FINN, E. J., *Fundamental University Physics*, Addison-Wesley, Reading, Mass., 1967, Vol. III.
  A very clear presentation of quantum mechanics is one of the many achievements of this fine book. Great stress is placed on the understanding of the physical principles.

BORN, Max, *Problems of Atomic Dynamics*, M.I.T. Press, Cambridge, Mass., 1926.
  Every serious student in physics should read this series of lectures given by Prof. Born at M.I.T.

MATTHEWS, P. T., *Introduction to Quantum Mechanics*, McGraw-Hill, New York, 1965.
  A very appropriate book for physics majors, written at the undergraduate level. Stress is placed on the mathematical formulation of the basic principles.

ROJANSKI, V., *Introductory Quantum Mechanics*, Prentice-Hall, Englewood Cliffs, N.J., 1946.
  One of the classical textbooks on wave mechanics. Very well written.

SCHIFF, L. I., *Quantum Mechanics*, McGraw-Hill, New York, 1949.
  The exposition is rigorous and concise and written at a higher level than is usual for undergraduate students majoring in physics. It is an excellent reference text.

SPOSITO, G., *An Introduction to Quantum Physics*, Wiley, New York, 1970.
  An excellent text written at the intermediate level.

19
Quantum Numbers I:
MAGNETIC MOMENTS

Max Karl Planck
(1858–1947)

Planck was born in Kiel, Germany, and received his Ph.D. from the University of Berlin in 1879. He was professor of theoretical physics at the University of Kiel (1885–1889) and later at the University of Berlin (1890–1892). Planck did research in thermodynamics and the theory of entropy; he also investigated the optical and electrical problems associated with the radiation of heat, and with the quantum theory. For his development of blackbody radiation, which from a historical point of view can be considered the beginning of quantum mechanics, he received the 1918 Nobel Prize.

19–1 THE ORBITAL QUANTUM NUMBER l
19–2 THE MAGNETIC QUANTUM NUMBER m_l
19–3 THE MAGNETIC MOMENT OF THE HYDROGEN ATOM

19-1 THE ORBITAL QUANTUM NUMBER l

In the last chapter, we applied the Schrödinger equation to the analysis of a simplified model of the hydrogen atom. One of the first results was a set of three independent wave equations in which each equation contained a function of only one of the coordinates of a spherical polar coordinate system. Second, for *bounded states* (those where the total energy of the electron is less than the local potential energy barrier), three quantized constants were obtained. These were named in analogy to the quantum numbers in the Bohr theory of the hydrogen atom. Finally, an expression for the allowed values of the energy for the bounded system, *energy eigenvalues*, was derived to give

$$E_n = \frac{-me^4}{32\pi^2 \varepsilon_0^2 \hbar^2} \frac{1}{n^2} = -\frac{13.6}{n^2} eV \qquad n = 1, 2, 3, \ldots \tag{19-1}$$

This result, in which n is the total quantum number, is the same as the corresponding result of the Bohr theory [Equation (12–14)].

Recall that in the Bohr theory the electron is considered as a particle in orbit around the proton and that the angular momentum of the electron is quantized arbitrarily to produce the quantum conditions for the atomic system. We must now discover the angular momentum conditions in the quantum mechanical results.

This may be readily done by direct inspection of the radial wave equation (18–9), which after collecting terms and describing the potential energy function by $V(r)$, is

$$\frac{1}{r^2}\frac{d}{dr}\left(r^2 \frac{dR}{dr}\right) + \frac{2m}{\hbar^2}\left[E - V(r) - \frac{\hbar^2}{2mr^2}l(l+1)\right]R = 0 \tag{19-2}$$

It is evident that the three terms in the square brackets, which multiply the radial function R, must all have the same units, namely energy. Writing the third of these terms as equal to an energy and solving for $\hbar^2 l(l+1)$, we obtain

$$\hbar^2 l(l+1) = r^2(2mE')$$
$$= r^2 p^2 \tag{19-3}$$

where E' is the kinetic energy and p is the linear momentum.*

---

* As in Equation (19–2), the first term inside the brackets is E (the total energy), the second

The quantity "radius times linear momentum" is just *angular momentum*, and it can now be expressed as

$$L = \text{angular momentum} = \hbar\sqrt{l(l+1)} \qquad (19\text{-}4)$$

This is the Schrödinger condition on the quantized orbital angular momentum carried by the electron wave in the bounded state of the hydrogen atom.

There are various ways in which to consider the subject of angular momentum in mechanics. Recall, for instance, that *momentum operators* have been defined as

$$\mathsf{p}_x = \frac{\hbar}{i}\frac{\partial}{\partial x}$$

The form for this operator is perfectly general in that the coordinate x may be any coordinate. The corresponding momentum is that component of the total momentum that is proper to that coordinate. Partial derivatives with respect to angular coordinates such as θ and ϕ signal the presence of angular momenta. Since these forms appear in the wave Equation (18-5), it is possible to visualize some sort of circulating wave system representing not only a distribution of *energy density*, but also a distribution of *angular momentum density*.

In the last chapter, it was shown that the possible values of the quantum number l include zero. Thus, a state of zero angular momentum, which was forbidden in the Bohr theory, is permitted in the Schrödinger result. Except for this, Equation (19-4) is quite close to the quantum conditions on angular momentum in the Bohr theory. In that treatment, $L = mvr = n\hbar$, where $n = 1, 2, 3, \ldots$ is the total quantum number of the electron in its orbit about the nucleus, whereas here l results from the properties of the *spherical harmonics* that describe the state of the total wave function in its distribution about the origin. In Bohr's theory, n is the only quantum number (l and m_l are excluded), but for large values of l Equation (19-4),

$$L = \hbar\sqrt{l(l+1)} \cong l\hbar$$

is similar to Bohr's postulate,

$$L = n\hbar$$

The only difference is that the value $n = 0$ is not allowed in Bohr's treatment, while $l = 0$ is permitted in the Schrödinger approach.†

---

term $V(r)$ is the potential energy, and the third term E' should be identified as a kinetic energy; this is why it is possible to make the substitution $2mE' = p^2$ in Equation (19-3).

† See, for example, M. Alonso and E. J. Finn, *Fundamental University Physics*, Addison-Wesley, Reading, Mass., 1968, Vol. III, pp. 122-123.

The following names are taken over from the nomenclature of optical spectroscopy as it was utilized by the early workers in atomic spectra:

VALUE OF l	STATE
0	s
1	p
2	d
3	f
4	g
etc.	etc.

The s state, where $l = 0$, is clearly one with zero angular momentum. Of the three wave functions for the hydrogen atom, only the radial one survives in this case, so the system has no dependence on either θ or ϕ. It is spherically symmetrical. As l takes on values different from zero, the total wave function departs from spherical symmetry, and the angular momentum appears in the system. Notice that the third term in the square brackets of Equation (19–2) has the same effect on the wave function R as does the Coulomb potential energy barrier $V(r)$. For this reason, the third term is often called the *angular momentum barrier*.

19–2 THE MAGNETIC QUANTUM NUMBER m_l

In the Schrödinger theory, the magnetic quantum number m_l, when multiplied by $\hbar$, represents the component of the angular momentum in the direction of an arbitrary coordinate such as the z axis. This restricts the angular momentum vector itself to a certain discrete set of orientations in that coordinate system that depends on the permissible values of m_l. This results in the so-called space quantization of the angular momentum. What physical significance is to be attached to this quantum number in the light of the Schrödinger wave mechanics? To arrive at an answer, we must introduce an *angular momentum operator*, which in spherical coordinates is

$$L_z = \frac{\hbar}{i} \frac{\partial}{\partial \phi} \qquad (19\text{–}5)$$

In classical mechanics, the angular momentum of a particle is defined by the vector equation

$$\mathbf{L} = \mathbf{r} \times \mathbf{p} \qquad (19\text{–}6)$$

where $\mathbf{r}$ is its vector distance from the origin and $\mathbf{p}$ is its linear momentum. Written in terms of rectangular coordinates, a *component conjugate* to the z axis is then given by

$$L_z = xp_y - yp_x \qquad (19\text{–}7)$$

where p_y and p_x are the components of the linear momentum **p** indicated by the subscripts. Replacing these linear momentum components by their corresponding quantum mechanical operators, we obtain

$$\boxed{\mathsf{L}_z = \frac{\hbar}{i}\left(x\frac{\partial}{\partial y} - y\frac{\partial}{\partial x}\right)} \tag{19-8}$$

This is, evidently, an operator for the z component of the angular momentum expressed in rectangular coordinates. This can be expressed in spherical polar coordinates by means of a direct coordinate transformation of Equation (19-8) to yield

$$\mathsf{L}_z = \frac{\hbar}{i}\frac{\partial}{\partial \phi} \tag{19-9}$$

which is the *angular momentum operator*, as was previously stated in Equation (19-5).

Just as the Hamiltonian operator, when applied to a wave function, yields a set of energy states (eigenvalues), which is expressed in the equation

$$\mathsf{H}\psi = E\psi \tag{19-10}$$

so also the angular momentum operator, when applied to a wave function, yields a set of z component angular momentum states (eigenvalues of the z component of the angular momentum)

$$\boxed{\mathsf{L}_z\psi = L_z\psi \quad \text{or} \quad \frac{\hbar}{i}\frac{\partial \psi}{\partial \phi} = L_z\psi} \tag{19-11}$$

This equation has solutions given by

$$\psi = f(r, \theta)e^{iL_z\phi/\hbar} \tag{19-12}$$

Recalling the arguments concerning the solutions of the azimuthal wave equation, Equation (18-13), we again say that ψ must be a continuous and single-valued function of the coordinates and have a period 2π in ϕ. This requires that

$$\boxed{L_z = m\hbar \quad m = 0, \pm 1, \pm 2, \pm 3, \ldots} \tag{19-13}$$

It can be shown that m here is the same constant m_l, the magnetic quantum number, as was obtained before [Equation (18-15)]. From Equations (19-12) and (19-13), it is evident that the eigenfunctions associated with the eigenvalues of L_z are given by $\Phi = Ae^{im\phi}$, A being a constant whose value can be obtained from the corresponding normalization condition $\int_0^{2\pi} \phi^*\phi \, d\phi = 1$. The eigenfunctions Φ were obtained in Equation (18-14) as solutions of the azimuthal equation. It specifies the permitted values of the z component of the orbital angular momentum:

$$L = \hbar\sqrt{l(l+1)} \qquad (19\text{-}14)$$

This is the same property as was found previously (Equation 19-4).

One of the many limitations of Bohr's theory was that it did not recognize states of zero angular momentum. Another limitation is that in Bohr's approach the quantization of L is postulated, while in Schrödinger's formulation it is obtained as a consequence of the wave equation and the properties of the spherical harmonic solutions.

The polar angle of the total angular momentum vector is now constrained to those values satisfying the condition

$$\cos\theta = \frac{L_z}{L} = \frac{m_l \hbar}{\hbar\sqrt{l(l+1)}} = \frac{m_l}{\sqrt{l(l+1)}} \qquad (19\text{-}15)$$

This is shown in Figures 19-1 and 19-2.

It is interesting to note that for very high values of angular momenta, where l becomes large compared to 1, the quantity $l(l+1)$ approaches l^2, and Equation (19-14) reduces to

$$L = l\hbar \qquad (19\text{-}16)$$

Here, the difference between successive values of angular momenta is small compared to the total angular momentum, and the spectrum of permitted values of L approaches a continuous distribution. The conditions on L_z, however, continue

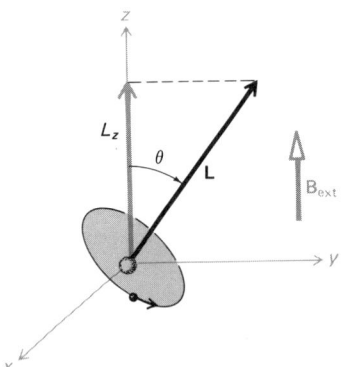

Figure 19-1 The orbital angular momentum **L** has a quantized component L_z in any designated z direction. When an external magnetic field $\mathbf{B}_{ext}$ is applied, the z axis is conveniently taken along its direction.

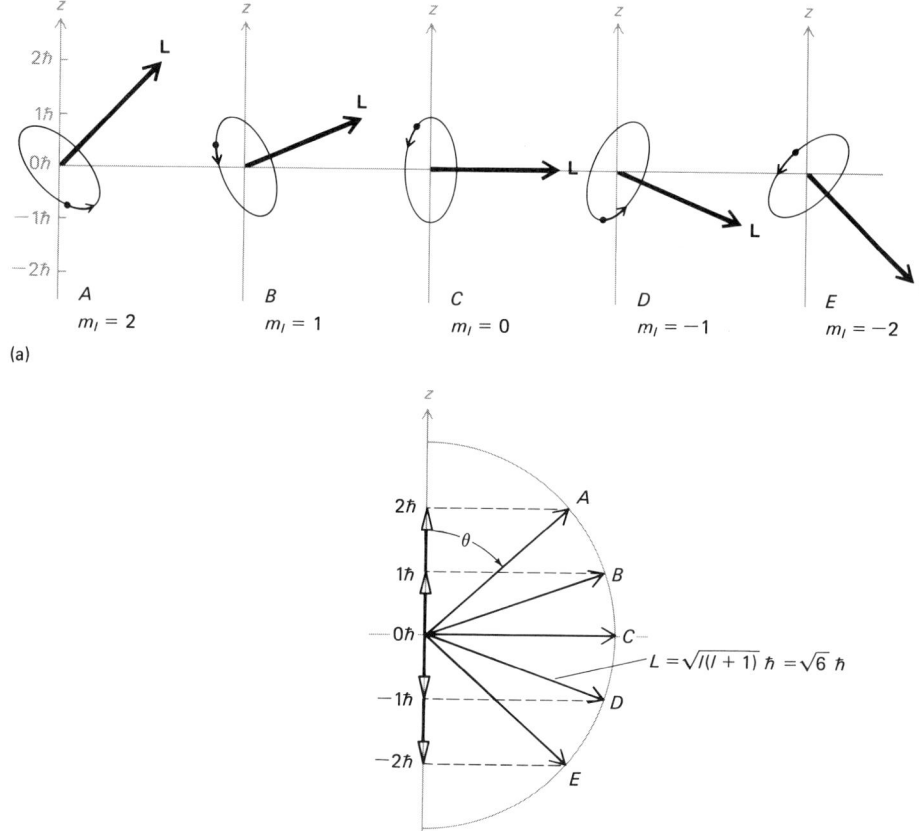

Figure 19-2 The angular momentum vector **L** is space quantized in (a). Only certain orientations with respect to a given z direction are permitted. When $l = 2$ [as in (b)], the space quantization of the orbital angular momentum $L = \sqrt{6}\hbar$ gives components $L_z = 0\hbar, \pm 1\hbar, \pm 2\hbar$.

to hold. The maximum values on m_l are $\pm l$. For large l and max $|m_l|$, Equation (19-15) becomes

$$\cos\theta = \frac{m_l}{\sqrt{l(l+1)}} = \pm\frac{l}{l} = \pm 1 \tag{19-17}$$

and essentially all orientations for the angular momentum vector between 0 and π are possible.

It should also be noted that in quantum mechanics, $\hbar$ is a natural unit of angular momentum.

19-3 THE MAGNETIC MOMENT OF THE HYDROGEN ATOM

An angular momentum associated with an electrical charge implies a rotating current of some sort. A rotating electrical current generates a magnetic field according to Maxwell's equations, and it is not surprising that the hydrogen atom possesses a permanent magnetic field of its own rather similar to that of a bar magnet. The strength of such a field is commonly expressed in terms of the magnetic moment. Let us consider the hydrogen atom as an electron circulating around the proton and calculate an expected magnetic moment for the system (see Figure 19-3).

If the frequency of rotation of the electron is f, then the equivalent current is

$$i = -ef \qquad (19\text{--}18)$$

and if m is the electron mass, R is the radius of rotation, and v is the linear speed of the electron, then its angular momentum is

$$L = (mv)R = m(2\pi f)R^2 \qquad (19\text{--}19)$$

where v has been replaced by $2\pi f R$. Now, if a current i flows in a closed loop of area A, the magnetic field it produces is almost identical to that of a magnet having a magnetic moment

$$\boxed{\mu_l = iA} \qquad (19\text{--}20)$$

where, for our case, $A = \pi R^2$.

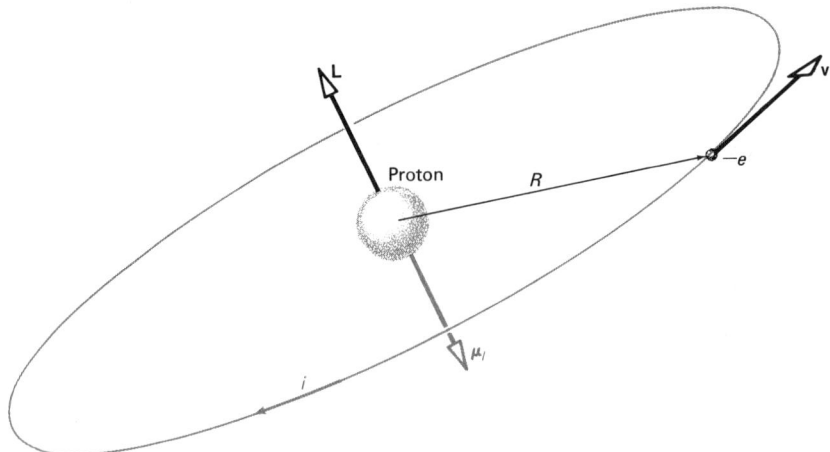

Figure 19-3   Schematic diagram of the hydrogen atom showing the relative direction of the orbital angular momentum **L** and the magnetic dipole moment μ_l.

The ratio determined by dividing the expression for the magnetic moment from Equation (19-20) by that for the angular momentum, Equation (19-19),

$$\frac{\mu_l}{L} = -\frac{ef\pi R^2}{m(2\pi f)R^2} = -\frac{e}{2m} = 8.8 \times 10^9 \text{ C/kg} \tag{19-21}$$

is known as the *gyromagnetic ratio*.

Recall that the orbital angular momentum of the electron is quantized and is equal to $\hbar\sqrt{l(l+1)}$. From Equation (19-21), then, the magnitude of the *magnetic dipole moment* associated with the hydrogen atom is

$$|\mu_l| = \frac{e}{2m} L = \frac{e\hbar}{2m}\sqrt{l(l+1)} \tag{19-22}$$

This is due to the orbital motion of the electron. Since the electrical charge on the electron is negative, the magnetic dipole moment vector is directed opposite to that of the angular momentum.

The unit of magnetic moment appearing in Equation (19-22),

$$\mu_B = \frac{e\hbar}{2m} = \beta = 9.27 \times 10^{-24} \text{ J } (wb/m^2)^{-1} \tag{19-23}$$

is a natural unit of the magnetic dipole moment known as the *Bohr electron magneton*. It will be seen later that additional magnetic dipole fields are associated with the electron and proton as they spin on their own axes, so Equation (19-22) does not tell the whole story.

When a magnetic dipole μ is placed in an external magnetic field $\mathbf{B}_{ext}$, it experiences a torque (Figure 19-4)

$$\tau = \mu \times \mathbf{B}_{ext} \tag{19-24}$$

which tends to align the dipole moment vector along that of the field. Maximum torque, according to Equation (19-24), occurs when the angle θ between μ and $\mathbf{B}_{ext}$ is 90°. The potential energy of a magnetic dipole at any angle θ relative to an external magnetic field is*

$$\Delta E_m = -\mu \cdot \mathbf{B}_{ext} = -\mu B_{ext} \cos\theta \tag{19-25}$$

---

* See, for example, D. Halliday and R. Resnick, *Physics for students of Science and Engineering*, Wiley, New York, 1969, p. 734.

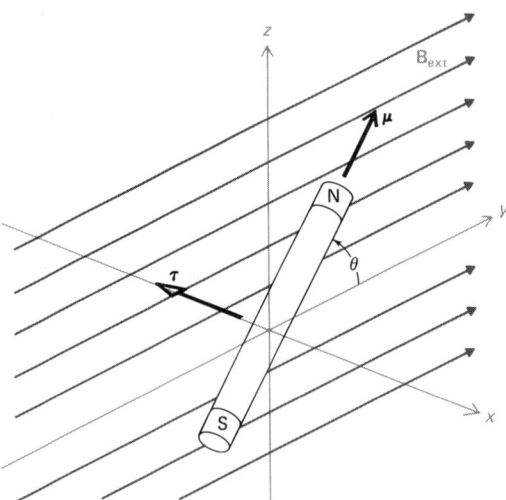

Figure 19-4 The torque τ acting on a magnetic dipole moment μ placed in an external magnet field $\mathbf{B}_{ext}$ is $\tau = \mu \times \mathbf{B}_{ext}$.

PROBLEMS

19-1 What is the value of the energy and the angular momentum of an electron in the hydrogen atom in the $3p$ state? The $4p$ state? (The notation $3p$ means $n = 3$, $l = 1$; $4s$ means $n = 4$, $l = 0$, and so on.)

19-2 Would you find an electron in the $2f$ state? Explain.

19-3 An electron is accelerated through 15,000 V and then allowed to circulate at right angles to a uniform magnetic field of 0.20 Wb/m². Determine the magnetic dipole moment produced by the circulating electron.

19-4 Determine the ratio of the angular momentum of a $3d$ electron as calculated by the Bohr theory and by the Schrödinger theory.

19-5 Use the equations $B = \mu_0 i/2R$ and $L = \sqrt{l(l+1)}\hbar = 2\pi fmR^2$ to determine the magnetic field set up by an electron in the first Bohr orbit. What is the effective current produced by an electron in the first Bohr orbit? Note that $R = n^2 r_1$, where $r_1 = 0.53$ Å is the Bohr radius.

19-6 Determine the maximum values of the angular momentum and the magnetic dipole moment of an electron in the M shell.

19-7 What is the maximum potential energy associated with the magnetic dipole moment of a $4p$ electron placed in an external magnetic field of 0.25 Wb/m²?

19-8 Calculate the orbital quantum number l for (a) an electron in the first Bohr orbit, where $r_1 = 0.53$ Å and $v_1 = 2.2 \times 10^6$ m/sec; (b) an insect of mass 1.0 mg riding on the end of a 2.0-cm-long minute hand of a clock.

19-9 A 5.0-eV proton circulates at right angles to a uniform magnetic field of magnetic induction 0.0063 Wb/m². (a) What is the orbital angular momentum (quantum number l) associated with the proton? (b) Determine the effective current of the proton. (c) Determine the magnetic dipole moment produced by the proton. (d) How many Bohr magnetons is this?

RECOMMENDED READING

ALONSO, M., *Fisica Atomica*, University of Havana Press, Havana, 1958, Chap. 4.
    A very complete study of atomic physics, at a slightly higher level than most of the intermediate books. Written in Spanish, but it can be read with very little knowledge of that language.

FERMI, E., *Notes on Quantum Mechanics*, University of Chicago Press, Chicago (1961).
    Contains an excellent group of articles on the basic principles of quantum mechanics.

FEYNMAN, R. P., LEIGHTON, R. B., and SANDS, L. M., *The Feynman Lectures on Physics*, Addison-Wesley, Reading, Mass., 1965, Vols. I, II.
    A standard reference book. Appropriate for every serious student in physics, undergraduate as well as graduate. Must be included in the library of every physicist.

HEISENBERG, W., *The Physical Principles of the Quantum Theory*, Dover, New York, 1930.
    A challenging book for students majoring in physics. Stress is placed on the physical meaning of the basic concepts.

MATHEWS, P. T., *Introduction to Quantum Mechanics*, McGraw-Hill, New York, 1963.
    Contains a complete coverage of the Schrödinger treatment of the hydrogen atom. Stress is on the mathematical approach.

SAXON, D. S., *Elementary Quantum Mechanics*, Holden-Day, San Francisco, 1964, Chap. 9.
    A very clear presentation of wave mechanics at the intermediate level. Chapter 9 is exceptionally good.

SPOSITO, G., *An Introduction to Quantum Mechanics*, Wiley, New York, 1970.
    Chapter 4, on the hydrogen atom, is very well presented. The emphasis is on the mathematical approach at the intermediate level.

20
Quantum Numbers II:
THE ZEEMAN EFFECT

Pieter Zeeman
(1865–1943)

Born in Zonnemaire, Zeeland, in the Netherlands, Zeeman received his education from the University of Leyden. He was professor of physics at the universities of Leyden and Amsterdam and was appointed director of the Physics Institute in Amsterdam in 1908. In 1896, while at Leyden, he discovered the Zeeman effect, which states that when a ray of light from a source placed in a magnetic field is examined spectroscopically, the spectral line is widened or occasionally doubled. For his research in this field he, with H. A. Lorentz, received the Nobel Prize in 1902.

20–1 AN ATOM IN AN EXTERNAL MAGNETIC FIELD
20–2 THE NORMAL ZEEMAN EFFECT
20–3 TOTAL NUMBER OF STATES

20-1 AN ATOM IN AN EXTERNAL MAGNETIC FIELD

In 1892 PIETER ZEEMAN, working in Leyden, the Netherlands, in optical spectroscopy and with strong magnetic fields, found that the bright yellow lines in the spectrum of a sodium flame were considerably broadened when the flame was placed between the poles of a powerful electromagnet. H. A. LORENTZ quickly analyzed this result from the point of view of his electron theory of matter and predicted that the light from the broadened line should be polarized. Observing the flame from a direction perpendicular to the field, Zeeman found that the edges of the lines were polarized parallel to the field and the central part perpendicular to the field. Improvements in spectroscopic techniques resulted in the resolution of the lines into three separate lines, polarized in this manner. This set of three lines became known as the *Lorentz triplets*.

It was soon found that most elements yielded much more complex patterns in their spectra when placed in magnetic fields, some showing many components with complex polarizations. These became known as examples of the *anomalous Zeeman effect* and were the subjects of much study in subsequent years. The separations of the component lines from one another in frequency, their relative intensities, and their polarizations as functions of temperature and magnetic field strength were all related problems. In 1912 F. PASCHEN and E. BACK found that the anomalous Zeeman effect for close doublets and triplets was changed into the normal pattern of Lorentz triplets in very high magnetic fields. (This normal pattern is usually called the *normal Zeeman effect*.) These effects are now understood in terms of the quantum theory of atomic structure and its interaction with external force fields.

In Figure 20-1, a hydrogen atom is pictured as being in a strong external magnetic field $\mathbf{B}_{\text{ext}}$. The magnetic dipole moment $\boldsymbol{\mu}_l$ of the electron current in its orbit around the proton interacts with $\mathbf{B}_{\text{ext}}$. The potential energy associated with this interaction is then [compare with Equation (19-25)]

$$\Delta E_m = -\boldsymbol{\mu}_l \cdot \mathbf{B}_{\text{ext}} = \mu_l B_{\text{ext}} \cos(\pi - \theta)$$

or

$$\boxed{\Delta E_m = \mu_l \mathbf{B}_{\text{ext}} \cos \theta} \qquad (20\text{-}1)$$

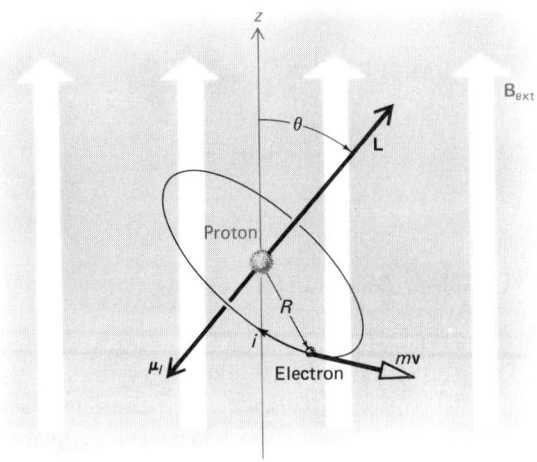

Figure 20–1 A hydrogen atom placed in an external magnetic field $\mathbf{B}_{ext}$ that determines the z direction. The orbital motion of the electron develops a magnetic dipole moment $\boldsymbol{\mu}_l$, which interacts with $\mathbf{B}_{ext}$ to produce a potential energy $\Delta E_m = -\boldsymbol{\mu}_l \cdot \mathbf{B}_{ext}$. $\mathbf{L}$ is the orbital angular momentum of the electron.

When $\boldsymbol{\mu}_l$ from Equation (19–22) is used, the potential energy due to this interaction may be written as

$$\Delta E_m = \frac{e}{2m} L B_{ext} \cos\theta = \frac{e}{2m} \mathbf{L} \cdot \mathbf{B}_{ext}$$

or, since $L = \hbar\sqrt{l(l+1)}$,

$$\Delta E_m = \frac{e\hbar}{2m} \sqrt{l(l+1)}\, B_{ext} \cos\theta = \mu_B \sqrt{l(l+1)}\, B_{ext} \cos\theta \qquad (20\text{--}2)$$

in which $\mu_B = e\hbar/2m$ is the Bohr magneton, Equation (19–23).

The total energy of the hydrogen atom in the magnetic field is now

$$E_{n,l,m_l} = -\underbrace{\frac{me^4}{32\pi^2\varepsilon_0^2\hbar^2}\frac{1}{n^2}}_{\substack{\text{due to Coulomb} \\ \text{interaction between} \\ \text{electron and proton}}} + \underbrace{\mu_B\sqrt{l(l+1)}\, B_{ext}\cos\theta}_{\substack{\text{due to interaction between} \\ \text{external magnetic field and} \\ \text{magnetic dipole moment of} \\ \text{the circulating electron}}}$$

Notice that the total energy is written with three quantum numbers, indicating that the energy also depends upon the magnetic quantum number m_l. This is seen by introducing Equation (19–15),

$$\cos\theta = \frac{m_l}{\sqrt{l(l+1)}}$$

and by writing the energy as

$$E_{n,l,m_l} = -\frac{me^4}{32\pi^2\varepsilon_0^2\hbar^2}\frac{1}{n^2} + \mu_B m_l B_{ext} \qquad (20\text{–}3)$$

This result now gives some insight into the effect of a magnetic field on the energy levels of the atom. For $B_{ext} = 0$ (no external field), the energy level is given by the first term only. As B_{ext} becomes large enough for the second term to contribute significantly, the various allowed values of m_l take effect. Recall that for a given n and a given l, there are $2l + 1$ values possible for m_l (0, ±1, ±2, ..., ±l). Each permitted value of m_l produces a different value for E_{n,l,m_l}, thus splitting the original energy level into $2l + 1$ levels.

EXAMPLE 20–1: For an atom with an electron in the $n = 4$ state, what will be the energy levels that result from placing the atom in an external magnetic field B_{ext}?

SOLUTION: The corresponding values of l and m_l when $n = 4$ are

l	0	1	2	3
m_l	0	0, ±1	0, ±1, ±2	0, ±1, ±2, ±3

and the possible energy levels from Equation (20–3) are represented in Table 20–1.

Table 20–1 Possible energy levels ($n = 4$) for a hydrogen atom placed in an external magnetic field

	s STATE ($l = 0$)			p STATE ($l = 1$)			d STATE ($l = 2$)			f STATE ($l = 3$)		
	n	l	m_l	n	l	m_l	n	l	m_l	n	l	m_l
										4	3	3
							4	2	2	4	3	2
				4	1	1	4	2	1	4	3	1
	4	0	0	4	1	0	4	2	0	4	3	0
				4	1	−1	4	2	−1	4	3	−1
							4	2	−2	4	3	−2
										4	3	−3
NUMBER OF ENERGY LEVELS		1			3			5			7	

According to Equation (20–3) and this simple picture, the separation between any two consecutive energy levels is constant and given by

$$\boxed{\mu_B B_{\text{ext}}} \quad (20\text{–}4)$$

20–2 THE NORMAL ZEEMAN EFFECT

The lifetime for an atom remaining in an excited state is very short, usually of the order of 10^{-8} sec. An atom leaves such a state by making a transition to a state of lower total energy with a different quantum number. The energy difference between the two states must be lost to the atom through some mechanism in order to obey the conservation law for energy. This is often done by the emission of a photon of frequency

$$\nu = \frac{E_i - E_f}{h} \quad E_f < E_i$$

where E_i and E_f are the total energies of the initial and final states, respectively. The transition may take place spontaneously and in a random fashion with respect to time, *so there is no way to predict just when the transition will occur*. However, the transition probability per unit time can be calculated by using the quantum mechanics. The results of such calculations include rules governing how the quantum numbers l and m_l may change in going from E_i to E_f. For example, the changes

$$\boxed{\begin{aligned} \Delta l &= l_i - l_f = \pm 1 \\ m_l &= (m_l)_i - (m_l)_f = 0 \text{ or } \pm 1 \end{aligned}} \quad (20\text{–}5)$$

are *selection rules* governing the situations in which the transition probability per unit time is very high (approaching unity), the so-called *allowed transitions*. Other selection rules may be derived for transition probabilities very much lower than this (of the order of 10^{-6} sec^{-1}) governing the (partially) *forbidden transitions*. The lifetimes of atoms in excited states that, for one reason or another, may decay only via a forbidden transition are understandably quite long compared to those for the allowed transitions. Such cases are commonly encountered in the spectra of the exceedingly rare, hot gas clouds in stellar atmospheres and in the nebulae produced by supernovae.

There is no selection rule governing the quantum number n other than the requirement that E_f be less than E_i.

As an example of the application of the allowed-transition selection rules, let us consider the possible transitions between the hydrogenic d state ($l = 2$) and the p state ($l = 1$). These transitions are represented schematically in Figure 20–2(a)

CHAPTER 20: QUANTUM NUMBERS II: THE ZEEMAN EFFECT · 271

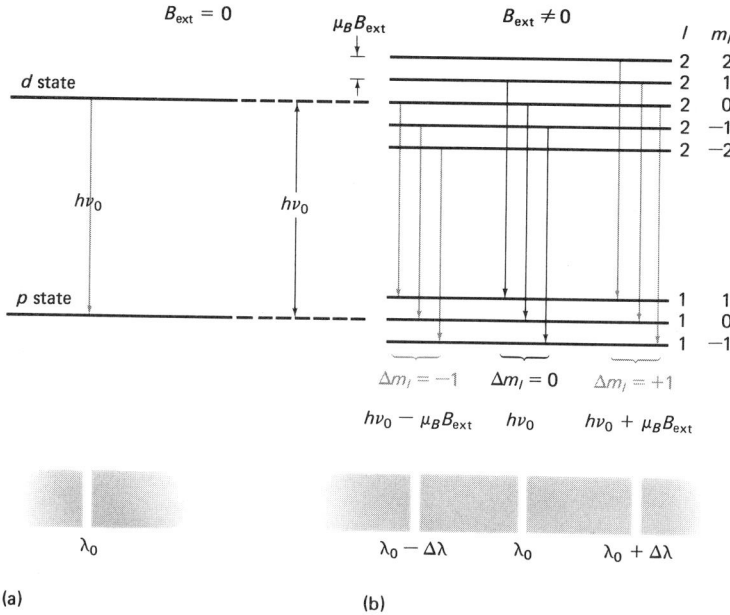

Figure 20–2 In the absence of an external magnetic field, only one spectral line results from the transition from the *d* state to the *p* state. When the field is applied, six transitions are possible although only three distinct energies or wavelengths arise.

for *zero external field* and in Figure 20–2(b) for an *applied external magnetic field*. It is seen that for $B_{ext} = 0$ only one transition is possible. The energy of the emitted photons is given by the Bohr formula

$$\Delta E = h\nu_0$$

with the corresponding spectral line having a wavelength

$$\lambda_0 = \frac{c}{\nu_0}$$

In this transition, the quantum number *l* changes from 2 to 1, so $\Delta l = 1$. With an external magnetic field applied, the former energy levels are split as indicated in Figure 20–2(b). The *d* state ($l = 2$) is split into five sublevels, because m_l can be $-2, -1, 0, +1,$ or $+2$; and the *p* state ($l = 1$) is split into three levels for m_l equal to $-1, 0,$ and $+1$.

The first group of three possible allowed transitions corresponding to the selection rules $\Delta l = 1$ and $\Delta m_l = -1$ involves the same energy release for each:

$$h\nu = h\nu_0 - \mu_B B_{ext} \tag{20-6}$$

The frequency and wavelength of the corresponding spectral line are

$$v = v_0 - \frac{\mu_B B_{ext}}{h}$$

and

$$\lambda = \frac{c}{v} = \frac{c}{v_0 - (\mu_B B_{ext}/h)}$$

The second group of transitions, those with selection rules $\Delta l = +1$, $\Delta m_l = 0$, are transitions that release energy hv_0. The frequency and wavelength of the corresponding spectral line are v_0 and $\lambda_0 = c/v_0$, respectively, which are the same as for the case when $B_{ext} = 0$.

Finally, the third group of transitions release a photon of energy

$$hv = hv_0 + \mu_B B_{ext} \tag{20-7}$$

with a frequency and wavelength of

$$v = v_0 + \frac{\mu_B B_{ext}}{h}$$

and

$$\lambda = \frac{c}{v} = \frac{c}{v_0 + (\mu_B B_{ext}/h)}$$

The original spectral line has thus been split into three components of wavelengths

$$\boxed{\frac{c}{v_0 - (\mu_B B_{ext}/h)} \qquad \frac{c}{v_0} \qquad \frac{c}{v_0 + (\mu_B B_{ext}/h)}}$$

Interpreting the Zeeman effect in this manner, that is, treating the orientation of a magnetic moment vector with respect to the direction of an applied magnetic field about which the vector precesses, clearly calls upon the space quantization of angular momentum. In much the same sense that the Franck–Hertz experiment provided some of the earliest direct evidence for the quantization of energy, so also has the Zeeman effect become one of the most direct confirmations for the concept of the *quantization of angular momentum*.

EXAMPLE 20–2: A spectrometer can resolve spectral lines in the visible region ($\lambda \cong 6000$ Å) when they are separated by $\Delta \lambda = 0.1$ Å. What will be the magnitude of the external magnetic field required to confirm experimentally the normal Zeeman effect?

SOLUTION: The energy separation between two consecutive spectral lines is $\Delta E = \mu_B B_{ext}$. The corresponding frequency separation is

$$\Delta v = \frac{\mu_B B_{ext}}{h}$$

which can also be written

$$|\Delta v| = \left| \Delta \left(\frac{c}{\lambda} \right) \right| = \frac{c \Delta \lambda}{\lambda^2}$$

Finally, the magnetic field required to produce a split in the spectral line that can just be resolved is

$$B_{ext} = \frac{hc \Delta \lambda}{\mu_B \lambda^2} = 0.059T \text{ wb/m}^2$$

where $\mu_B = 9.3 \times 10^{-24}$ J/Wb/m², $\lambda = 6.0 \times 10^{-7}$ m, and $\Delta \lambda = 1.0 \times 10^{-11}$ m.

20–3 TOTAL NUMBER OF STATES

Every combination of the three quantum numbers n, l, and m_l, defines a *state of the atom*. Since the general solution of the Schrödinger equation for the hydrogen atom,

$$\psi = R(r)\Theta(\theta)\Phi(\phi)$$

in which

$$R(r) = e^{-nrl} r^l L_{nl}(r)$$
$$\Theta(\theta) = P_{l, m_l}(\cos \theta)$$
$$\Phi(\phi) = A e^{lm_l \phi}$$

shows a dependence on the three quantum numbers n, l, and m_l, it can be concluded that every state is characterized by a corresponding eigenfunction

$$\psi_{n, l, m_l}$$

When the atom is not placed in an external magnetic field, the eigenvalues of the energy are

$$E_n = -\frac{me^4}{32\pi^2 \varepsilon_0^2 \hbar^2} \frac{1}{n^2}$$

and this energy is independent of l and m_l. As a consequence in this particular case, there is *not* a one-to-one correspondence between eigenvalues and eigenfunctions. The Schrödinger equation will have two or more solutions for one value of the energy, and the solutions in this case are said to be *degenerate*. When there is only one independent solution for a given eigenvalue of energy, the solution is *nondegenerate*.

EXAMPLE 20–3: Show that for a given principal quantum number n, there are n^2 possible states.

SOLUTION: For a given n, the orbital quantum number l can have the following n values:

$$l = 0, 1, 2, \ldots, (n-1)$$

For every value of l, the magnetic quantum number m_l has the following $2l + 1$ values:

$$m_l = 0, \pm 1, \pm 2, \ldots, \pm l$$

The total number of eigenfunctions or possible states is

$$\text{total number of states} = \sum_{l=0}^{l=n-1} (2l + 1) = 1 + 3 + 5 + \cdots + (2n - 1)$$

This is an arithmetic progression whose sum can be computed by

$$s = \left(\frac{a + u}{2}\right) n \tag{20-8}$$

in which a is the value of the first term, u is the value of the last term, and n is the total number of terms. Application of Equation (20–8) then shows that

$$\boxed{\text{total number of states} = \left(\frac{1 + (2n - 1)}{2}\right) n = n^2}$$

EXAMPLE 20–4: Make a table of all possible states and eigenfunctions for $n = 3$.

SOLUTION: For $n = 3$, the possible values of l and m_l are

l	0	1	2
m_l	0	0, ±1	0, ±1, ±2

Table 20–2 shows the possible combinations. The total number of states is $n^2 = 3^2 = 9$.

Table 20-2 States and eigenfunctions for $n = 3$

STATES			EIGENFUNCTIONS, ψ_{n,l,m_l}	NUMBER OF STATES, N
n	l	m_l		
3	0	0	$\psi_{3,0,0}$	1
3	1	-1	$\psi_{3,1,-1}$	
3	1	0	$\psi_{3,1,0}$	3
3	1	$+1$	$\psi_{3,1,1}$	
3	2	-2	$\psi_{3,2,-2}$	
3	2	-1	$\psi_{3,2,-1}$	
3	2	0	$\psi_{3,2,0}$	5
3	2	1	$\psi_{3,2,1}$	
3	2	2	$\psi_{3,2,2}$	
			Total	9

PROBLEMS

20-1 Make a diagram illustrating the split of the energy levels in the normal Zeeman effect for the states s, p, d, and f of the hydrogen atom.

20-2 Combine Equations (20–1) and (20–3) to show that the components of the magnetic dipole moment μ_l along the z axis are given by

$$\mu_{l_z} = \mu_B m_l$$

which are the only ones that can be observed and measured experimentally.

20-3 According to Problem 20-2, the z components of μ_l are integral multiples of the Bohr magneton. Hence, μ_B is a natural unit for the determination of μ_{l_z}. (a) For $n = 3$, how many values of μ_{l_z} are possible? (b) Is μ_l space-quantized? (c) How many orientations of μ_l are possible for the d state?

20-4 The nomenclature s, p, d, f for the states $l = 0, 1, 2, 3$, respectively, originated with the early spectroscopists, who called the corresponding series by the names sharp, principal, diffuse, and fundamental. Draw an energy level diagram in which the allowed transitions in the normal Zeeman effect for the principal series are illustrated.

20-5 The lack of information about the y and z components of the orbital angular momentum **L** of the electron in the hydrogen atom can be illustrated by drawing a vector **L** of magnitude $|\mathbf{L}| = \sqrt{l(l+1)}\hbar$ at the origin of the (x, y, z) reference frame, in which the z axis is taken along the direction of

an external magnetic field B_{ext}. Recall from Equation (19–13) that $L_z = m_l \hbar$.
(a) Show that the projection of **L** onto the xy plane is given by

$$L_{xy} = \sqrt{(l^2 + l - m_l^2)}\hbar$$

(b) Compute the different possible values of L_z and L_{xy} for the f state.
(c) Draw to scale a diagram for part (b).

20–6 (a) For Problem 20–5, draw a diagram in which the lack of information about L_x and L_y can be pictured as a conical surface of revolution about the z axis with the vector **L** shown precessing about B_{ext}.
(b) Compute the different values of the polar angle θ from the equation $L_z = L \cos \theta = m_l \hbar$ for the d state.

20–7 The normal Zeeman effect takes place when the atom gas is under the action of a very intense external magnetic field. The orbital angular momentum **L** precesses about B_{ext}. Analyze the normal Zeeman effect for the allowed transitions in the sharp, principal, and diffuse series.

20–8 Compute the separation between the consecutive states in the normal Zeeman effect (a) in units of energy (electron volts and joules), (b) in units of frequency (sec^{-1}), and (c) in units of length (Å and meters). Assume that $B_{ext} = 1.2$ Wb/m².

20–9 A hydrogen atom gas is placed in an intense magnetic field $B_{ext} = 0.8$ Wb/m². Compute the separation between two consecutive states (normal Zeeman effect) when $\lambda = 5000$ Å. Give the results in units of wavelength and frequency.

20–10 In reference to Problems 20–5 and 20–6, the fact that the magnitude of **L**, $|\mathbf{L}| = \sqrt{l(l + 1)}\hbar$, and the z component, $L_z = m_l \hbar$, can be measured simultaneously indicates that there is no uncertainty relation between **L** and L_z. However, if L_z is known, L_x and L_y cannot be known simultaneously. Only $L_{xy} = \sqrt{l^2 + l - m_l^2}\hbar$ is known. L_z and the L_{xy} orientation, given by the azimuth angle ϕ, cannot be known *simultaneously* with absolute precision. Hence, L_z and ϕ are also *conjugate variables*, like linear momentum and position. Therefore, in addition to the uncertainty relationships [see R. T. Birge, *Rev. Mod. Phys.* **13**, 233 (1941)] $\Delta p \, \Delta x \geq h$ and $\Delta E \, \Delta t \geq h$ it is possible to add $\Delta L_z \, \Delta \phi \geq h$. Compute the different uncertainties in $\Delta \phi$ for the d state.

RECOMMENDED READING

Beiser, A., *Concepts of Modern Physics*, McGraw-Hill, New York, 1963, Chap. 8.
Marks, Hans, and Olson, N. T., *Experiments in Modern Physics*, McGraw-Hill, New York, 1966, pp. 180–186.

PAULING, L., and GOUDSMIT, S., *The Structure of Line Spectra*, McGraw-Hill, New York, 1930.

SHAMOS, M. H., *Great Experiments in Physics*, Holt, Rinehart & Winston, New York, 1959, pp. 301–302.

TIPLER, P. A., *Foundations of Modern Physics*, Worth, New York, 1969, pp. 347–355.
An outstanding presentation of the whole subject of modern physics. Beautiful illustrations. This is certainly a book written very carefully and with devotion.

WEIDNER, R. T., and SELLS, R. L., *Elementary Modern Physics*, Allyn & Bacon, Boston, 1967, pp. 225–235.
A very popular standard book, written at the undergraduate introductory level. The main feature is its excellent selection of problems.

WHITE, H. E., *Introduction to Atomic Spectra*, McGraw-Hill, New York, 1934.

WHITE, H. E., *Introduction to Atomic and Nuclear Physics*, Van Nostrand Reinhold, New York, 1964, pp. 150–152.
A very elementary presentation of the normal Zeeman effect.

21
The Wave Functions of the Hydrogen Atom

Harold Clayton Urey
(1893–)

Born in Walkerton, Indiana, Urey received his Ph.D. from the University of California in 1923. He was professor of chemistry at Johns Hopkins University (1924–1929), at Columbia University (1929–1945), and at the University of Chicago (1945–1958). His research has been in various fields, including the separation of isotopes, absorption spectra, and molecular structure. Urey is also active in the study of the origin of the solar system and atomic and nuclear structure. In 1934 he received the Nobel Prize in chemistry for the discovery of deuterium.

21–1 THE WAVE FUNCTIONS OF THE HYDROGEN ATOM
21–2 THE RADIAL PROBABILITY DISTRIBUTION
21–3 ANGULAR PROBABILITY DEPENDENCE

21-1 THE WAVE FUNCTIONS OF THE HYDROGEN ATOM

We have seen that the Schrödinger equation for the hydrogen atom,

$$H\psi = E\psi$$

has the general solution

$$\psi_{n,l,m_l} = R_{n,l}\Theta_{lm_l}\Phi_{m_l} = [e^{-nr}r^l L_{nl}(r)][P_{l,m_l}(\cos\theta)](Ae^{im_l\phi}) \quad (21\text{-}1)$$
$$\text{radial part} \qquad \text{polar part} \quad \text{azimuthal part}$$

where $L_{n,l}(r)$ are called the *associated Laguerre polynomials*, and $P_{l,m_l}(\cos\theta)$ are *Legendre polynomials*.* In these equations, n, l, and m_l are the principal orbital and magnetic quantum numbers, respectively, and take on the following ranges of values:

$$n = 1, 2, 3, \ldots$$

$$l = 0, 1, 2, \ldots, (n-1) \qquad n \text{ values}$$

$$m_l = 0, \pm 1, \pm 2, \ldots, \pm l \qquad (2l+1) \text{ values}$$

Every allowed combination of these three numbers defines a *state* of the atom, and for every state there is a corresponding eigenfunction ψ_{nlm_l}. The function ψ_{nlm_l} is, in general, complex. If $\psi^*_{nlm_l}$ is the complex conjugate of this function, then the real quantity

$$p = \psi^*_{nlm_l}\psi_{nlm_l} \quad (21\text{-}2)$$

represents the probability per unit volume, or the *probability density*, of finding an electron at a given point in space relative to the central proton nucleus. Taking

---

* The L_{nl} polynomials are given by $L_{nl} = (A_0 + A_1 r + A_2 r^2 + \cdots + A_{n-l-1}r^{n-l-1})$. For more information about the Laguerre and Legendre polynomials and functions, see M. R. Spiegel, *Mathematical Handbook of Formulas and Tables*, McGraw-Hill, New York, 1968, pp. 149–150 and 153–155.

$dV = r^2 \sin\theta \, dr \, d\theta \, d\phi$ as the element of volume in spherical coordinates, then we may require that

$$\int_{\phi=0}^{2\pi} \int_{\theta=-\pi}^{+\pi} \int_{r=0}^{\infty} \psi_{nlm_l}^* \psi_{nlm_l} \, dv = 1$$

as a normalization condition, stating that the electron is, in fact, to be found somewhere about the proton. Multiplying the probability density by the electronic charge then yields a charge density distribution for the atom.

21–2 THE RADIAL PROBABILITY DISTRIBUTION

In general, the probability density distribution $\psi^*\psi$ for locating the particle depends on the three coordinates r, θ, and ϕ as well as the quantum numbers n, l, and m_l. The distributions are different for the different states. It is informative to investigate the distribution of density as a function of the length of the radius vector. The radial part of the wave function is, from Equation (21–1),

$$R_{nl} = e^{-nr} r^l L_{nl}(r)$$

This function is real and independent of the angular coordinates θ and ϕ and from the magnetic quantum number m_l. The probability per unit volume for finding the particle at a given point along the radius is

$$\boxed{p_r = R_{nl}^* R_{nl}} \qquad (21\text{–}3)$$

The element of volume dependent only on the radius is a spherical shell enclosed by the two concentric spheres of radii r and $r + dr$, respectively. This volume is $dV_r = 4\pi r^2 \, dr$ (Figure 21–1), and the *radial probability of finding the particle in dV_r* is

$$\boxed{P_r = R_{nl}^* R_{nl} \, dV_r = 4\pi r^2 R_{nl}^2 \, dr = p_r \, dV_r} \qquad (21\text{–}4)$$

or

$$P_r = e^{-2nr} r^{2(l+1)} L_{nl}^2(r) 4\pi \, dr$$

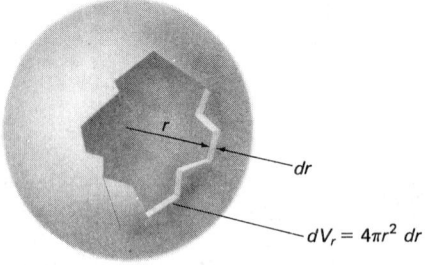

Figure 21–1 Element of volume of a spherical shell.

CHAPTER 21: THE WAVE FUNCTIONS OF THE HYDROGEN ATOM · 281

The radial probability is the probability of finding the electron between the two concentric spheres of radii r and $r + dr$. The charge density at this radius is just the electronic charge multiplied by this probability.

Figure 21-2 is a graph of the radial probability P_r versus the radial distance measured in units of the Bohr radius $r_1 = 0.53$ Å. For the 1s state ($n = 1, l = 0$), Equation (21-4) becomes

$$P_r = R_{10}^* R_{10}\, dV_r = R_{10}^2\, dV_r = e^{-2r} r^2 A_0^2 4\pi\, dr$$

Note that when $r = 0$, $P_r = 0$; and the maximum probability is obtained when $r = r_1 = 0.53$ Å, which is in agreement with the radius of the orbit for the ground state as computed by the Bohr theory. As r approaches $5r_1$, P_r becomes very small and approaches zero.

The 2s state ($n = 2, l = 0$) has a probability

$$P_r = e^{-4r} r^2 (A_0 + A_1 r)^2 4\pi\, dr$$

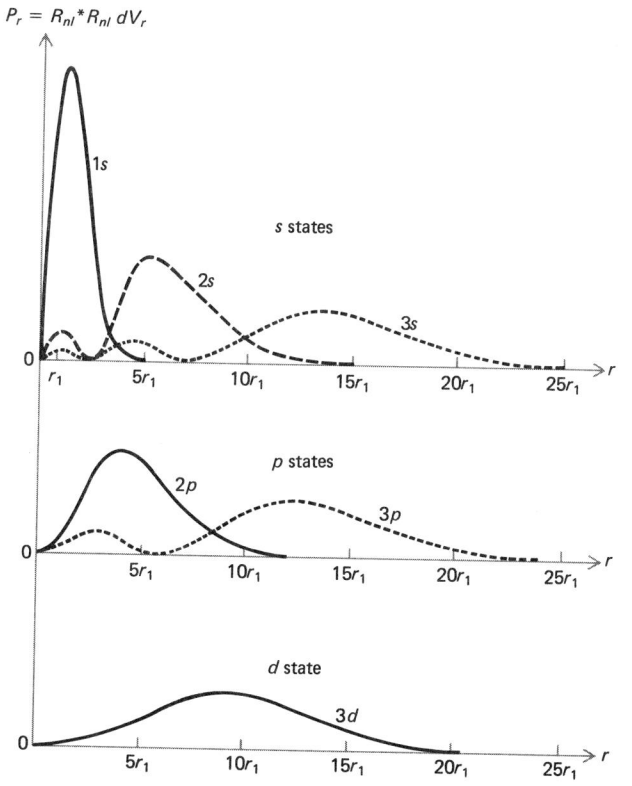

Figure 21-2 Radial probability $P_r = R_{nl}^* R_{nl}\, dV_r$ versus distance r in units of the Bohr radius $r_1 = 0.53$ Å. Vertical scale is in arbitrary units.

and as before $P_r = 0$ for $r = 0$. Now there are two maxima, one at $r = r_1$ and the other at $r \cong 5r_1$, with the second being larger than the first.

The probability function for the 3s state ($n = 3, l = 0$) is

$$P_r = e^{-6r}r^2(A_0 + A_1r + A_2r^2)^2 4\pi \, dr$$

and three maxima are now found to occur around r_1, $4r_1$, and $12r_1$, with the latter being the largest.

The electrons found in the 2p state have a maximum radial probability at about $4r_1$, and those in the 3p state have maxima at about $2r_1$, and $12r_1$. Notice that in each case there is a maximum more pronounced than the others. The probability of locating the electrons at one of these distances along the radius is greater here, and we have the mechanism for locating the electrons in certain shells about the nucleus. Unlike the Bohr theory, however, there are other positions where there is a definite probability of locating the electron.

In each case, the probability of finding the electron at $r = 0$ is zero. The constants $A_0, A_1, A_2, \ldots$ can be found by using the normalization condition*:

$$\frac{1}{4\pi}\int_{r=0}^{r=\infty} R_{nl}^* R_{nl} \, dV_r = \frac{1}{4\pi}\int_{r=0}^{r=\infty} R_{nl}^2 \, dV_r = 1 \qquad (21\text{-}5)$$

21-3 ANGULAR PROBABILITY DEPENDENCE

The probability per unit volume of finding the electron at a particular point in space is

$$p = R_{nl}^* R_{nl} \Theta_{lm_l}^* \Theta_{lm_l} \Phi_{m_l}^* \Phi_{m_l} \qquad (21\text{-}6)$$

but since

$$\Phi_{m_l}^* \Phi_{m_l} = (A^* e^{-im_l\phi})(A e^{im_l\phi}) = A^* A = \text{constant}$$

the probability density in this case is not dependent on ϕ. The probability density is determined by the product of the radially dependent probability $p_r = R_{nl}^* R_{nl}$ and the θ-dependent probability

$$p_\theta = \Theta_{lm_l}^* \Theta_{lm_l} \qquad (21\text{-}7)$$

---

* The reason for the factor $1/4\pi$ in Equation (21-5) is that while $(R_{nl}^* R_{nl})$ represents the probability per unit volume along the radius, $(1/4\pi)(R_{nl}^* R_{nl})$ represents the probability per unit volume per unit solid angle (in sterradians), and this is the one that must be normalized.

CHAPTER 21: THE WAVE FUNCTIONS OF THE HYDROGEN ATOM · 283

Since Θ_{lm_l} depends on l and m_l as well as θ, the effect of p_θ is different for the various states as determined by the quantum numbers l and m_l. Table 21-1 lists several of the normalized zenithal functions Θ_{lm_l} for different combinations of l and m_l. Because $\Theta_{lm_l}*\Theta_{lm_l}$ is independent of ϕ, a plot of $p_\theta = \Theta_{lm_l}*\Theta_{lm_l}$ versus θ in each case exhibits a symmetry of revolution with respect to the z axis.

Table 21-1 Normalized functions $\Theta_{lm_l} = P_{lm_l}(\cos\theta)$ for various states as determined by l and m_l

STATE	l	m_l	Θ_{lm_l}
s	0	0	$\Theta_{0,0} = \dfrac{1}{\sqrt{2}}$
p	1	0	$\Theta_{1,0} = \sqrt{\tfrac{3}{2}}\cos\theta$
p	1	±1	$\Theta_{1,\pm 1} = \sqrt{\tfrac{3}{4}}\sin\theta$
d	2	0	$\Theta_{2,0} = \sqrt{\tfrac{5}{8}}(3\cos^2\theta - 1)$
d	2	±1	$\Theta_{2,\pm 1} = \sqrt{\tfrac{15}{4}}\sin\theta\cos\theta$
d	2	±2	$\Theta_{2,\pm 2} = \sqrt{\tfrac{15}{16}}\sin^2\theta$

There is only one possible energy state for the s state, that is, the state for $l = 0$, $m_l = 0$. The Legendre polynomial from Table 21-1 is $\Theta_{0,0} = 1/\sqrt{2}$, and $p_\theta = \Theta_{0,0}*\Theta_{0,0} = \tfrac{1}{2}$. When p_θ is plotted against θ in polar coordinates, the graph is a sphere of radius $\tfrac{1}{2}$ as shown in Figure 21-3(a). The straight line from the origin to the curve has a length equal to $\Theta_{lm_l}^*\Theta_{lm_l}$ and represents the probability that the electron is in the direction of that line. The diagrams in Figure 21-4 show the total probability density determined by the angular probability p_θ as it modulates or affects the radial probability p_r.

In the p state, when $m_l = 0$ the Legendre polynomial is $\Theta_{1,0} = \sqrt{\tfrac{3}{2}}\cos\theta$ with the θ-dependent probability density given by $p_\theta = \tfrac{3}{2}\cos^2\theta$. Figures 21-3(a) and 21-4(b) show the probability distribution as two lobes with a maximum at $\theta = 0$ and a minimum at $\theta = \pm\pi/2$. These lobes are symmetrical about the z axis, and the most probable position of the electron is indicated by the density of points in Figure 21-4(b). Also in the p state, when $m_l = \pm 1$, the probability distribution $p_\theta = \tfrac{3}{4}\sin\theta$ is again two lobes oriented 90° away from the $m_l = 0$ case. The probability is a maximum when $\theta = \pm\pi/2$ and zero in the vertical direction when $\theta = 0$ or π.

The Bohr theory makes no provision for locating the electron in the positions indicated by Figures 21-2, 21-3, and 21-4. There is some correspondence, however, for the cases $l = n - 1$. Notice from Figure 21-2 that the most probable location of the electron corresponds closely to the radius determined by the Bohr theory.

284 · PART THREE: THE ATOM

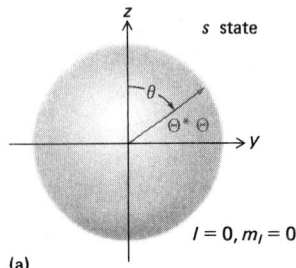

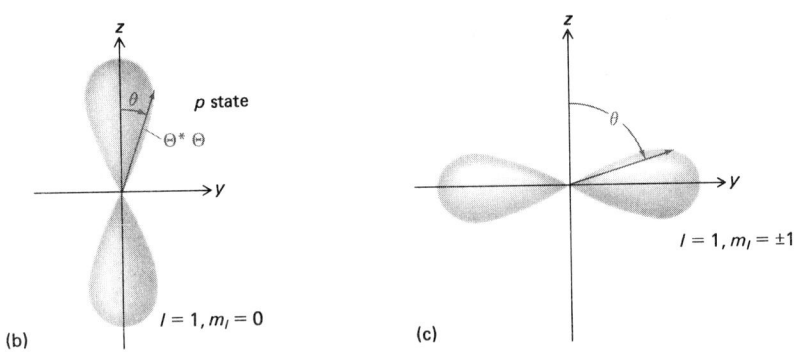

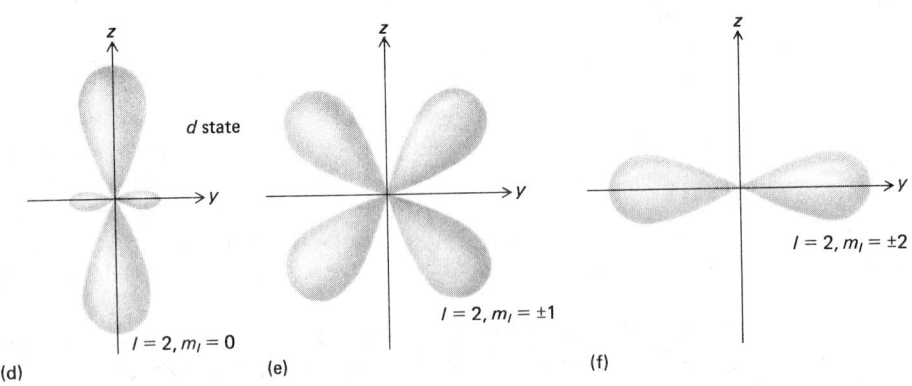

Figure 21-3 Probability density $p_\theta = \Theta^*\Theta$ is plotted against θ. Since p_θ is independent of ϕ, there is a symmetry of revolution about the z axis.

CHAPTER 21: THE WAVE FUNCTIONS OF THE HYDROGEN ATOM · 285

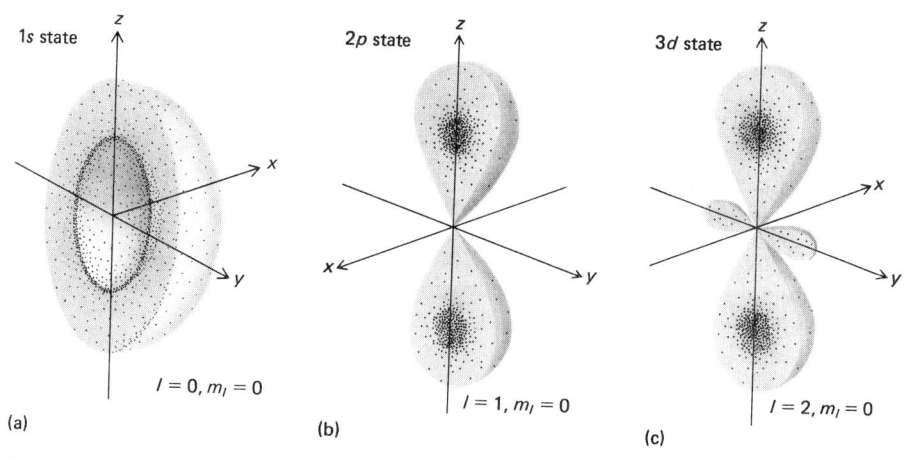

Figure 21-4 The probability density distribution in three dimensions of ψ^2_{nlm} is represented by the density of points. The figures are obtained by combining Figures 21-2 and 21-3.

PROBLEMS

Table 21-2 First normalized radial functions R_{nl} for the hydrogen atom[a]

n	l	R_{nl}
1	0	$2\left(\dfrac{1}{r_1}\right)^{3/2} e^{-r/r_1}$
2	0	$\dfrac{1}{2\sqrt{2}}\left(\dfrac{1}{r_1}\right)^{3/2}\left(2 - \dfrac{r}{r_1}\right) e^{-r/2r_1}$
2	1	$\dfrac{1}{2\sqrt{6}}\left(\dfrac{1}{r_1}\right)^{3/2}\left(\dfrac{r}{r_1}\right) e^{-r/2r_1}$
3	0	$\dfrac{1}{9\sqrt{3}}\left(\dfrac{1}{r_1}\right)^{3/2}\left(6 - \dfrac{4r}{r_1} + \dfrac{4r^2}{9r_1}\right) e^{-r/3r_1}$
3	1	$\dfrac{1}{9\sqrt{6}}\left(\dfrac{1}{r_1}\right)^{3/2}\left(\dfrac{2r}{3r_1}\right)\left(4 - \dfrac{2r}{3r_1}\right) e^{-r/3r_1}$
3	2	$\dfrac{1}{9\sqrt{30}}\left(\dfrac{1}{r_1}\right)^{3/2}\left(\dfrac{2r}{3r_1}\right)^2 e^{-r/3r_1}$

[a] r_1 is the Bohr radius.

21-1 Construct a table of the probabilities of finding the electron between two concentric spheres of radii r and $r + dr$, according to Equation (21-4).

21-2 Find the probabilities of locating an electron in an interval dr centered around r, in the following cases: (a) $n = 1, l = 0$; (b) $n = 2, l = 0$; (c) $n = 3, l = 0$. Compare these results with Figure 21-2.

21-3 (a) Show that for the state $n = 1, l = 0$, the maximum probability of finding the electron for $r = r_1$ is in agreement with the results of the Bohr theory.
(b) Show also that for the same state, $P_r = 0$ when $r \to 0$ or when $r \to \infty$.

21-4 (a) Using Table 21-2, show that for the state $n = 2, l = 0$, there are two maxima for P_r centered approximately about $r = 0.7r_1$ and $r = 5r_1$. Also show that P_r is larger for the second case.
(b) For the state $n = 3, l = 0$, show that there are three maxima for P_r centered approximately about $r = 0.6r_1$, $r = 4r_1$, and $r = 13r_1$.
(c) For the states mentioned above, show that $P_r \to 0$ when $r \to 0$ and when $r \to \infty$. Is this a general property?

21-5 Use the results in Table 21-1 and find the zenithal probability $p_\theta = \Theta^*\Theta$ when $\theta = \pi/4$ in the following cases: (a) $l = 0, m_l = 0$; (b) $l = 1, m_l = 1$; (c) $l = 2, m_l = -1$; (d) $l = 2, m_l = 2$.

21-6 By combining the results of Tables 21-1 and 21-2, find the probability densities $\psi^*\psi$ in the following cases: (a) $n = 1, l = 0, m_l = 0$; (b) $n = 2, l = 0, m_l = 0$; (c) $n = 2, l = 1, m_l = 0$; (d) $n = 2, l = 1, m_l = -1$.

RECOMMENDED READING

EDMONDS, A. R., *Angular Momentum in Quantum Mechanics*, Princeton University Press, Princeton, N.J., 1957, Chap. 2.

FONG, P., *Elementary Quantum Mechanics*, Addison-Wesley, Reading, Mass., 1968, Chap. 8.

SAXON, D. S., *Elementary Quantum Mechanics*, Holden-Day, San Francisco, 1964, Chap. 10.

SCHIFF, L. I., *Quantum Mechanics*, McGraw-Hill, New York, 1949, pp. 69–91.
A very well written book at the upper undergraduate or junior graduate level.

SPOSITO, G., *An Introduction to Quantum Physics*, Wiley, New York, 1970, Chap. 4.

WHITTAKER, E. T., and WATSON, G. N., *A Course in Modern Analysis*, 4th ed., Cambridge University Press, London, 1935.
Chapter 15 contains a very complete study of the Legendre polynomials.

WHITTAKER, E. T., and WATSON, G. N., *Modern Analysis*, Cambridge University Press, London, 1946, pp. 323–325.

22
Electron Spin

Otto Stern
(1888–1969)

Stern was born in Upper Silesia, Germany, and earned his doctorate from the University of Breslau. He later taught with Einstein at the universities of Prague and Zurich. In 1933 he left Nazi Germany and became research professor at the Carnegie Institute of Technology (now Carnegie-Mellon University) and then professor emeritus. Stern's research confirmed the existence of the magnetic moment in atoms, and he discovered the magnetic moment of the proton. In 1943 he received the Nobel Prize in physics for developing the molecular beam technique with which he conducted his study of atomic particles.

22–1 INTRINSIC SPIN
22–2 SPIN ANGULAR MOMENTUM
22–3 THE STERN–GERLACH EXPERIMENT
22–4 ENERGY OF THE SPIN-ORBIT INTERACTION—FINE STRUCTURE

22-1 INTRINSIC SPIN

In the previous chapters, we have seen that quantum mechanics greatly extends our model of the atom over that previously suggested by Bohr. This new approach based on quantum mechanics gives a somewhat complete and satisfying picture of the hydrogen atom. However, experience tells us that new experiments and new ideas will cause us again to change our model or even develop a new model. In this chapter, the electron is no longer pictured as just a point of charge in space. It will be seen that it has certain physical attributes, which, when understood, lead to the explanation of some of the fine details of atomic spectra.

The strongest emission of $^{23}_{11}$Na is from a $3p \rightarrow 3s$ transition, and in a simple spectrum it appears as a single strong line. However, on close inspection it is found that it is not a simple line, but a doublet consisting of two wavelengths, 5890.12 Å and 5896.26 Å. This splitting of a single spectral line into two distinct but closely spaced lines is known as the *fine structure* of the atomic spectra.

In Chapter 20, a kind of fine structure was observed when an atom was placed in an external magnetic field. This interaction of the electrons within the atom with the magnetic field is the Zeeman effect. The fine structure, like that in the sodium doublet, differs from the Zeeman effect in that it does not require the presence of an external magnetic field, and the splitting between any two consecutive levels is different.

In order to explain this splitting, an addition to the model of the atom as it has been visualized this far must be made. A. H. Compton, in 1921, had suggested that the electron might be a spinning particle. Then, in 1925, SAMUEL GOUDSMIT and GEORGE UHLENBECK, both graduate students at Leiden University, theoretically explained the splitting of the line spectra by assuming that the electron had a rotational motion about an axis within itself. Associated with the spinning electron is an *intrinsic magnetic dipole moment* μ_s and an angular momentum L_s called the *spin angular momentum*. The angular momentum L of the electron about the nucleus can be compared with the rotation of the earth about the sun as it goes through the seasons of the year, while the spin angular momentum can be compared to the rotation of the earth on its axis as it produces day and night. In 1928 Dirac developed a relativistic quantum theory of the electron, and the effect of this intrinsic spin of the electron was a natural consequence of this theory.

A passenger riding on an electron as it orbits a proton would observe the proton orbiting the electron in an orbit of the same size as it is in Figure 22–1. This motion of the proton about the electron produces a magnetic field as shown in the figure at the site of the electron. In the figure, **L** is the orbital angular momentum of the electron around the proton and **L**$_s$ is the spin angular momentum of the electron. The orbital motion of the electron produces a magnetic dipole moment μ_l at the proton and the intrinsic spinning of the electron, like a spinning negatively charged sphere, produces the magnetic dipole moment μ_s (see Figure 22–2).

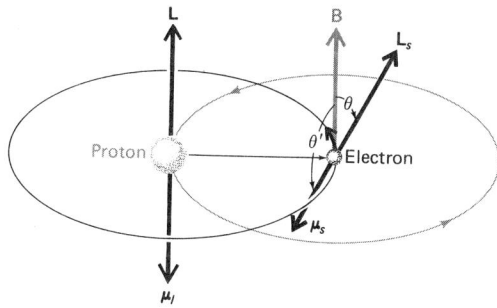

Figure 22–1 A passenger riding on an electron as it orbits a proton would see the proton in an orbit of the same size about the electron. The motion of this proton about the electron produces a magnetic field **B** at the site of the electron.

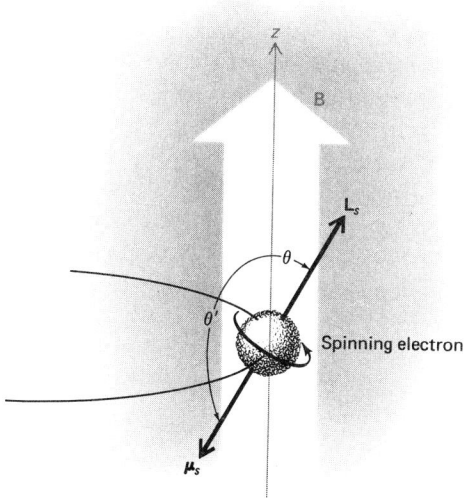

Figure 22–2 The magnetic dipole moment μ_s produced by the intrinsic spin of the electron interacts with the magnetic field **B** produced by the proton as it moves about the electron.

The magnetic field produced by the proton as it circles the electron interacts with the magnetic dipole moment μ_s from the spinning electron. This results in an additional energy term,

$$\Delta E_s = -\boldsymbol{\mu}_s \cdot \mathbf{B} = -\mu_s B \cos \theta' = \mu_s B \cos \theta \qquad (22\text{-}1)$$

which must be added to the energy from the electrostatic interaction of the electron and proton.

This interaction is called the *spin-orbit interaction*, because it relates the magnetic dipole moment μ_s, from the spin of the electron, and the magnetic field **B** due to the orbital motion of the electron. This spin-orbit interaction is the source of additional energy levels, which produce the fine structure of the spectrum. In the absence of an applied external magnetic field then, the total energy of the atom is

$$\begin{aligned} E &= -\frac{me^4}{32\pi^2\varepsilon_0^2\hbar^2}\frac{1}{n^2} + \Delta E_s \\ &= -\frac{me^4}{32\pi^2\varepsilon_0^2\hbar^2}\frac{1}{n^2} + \mu_s B \cos \theta \end{aligned} \qquad (22\text{-}2)$$

22-2 SPIN ANGULAR MOMENTUM

With the addition of the spinning electron to the model of the atom, another quantum number must be introduced. Dirac's relativistic quantum treatment of the electron introduced the *spin angular momentum quantum number* (or just the spin quantum number) s, and it has since been experimentally verified by study of the fine structure of atomic spectra. The spin angular momentum is quantized according to

$$L_s = \sqrt{s(s+1)}\,\hbar \qquad (22\text{-}3)$$

in which the *spin angular momentum quantum number s has only the single unique value* $\tfrac{1}{2}$. The only possible value of L_s is

$$L_s = \sqrt{\tfrac{1}{2}(\tfrac{1}{2} + 1)}\,\hbar = \sqrt{3}\,\frac{\hbar}{2}$$

The number s is called a quantum number, but it is not used along with the other quantum numbers since it has only the value $\tfrac{1}{2}$ and does not really distinguish one state from another. The only value the spin angular momentum can have is $L_s = \sqrt{3}\hbar/2$ and for this reason it is as much a fundamental character of the electron as the mass or the charge.

Again, similar to the Zeeman effect, if the direction along the magnetic field

B is chosen as the z direction (Figure 22–2), then the component of $\mathbf{L}_s$ in the z direction is quantized and given by

$$L_{sz} = L_s \cos \theta = m_s \hbar \qquad (22\text{–}4)$$

where $m_s = \pm\frac{1}{2}$ and is called the *magnetic spin quantum number*. Frequently, $m_s = \frac{1}{2}$ is designated as *spin-up* (↑) and $m_s = -\frac{1}{2}$ as *spin-down* (↓).

Table 22–1 lists the complete set of quantum numbers. Since s is always $\frac{1}{2}$, the state of the system can be completely designated by the quantum numbers n, l, m_l, m_s. Note that for every l there are $2l + 1$ values of m_l, and for every s there are $2s + 1 = 2$ possible values of m_s.

Table 22–1

QUANTUM NUMBERS	VALUES	NUMBER OF POSSIBLE VALUES
principal, n	1, 2, 3, …	any number
orbital, l	0, 1, 2, …, $(n-1)$	n
magnetic, m_l	0, ±1, ±2, …, ±l	$2l + 1$
spin, s	$\frac{1}{2}$	1
magnetic spin, m_s	$\pm\frac{1}{2}$	$2s + 1 = 2$

There are only two possible orientations of the spin angular momentum in a given direction, such as that of the magnetic field produced by a proton circulating about an electron. The two possible values shown in Figure 22–3 from Equation (22–4) are $L_{sz} = \pm\frac{1}{2}\hbar$. This restriction is known as the *space quantization* of the spin angular momentum.

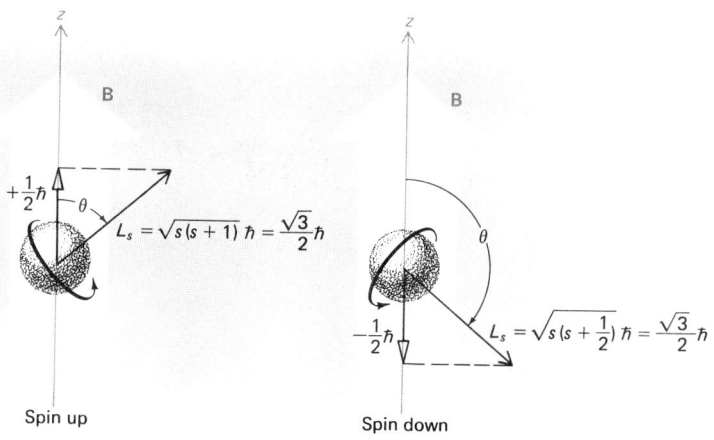

Figure 22–3 The spin angular momentum of the electron is space quantized, since it has only two allowed orientations with respect to a magnetic field **B**. One orientation has a component in the direction of the field of $L_{sz} = \frac{1}{2}\hbar$ and the other has a component $L_{sz} = -\frac{1}{2}\hbar$.

22-3 THE STERN–GERLACH EXPERIMENT

The remarkable idea that the spin angular momentum of electrons is space quantized in a magnetic field is a subtle one, and at first glance it would seem difficult to verify experimentally. In 1920 OTTO STERN, from his work with beams of atomic and molecular particles, designed an experiment in collaboration with WALTER GERLACH, which became a direct experimental verification of the space quantization of L_s.

In the Stern–Gerlach experiment shown in Figure 22–4, silver is boiled in an oven, and atoms of silver stream from an opening in the oven into an evacuated region where they travel in straight lines through collimating slits. These atoms then pass through a very inhomogeneous magnetic field between the shaped poles of a magnet and fall onto a photographic plate. Classical theory predicts that a single smeared pattern should be produced on the photographic plate. Instead, the stream of silver atoms splits into two distinct lines after passing through the inhomogeneous magnetic field.

As the shells of electrons about the nucleus are filled, the electrons are alternately aligned such that each new pair consists of one electron with *spin-up* and one electron with *spin-down*. As a result, each completed shell has a resultant spin angular momentum of zero. In a normal silver atom, all of the shells are filled with the exception of the outer shell, which contains a single electron. This electron has an orbital quantum number $l = 0$, and the total angular momentum is*
$J = \sqrt{j(j+1)}\hbar$ or $J = \sqrt{\frac{1}{2}(\frac{1}{2}+1)}\hbar = \frac{1}{2}\sqrt{3}\hbar$, since $j = \frac{1}{2}$. When $l > 0$, then $j = l \pm s$.*

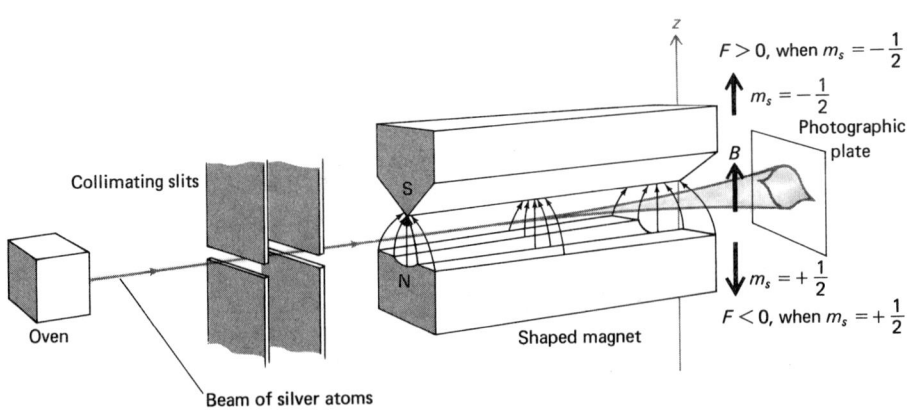

Figure 22–4 The Stern–Gerlach experiment, in which a beam of silver atoms passes between the poles of a shaped magnet.

---

* The quantum number j is often called the *total angular momentum quantum number*. See, for example, M. Alonso and E. J. Finn, *Fundamental University Physics*, Addison-Wesley, Reading, Mass., 1968, Vol. 3, pp. 137–139.

The spin angular momentum is quantized according to $L_s = \sqrt{s(s+1)}\hbar$, and with respect to the z direction determined by the magnetic field it has only components (Figure 22-3)

$$L_{sz} = L_s \cos \theta = m_s \hbar$$

A spinning electron creates a magnetic dipole moment μ_s [Equation (22-4)] that under the action of the magnetic field **B** experiences a torque

$$\tau = \mu_s \times \mathbf{B}$$

that tends to align the dipole with the applied field **B**. The energy associated with this magnetic dipole is*

$$\boxed{\Delta E = -\mu_s \cdot \mathbf{B} = -\mu_s B \cos \alpha} \tag{22-5}$$

Figure 22-5 shows a magnetic dipole in an *inhomogeneous magnetic field*. In addition to the rotational effect given by the torque, there is a net translational force†

$$\boxed{F = -\frac{d}{dz}(\Delta E)} \tag{22-6}$$

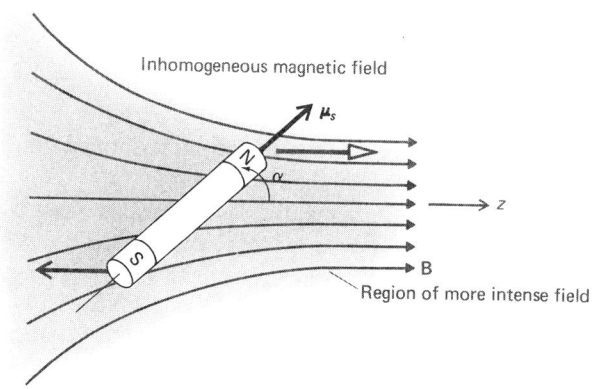

Figure 22-5 A magnet somewhat aligned with a magnetic field is pulled toward the region of the more intense field. A dipole not aligned will be pushed away from the region of intense field.

---

* Since μ_s is related to L_s, and **B** is due to the orbital motion of the proton as viewed from the electron (Figure 22-1), it can be said that the energy $\Delta E = -\mu_s \cdot \mathbf{B}$ is associated with the spin-orbit interaction.

† See, for example, G. Sposito, *An Introduction to Quantum Physics*, Wiley, New York, 1970, p. 11.

which pushes the dipoles that are somewhat aligned with the field in the direction of the most intense part of the field; or for those antialigned, they are pushed away from the intense portion of the field. The energy associated with the interaction of L_s and the external field **B** is

$$\Delta E = m_s 2 \left(\frac{e\hbar}{2m}\right) B = 2m_s \mu_B B \qquad (22\text{-}7)$$

where $\mu_B = e\hbar/2m$ is the Bohr magneton. This equation will be derived in Section 22–4.

The net force acting on the atoms is

$$F = -\frac{d}{dz}(\Delta E) = -m_s 2 \left(\frac{e\hbar}{2m}\right)\frac{dB}{dz} \qquad (22\text{-}8)$$

Those atoms with $m_s = +\tfrac{1}{2}$ are displaced downward in Figure 22–4, since from Equation (22-8), $F < 0$ because $dB/dz > 0$ and **B** increases in the $+z$ direction as indicated. For atoms with $m_s = -\tfrac{1}{2}$, $F > 0$, and the atoms are displaced upward.

In general, as the total angular momentum is $J = \sqrt{j(j+1)}\hbar$, the number of lines in the Stern–Gerlach pattern is

$$\text{multiplicity} = 2j + 1 \qquad (22\text{-}9)$$

Thus, in addition to silver atoms, any atom with a single outer electron in some state $l = 0$ should show a pattern of two distinct lines. With apparatus of increased accuracy, the Stern–Gerlach experiment was repeated with hydrogen, copper, gold, sodium, and potassium, and all showed the double pattern.

It might be asked if the Stern–Gerlach patterns could have been produced by spinning particles other than electrons, such as protons and neutrons in the nucleus. Theory states that the magnitude of the magnetic dipole is inversely proportional to the mass of the particle. The mass of the proton is about 2000 times that of the electron, and the magnetic dipole should accordingly be 2000 times smaller. This small magnetic dipole was also later measured by Frisch, Estermann, and Stern to distinguish it from the dipole of the electron spin.

22–4 ENERGY OF THE SPIN-ORBIT INTERACTION— FINE STRUCTURE

By experimental evidence from the Stern-Gerlach experiment and the analysis of the fine structure of atomic spectra, the gyromagnetic ratio of the electron spin has been found to be approximately twice the corresponding value of the orbital gyromagnetic ratio from Equation (19–21). The spin magnetic moment from experiment is found to be*

$$\mu_s = 2.0024 \left(\frac{e}{2m}\right) L_s$$

or the spin gyromagnetic ratio is approximately

$$\frac{\mu_s}{L_s} \cong 2\left(\frac{e}{2m}\right) \qquad (22\text{–}10)$$

The magnetic potential energy associated with the spin-orbit interaction is, from Equation (22–1),

$$\Delta E_s = \mu_s B \cos \theta$$
$$= 2\left(\frac{e}{2m}\right) L_s (\cos \theta) B$$

Then, from Equation (22–4), we can write

$$\Delta E_s = 2\left(\frac{e\hbar}{2m}\right) m_s B$$

or, in terms of the Bohr magneton μ_B,

$$\Delta E_s = 2\mu_B m_s B = \pm \mu_B B \qquad (22\text{–}11)$$

since $m_s = \pm\frac{1}{2}$. Now the total energy of the hydrogen atom (in the absence of an external magnetic field) is

$$E_{n,l,m_s} = -\frac{me^4}{32\pi^2 \varepsilon_0^2 \hbar^2} \frac{1}{n^2} + 2\mu_B m_s B \qquad (22\text{–}12)$$

In Figure 22–6, the single line of frequency v that results from the transition from the n state to the $n = 1$ ground state is revealed to be, by resolution of the fine structure, two lines separated in energy by

$$E_s = [E_n + 2\mu_B(\tfrac{1}{2})B] - [E_n + 2\mu_B(-\tfrac{1}{2})B] = 2\mu_B B(\tfrac{1}{2} + \tfrac{1}{2}) = 2\mu_B B$$

$$(22\text{–}13)$$

---

* The same result can be obtained on theoretical grounds following Dirac's relativistic theory of the electron. See, for example, L. I. Schiff, *Quantum Mechanics*, McGraw-Hill, New York, 1949, Chap. 12, and in particular pp. 319–327.

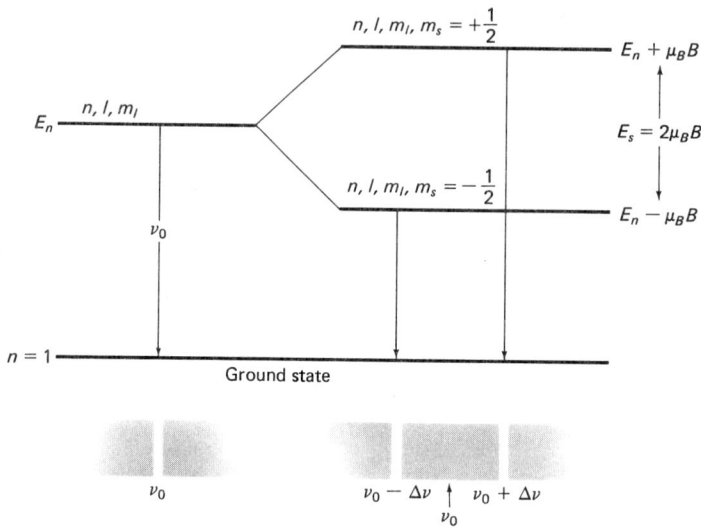

Figure 22-6 The single line ν_0 resulting from the transition from the n state to the $n = 1$ ground state is, when the fine structure is revealed, two lines separated in energy by $2\mu_B B$.

Since $m_s = +\frac{1}{2}$ or $-\frac{1}{2}$, the line is said to have a multiplicity of $2s + 1 = 2 \times \frac{1}{2} + 1 = 2$ values, that is, each state designated by n, l, m_l is split into two substates, one for $m_s = +\frac{1}{2}$ and the other for $m_s = -\frac{1}{2}$.

Notice that there is one exception. When an electron is in the s state, $l = 0$, and as a result the magnetic field at the site of the electron is zero. Since $B = 0$, ΔE_s must also be zero, and there is *no* fine structure splitting for this state. This is a degenerate state because the energy is the same whether $m_s = +\frac{1}{2}$ or $-\frac{1}{2}$, but this degeneracy can be removed if the atom is placed in an external magnetic field.

For the quantum numbers n, l, m_l it was shown previously that there are n^2 possible states. Now, with the multiplicity of $2s + 1 = 2$ states, the total number of states is $2n^2$.

EXAMPLE 22-1: Calculate the wavelength separation of the fine structure resulting from the spin-orbit interaction within the hydrogen atom.

SOLUTION: The magnetic field resulting from the orbital motion of the electron around the nucleus (or the proton around the electron) is, from the Biot-Savart law,

$$B = \frac{\mu_0}{4\pi} \frac{2\pi i}{R} = \frac{\mu_0}{4\pi} \frac{2\pi e f}{R}$$

For an atom in the ground state, $R = 0.53$ Å, $f = 7 \times 10^{15}$ sec^{-1}, and B is

$$B = 10^{-7} \frac{2 \times 3.14 \times (1.60 \times 10^{-1} \text{C}) \times (7 \times 10^{15} \text{ sec}^{-1})}{53 \times 10^{-12} \text{ m}} \cong 14 \text{ Wb/m}^2$$

The energy of separation of the split lines is

$$E_s = \frac{hc}{\lambda} - \frac{hc}{\lambda'} = \frac{hc \, \Delta\lambda}{\lambda\lambda'} \cong \frac{hc \, \Delta\lambda}{\lambda^2}$$

and, from Equation (22–13),

$$E_s = 2\mu_B B = 2 \times 9.3 \times 10^{-24} \frac{\text{J}}{\text{Wb/m}^2} \times 14 \text{ Wb/m}^2 \cong 28 \times 10^{-23} \text{ J}$$

If $\lambda \cong \lambda' = 4000$ Å $= 4.0 \times 10^{-7}$ m, then

$$\Delta\lambda = \frac{\lambda^2 E_s}{hc} = \frac{(4 \times 10^{-7} \text{ m})^2 \times (28 \times 10^{-23} \text{ J})}{6.63 \times 10^{-34} \text{ J-sec} \times 3 \times 10^8 \text{ m/sec}} \times 10^{10} \text{ Å/m}$$

$$= 1.12 \text{ Å}$$

This difference in wavelengths can be readily observed by a high-resolution spectrometer or an interferometer.

PROBLEMS

22–1 (a) Show that the z components of μ_s are given by $\mu_{sz} = -2m_s\mu_B$, where the z axis is taken in the direction of the magnetic field **B** set up by the proton in its orbital motion around the electron (as viewed from a reference frame attached to the electron). This is the only component of the spin dipole moment that can be observed and measured experimentally.
(b) Does this result show that μ_B is a natural unit with which to measure the spin dipole moment?
(c) Does it prove that μ_s is also space-quantized?
(d) Draw a diagram to scale to show the space quantization of μ_s.

22–2 (a) Show that the projection of μ_s onto the xy plane is given by $\mu_s(xy) = 2\mu_B\sqrt{s(s+1) - m_s^2}$. Although μ_s is known, μ_{sx} and μ_{sy} cannot be measured simultaneously with absolute precision.
(b) From the result of part (a), show that μ_{sz} and the azimuth angle ϕ, which gives the orientation of $\mu_s(xy)$ in the xy plane, cannot be determined with absolute precision simultaneously. (The conjugate variables μ_{sz} and ϕ obey Heisenberg's uncertainty principle $\Delta(\mu_{sz})\Delta\phi \geq h(m/e)$.
(c) Check the units for the uncertainty equation in part (b).

(d) Show this uncertainty graphically with a diagram in which μ_s is precessing about the z axis, generating a conical surface of revolution with an axis taken along **B**.

(e) Draw to scale the energy level diagrams that show the splitting of the energy levels (fine structure) for the following cases: $l = 2, m_l = 0$; $l = 2, m_l = \pm 1$; $l = 2, m_l = \pm 2$; $l = 1, m_l = 0$; $l = 1, m_l = \pm 1$.

(f) Show that there is no splitting for the case $l = 0$ unless there is an external magnetic field applied.

22-3 (a) Show that μ_s is also given by the equation $\mu_s = 2\mu_B \sqrt{s(s+1)}$

(b) Find the numerical values of μ_s, μ_{sz}, and $\mu_s(xy)$ for the following states: $m_l = 2, m_s = \frac{1}{2}$; $m_l = 2, m_s = -\frac{1}{2}$; $m_l = 1, m_s = \frac{1}{2}$; $m_l = 1, m_s = -\frac{1}{2}$.

22-4 Compute the separation of energy levels in the hydrogen atom (fine structure splitting) between the states $n = 2, l = 1, m_l = 0, m_s = \frac{1}{2}$ and $n = 1, l = 0, m_l = 0, m_s = -\frac{1}{2}$ (a) in units of energy (electron volts and joules); (b) in units of frequency; (c) in units of length (angstroms and meters). Assume that the atom is in the ground state postulated by the Bohr approach and $R = r_1 = 0.53$ Å and $f_1 = 7 \times 10^{15}$ sec^{-1}.

22-5 (a) In the Bohr theory, the speed of the electron in the nth orbit is given by

$$v_n = \frac{e^2}{4\pi\varepsilon_0 \hbar} \frac{1}{n}$$

Derive this equation from the principles of the Bohr theory described in Chapter 12.

(b) The radii of the allowed orbits, according to Bohr's treatment, are given by $R = n^2 r_1$. Use this information and calculate the energy separation due to fine structure splitting for the state $n = 2, l = 1, m_l = 0$. Assume that the atom is in the first excited state $n = 2$.

(c) Calculate the wavelength λ of the emitted photon when a transition takes place from $n = 2, l = 1, m_l = 0, m_s = \frac{1}{2}$ to $n = 1, l = 0, m_l = 0, m_s = -\frac{1}{2}$.

22-6 (a) In reference to Problem 22-5, show that

$$\frac{v_n}{c} \cong \frac{7.33 \times 10^{-3}}{n}$$

(b) If the relativistic effects of this speed are taken into account, the allowed energies in Bohr's treatment are given by

$$E_n = -\frac{me^4}{32\pi^2 \varepsilon_0^2 \hbar^2} \frac{1}{n^2} + \Delta E_n$$

where

$$\Delta E_n = \frac{1}{4}\frac{v^2}{c^2}\left(-\frac{me^4}{32\pi^2\varepsilon_0^2\hbar^2}\right)$$

Calculate the relativistic correction in electron volts for $n = 100$ and for $n = 1$. Can this correction be considered negligible?

(c) More refined computations show that

$$\Delta E_n = \left(-\frac{me^4}{32\pi^2\varepsilon_0^2\hbar^2}\right)\frac{\alpha^2}{n}\left(\frac{3}{4n} - \frac{1}{l+\frac{1}{2}}\right)$$

where $\alpha = v_1/c = e^2/4\pi\varepsilon_0\hbar c$ is called the fine structure constant. In terms of α, calculate ΔE_n for the state $n = 2$, $l = 1$.

22-7 Show that the fine structure constant (Problem 22-6) is an absolute number and that its value is approximately 1/137.

22-8 The energy associated with the fine structure splitting is given by

$$\Delta E_s = -\boldsymbol{\mu}_s \cdot \mathbf{B}$$

but $\boldsymbol{\mu}_s$ is antiparallel to $\mathbf{L}_s$ and $\mathbf{B}$ is parallel to $\mathbf{L}$, hence we can also write

$$\Delta E_s = b\mathbf{L}_s \cdot \mathbf{L}$$

where b is a constant. This is another expression to show that this energy is due to spin-orbit interaction. Show that (a) $\mathbf{L}_s \cdot \mathbf{L} = \frac{1}{2}(J^2 - L^2 - L_s^2)$, where $\mathbf{J} = \mathbf{L} + \mathbf{L}_s$; (b) $\mathbf{L}_s \cdot \mathbf{L} = \frac{1}{2}[j(j+1) - l(l+1) - s(s+1)]\hbar^2$. (c) Taking into account that $j = l + \frac{1}{2}$ for the case spin-up and $j = l - \frac{1}{2}$ for spin-down, prove that

$$\Delta E_s = \frac{1}{2}b\hbar^2(2l+1)$$

(d) By substituting for the value of b, it can be shown that

$$\Delta E_s = \left(-\frac{me^4}{32\pi^2\varepsilon_0^2\hbar^2}\right)\frac{\alpha^2}{n^3 l(l+1)}$$

Compute E_s for the states $n = 100$, $l = 1$, and $n = 2$, $l = 1$. What can you conclude about whether the correction is appreciable or not?

22-9 (a) Show that the projection of the spin angular momentum onto the xy plane is

$$L_s(xy) = \sqrt{s(s+1) - m_s^2}\,\hbar$$

(b) The lack of information about L_{sx} and L_{sy} shows that we cannot know simultaneously with absolute precision $L_{sz} = m_s\hbar$ and the orientation of $L_s(xy)$ as given by the azimuth angle ϕ. L_{sz} and ϕ are conjugate variables, and their uncertainties obey Heisenberg's principle,

$$\Delta L_{sz} \, \Delta \phi \geqq h$$

Show that the above relationship is dimensionally correct.

(c) Compute the magnitudes of $\mathbf{L}_s$, L_{sz}, and $L_s(xy)$ for the states $n = 2$, $l = 1, m_l = 0, m_s = \frac{1}{2}; n = 2, l = 1, m_l = 0, m_s = -\frac{1}{2}$.

RECOMMENDED READING

ALONSO, M., and FINN, E. J., *Fundamental University Physics*, Addison-Wesley, Reading, Mass., 1968, Vol. 3, Chap. 3.

DYSON, F. J., *Advanced Quantum Mechanics*, Cornell University Press, Ithaca, New York, 1951.
  A series of lectures on relativistic quantum mechanics given by Prof. Dyson at Cornell University in the fall of 1951. See Chapter 2. Even though this is an advanced treatment, the serious student should take a look at Chapter 3, where one of the greatest triumphs of the Dirac theory (the theoretical derivation of the electron spin dipole moment) is studied.

SPOSITO, G., *An Introduction to Quantum Mechanics*, Wiley, New York, 1970.
  For a discussion of the Stern–Gerlach experiment, see pp. 10, 130, and 200. A study of Dirac's approach to the problem of the magnetic dipole moment of the electron is made on pp. 348–352.

VON NEUMANN, J., *Mathematical Foundations of Quantum Mechanics*, translated from the German by R. T. Boyer, Princeton University Press, Princeton, N.J., 1955.
  Another book at the graduate level. Heisenberg's principle is discussed at length (pp. 231–247).

WEIDNER, R. T., and SELLS, R. L., *Elementary Modern Physics*, Allyn & Bacon, Boston, 1964, pp. 235–236.

23
Atomic and Molecular Spectra

Gerhard Herzberg

(1904–)
A native of Hamburg, Germany, Herzberg studied at Darmstadt Institute of Technology, where he was later professor of physics (1930–1935). He held positions at the University of Saskatchewan and at Yerkes Observatory at the University of Chicago. Active in research on atomic and molecular spectroscopy, he successfully applied his spectroscopic techniques to identify certain molecules in planetary atmosphere, comets, and interstellar space. This work was of such fundamental importance to chemical physics and quantum theory that he was awarded the Nobel Prize in chemistry in 1971.

23–1 TOTAL ANGULAR MOMENTUM
23–2 ATOMIC SPECTRA
23–3 MOLECULAR SPECTRA
23–4 LASERS

23-1 TOTAL ANGULAR MOMENTUM

An electron in orbit about the nucleus of an atom has an orbital angular momentum **L**, an intrinsic spin angular momentum $\mathbf{L}_s$, and a total angular momentum, by vector addition of these moments, of

$$\mathbf{J} = \mathbf{L} + \mathbf{L}_s \qquad (23\text{-}1)$$

In the same manner that $L = \sqrt{l(l+1)}\hbar$ and $L_s = \sqrt{s(s+1)}\hbar$, quantum theory shows the total angular momentum vector to be quantized according to

$$J = \sqrt{j(j+1)}\hbar \qquad (23\text{-}2)$$

The total angular momentum quantum number j can assume the values $j = l + s$ or $j = l - s$. Note that since $s = \frac{1}{2}$, j always has half-integer values.

EXAMPLE 23-1: Determine the values of the total angular momentum for a $3d$ electron.

SOLUTION: The $3d(n = 3, l = 2)$ electron has an orbital angular momentum

$$L = \sqrt{2(3)}\hbar = \sqrt{6}\hbar$$

and a spin angular momentum

$$L_s = \sqrt{\frac{1}{2}\left(\frac{3}{2}\right)}\hbar = \frac{\sqrt{3}}{2}\hbar$$

The total angular momentum quantum number is thus

$$j = l + s = 2 + \tfrac{1}{2} = \tfrac{5}{2}$$

or

$$j = l - s = 2 - \tfrac{1}{2} = \tfrac{3}{2}$$

and the total angular momentum vector has a magnitude

$$J = \sqrt{\frac{5}{2}\left(\frac{5}{2} + 1\right)}\hbar = \frac{\sqrt{35}}{2}\hbar$$

or

$$J = \sqrt{\frac{5}{2}\left(\frac{5}{2} - 1\right)}\hbar = \frac{\sqrt{15}}{2}\hbar$$

The two possible orientations of **J** are illustrated in Figure 23–1.

Also similar to momenta **L** and **L**$_s$, the component of **J** in a z direction determined by some external field is quantized as

$$\boxed{J_z = m_j\hbar} \tag{23-3}$$

with $m_j = \pm\frac{1}{2}$.

Since all the angular momenta **L**, **L**$_s$, and **J** are quantized at the same time, there are only certain allowed orientations of these vectors relative to each other. The spin of the electron is always at some angle relative to the orbital angular momentum and is never aligned parallel to the orbital angular momentum or the total angular momentum.

If there is no external magnetic field, only the magnetic fields from the electron orbital motion and spin interact. *There is no external torque, and the total angular momentum is conserved.* These internal torques cause the precession of **L** and **L**$_s$ about **J** as shown in Figure 23–2(a). However, in the presence of an applied external magnetic field, **J** *precesses about* **B** *while* **L** *and* **L**$_s$ *continue to precess about* **J** *as*

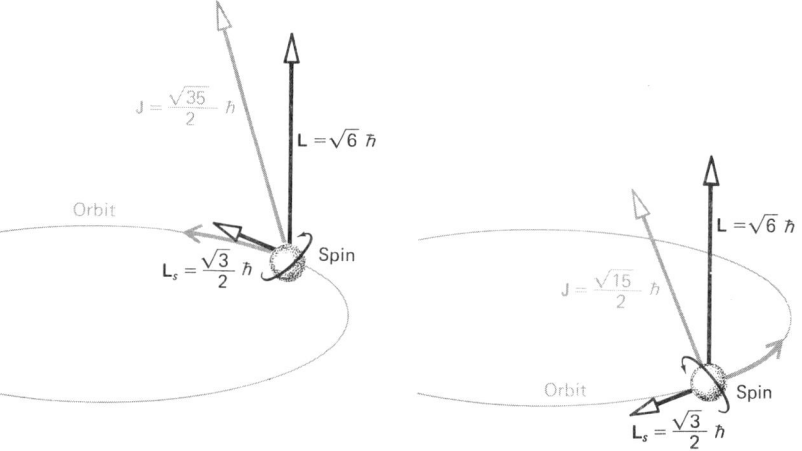

Figure 23–1 Addition of the orbital angular momentum **L** and the spin angular momentum **L**$_s$ of a $3d$ electron to form the total angular momentum vector **J** according to quantum mechanics.

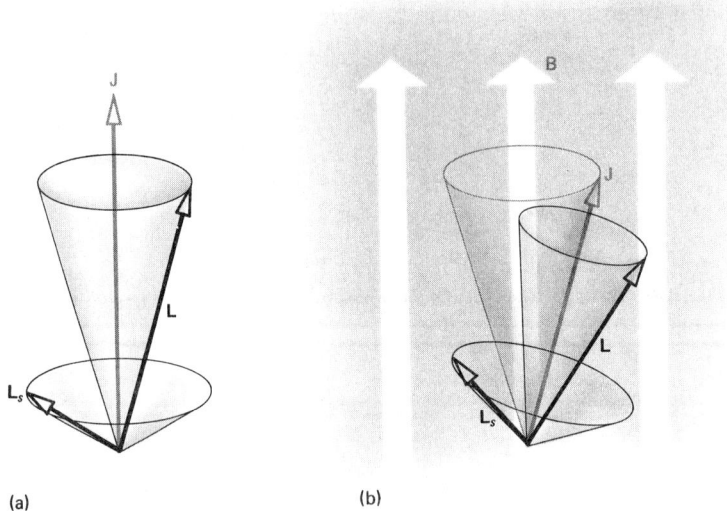

Figure 23-2 (a) In the absence of an external magnetic field, there are no external torques, total angular momentum **J** is conserved, and **L** and $\mathbf{L}_s$ precess about the fixed vector **J**. (b) In the presence of an applied magnetic field **B**, **J** precesses about an axis in the direction of **B**.

depicted in Figure 23–2(b). The interaction of **J** with an external weak magnetic field gives rise to an additional small splitting known as the *anomalous Zeeman effect*.

23-2 ATOMIC SPECTRA

A spectroscopic notation using capital letters for the corresponding orbital angular momentum quantum number

$$l = 0\ 1\ 2\ 3\ 4\ 5\ldots$$
$$S\ P\ D\ F\ G\ H\ldots$$

is used to specify the energy state of each electron. The concise, shorthand notation

$$\boxed{n^{2s+1}D_j}$$

where

n represents the principal quantum number
$2s + 1$ represents multiplicity
D represents $l = 2$
$j = l \pm s$ represents the total angular momentum quantum number
$l = 2$ represents the orbital angular momentum quantum number

could designate a 3d electron by

$$3^2D_{5/2} \quad \text{or} \quad 3^2D_{3/2}$$

where

$$n = 3$$
$$2s + 1 = 2(\tfrac{1}{2}) + 1 = 2$$
$$j = 2 + \tfrac{1}{2} = \tfrac{5}{2} \quad \text{or} \quad 2 - \tfrac{1}{2} = \tfrac{3}{2}$$
$$l = 2$$

The alkali metals, such as lithium, sodium, and potassium, have a single valence electron outside completely filled subshells. The nuclear charge $+Ze$ is shielded by a charge of $-(Z-1)e$ from the electrons in the closed subshells, and the effective charge seen by the valence electrons is $Ze - (Z-1)e = +e$. Thus, the effective charge of the valence electron is $Ze - (Z-1)e = +e$. The valence electron in the alkali metals therefore behaves much like the orbiting electron in the hydrogen atom. Only the quantum numbers of the relatively simple spectra are similar to those of the hydrogen atom.

The neutral sodium atom $^{23}_{11}\text{Na}$ has ten electrons in completely filled subshells and a single valence electron with a ground state energy $3^2S_{1/2}$. These valence electrons can be raised to higher energy states by bombarding them with electrons, by heating, or by allowing them to absorb radiant energy. The atoms in these excited energy states return to the state of lower energy by giving off photons whose energies are given by Bohr's formula, $h\nu = E_i - E_f$. The energy level diagram in Figure 23–3 shows that the excited levels of the valence electrons of the many atoms of sodium will be in various excited states. The energy level diagram shows some of the possible excited states and the most prominent allowed transitions with the associated wavelengths in angstroms. The selection rule that controls the allowed transitions is

$$\boxed{\Delta l = \pm 1}$$

The most prominent feature of the sodium spectrum is the pair of intense yellow lines that represent the transitions

$$3^2P_{3/2} \to 3^2S_{1/2} \quad (\lambda = 5889.9 \text{ Å})$$
$$3^2P_{1/2} \to 3^2S_{1/2} \quad (\lambda = 5895.9 \text{ Å})$$

There are other allowed transitions not shown in the diagram, but they are quite weak.

The spin-orbit interaction described in Section 22–4 splits the energy levels into two, one with $j = l + \tfrac{1}{2}$ and the other with $j = l - \tfrac{1}{2}$. (Although the S terms with $l = 0$ and $j = \tfrac{1}{2}$ are customarily designated $^2S_{1/2}$, with the multiplicity

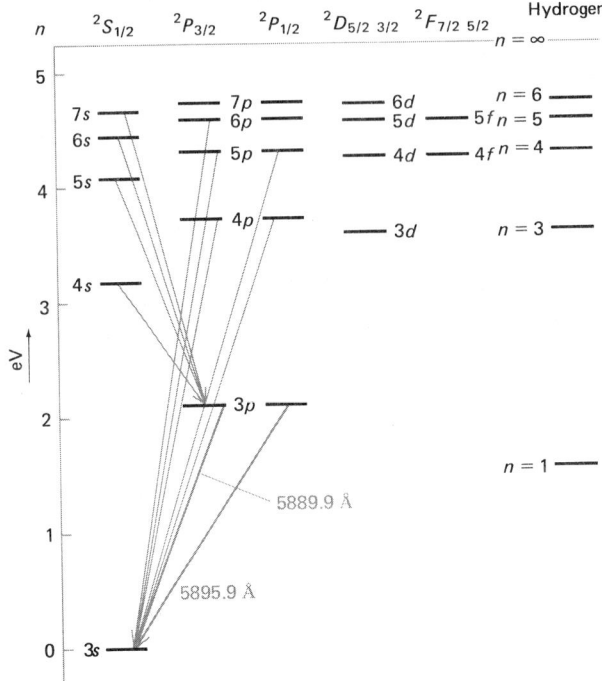

Figure 23-3 Energy level diagram for sodium, with the energy levels for hydrogen included for comparison.

superscript of 2 indicating a doublet, it is really only a single energy level.) In the energy level diagram, note that all of the states other than the S states are really *doublet states*, since for the P state $j = \frac{1}{2}$ or $\frac{3}{2}$ and the states are

$$^2P_{1/2} \quad ^2P_{3/2}$$

for the D state, $j = \frac{3}{2}$ or $\frac{5}{2}$ and

$$^2D_{3/2} \quad ^2D_{5/2}$$

and so on. In the figure, the differences in energy between the ^2D levels and ^2F levels are so small that they are not shown.

The energy levels of hydrogen are shown in Figure 23-3 for comparison with the sodium energy levels. For given values of n, the electron is found closer to the nucleus for small values of l for both sodium and hydrogen. Those electrons of sodium with small values of l thus penetrate the inner core of electrons and pass closer to the nucleus, and consequently they are less shielded from the nuclear charge. They have a more negative total energy and the states associated with small angular momentum are displaced downward from the same states in hydrogen. For large values of l, the energy levels of sodium and hydrogen approach the same values.

23–3 MOLECULAR SPECTRA

The energy of an isolated molecule is

$$E = E_e + E_v + E_R + E_t \qquad (23\text{–}4)$$

where the total energy is made up of the electronic energy E_e, the vibrational energy E_v, the rotational energy E_R, and the translational energy E_t. These energies, with the exception of the translational energy, are quantized and specified by quantum numbers. As in the case of atomic spectra, radiation of energy $\Delta E = h\nu$ is absorbed or emitted when a transition from one energy state to another occurs. Transitions can take place that involve simultaneous changes of electronic, vibrational, and rotational energy.

The simplest picture of a molecule is the diatomic molecule pictured as a rigid rotator in the form of a dumbbell that rotates about the center of mass of the two atoms as shown in Figure 23–4. The moment of inertia of the system is

$$I = \mu r^2 \qquad (23\text{–}5)$$

where $\mu = m_1 m_2/(m_1 + m_2)$ is the reduced mass and $r = r_1 + r_2$. The angular momentum of the rotating diatomic molecule is quantized according to

$$L_{\text{rot}} = \sqrt{J_R(J_R + 1)}\hbar \qquad (23\text{–}6)$$

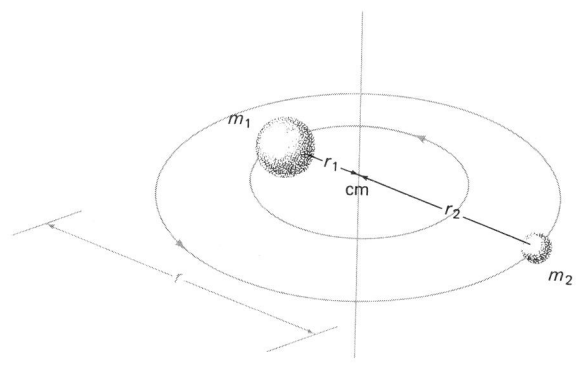

Figure 23–4 In the dumbbell model of the diatomic molecule, the atom rotates about a common center of mass.

where the rotational quantum number J_R has the possible integral values $J_R = 0, 1, 2, 3, \ldots$. The rotational kinetic energy of angular momentum $L = I\omega$ is

$$E_{rot} = \tfrac{1}{2}I\omega^2 = \frac{(I\omega)^2}{2I} = \frac{L^2}{2I_0}$$

and is quantized according to

$$E_{rot} = J_R(J_R + 1)\frac{\hbar^2}{2I} \tag{23-7}$$

The difference in energy between two successive quantum numbers

$$E_{rot\,(J_R+1)} - E_{rot\,(J_R)} = \Delta E_{rot} = \frac{\hbar^2}{I}(J_R + 1) \tag{23-8}$$

shows the rotational molecular spectrum to be a series of equally spaced lines each separated in energy by $(\hbar^2/I)(J_R + 1)$. The photons emitted in these transitions with a selection rule $\Delta J_R = \pm 1$ have an energy

$$h\nu_R = \frac{\hbar^2}{I}(J_R + 1) \tag{23-9}$$

The spectral lines due to these transitions are found in the far infrared and microwave portions of the spectrum.

When the diatomic molecule receives sufficient excitation, it can vibrate as well as rotate. For the lowest values of energy, the energies resemble those for a harmonic oscillator, with the quantized energy given by

$$E_v = (v + \tfrac{1}{2})h\nu_0 \tag{23-10}$$

where $v = 0, 1, 2, \ldots$. Since the lowest vibrational energy state is equal to $\tfrac{1}{2}h\nu_0$, the vibrational transitions never occur alone, but always with rotational transitions. As a consequence, these transitions give rise to a group of lines in the infrared spectrum known as the *vibration-rotation band*. A typical vibration-rotation molecular spectrum of HBr, shown in Figure 23–5, is made by sending radiation from a source of continuous infrared wavelengths through a cell containing HBr and recording the absorption spectra.

When there is enough available energy, the electronic states of the molecules can be excited. This gives rise to the most general type of transitions, *rotation-vibration electronic transitions*. These transitions produce photons of energy

$$h\nu = h\nu_e + (E_v'' - E_v') + (E_{rot}'' - E_{rot}')$$

$$= h\nu_e + \tfrac{1}{2}h\nu_0 + J_R(J_R + 1)\frac{\hbar^2}{I} \tag{23-11}$$

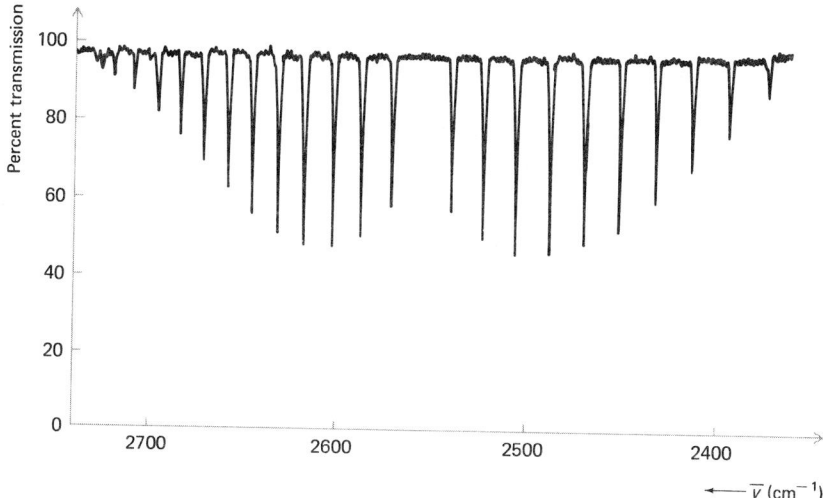

Figure 23–5 The rotational structure of the $v = 0$ to $v = 1$ transition of HBr gas. Here $\bar{v} = 1/\lambda$ is the so-called wave number and means the number of waves per unit length. Spectroscopists usually measured $\bar{v}$ in waves per centimeter or reciprocal centimeters (cm^{-1}). (From Gordon Barrow, *Introduction to Molecular Spectroscopy*, McGraw-Hill, New York, 1962. Used with permission.)

It is the outermost electrons of the atoms composing the molecules that are involved, and these transitions are energetic enough that they produce electronic bands in the visible and ultraviolet regions. The relative arrangement of these molecular energies is shown in Figure 23–6. For each electronic state there are the associated vibration states, and for each vibration state there are associated rotation levels. Since the selection rules forbid transitions of $\Delta J_R = 0$, the two branches shown in Figure 23–6 represent transitions $\Delta J_R = -1$ and $\Delta J_R = +1$.

23–4 LASERS

In 1919 Einstein, in a paper on blackbody radiation, pointed out that:

1. Atoms in an excited state can return to their ground state by spontaneous emission of photons of the corresponding energy.
2. Atoms in a lower energy state can be raised to a higher energy state by absorption of photons of the correct energy.
3. Emission of a photon from atoms in an excited state can be increased or stimulated when an excited atom is hit by a quantum of light of the same energy.

The first two types of transitions are just those that produce emission or absorption spectra. It is the stimulated emission that interests us here.

310 · PART THREE: THE ATOM

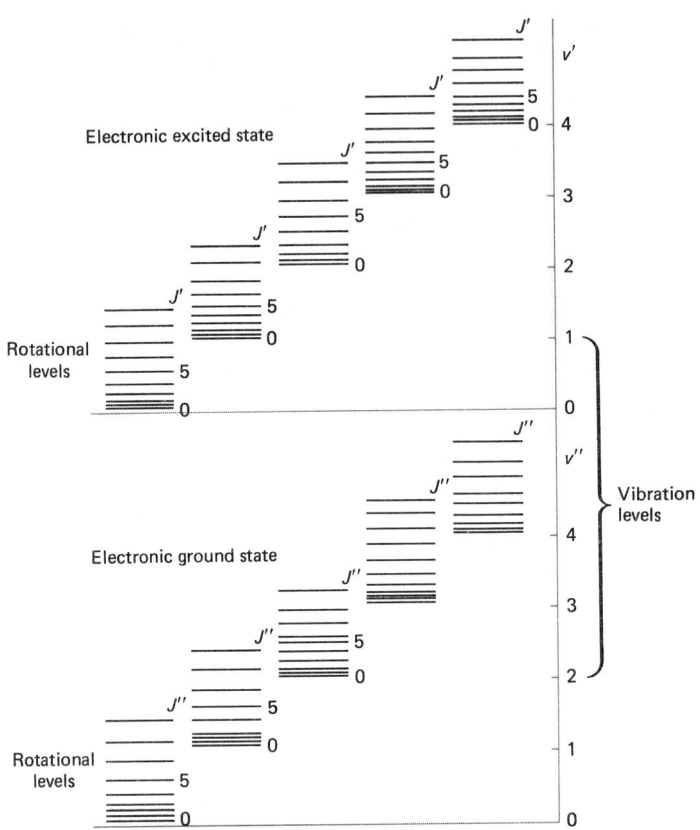

Figure 23-6   Vibrational and rotational levels of two electronic states of a molecule.

Stimulated emission means that because more quanta are now coming from atoms in a given excited state, the average time these atoms spend in this state is decreased. Since the stimulating photon is not absorbed by this process, there will be two photons leaving the atom, as pictured schematically in Figure 23-7. Both of the photons will be traveling in the same direction as the original photon, and they will be coherent because they are in phase and have the same frequency. These two photons can now stimulate other excited atoms, and eventually an avalanche of concentrated, coherent quanta is produced.

If enough atoms can be kept in the proper excited state, and if enough quanta of the proper frequency can be found to produce the stimulated emission in a very short time, a "burst" of quanta will be produced. In 1952 CHARLES H. TOWNES first demonstrated the possibility of producing and maintaining a larger number of atoms in excited states than in the ground state. This is known as *population inversion*, and the first method for producing it was a microwave device called a MASER, which is an acronym for *M*icrowave *A*mplification by the *S*timulated

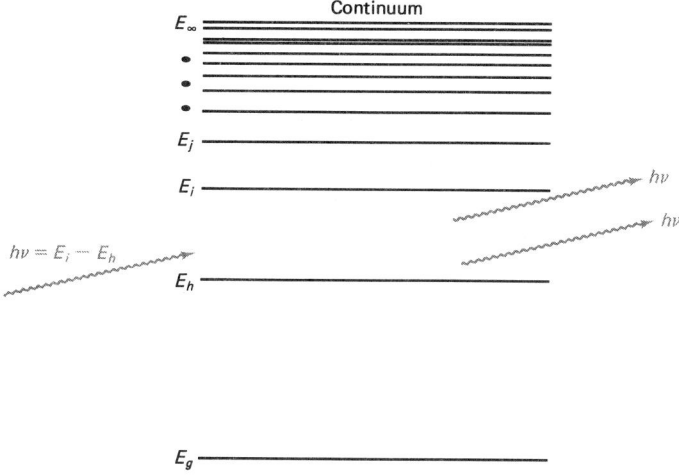

Figure 23-7 A photon of energy $h\nu = E_i - E_n$ will stimulate the emission of a second photon of the same energy.

*E*mission of *R*adiation. Later devices making use of the phenomena in the visible region were labeled LASERS (the L from *L*ight); and in the infrared region they are labeled, appropriately enough, IRASERS* (the IR from *I*nfrared).

Of the several methods of inverting the population from the ground state to the excited state, the two most commonly employed are electrical discharge in a gaseous discharge tube and a flash lamp, which produces a very intense source of light on a crystal such as ruby or neodymium. This process of inverting the population of the atom is known as *pumping*. The discharge produces atoms in many different excited states, some of which will be heavily populated and give up photons of the frequency for lasing ν_L. In the case of the flash lamp, photons in a continuous range of frequencies are given off, but among these are again photons of the proper frequency.

If the devices were only pumped, there would be little if any lasing because there are many competing processes that would soon absorb all of the photons of frequency ν_L. To produce stimulated emission and hence an avalanche of photons of frequency ν_L, the gas laser in Figure 23-8 uses mirrors that reflect as much as 95% of the radiation back into the tube. The *Brewster windows* shown are not necessary for lasing, but they are frequently used since they make more efficient use of the radiation available, and they cause the lasing radiation to be polarized. Thus, photons of the proper frequency ν_L are sent back into the tube to produce a cascade of similar photons. When enough photons are produced so that more photons ν_L are produced then absorbed, lasing begins. The output mirror is gener-

---

* See, for example, *Phys. Today*, July 1970, p. 55.

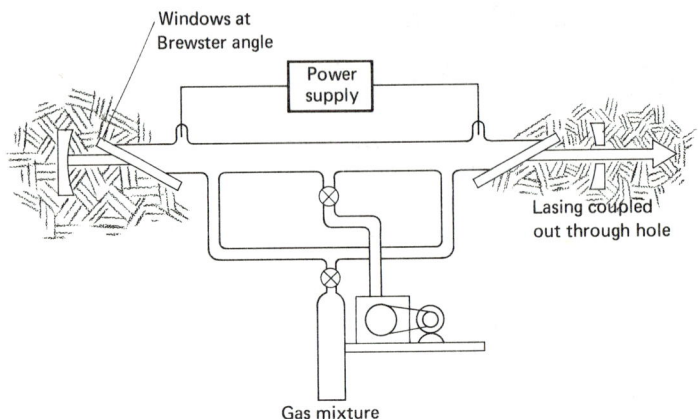

Figure 23-8   Typical apparatus for a gas laser such as a CO_2 laser.

ally a dielectrically coated mirror that reflects most of the radiation back into the tube but also allows part of it to leave the laser. Sometimes a mirror with a small hole drilled in the center is also used as the output mirror.

In Figure 23-9, there are no mirrors required because the ends of the ruby crystals themselves are polished and coated to act as reflectors.

In ruby lasers, a tremendous surge of energy is put into the flash lamp, which in turn gives off a "burst" of very intense light about the crystal and causes it to lase. This process is then repeated, and the crystal lases with each "burst" of light from the flash lamp. Ruby lasers are, therefore, known as *pulsed lasers*, because they produce very powerful lasing radiation in very short intervals. The gaseous lasers, such as He-Ne or CO_2, can be pumped into excited states continuously and are *continuous wave* (cw) lasers.

The CO_2 laser is typical of the continuous wave lasers. Many different gases, including CO_2, N_2O, H_2O, Ne and A, have been made to lase because they contain a particular energy level that can be populated well. The CO_2 molecule is linear, and as such, has both quantized rotational energy and vibrational energy. Figure 23-10 shows the three modes of vibration for a linear molecule such as CO_2. Associated with vibrations causing symmetric stretching of the molecule is a quantum number v_1. Quantum number v_2 is associated with the bending modes of

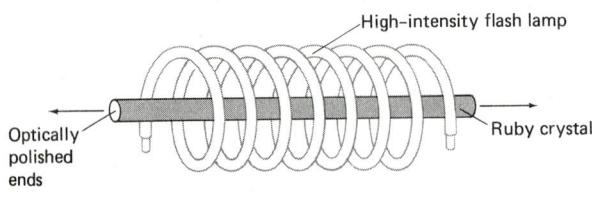

Figure 23-9   Ruby crystal pumped by flash lamp.

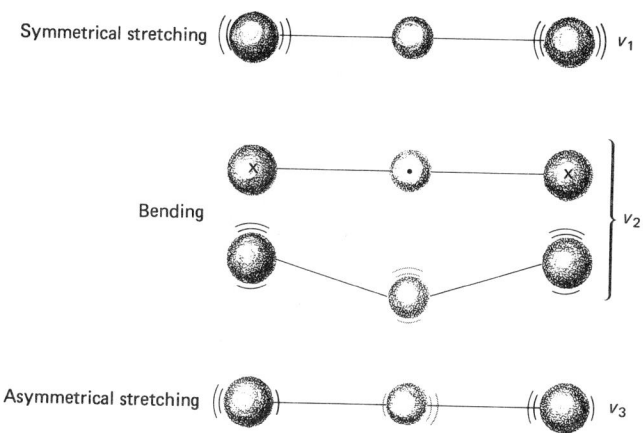

Figure 23-10 Possible modes of vibration of a linear molecule such as CO_2.

vibration, which are degenerate since this can occur in two different ways. Asymmetric stretching is designated by the quantum number v_3. The degeneracy of quantum number v_2 is broken up by centrifugal stretching as the molecule rotates and whirls and the atoms stretch farther apart. The quantum number associated with this is l (not to be confused with the orbital angular momentum quantum number of the electrons). The vibrational energy state of a molecule with quantum numbers $v_1 = 0, v_2 = 1, v_3 = 1, l = 0$ would then be written $v_1 v_2{}^l v_3$ or $(01^0 1)$.

Some of the important vibrational energy levels of CO_2 are indicated in Figure 23-11, with only the rotational energy levels associated with each vibration level

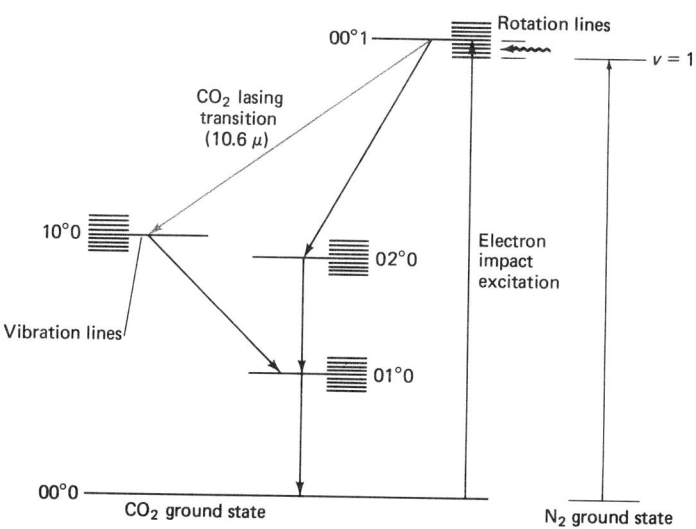

Figure 23-11 Energy level diagram showing pertinent levels for a CO_2 laser.

indicated. The ground level (00^00) is excited to the (00^01) level, which has a long lifetime and allows many molecules to be in this level. Then, as the molecule relaxes to return to the ground state, it goes to the (10^00) level, then to the (01^00) level, and finally back to the ground level (00^00). There are several other paths by which it might come back to the ground level, but in this case the transition from the (00^01) level to the (10^00) level is the *lasing transition*. This produces an infrared wavelength of 10.6 μ (about 20 times the wavelength of visible light). Helium is used in CO_2 lasers because it takes atoms by collision out of the (10^00) level and allows even more (00^01)–(10^00) transitions. Nitrogen is also used, because it has an energy level close to the (00^01) CO_2 level and, by energy transfer, helps increase the population of the (00^01) level.

Most research to date has been on the development of lasers, new types of lasers, and laser techniques. More recently, research on lasers has concentrated on a variety of methods of using lasers in practical applications.

PROBLEMS

23-1 Use the law of cosines and show that the angle between L and L_s can be obtained from

$$\cos(L, L_s) = \frac{j(j+1) - l(l+1) - s(s+1)}{2\sqrt{s(s+1)}\sqrt{l(l+1)}}$$

23-2 What is the expression for the cosine of Problem 23-1 when $j = l + \frac{1}{2}$? $j = l - \frac{1}{2}$?

23-3 Use the equation in Problem 23-1 to determine the possible angles between the spin angular momentum and the orbital angular momentum for a $3d$ electron.

23-4 The electron in a hydrogen atom is in the $4f$ state. Determine (a) the possible values of the total angular momentum, and (b) the corresponding components in the z direction of the total angular momentum.

23-5 On an energy level diagram, show the possible transitions for the valence electron of sodium in the $6^2P_{1/2}$ state to return to the ground state.

23-6 Sketch in some detail the allowed $6F$–$6D$ transitions for sodium.

23-7 From Figure 23-3 and Equation (22-13), determine the difference in energy in electron volts between the $3^2P_{3/2}$ and $3^2P_{1/2}$ states.

23-8 A source of the sodium spectrum is placed in a strong magnetic field of 1.5 tesla. What will be the wavelengths of the Zeeman lines associated with the strong $3^2P_{3/2} \rightarrow 3^2S_{1/2}$ and $3^2P_{1/2} \rightarrow 3^2S_{1/2}$ transitions?

CHAPTER 23: ATOMIC AND MOLECULAR SPECTRA · 315

23-9 The atoms of CsI are separated by 3.31 Å. Calculate the reduced mass and the rotational inertia of this molecule.

23-10 The hydrogen and chlorine atoms are separated by 1.25 Å in the HCl molecule; $h\nu_0$ is 0.369 eV. Calculate a few of the lowest vibration-rotation lines of HCl.

RECOMMENDED READING

BARROW, G. N., *Introduction to Molecular Spectroscopy*, McGraw-Hill, New York, 1962.
HERZBERG, G., *Atomic Spectra and Atomic Structure*, Dover, New York, 1944.
MCELROY, J., "The CO_2 Laser," *Electronics World*, May 1968.
PATEL, C. K. N., "High-Power Carbon Dioxide Lasers," *Sci. Am.*, August 1968.
PAULING, L., and GOUDSMIT, S., *The Structure of Line Spectra*, McGraw-Hill, New York, 1930.

24
The Exclusion Principle

Wolfgang Pauli
(1900–1958)

A native of Vienna, Pauli received his Ph.D. from the University of Munich in 1921. He was professor at the Federal Institute of Technology in Zurich (1928) and was appointed to the chair of theoretical physics at the Institute for Advanced Studies at Princeton (1940). His contributions to quantum theory and the theory of other fields are numerous. For his discovery of the exclusion principle (Pauli principle), which governs the electronic configuration of the outer shells of atoms, he received the 1945 Nobel Prize.

24–1   THE EXCLUSION PRINCIPLE
24–2   TWO-ELECTRON ATOMS
24–3   THE PERIODIC TABLE

24-1 THE EXCLUSION PRINCIPLE

The assignment of the various quantum numbers to electrons is more than a means of classifying the electrons. A simple, yet powerful law of physics known as the *Pauli principle*, or the *exclusion principle*, states that *there cannot be more than one electron in a given energy state when this energy state is completely described by the four quantum numbers n, l, m_l, m_s.* During the summer of 1922, WOLFGANG PAULI went to Copenhagen to work with Bohr on the explanation of the anomalous Zeeman effect. The exclusion principle evolved from his efforts to classify the energy levels of an electron when it is placed in a strong magnetic field. The concept was completed when Uhlenbeck and Goudsmit proposed the idea of electron spin and the quantum number m_s was added.

The exclusion principle was formulated in the same period that the Schrödinger and Heisenberg quantum theories were being developed, and it was quickly incorporated as an essential component of the new quantum physics. In theory, quantum mechanics can predict all of the properties of the chemical elements, but in practice the complexity of the mathematics involved makes it a formidable task. One of the great achievements of the quantum theory is the use of the exclusion principle for a simple, complete explanation of the properties of the elements by means of the periodic table.

To specify completely the energy state of an atom, the four quantum numbers needed are n, l, and m_l, which can be derived from the Schrödinger equation, and the spin quantum number m_s, which comes from Dirac's relativistic theory of the electron.

Another way of stating the Pauli exclusion principle is that *no two electrons in a given atom may have the same set of quantum numbers n, l, m_l, m_s.* In the classical theory, the most stable state of the electrons is when they are at their lowest energy state, the ground state. Thus, from hydrogen on up to the heaviest elements, all of the electrons should be in the first orbit, with the orbits being smaller for the heavier atoms. As a consequence, each individual atom should differ in size as indicated in Figure 24-1, but in reality this does not happen. All of the atoms are approximately equal in size.

The possible energy states of the hydrogen atom according to the scheme of the exclusion principle are illustrated in Figure 24-2. Each arrow represents an energy state of an electron with either spin-up or spin-down; and although there appear to be positions where energy levels are the same, this degeneracy is removed and

318 · PART THREE: THE ATOM

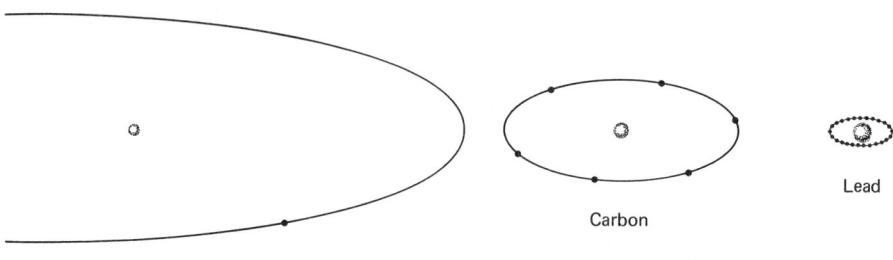

Figure 24-1 If the exclusion principle did not apply, atoms would vary greatly in size. Actually, all atoms are of approximately the same size. (Idea for diagram courtesy of Dr. J. Calame.)

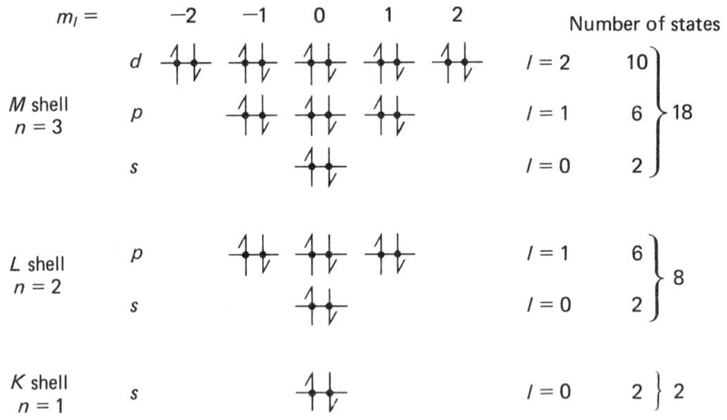

Figure 24-2 Possible energy states of the single electron of the hydrogen atom according to the exclusion principle. For each set of quantum numbers n, l, m_l there are two states, one for $m_s = \frac{1}{2}$ and one for $m_s = -\frac{1}{2}$.

each electron is distinct because each has a different spin orientation. Notice that the grouping according to the quantum numbers has arranged the energy levels naturally into shells and subshells. The ground state for hydrogen then has quantum numbers $n = 1$, $l = 0$, $m_l = 0$, and $m_s = \pm\frac{1}{2}$. An electron in the shaded energy state at the top of the figure is designated

$$3^2 D_{5/2}$$

with $n = 3$, $2s + 1 = 2$, $D \rightarrow l = 2$, $j = l + s = 2 + \frac{1}{2}$.

24-2 TWO-ELECTRON ATOMS

The alkali metals have a single electron outside a closed shell of electrons and can be treated somewhat like the hydrogen atom. The alkaline earth elements, beryllium, magnesium, calcium, strontium, barium, and radium, have two electrons outside a closed shell and are treated like the helium atom.

The helium atom, with two protons and two neutrons in the nucleus, has two orbiting electrons. If $\mathbf{L}_1$ and $\mathbf{L}_2$ are the respective orbital angular momenta, then the total orbital angular momentum is

$$\mathbf{L} = \mathbf{L}_1 + \mathbf{L}_2 = \sqrt{L_T(L_T + 1)}\hbar \qquad (24\text{-}1)$$

where L_T is the total orbital angular momentum quantum number. Similarly, the respective spins are $\mathbf{L}_{s_1}$ and $\mathbf{L}_{s_2}$, with a total spin angular momentum

$$\mathbf{L}_s = \mathbf{L}_{s_1} + \mathbf{L}_{s_2} = \sqrt{S(S + 1)}\hbar \qquad (24\text{-}2)$$

where S is the total spin angular momentum quantum number. The total angular momentum of the electron is

$$\mathbf{J} = \mathbf{L} + \mathbf{L}_s = \sqrt{J_T(J_T + 1)}\hbar \qquad (24\text{-}3)$$

In these equations L_T, S, and J_T are the quantum numbers representing, respectively, total orbital momentum, spin angular momentum, and total angular momentum.

In the ground state K shell, the electrons have the following quantum numbers:

electron 1 $n_1 = 1$, $l_1 = 0$, $m_{l_1} = 0$, $m_{s_1} = \frac{1}{2}$

electron 2 $n_2 = 1$, $l_2 = 0$, $m_{l_2} = 0$, $m_{s_2} = \frac{1}{2}$

which contradicts the exclusion principle because the two sets of numbers are the same. Since it is prohibited that two electrons have the same energy state and move in the same orbit, each electron in Figure 24–3, and hence each orbit, is distinguished by one electron with spin-up and one with spin-down. The two spin angular momenta are antiparallel, and

$$\mathbf{L}_s = \mathbf{L}_{s_1} + \mathbf{L}_{s_2} = 0$$

by vector addition. Then $L_s = \sqrt{S(S + 1)}\hbar = 0$, the total spin quantum number $S = 0$, and the multiplicity is now $2S + 1 = 1$. Also, because $l_1 = l_2 = 0$,

$$\mathbf{L} = \mathbf{L}_1 + \mathbf{L}_2 = 0$$

and as $L = \sqrt{L_T(L_T + 1)}\hbar = 0$, the total orbital angular momentum quantum number is $L_T = 0$.

The ground energy state of the atom can now be written in spectroscopic notation as

$$1^1S_0$$

since $n = 1$, $2S + 1 = 1$, $L_T = 0$, and $J_T = L_T + S = 0$. This is a singlet state, since $\mathbf{L}_s = 0$ and the total angular momentum can have only the single value

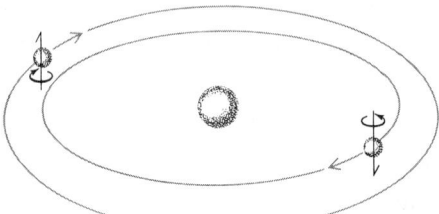

Figure 24-3 The exclusion principle prohibits two electrons from moving in the same orbits. The two orbits are distinguished by one electron with spin-up and the second with spin-down.

$J = L$. In the same fashion, the states $^1P_1$, $^1D_2$, and $^1F_3$ are also singlet states.

For two-electron systems with spins parallel, $S = s_1 + s_2 = \frac{1}{2} + \frac{1}{2} = 1$, and the multiplicity $2S + 1 = 3$ indicates a triplet state because the total angular momentum quantum number can have the values $J_T = L_T + 1$, $J_T = L_T$, and $J_T = L_T - 1$. Thus, for $L_T = 1$ we have triplet states $^3P_2$, $^3P_1$, $^3P_0$; for $L_T = 2$, we have $^3D_3$, $^3D_2$, $^3D_1$; and so on.

24-3 THE PERIODIC TABLE

In 1869 the Russian chemist DMITRI IVANOVICH MENDELEEFF arranged the chemical elements according to their atomic weights and physical properties. Although elements such as the rare earths did not fit well into his scheme, he was able to develop a somewhat successful periodic table of elements. It was not until the discoveries of many missing elements and the concept of the exclusion principle that the periodic table was put into complete and final order.

The following notation is used in Table 24-1 to show the configuration of the electrons for the ground states of the elements:

$1s^2$

where

$n = 1$ (both electrons are in the K shell),
s indicates that $l = 0$ for both electrons, and
2 indicates that there are two electrons, which are in the respective states:

$n_1 = 1, l_1 = 0, m_{l_1} = 0, m_{s_1} = \frac{1}{2}$

and

$n_2 = 1, l_2 = 0, m_{l_2} = 0, m_{s_2} = -\frac{1}{2}$

For each value of l there are $2l + 1$ values of m_l and two values each of m_s. Consequently, there are $2(2l + 1)$ electrons in each completed subshell and, as previously discussed, each shell contains a maximum of $2n^2$ electrons. When a subshell of an atom in the ground state is completed with the number of electrons allowed by the exclusion principle, the total orbital and spin angular momenta are zero.

A comparison of Table 24-1 and the periodic table (Table 1 in the Appendix) shows how the arrangement of the electrons in the shells and subshells accounts for groups of elements with similar chemical and physical properties. The periodic table is arranged with columns forming eight groups, which contain elements with similar properties. The seven horizontal rows are labeled periods and consist of a transition from chemically active metals to the inert gases.

When a shell is completely filled, the electrons within this shell are very tightly bound and do not take part in chemical reactions. The elements with these closed shells are the inert gases in group VIII. Originally, there were only seven groups in the periodic table, but the discovery of the inert gases (helium, neon, argon, krypton, xenon, and radon) was accommodated by including the entire family in the periodic table by simply adding another column and calling it group VIII.

All of the elements of group I are hydrogenlike elements, since they are composed of a closed *core* that is tightly bound and a single electron of valence +1. These valence electrons very readily combine with other elements, and the elements of group I are characterized by being very active chemically.

Most of the rare earth elements were unknown at the time Mendeleeff organized the periodic table. These elements, because they are chemically very similar, form a somewhat separate *rare earth* or *lanthanide* series within the periodic table, beginning with atomic number 57, lanthanum, and continuing to 71, luthelium. The *actinide series* also forms a somewhat independent set of elements running from atomic number 89, actinium, to man-made elements of high atomic number such as neptunium, plutonium, and americium. The latest man-made elements are atomic number 104, rutherfordium, and 105, hahnium, which were discovered in 1970 at the Lawrence Radiation Laboratory, Berkeley, Calif. These man-made elements of high atomic number are difficult both to produce and to identify. (Hahnium was originally produced at the rate of about six atoms per hour.)

If electrons obey the Pauli exclusion principle, what about other particles such as neutrons, protons, or atoms as a whole? The basic elements of the nucleus, neutrons and protons, have been found to have a spin and a magnetic moment, and they are regulated by a set of quantum numbers. Neutrons and protons are restricted to move in limited numbers in quantum orbits similar to electrons, although the nuclear energy levels are different.

However, all particles do not obey the Pauli exclusion principle. It has been found that particles with spin $s = \frac{1}{2}$ obey the exclusion principle and *Fermi statistics*. These are known as *fermions* and include neutrinos and atoms containing an odd number of particles. Other particles that do not comply with the exclusion principle obey the laws of *Bose-Einstein statistics*. These are called *bosons* and

Table 24–1 Electron configurations for the ground states of the elements

	K	L		M			N				O			P			Q
	1s	2s	2p	3s	3p	3d	4s	4p	4d	4f	5s	5p	5d	5f	6s	6p	6d	7s
1 H	1																	
2 He	2																	
3 Li	2	1																
4 Be	2	2																
5 B	2	2	1															
6 C	2	2	2															
7 N	2	2	3															
8 O	2	2	4															
9 F	2	2	5															
10 Ne	2	2	6															
11 Na	2	2	6	1														
12 Mg	2	2	6	2														
13 Al	2	2	6	2	1													
14 Si	2	2	6	2	2													
15 P	2	2	6	2	3													
16 S	2	2	6	2	4													
17 Cl	2	2	6	2	5													
18 A	2	2	6	2	6													
19 K	2	2	6	2	6		1											
20 Ca	2	2	6	2	6		2											
21 Sc	2	2	6	2	6	1	2											
22 Ti	2	2	6	2	6	2	2											
23 V	2	2	6	2	6	3	2											
24 Cr	2	2	6	2	6	5	1											
25 Mn	2	2	6	2	6	5	2											
26 Fe	2	2	6	2	6	6	2											
27 Co	2	2	6	2	6	7	2											
28 Ni	2	2	6	2	6	8	2											
29 Cu	2	2	6	2	6	10	1											
30 Zn	2	2	6	2	6	10	2											
31 Ga	2	2	6	2	6	10	2	1										
32 Ge	2	2	6	2	6	10	2	2										
33 As	2	2	6	2	6	10	2	3										
34 Se	2	2	6	2	6	10	2	4										
35 Br	2	2	6	2	6	10	2	5										
36 Kr	2	2	6	2	6	10	2	6										
37 Rb	2	2	6	2	6	10	2	6			1							
38 Sr	2	2	6	2	6	10	2	6			2							
39 Y	2	2	6	2	6	10	2	6	1		2							
40 Zr	2	2	6	2	6	10	2	6	2		2							
41 Nb	2	2	6	2	6	10	2	6	4		1							
42 Mo	2	2	6	2	6	10	2	6	5		1							
43 Tc	2	2	6	2	6	10	2	6	5		2							
44 Ru	2	2	6	2	6	10	2	6	7		1							
45 Rh	2	2	6	2	6	10	2	6	8		1							
46 Pd	2	2	6	2	6	10	2	6	10									
47 Ag	2	2	6	2	6	10	2	6	10		1							
48 Cd	2	2	6	2	6	10	2	6	10		2							
49 In	2	2	6	2	6	10	2	6	10		2	1						
50 Sn	2	2	6	2	6	10	2	6	10		2	2						
51 Sb	2	2	6	2	6	10	2	6	10		2	3						
52 Te	2	2	6	2	6	10	2	6	10		2	4						
53 I	2	2	6	2	6	10	2	6	10		2	5						

Table 24–1 (Continued)

	K	L		M			N				O				P			Q
	1s	2s	2p	3s	3p	3d	4s	4p	4d	4f	5s	5p	5d	5f	6s	6p	6d	7s
54 Xe	2	2	6	2	6	10	2	6	10		2	6						
55 Cs	2	2	6	2	6	10	2	6	10		2	6			1			
56 Ba	2	2	6	2	6	10	2	6	10		2	6			2			
57 La	2	2	6	2	6	10	2	6	10		2	6	1		2			
58 Ce	2	2	6	2	6	10	2	6	10	2	2	6			2			
59 Pr	2	2	6	2	6	10	2	6	10	3	2	6			2			
60 Nd	2	2	6	2	6	10	2	6	10	4	2	6			2			
61 Pm	2	2	6	2	6	10	2	6	10	5	2	6			2			
62 Sm	2	2	6	2	6	10	2	6	10	6	2	6			2			
63 En	2	2	6	2	6	10	2	6	10	7	2	6			2			
64 Gd	2	2	6	2	6	10	2	6	10	7	2	6	1		2			
65 Tb	2	2	6	2	6	10	2	6	10	9	2	6			2			
66 Dy	2	2	6	2	6	10	2	6	10	10	2	6			2			
67 Ho	2	2	6	2	6	10	2	6	10	11	2	6			2			
68 Er	2	2	6	2	6	10	2	6	10	12	2	6			2			
69 Tm	2	2	6	2	6	10	2	6	10	13	2	6			2			
70 Yb	2	2	6	2	6	10	2	6	10	14	2	6			2			
71 Lu	2	2	6	2	6	10	2	6	10	14	2	6	1		2			
72 Hf	2	2	6	2	6	10	2	6	10	14	2	6	2		2			
73 Ta	2	2	6	2	6	10	2	6	10	14	2	6	3		2			
74 W	2	2	6	2	6	10	2	6	10	14	2	6	4		2			
75 Re	2	2	6	2	6	10	2	6	10	14	2	6	5		2			
76 Os	2	2	6	2	6	10	2	6	10	14	2	6	6		2			
77 Ir	2	2	6	2	6	10	2	6	10	14	2	6	7		2			
78 Pt	2	2	6	2	6	10	2	6	10	14	2	6	9		1			
79 Au	2	2	6	2	6	10	2	6	10	14	2	6	10		1			
80 Hg	2	2	6	2	6	10	2	6	10	14	2	6	10		2			
81 Tl	2	2	6	2	6	10	2	6	10	14	2	6	10		2	1		
82 Pb	2	2	6	2	6	10	2	6	10	14	2	6	10		2	2		
83 Bi	2	2	6	2	6	10	2	6	10	14	2	6	10		2	3		
84 Po	2	2	6	2	6	10	2	6	10	14	2	6	10		2	4		
85 At	2	2	6	2	6	10	2	6	10	14	2	6	10		2	5		
86 Rn	2	2	6	2	6	10	2	6	10	14	2	6	10		2	6		
87 Fr	2	2	6	2	6	10	2	6	10	14	2	6	10		2	6		1
88 Ra	2	2	6	2	6	10	2	6	10	14	2	6	10		2	6		2
89 Ac	2	2	6	2	6	10	2	6	10	14	2	6	10		2	6	1	2
90 Th	2	2	6	2	6	10	2	6	10	14	2	6	10		2	6	2	2
91 Pa	2	2	6	2	6	10	2	6	10	14	2	6	10	2	2	6	1	2
92 U	2	2	6	2	6	10	2	6	10	14	2	6	10	3	2	6	1	2
93 Np	2	2	6	2	6	10	2	6	10	14	2	6	10	4	2	6	1	2
94 Pu	2	2	6	2	6	10	2	6	10	14	2	6	10	5	2	6	1	2
95 Am	2	2	6	2	6	10	2	6	10	14	2	6	10	6	2	6	1	2
96 Cm	2	2	6	2	6	10	2	6	10	14	2	6	10	7	2	6	1	2
97 Bk	2	2	6	2	6	10	2	6	10	14	2	6	10	8	2	6	1	2
98 Cf	2	2	6	2	6	10	2	6	10	14	2	6	10	10	2	6		2
99 E	2	2	6	2	6	10	2	6	10	14	2	6	10	11	2	6		2
100 Fm	2	2	6	2	6	10	2	6	10	14	2	6	10	12	2	6		2
101 Md	2	2	6	2	6	10	2	6	10	14	2	6	10	13	2	6		2
102 No	2	2	6	2	6	10	2	6	10	14	2	6	10	14	2	6		2
103 Lw	2	2	6	2	6	10	2	6	10	14	2	6	10	14	2	6	1	2
104 RF																		
105 Ha																		

Elements 57–71 constitute the Lanthanide Series. Elements 89–103 constitute the Actinide Series.

include photons, α particles, and atoms composed of an even number of protons and neutrons.

The protons and neutrons within the nucleus are also arranged in definite shells and subshells, much like those for the orbiting electrons. However, the energy interactions within the nucleus are quite complex, because both nuclear forces and Coulomb forces are involved, and because two kinds of particles—protons and neutrons—are involved. For these reasons, these shells are much more complex than atomic shells. They will be discussed at greater length in Chapter 26.

RECOMMENDED READING

GAMOW, George, "The Exclusion Principle," *Sci. Am.*, July 1959.

HAMPEL, Clifford A. (Ed.), *The Encyclopedia of the Chemical Elements*, Van Nostrand Reinhold, New York, 1968.

FOUR The Nucleus

> When the oscillator switch was opened this time, the counter was turned on, and click-click-click-click-click. We were observing induced radioactivity within less than half-hour after hearing of the Curie-Joliot results.
>
> M. STANLEY LIVINGSTON
> "History of the Cyclotron,"
> *Physics Today,* October 1959

Livingston is relating the excitement he shared with Lawrence and co-workers when they induced radioactivity with an early cyclotron. Such moments of excitement, which occurred only after much work and many trials and failures, eventually gave physicists insight into the energy contained within the nucleus of the atom. Man's later accomplishments in controlling and harnessing this energy, thus ushering in the atomic age, are well-known history.

25
The Nucleus

Hideki Yukawa
(1907–)

Born in Tokyo, Yukawa received his Doctor of Science degree from Osaka University in 1938, where he has taught since 1939. In 1935 Yukawa postulated a new type of field force, attempting to explain how particles are held together in an atomic nucleus. He described an intermediate mass particle (meson) that served as carrier of the nuclear force. The hypothesis was later confirmed, revealing that different types of mesons existed, including the pi meson, whose existence he had predicted. For his work in nuclear forces, he received the Nobel Prize in physics in 1949.

25–1 THE NUCLEAR ATOM
25–2 NUCLEAR FORCES
25–3 SOME PROPERTIES OF THE NUCLEUS
25–4 NUCLEAR BINDING ENERGY

25-1 THE NUCLEAR ATOM

By 1911 ERNEST RUTHERFORD's scattering experiments (Chapter 11) had shown that the atom consists of a very small nucleus ($\sim 10^{-14}$ m in diameter) and electrons orbiting about this nucleus. Compared to the dimensions of the nucleus, the orbits of the electrons were at great distances from the nucleus, and since most of the mass was formed in the nucleus, the atom was mostly empty space. We recall that the Rutherford model of the atom was a dynamic one, because electrostatic attraction of the positive nucleus and the negative electrons would have soon caused collapse in a static model.

As we have seen, although the Rutherford model was based on classical mechanics and the later Bohr model was based on some early concepts of quantum mechanics, it was the quantum theory of Schrödinger and Heisenberg that correctly described atomic structure. Thus quantum mechanics is again the model that must be used to describe the workings of the nucleus. However, the problem that we must consider here is much more complex, and a complete, unified nuclear theory has not yet been formulated. Physicists have adopted two models of the nucleus that we consider here—the *liquid drop model* and the *shell model* (other models are also used in differing contexts*). We approach these models from an experimental point of view to simplify their study.

Atomic physics is mainly a study of the electrons located in shells and subshells about the nucleus. The energies involved in the release or acceptance of electrons by the atom or in the transitions of electrons from one stationary state to another are of the order of several electron volts. Let us recall that only about 13.6 eV of energy are needed to remove from the atom an electron in the ground energy state of hydrogen. Because only the shell electrons are involved, this sort of chemistry can properly be called atomic chemistry.

EXAMPLE 25–1: How much energy per molecule is involved in the formation of water?

SOLUTION: The heat of formation of water is 68.3 kcal/mole; and since 1.0 kcal = 2.62×10^{22} eV and 1.0 mole of water contains 6.02×10^{23} molecules, the heat

---

* For the study of other, more sophisticated nuclear models, such as the optical model, the collective model, etc., see M. A. Preston, *Physics of the Nucleus*, Addison-Wesley, Reading, Mass., 1963, Chaps. 12, 13, and 18.

of formation of water per molecule is

$$Q = \frac{68.3 \times 2.62 \times 10^{22}}{6.02 \times 10^{23}} = 2.97 \text{ eV/molecule}$$

The simplest nuclear structure is the deuteron, which has a nucleus composed of a proton and a neutron held together by a very strong attractive nuclear force. To separate the constituents of a deuteron nucleus, 2.24 MeV of energy is required. In a nuclear reaction such as the fission of uranium-235 into krypton-89 and barium-144, a tremendous amount of energy, on the order of 200 MeV per uranium atom, is released. The chemistry involved in these reactions is appropriately called nuclear chemistry.

The particles making up the nucleus, the protons and neutrons, are referred to as *nucleons* when they are part of the nucleus. A species of nucleus, known as a *nuclide*, is represented schematically by

$$^A_Z B_N \tag{25-1}$$

where

Z, the *atomic number*, indicates the number of protons*
N, the *neutron number*, indicates the number of neutrons
A = N + Z, the *mass number*, indicates the total number of protons plus neutrons

As an example, the chlorine nucleus $^{35}_{17}Cl_{18}$ has $Z = 17$ protons, $N = 18$ neutrons, and $A = 17 + 18 = 35$ nucleons. Often the number N, which is somewhat redundant, is omitted, and the nucleus is written $^{35}_{17}Cl$.†

Isotopes are nuclei with the same atomic number Z but different mass numbers. The nuclei $^{28}_{14}Si$, $^{29}_{14}Si$, $^{30}_{14}Si$, and $^{32}_{14}Si$ are all isotopes of silicon. Those nuclei with the same mass number A but different atomic number Z are called *isobars*. The nuclei $^{16}_{8}O$ and $^{16}_{7}N$ are examples of isobars. Nuclei with an equal number of neutrons, that is, with the same N, are *isotones*. Some isotones are $^{14}_{6}C_8$, $^{15}_{7}N_8$, and $^{16}_{8}O_8$.

All of the nuclides listed in Table 6 in the Appendix include isotopes. For example, the nuclide chlorine-17 has nine isotopes ranging from $^{32}_{17}Cl$ to $^{39}_{17}Cl$, but the most abundant in nature are $^{35}_{17}Cl$ (abundance 75.53%) and $^{37}_{17}Cl$ (abundance 24.47%). The corresponding atomic masses in atomic mass units are $^{35}_{17}Cl$ (34.98 amu) and $^{37}_{17}Cl$ (36.98 amu). As found in nature, chlorine is a mixture largely of these two isotopes, and the average atomic mass of chlorine is

$$A_{Cl} = \frac{(75.53 \times 34.98) + (24.47 \times 36.98)}{75.53 + 24.47} = 35.46$$

---

* The atomic number is sometimes called the *proton number*.
† This symbolic representation has been recommended by the A.A.P.T. (American Association of Physics Teachers). Physicists engaged in research usually skip the atomic and neutron numbers and write simply Cl-35.

The number 35.46 is often called the *atomic weight* of chlorine, although this is something of a misnomer.

The most abundant element, hydrogen, is found with the following abundances of isotopes:

ISOTOPE	SYMBOL	ATOMIC MASS (amu)	ABUNDANCE (%)
hydrogen	$_1^1H$	1.007825	99.99
deuterium	$_1^2H$	2.014102	0.01
tritium	$_1^3H$	3.016049	traces

The average atomic mass of hydrogen is then

$$\frac{(99.99 \times 1.0078) + (0.01 \times 2.0141)}{99.99 + 0.01} = 1.0079$$

25-2 NUCLEAR FORCES

According to Coulomb's law, the positively charged protons closely spaced within the nucleus should repel each other very strongly, and the protons should fly apart. It is difficult to explain the stability of the nucleus unless it is assumed that the nucleons are under the influence of some very strong attractive forces. These forces, classified as *strong interactions*, were studied extensively over a long period of time by the Japanese physicist HIDEKI YUKAWA. In 1935 he announced the main characteristics of the nuclear forces and postulated the existence of a particle called a *pion*, which has a rest mass of $270m_e$ (270 times the mass of the electron). The pion played an integral part in the explanation of nuclear forces. For his contributions to the understanding of these nuclear forces, Prof. Yukawa was awarded the Nobel Prize in physics in 1949.

Yukawa attributed the following characteristics to the nuclear forces:

1. Nuclear forces are effective only at *short ranges*.
2. Nuclear forces are *charge-independent*.
3. Nuclear forces are the *strongest* known forces in nature.
4. Nuclear forces are readily *saturated* by surrounding nucleons.

Let us analyze these characteristics in more detail.

Short Range. From the results of scattering experiments, it is found that nuclear forces are appreciable only when the distance between two nucleons is of the order of 10^{-15} m or less. If a nucleus is bombarded by protons and if the range of nuclear forces is of the same order of magnitude as the repulsive Coulomb forces, then no matter how close the protons approach the nucleus they will be affected by both types of forces, and the scattering distribution of protons will be different from that corresponding to a pure Coulomb scattering.

Incident protons that do not pass too close to the nucleus are scattered by the action of electrical repulsive forces. However, if the energy of the incoming protons is large enough to overcome the repulsive effect of the Coulomb forces, they can pass very close to the nucleus and fall in the range of the attractive nuclear forces. The distribution of the scattered protons in this case arises from scattering largely by the strong, attractive nuclear forces, and the distribution is quite distinct from Coulomb scattering.

There is also some evidence to suggest that at extremely short distances (0.5 fermi*), *nucleons repel each other* (see Figure 25–1).

Figure 25–2 shows a proton approaching a nucleus. When the proton is very far from the nucleus, its total energy is kinetic. However, as it approaches closer

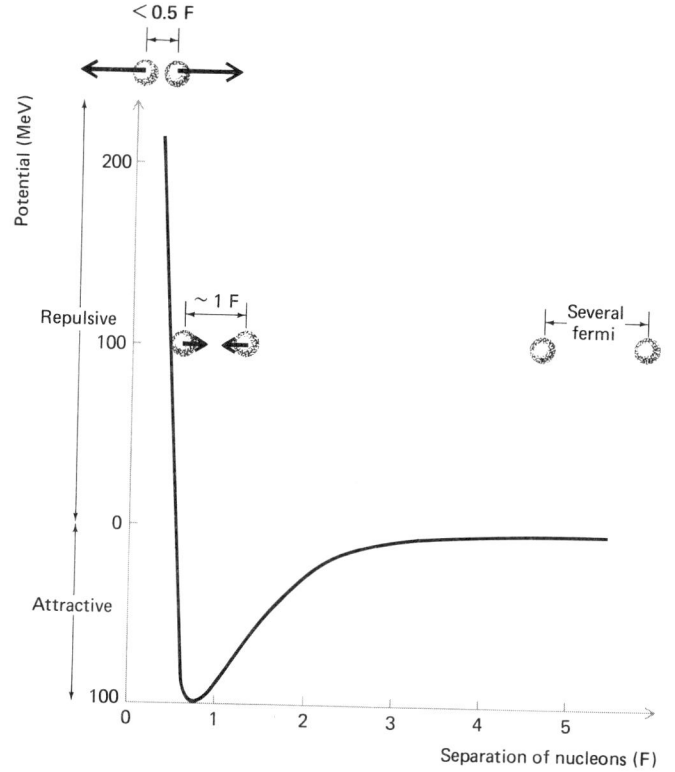

Figure 25–1 The potential of the nuclear force shows it to be strongly attractive at a little less than 1 F. Experimental evidence suggests that at shorter distances it is repulsive.

* The very small nuclear dimensions are often written in terms of the unit of length called the fermi (F), where 1 F = 10^{-15} m.

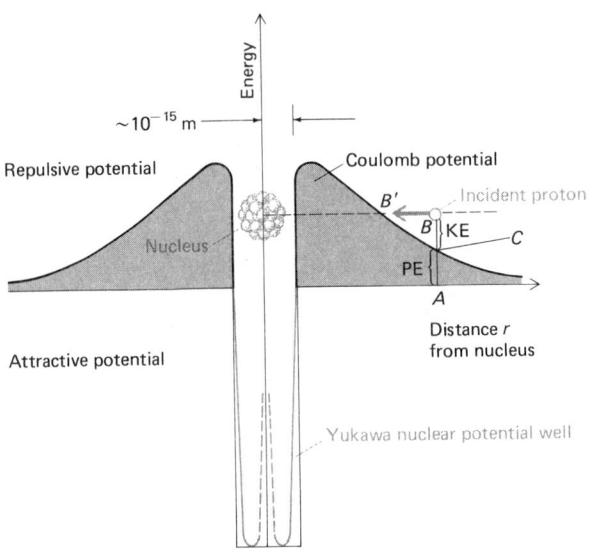

Figure 25-2 Yukawa potential of short-range nuclear forces compared to the Coulomb barrier.

to the nucleus, it is repulsed by the positively charged nucleus and experiences a repulsive Coulomb potential

$$eV = k\frac{Ze^2}{r} \tag{25-2}$$

Along with an increase in potential energy, there is a decrease in kinetic energy. From Figure 25-2, the total energy is given by

$$AB = AC + CB$$

where

AB = total energy

$AC = eV = \dfrac{kZe^2}{r}$ potential energy

$CB = KE$ kinetic energy

and since the curve shows that $dV/dr < 0$, the electrical force

$$F_E = -e\frac{dV}{dr} > 0$$

is repulsive.

As pictured in Figure 25–2, the total energy of the incident proton is less than the height of the Coulomb potential barrier. In terms of classical mechanics, when the proton reaches position B and the kinetic energy just equals the potential energy, it classically will "collide" and bounce back, unable to penetrate the barrier and become attracted by the short-range nuclear forces.

Quantum mechanics explains by the tunnel effect (Chapter 15) how the approaching proton has a finite probability of tunneling through the potential barrier and "falling" in the *Yukawa potential well* associated with nuclear forces

$$V = -g^2 \frac{e^{-ar}}{r} \qquad (25\text{-}3)$$

where $a = mc/\hbar$, m is the mass of the proton, and g is a constant related to the nucleus.

Charge Independence. Experimental evidence has shown that the interaction between any two nucleons is independent of the charge. To a high degree of accuracy, the interactions among the nuclear forces between protons and neutrons, neutrons and neutrons, and protons and protons exclusive of Coulomb forces have been found to be the same.

Strong Forces. The forces between the nucleons, the *strong interactions*, are the strongest forces found in nature. The weaker gravitational and electromagnetic interacting forces were observed long before the nuclear forces, because they are associated with macroscopic bodies such as the gravitational forces between the planets and the sun and the electrical forces associated with charged bodies.

Saturation Effect. Nuclear forces are the only forces in nature that exhibit saturation effects. The ability of nuclear forces to act on other particles reaches a saturation point when a nucleon is completely surrounded by other nucleons. Other nucleons located outside the surrounding nucleons do not "feel" the interaction of the surrounded nucleon.

25–3 SOME PROPERTIES OF THE NUCLEUS

The nucleus, composed of neutrons (which are electrically neutral) and protons, carries a net electrical charge of $+Ze$, where the atomic number Z gives the number of protons. It would seem that the mass of the nucleus should be

$$\text{assumed nuclear mass} = Zm_p + Nm_n$$

where m_p and m_n are the respective proton and neutron masses and N is the neutron number. Measurements by mass spectrometer, however, show that

$$\text{real nuclear mass} < Zm_p + Nm_n$$

The difference in masses,

$$Zm_p + Nm_n - \text{real nuclear mass} = M_D \quad (25\text{-}4)$$

is called the *mass defect*.

The theoretical explanation for this mass defect is based on Equation (6–18), the mass-energy conversion equation

$$\Delta E = (\Delta m)c^2 \quad (25\text{-}5)$$

When the Z protons and the N neutrons combine to form a stable nucleus, some of the mass (Δm) disappears in the form of energy released (usually in the form of γ-ray energy).

When the nucleus is bombarded with protons, the protons must penetrate the Coulomb barrier to get close enough to "feel" the nuclear forces. Experiments using charged particles scattered from the nucleus arrive at an expression for the radius of the nucleus given by

$$R = r_0 A^{1/3} \quad (25\text{-}6)$$

where A is the mass number and $r_0 = 1.2 \times 10^{-15}$ m = 1.2 F.

The Coulomb barrier produces no resistance when neutrons are scattered from the nucleus. The neutrons then "feel" the nuclear forces at a somewhat larger distance than do charged particles, and for this case

$$r_0 = 1.5 \times 10^{-15} \text{ m} = 1.5 \text{ F}.$$

In working problems, we shall compromise by using a value of $r_0 = 1.3$ F.

The *nuclear density* ρ_N can be calculated approximately from

$$\rho_N = \frac{\text{nuclear mass}}{\text{nuclear volume}} \quad (25\text{-}7)$$

The nuclear mass is approximately

$$\text{nuclear mass} \cong A m_N$$

where A is the mass number and m_N is roughly the mass of the nucleon, $m_N = 1.67 \times 10^{-27}$ kg. The nuclear volume is

$$\text{nuclear volume} = \tfrac{4}{3}\pi R^3 = \tfrac{4}{3}\pi (r_0 A^{1/3})^3$$

or

$$\text{nuclear volume} = (\tfrac{4}{3}\pi r_0^3) A \quad (25\text{-}8)$$

Since the volume of a single nucleon when $A = 1$ is $\tfrac{4}{3}\pi r_0^3$, Equation (25-8) shows that the nuclear volume is directly proportional to the number of nucleons A. The nuclear density is then

$$\rho_N = \frac{m_N A}{\frac{4}{3}\pi r_0^3 A} = \frac{m_N}{\frac{4}{3}\pi r_0^3} \tag{25-9}$$

When numerical values are substituted in Equation (25-9), the nuclear density is

$$\boxed{\rho_N \cong 2 \times 10^{17} \text{ kg/m}^3 = 2 \times 10^{14} \text{ tons/m}^3}$$

an incredibly large number.

The atomic radius is 10^4 times the nuclear radius, and the *atomic density* is

$$\boxed{\rho_A \cong \frac{2 \times 10^{17}}{(10^4)^3} = 2 \times 10^5 \text{ kg/m}^3}$$

a much smaller number.

Finally, the density of matter in bulk is considerably less than the atomic or nuclear densities. The density of water is

$$\boxed{\rho_W = 1 \text{ g/cm}^3 = 10^3 \text{ kg/m}^3}$$

25-4 NUCLEAR BINDING ENERGY

When the Z protons and N neutrons combine to make a nucleus, some of the mass (Δm) disappears because it is converted into an amount of energy $\Delta E = (\Delta m)c^2$. This energy is called the *binding energy* BE of the nucleus. To disrupt a stable nucleus into its constituent protons and neutrons, the minimum energy required is the binding energy. The binding energy is then

$$\boxed{\text{BE} = (Zm_p + Nm_n)c^2 - M_n c^2} \tag{25-10}$$

where M_n is the nuclear mass and each mass term has been multiplied by c^2 to express the equation in terms of energy.

Because tables of nuclides such as Table 6 in the Appendix, are tabulated in terms of atomic masses rather than nuclear masses, Equation (25-10) will be modified accordingly. First, the nuclear mass can be computed from

$$M_n = M_a - Zm_e \tag{25-11}$$

where M_n represents the atomic mass corresponding to the specified nuclear mass and Zm_e is the total mass of the orbiting electrons. The binding energy of electrons has been neglected because it is quite small compared to the nuclear binding energies.

Second, the proton mass can be found from

$$m_p = m_H - m_e \qquad (25\text{-}12)$$

where m_H is the atomic mass of the hydrogen atom. The electron binding energy of 13.6 eV is also negligible here.

The binding energy equation now becomes

$$BE = Z(m_H - m_e)c^2 + Nm_n c^2 - (M_a - Zm_e)c^2 \qquad (25\text{-}13)$$

which simplifies to

$$BE = (Zm_H + Nm_n)c^2 - M_a c^2 \qquad (25\text{-}14)$$

It is frequently more convenient to express the binding energy in terms of mass units rather than energy units. In this case, the c^2 factor is dropped and Equation (25-14) becomes

$$\boxed{BE = (Zm_H + Nm_n) - M_a} \qquad (25\text{-}15)$$

BE is now in atomic mass units.

If $BE > 0$, the nucleus is stable and energy must be supplied from the outside to disrupt it into its constituents. If $BE < 0$, the nucleus is unstable and it will disintegrate by itself.

EXAMPLE 25-2: Compute the binding energy for $^{16}_{8}O$.

SOLUTION: From Equation (25-15),

$$Zm_H = 8 \times 1.007825 = 8.062600 \text{ amu}$$
$$Nm_n = 8 \times 1.008665 = 8.069320 \text{ amu}$$
$$16.131920 \text{ amu}$$

atomic mass of $^{16}_{8}O = 16.000000$ amu

$$BE = +0.131920 \text{ amu}$$

The plus sign indicates that the nucleus is stable.

Since 1 amu = 931.48 MeV, the binding energy in MeV is

$$BE = 0.131920 \times 931.48 = 122.8 \text{ MeV}$$

The binding energy per nucleon is 123 MeV/16 = 7.68 MeV/nucleon.

EXAMPLE 25-3: Compute the *separation energy* SE to remove one proton from $^{16}_{8}O$. The separation energy SE is the minimum energy that must be supplied to remove the least tightly bound nucleon from the nucleus. It is the binding energy of the least tightly bound nucleon.

SOLUTION: When a single proton is removed from the $^{16}_{8}$O nucleus, a $^{15}_{7}$N nucleus remains:

$$^{16}_{8}\text{O} \rightarrow {}^{15}_{7}\text{N} + {}^{1}_{1}p$$

The computation of the separation energy is made by following the mass-energy conservational principle (Chapter 6). The separation energy is then

$$\text{SE} = (\text{atomic mass of } {}^{15}_{7}\text{N} + m_\text{H}) - (\text{atomic mass of } {}^{16}_{8}\text{O})$$

Numerical values in this equation give

$$\begin{aligned}
\text{atomic mass of } {}^{15}_{7}\text{N} &= 15.000108 \text{ amu} \\
m_\text{H} &= \underline{1.007825 \text{ amu}} \\
&\ 16.007933 \text{ amu} \\
\text{atomic mass of } {}^{16}_{8}\text{O} &= 16.000000 \text{ amu} \\
\text{separation energy} &= 0.007933 \text{ amu}
\end{aligned}$$

and in MeV the separation energy is

$$\text{SE} = 0.007933 \times 931.48 = 7.40 \text{ MeV}$$

Can you suggest why the binding energy of the single proton is less than the binding energy per nucleon calculated in Example 25-2?

PROBLEMS

25-1 In a scattering experiment, gold nuclei $^{197}_{79}$Au are bombarded with α particles $^{4}_{2}$He. Calculate the nuclear radius of the gold nuclei.

25-2 In an experiment similar to the one described in Problem 25-1, the gold nuclei are bombarded by neutrons. Compute the nuclear radius of the gold nuclei for these neutral particles.

25-3 In Problem 25-1, if it is known that the incident α particles have a kinetic energy of 7.68 MeV, calculate (a) the height of the Coulomb barrier in MeV, and (b) the distance between the α particle and the gold nucleus when the α particle "collides" with the Coulomb potential barrier.

25-4 An α particle is inside a $^{226}_{88}$Ra nucleus. Compute the *height* of the potential barrier in MeV. If the radium nucleus emits a 4.78-MeV α particle, calculate the closest distance the α particle can approach a gold nucleus ($Z = 79$, $A = 197$).

25-5 Calculate the nuclear binding energy in MeV for a deuteron, $^{2}_{1}$H.

25-6 Calculate the nuclear binding energy for the isotopes $^{8}_{4}$Be and $^{9}_{4}$Be and decide which one is more stable.

25-7 Two nuclides are called *mirror nuclides* when one nuclide can be transformed into the other by interchanging their respective proton and neutron numbers. Determine the difference in the total nuclear binding energy of the mirror nuclides $^{11}_{5}B$ and $^{11}_{6}C$.

25-8 Compute the nuclear binding energy per nucleon for the following nuclides: (a) $^{6}_{4}Be$, (b) $^{8}_{5}B$, (c) $^{20}_{10}Ne$, and (d) $^{56}_{26}Fe$.

25-9 What is the separation energy for a neutron from $^{4}_{2}He$?

25-10 Assume that the radius of the hydrogen atom is 0.53 Å, which corresponds to its ground state, and show that the nuclear density of $^{1}_{1}H$ is about 10^{14} larger than the atomic density.

25-11 The nuclear binding energy of nitrogen-14 is 104.631 MeV. What is this in terms of atomic mass units? What fraction of the mass of a proton is this?

RECOMMENDED READING

ALONSO, M., and FINN, E. J., *Fundamental University Physics*, Addison-Wesley, Reading, Mass., 1968, Vol. III.

ARYA, A. P., *Fundamentals of Nuclear Physics*, Allyn & Bacon, Boston, 1966.

BETHE, H. A., "What Holds the Nucleus Together?" *Sci. Am.*, September 1953.

CHEW, G. F., GELL-MANN, M., and ROSENFELD, A. R., "Strongly Interacting Particles," *Sci. Am.*, February 1964.

ENGE, H., *Introduction to Nuclear Physics*, Addison-Wesley, Reading, Mass., 1966.

HOFSTADER, R., "The Atomic Nucleus," *Sci. Am.*, July 1956.

MARSHAK, R. E., "The Nuclear Force," *Sci. Am.*, March 1960.

MAYER, M. G., "The Structure of the Nucleus," *Sci. Am.*, March 1951.

26
Models of the Nucleus

Maria Goeppert-Mayer
(1906–)

A native of Kattowitz, Poland, Mayer received her Ph.D. from the University of Göttingen in 1930. She immigrated to the United States in 1930, where she was associated with the Argonne National Laboratory (1946–1960) and the Enrico Fermi Institute of Nuclear Studies (1946–1959). She is now at the University of California. Mayer, E. P. Wigner, and J. Hans Jensen theorized that the protons and neutrons of the atomic nucleus are arranged in shells much as are the electrons in the outer atom. For their work they received the Nobel Prize in 1963.

26–1 PHOTODISINTEGRATION—NUCLEAR STABILITY
26–2 SPIN ANGULAR MOMENTUM
26–3 ELECTRONS IN THE NUCLEUS?
26–4 LIQUID DROP MODEL
26–5 SHELL MODEL

26-1 PHOTODISINTEGRATION—NUCLEAR STABILITY

Compared to chemical energies, the energies binding the nucleons together within the nucleus are quite large. Nevertheless, when atomic particles such as neutrons, protons, and α particles, or high-energy photons, collide with the nucleus they can produce changes in the nucleus. There are a wide variety of events that can occur. Sometimes the incident particle enters the nucleus, where it remains while a second particle, or even several other particles, are ejected. The incident particle can even fracture the nucleus into several pieces, or it may simply bounce off in an elastic collision. The total stability of the nucleus is measured in terms of the energy of the incident particle and the binding energies of the particles making up the nucleus.

Other than the single proton of the hydrogen atom, the next simplest nuclear structure is the deuterium nucleus, or deuteron, which contains a proton and a neutron. In the disintegration of the deuteron,

$$^2_1H \rightarrow {}^1_1H + {}_0n^1$$

the binding energy in atomic mass units is found from

$$BE = Zm_H + Nm_n - \text{(atomic mass of } {}^2_1H\text{)}$$

or since in this case $Z = 1$ and $N = 1$,

$$BE = m_H + m_n - \text{(atomic mass of } {}^2_1H\text{)}$$

and we have, by substitution,

$$\begin{aligned}
1.007825 \text{ amu} &= m_H \\
+1.008665 \text{ amu} &= m_n \\
\hline
2.016490 \text{ amu} & \\
-2.014102 \text{ amu} &= \text{atomic mass of } {}^2_1H \\
\hline
BE = 0.002388 \text{ amu} &
\end{aligned}$$

or as more commonly expressed in MeV, $BE = 2.22$ MeV.

If m_d is the nuclear mass of the deuteron, the binding energy in MeV can be computed from

$$BE + m_d c^2 = m_p c^2 + m_n c^2 \tag{26-1}$$

Graphically, these relations are shown in Figure 26-1.

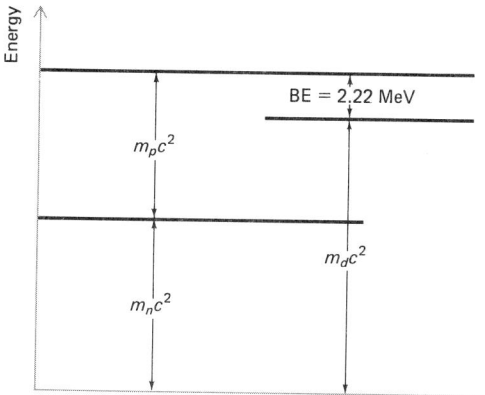

Figure 26-1 Binding energy of the deuteron compared to the rest mass of the constituent proton and neutron.

Experimentally, the binding energy of the deuteron d can be determined by *photodisintegration*, where monoenergetic photons, γ rays, are absorbed by the nucleus and the nucleus disintegrates through the process

$$\gamma + d \rightarrow p + n \qquad (26\text{-}2)$$

The mass-energy balance of Equation (26-2) is

$$(h\nu) + m_d c^2 = m_p c^2 + m_n c^2 + K_p + K_n \qquad (26\text{-}3)$$

where K_p and K_n are the respective kinetic energies of the released proton and neutron, and the deuteron is initially at rest. The minimum or threshold energy of the incident photon $h\nu_0$ is found from Equation (26-3) by letting $K_p = K_n = 0$. The threshold energy is then

$$(h\nu_0)_\gamma + m_d c^2 = m_p c^2 + m_n c^2 \qquad (26\text{-}4)$$

and evidently

$$h\nu_0 = \text{BE}$$

The inverse photonuclear effect is the production of γ-ray photons with an energy of 2.22 MeV from the synthesis of protons and neutrons at rest. This process is

$$\underset{\text{at rest}}{p + n} \rightarrow d + \gamma \qquad (26\text{-}5)$$

Table 26-1 shows how the 272 stable nuclei found in nature are classified according to even and odd numbers of protons and neutrons.

Table 26-1

PROTONS	NEUTRONS	STABLE NUCLIDES
even	even	160
even	odd	56
odd	even	52
odd	odd	4
		272

The combination of an even number of protons and an even number of neutrons composing the nucleus is evidently preferred by nature for stable nuclides. The odd-odd combination of stable nuclides is found only in the light elements. The number of even-odd combinations is about the same.

A plot of the number of neutrons versus the number of protons for the stable nuclides is shown in Figure 26-2. Notice that for $Z < 20$, the stability line is a

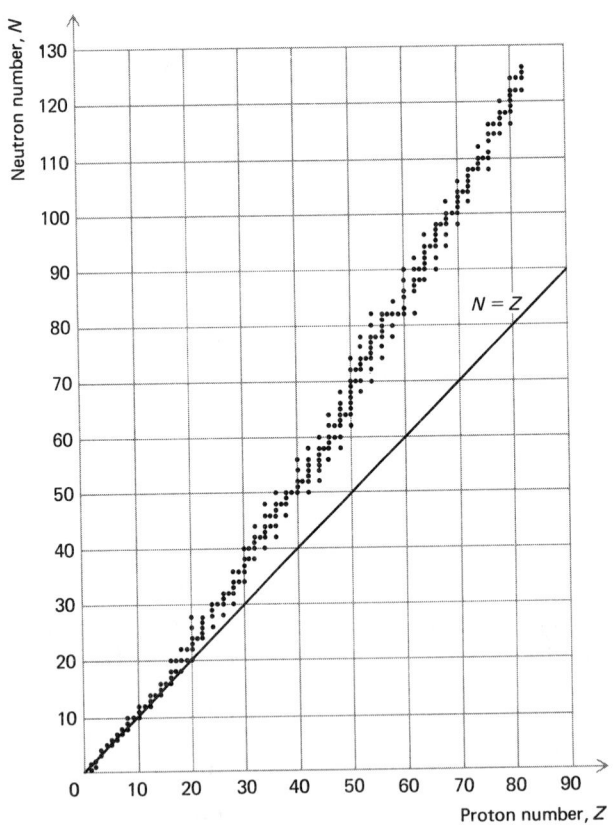

Figure 26-2 Stability curve for nuclides. The neutron number of each nuclide is plotted versus its proton number.

straight line with $Z = N$. For the heavier nuclides $Z > 20$, $N > 20$, the stability curve bends in the direction of $N > Z$. For example, $^{48}_{20}\text{Ca}_{28}$ has $N = 28$, $Z = 20$; for larger values of Z the tendency is more pronounced, as in the case of $^{232}_{91}\text{Pa}_{141}$, which has $N = 141$, $Z = 91$.

Evidently, for large values of Z the Coulomb electrostatic repulsion becomes important, and the number of neutrons must be greater to compensate this repulsive effect.

26–2 SPIN ANGULAR MOMENTUM

The proton and neutron, like the electron, each have an intrinsic spin; the *spin angular momentum* is computed by

$$L_s = \sqrt{I(I + 1)}\hbar \tag{26-6}$$

where the quantum number I, commonly called the spin, is equal to $\frac{1}{2}$. The spin angular momentum then has a value

$$L_s = \frac{\sqrt{3}}{2}\hbar$$

In the presence of an external magnetic field B_{ext}, the spin angular momentum is space-quantized with respect to the field. With the z direction chosen to be in the direction of the field B_{ext}, the spin angular momentum has only the components

$$L_{sz} = \pm \tfrac{1}{2}\hbar = L_s \cos \theta \tag{26-7}$$

The space quantization of the spin angular momentum of a proton or neutron is illustrated in Figure 26–3.

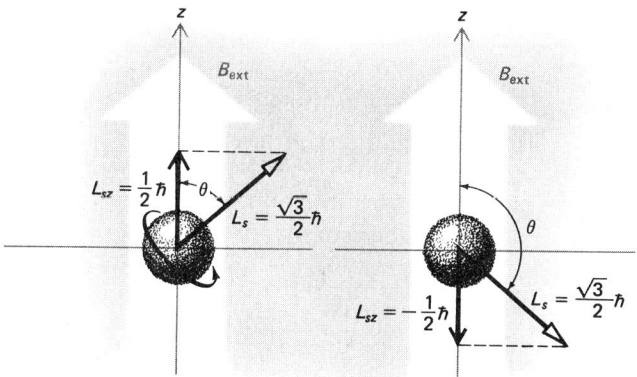

Figure 26–3 Space quantization of the spin angular momentum of a proton or neutron.

In addition to the spin angular momentum, the protons and neutrons in the nucleus have an *orbital angular momentum*. The *resultant angular momentum* of the nucleus is obtained by adding the spin and orbital angular momenta of all the nucleons within the nucleus. The total angular momentum of a nucleus is given by

$$\boxed{L_N = \sqrt{I_N(I_N + 1)}\,\hbar} \qquad (26\text{–}8)$$

where the quantum number I_N, usually called the *nuclear spin*, has values of $I_N = 0, 1, \ldots$ or half-integer values $\tfrac{1}{2}, \tfrac{3}{2}, \ldots$.

The *nuclear angular momentum* is space-quantized with respect to an externally applied magnetic field. The z component, the component in the direction of the applied field, is

$$L_{Nz} = L_N \cos\theta = m_I \hbar \begin{cases} m_I = \pm I_N, \pm(I_N - 1), \ldots, \pm\tfrac{1}{2} \\ \quad \text{if } I_N \text{ is a half-integer} \\ m_I = \pm I_N, +(I_N - 1), \ldots, 0 \\ \quad \text{if } I_N \text{ is an integer} \end{cases} \qquad (26\text{–}9)$$

As an example, $^{43}_{20}\text{Ca}$ has a nuclear spin $I_N = \tfrac{7}{2}$ and a total angular momentum $L_N = \sqrt{\tfrac{7}{2}(\tfrac{7}{2} + 1)}\,\hbar = (\sqrt{63}/2)\hbar$. There are a total of $2I_N + 1 = 8$ possible orientations of the nuclear angular momentum in an applied magnetic field

$$L_{Nz} = \pm\tfrac{7}{2}\hbar,\ \pm\tfrac{5}{2}\hbar,\ \pm\tfrac{3}{2}\hbar,\ \pm\tfrac{1}{2}\hbar$$

The spinning electron has an associated magnetic dipole moment of 1 Bohr magneton, $e\hbar/2m_e$. Similarly, there is associated with a nucleon a *nuclear magneton* given by

$$\boxed{\mu_N = \frac{e\hbar}{2m_p}} \qquad (26\text{–}10)$$

where m_p is the mass of the proton, which is 1836 times the mass of the electron. The nuclear magneton is thus proportionately 1836 times smaller than the Bohr magneton, having a value of 5.050×10^{-27} J/T.

In the presence of an external magnetic field B_{ext}, if the z axis is taken along B_{ext}, the z components (the only ones that can be observed) of the magnetic dipole moment of the proton and the neutron are, respectively,

$$\begin{aligned} \mu_{pz} &= +2.79\,\frac{e\hbar}{2m_p} = +2.79\mu_N \\ \mu_{nz} &= -1.91\,\frac{e\hbar}{2m_p} = -1.91\mu_N \end{aligned} \qquad (26\text{–}11)$$

We should note that these values are not well understood from the theoretical point of view. Physicists have found that it is especially hard to understand how the neutron, which is a neutral particle, can have a magnetic moment. The negative sign associated with the z component of the magnetic moment of the neutron indicates that it is in the direction opposite to its observable angular momentum.

The spin and observable magnetic dipole moments of several nuclides are listed in Table 26–2.

26-3 ELECTRONS IN THE NUCLEUS?

The question of whether there are electrons in the nucleus or not is intriguing because Thomson originally pictured the atom as one big positive mass embedded throughout with small electrons. The atom, as Thomson saw it, had no nucleus at all. Then the efforts of Rutherford and Bohr revised the picture of the atom, and it was seen to have a massive, positively charged nucleus about which and some distance away, revolved electrons. However, until 1931 it was generally believed that electrons did exist in the nucleus, and that the net positive charge of the nucleus is due to an excess number of nuclear protons over the number of nuclear electrons. For example, the nucleus of deuterium was supposed to consist of two protons and one nuclear electron, and the excess charge of $+1e$ was compensated in the neutral atom by the one electron in the outer shell. However, with the discovery of the neutron by Chadwick in 1932, a new theory was proposed in which the presence of nuclear electrons was denied. According to this theory, the nucleus contains only protons and neutrons and no electrons. Several different experiments have since been performed to verify that electrons do not exist in the nucleus.

If electrons did exist in the nucleus, the magnetic moment of a nucleus would be of the order of the Bohr magneton. Experimental evidence expressed in Equations (26–10) and (26–11) show the magnetic moments to be a factor of a thousand or so smaller than this.

Many experiments have shown the deuterium nucleus to have a spin $+1$. Figure 26–4 shows the possible spins of a nucleus with electrons and the possible spins without electrons. The figure shows that any combination of allowed orientations of the spins of the electrons in the nucleus will give a half-integer value for the spin, which is in contradiction to experimental evidence.

Rutherford's scattering experiments showed the size of the nucleus to be of the order of 10^{-14} m. The uncertainty in position of an electron in the nucleus is

$$\Delta x = 10^{-14} \text{ m}$$

and from the uncertainty principle the uncertainty in momentum is

$$\Delta p = \frac{\hbar}{\Delta x} = 1.1 \times 10^{-20} \text{ kg-m/sec}$$

Table 26-2

NUCLIDE	SPIN	MAGNETIC DIPOLE MOMENT IN NUCLEAR MAGNETONS
$^{1}_{0}n$	$\frac{1}{2}$	-1.9131
$^{1}_{1}H$	$\frac{1}{2}$	2.7927
$^{2}_{1}H$	1	0.8574
$^{3}_{2}He$	$\frac{1}{2}$	-2.1275
$^{3}_{1}H$	$\frac{1}{2}$	2.9789
$^{4}_{2}He$	0	0
$^{7}_{3}Li$	$\frac{3}{2}$	3.2563
$^{12}_{6}C$	0	0
$^{13}_{6}C$	$\frac{1}{2}$	0.7024
$^{14}_{7}N$	1	0.4036
$^{15}_{7}N$	$\frac{1}{2}$	-0.2831
$^{16}_{8}O$	0	0
$^{17}_{8}O$	$\frac{5}{2}$	-1.8937
$^{21}_{10}Na$	$\frac{3}{2}$	-0.6618
$^{27}_{13}Al$	$\frac{5}{2}$	3.6414
$^{35}_{17}Cl$	$\frac{3}{2}$	0.8218
$^{40}_{20}Ca$	0	0
$^{43}_{20}Ca$	$\frac{7}{2}$	-1.3172
$^{56}_{26}Fe$	0	1.06
$^{57}_{26}Fe$	$\frac{1}{2}$	0.0905
$^{60}_{27}Co$	5	3.8100
$^{63}_{29}Cu$	$\frac{3}{2}$	2.2260
$^{79}_{35}Br$	$\frac{3}{2}$	2.1060
$^{88}_{38}Sr$	0	0
$^{93}_{41}Nb$	$\frac{9}{2}$	6.1670
$^{103}_{45}Rh$	$\frac{1}{2}$	0
$^{114}_{48}Cd$	0	0
$^{127}_{53}I$	$\frac{5}{2}$	2.8080
$^{155}_{64}Gd$	$\frac{3}{2}$	-0.2700
$^{175}_{71}Lu$	$\frac{7}{2}$	2.2300
$^{176}_{71}Lu$	7	3.1800
$^{177}_{71}Lu$	$\frac{7}{2}$	2.2400
$^{180}_{72}Hf$	0	0
$^{185}_{75}Re$	$\frac{5}{2}$	3.1716
$^{208}_{82}Pb$	0	0
$^{209}_{83}Bi$	$\frac{9}{2}$	4.0802
$^{227}_{83}Ac$	$\frac{3}{2}$	1.1
$^{233}_{92}U$	$\frac{5}{2}$	0.54
$^{235}_{92}U$	$\frac{7}{2}$	0.35
$^{238}_{92}U$	0	0
$^{241}_{94}Pu$	$\frac{5}{2}$	-0.730

SOURCE: William H. Sullivan, *Trilinear Chart of Nuclides,* U.S. Government Printing Office, Washington, D.C., 1957.

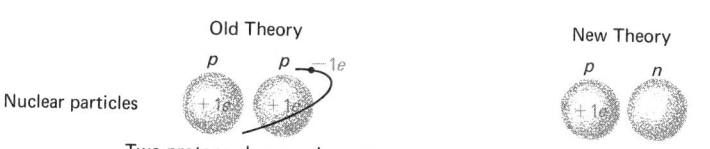

Figure 26-4 Experimental evidence shows that the spin of the deuteron is +1, which contradicts the old theory that electrons exist in the nucleus.

The momentum of the electron must be at least this large, and from Equation (6–27) the kinetic energy is

$$K \cong pc = 3.3 \times 10^{-12} \text{ J}$$

$$= 20 \text{ MeV}$$

The energies of electrons associated with unstable atoms are only of the order of several electron volts according to experimental verifications, and this fact is in contradiction with the theoretical expectation of the uncertainty principle.

26–4 LIQUID DROP MODEL

Since there is no one complete theory that accounts for all of the nuclear properties, several models of the nucleus have been developed to describe different particular properties. Here we shall first examine the liquid drop model. In a liquid, the binding energies of two atoms in a molecule are independent of other atoms in the same molecule. This situation is similar to that of the saturation effect in nuclear forces, where the binding energy per nucleon remains practically constant for mass numbers $A \geqq 20$.

In 1935 C. V. Weizacker, a German physicist, proposed the semi-empirical nuclear binding energy formula for a nucleus (Z, N, A)

$$\text{BE}_{\text{(nuclear)}} = aA - bA^{2/3} - \frac{cZ(Z-1)}{A^{1/3}} - \frac{d(N-Z)^2}{A} \pm \frac{\delta}{A^{3/4}} \text{ (MeV)}$$

(26–12)

with the constants having values $a = 15.8$, $b = 17.8$, $c = 0.71$, and $d = 23.7$. The constant δ is chosen according to

δ	Z	N	A
34	even	even	even
0	even	odd	odd
0	odd	even	odd
34	odd	odd	even

The first term, $E_v = aA$, is the *volume effect*. The larger the total number of nucleons A, the more difficult it will be to remove the individual proton and neutrons from the nucleus. The binding energy is directly proportional to the total number of nucleons in the same sense that the amount of heat energy (the heat of vaporization) necessary to change a liquid to a vapor state is proportional to the mass of the liquid.

The nucleons at the surface of the nucleus are not completely surrounded by other nucleons. The *surface effect*, expressed in the second term of Equation (26–12),

$$E_2 = bA^{2/3}$$

is similar to the surface tension in liquids. Since this tends to reduce the effect of the binding energy, it appears as a negative quantity in the binding energy equation.

Another effect that dilutes the binding energy is the *Coulomb electrostatic repulsion*. Because each charged particle in the nucleus repulses all of the other charged particles, there, this energy is directly proportional to the possible number of combinations for a given proton number Z, which is $[Z(Z - 1)]/2$, as shown in Figure 26–5.

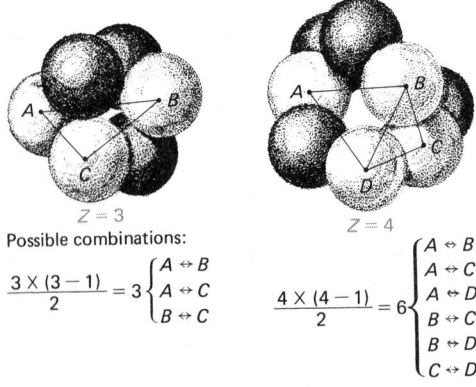

Figure 26–5 The number of possible combinations of interactions between charged particles within the nucleus is $[Z(Z - 1)]/2$.

The energy of interaction between protons is also inversely proportional to the distance of separation R. The energy associated with Coulomb electrostatic repulsion is

$$E_c = K\frac{Z(Z-1)}{R} = K\frac{Z(Z-1)}{r_0 A^{1/3}} = \frac{cZ(Z-1)}{A^{1/3}}$$

where R has been replaced by the radius of the nucleus $r_0 A^{1/3}$. Because this effect is disruptive, it appears as a negative energy term.

The term

$$E_a = \frac{d(N-Z)^2}{A}$$

originates from the *lack of symmetry* between the number of protons and the number of neutrons in the nucleus. The maximum stability of a nucleus occurs when $N = Z$. Any departure from this introduces an asymmetry $N - Z = A - 2Z$, which results in a decrease in stability; for this reason this term is also negative.

A pure corrective term, called the *pairing term*,

$$E_p = \pm \frac{\delta}{A^{3/4}}$$

is added if the mass number A is even. This provides agreement with the semi-empirical formula and laboratory results. If A is odd, no correction is needed.

The binding energy per nucleon, from Equation (26–12), is

$$\boxed{\frac{\text{BE}}{A} = a - \frac{b}{A^{1/3}} - \frac{cZ(Z-1)}{A^{4/3}} - \frac{d(N-Z)^2}{A^2} \pm \frac{\delta}{A^{7/4}}}$$

These binding energies per nucleon appear in Figure 26–6, which shows the binding energy per nucleon in MeV plotted against the mass number.

With the exception of a few irregularities, such as $^4_2$He, $^{12}_6$C, and $^{16}_8$O, the curve is relatively smooth. The curve rises sharply for small values of A, but for values of $A \geq 30$ the binding energy is close to 8 MeV/nucleon. The contributions of the various effects in Weizacker's empirical formula are represented schematically in the graph of Figure 26–7.

The liquid drop model is successful in explaining the division of the nucleus into two or more parts when nuclear fission occurs. However, the model is not adequate to describe the excited energy states known to exist in the nucleus. For this reason, another nuclear model must be developed.

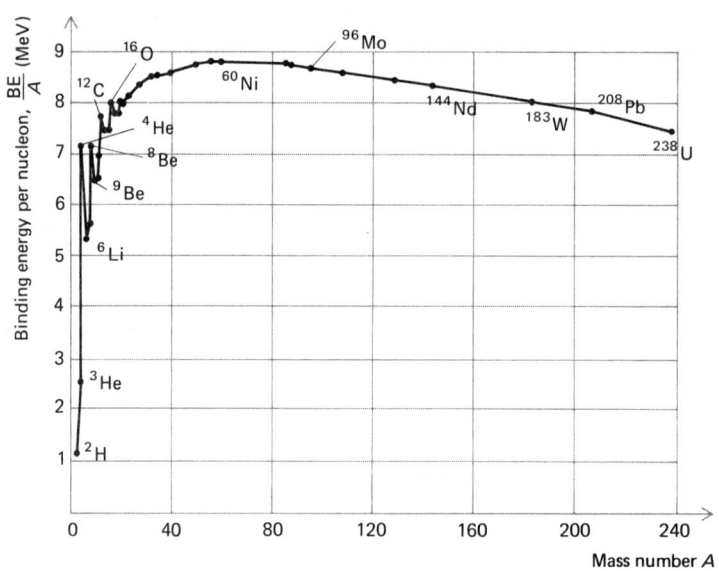

Figure 26–6 Average binding energy per nucleon for the stable nuclides.

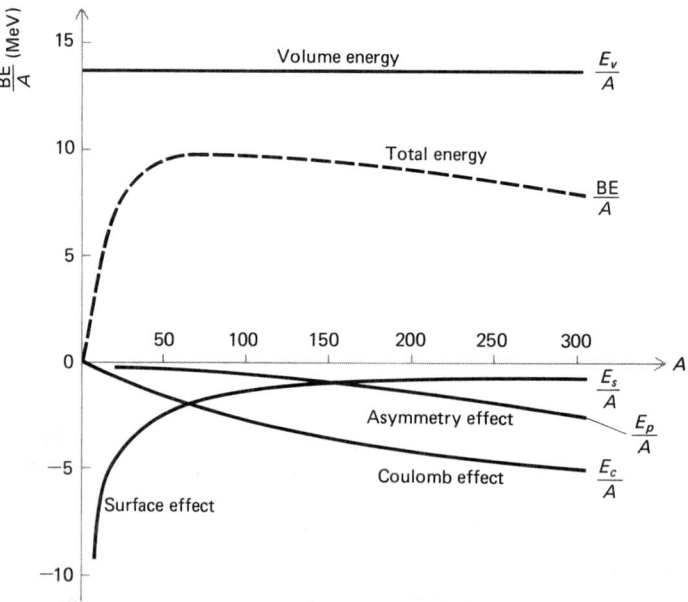

Figure 26–7 Contributions to the binding energy per nucleon.

26-5 SHELL MODEL

Experimentally, it is well established that many nuclear properties vary periodically in a manner similar to the way in which atomic properties vary according to the periodic table of elements. For example, the most stable nuclei occur when the number of protons (Z) or the number of neutrons ($N = A - Z$) are equal to one of the following numbers,

$$2, 8, 20, 50, 82, 126, \ldots$$

which are called *magic numbers*.

These magic numbers have been explained by the *shell model* of the nucleus, which proposes that protons and neutrons form closed shells similar to electronic shells.

If the curve of binding energy per nucleon in Figure 26–6 is magnified, it will show "peaks" in which the binding energy per nucleon has a larger value than for neighboring nuclides. These peaks occur for nuclides $^4_2$He and $^{16}_8$O, with other peaks at $^{88}_{38}$Sr, $^{120}_{50}$Sn, $^{140}_{58}$Sn, and $^{208}_{82}$Pb.

The magic numbers also appear when the number of stable nuclides is compared to proton or neutron number, as in Figures 26–8 and 26–9. Both graphs show greater numbers of nuclides at proton and neutron numbers of 20, 50, and 82. Notice that for the isotones in Figure 26–9, $N = 28$ has a peak. This is called a *semimagic* number.

The periodicity of the chemical properties arranged in Mendeleeff's table of elements (Chapter 24) found a straightforward explanation by assuming that the electrons are distributed in *shells* and *subshells*. It was then that J. H. BARTLETT in 1932 proposed that a shell structure in nuclei could explain nuclear stability. He

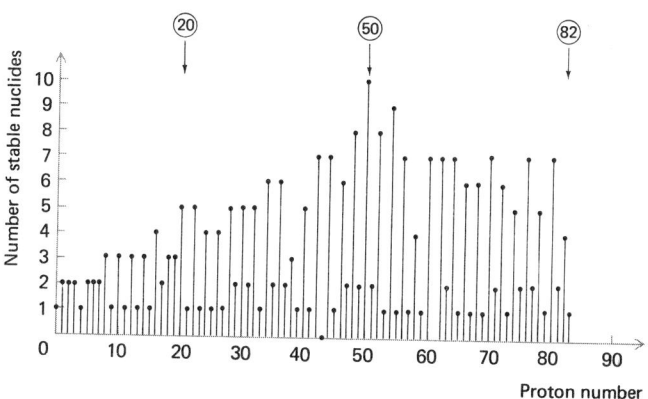

Figure 26–8 Number of stable nuclides per element as a function of proton number. [From B. H. Flowers, in O. R. Frisch (Ed.), *Progress in Nuclear Physics*, Vol. 2, Pergamon, Oxford, 1952. Used with permission.]

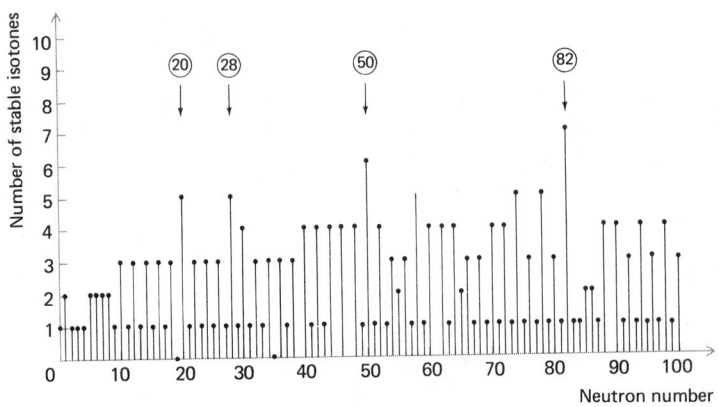

Figure 26-9 Number of stable nuclides per element as a function of neutron number. [From B. H. Flowers, in O. R. Frisch (Ed.), *Progress in Nuclear Physics*, Vol. 2, Pergamon, Oxford, 1952. Used with permission.]

observed that the pattern of composition of naturally occurring isotopes undergoes a change at $^{16}_{8}$O and another change at $^{36}_{18}$A. Between $^{4}_{2}$He and $^{16}_{8}$O, all the stable nuclei are built up according to the scheme $^{4}_{2}$He $+ n + p + n + p + \ldots$, and from $^{16}_{8}$O to $^{36}_{18}$A the pattern is $^{16}_{8}$O $+ n + n + p + p + n + n + \ldots$. It was then proposed that, without violating the Pauli exclusion principle for neutrons and protons, it was possible to put two neutrons and two protons in the s shell with an orbital angular momentum of zero. The next shell, the p shell with orbital angular momentum equal to 1, has room for six protons and six neutrons. When the s and p shells are first complete, the $^{16}_{8}$O nuclide is formed. The next shell is the d shell, with orbital angular momentum equal to 2. This shell has room for ten protons and ten neutrons. The $^{36}_{18}$A nuclide is formed when the s, p, and d shells are filled. Generally, the number of protons or neutrons in a shell is given by $2(2l + 1)$, where $l = 0, 1, 2, \ldots$.

According to the shell model, the total angular momentum of a nucleus (called the nuclear spin) is given by

$$\text{nuclear spin} = \sum \begin{pmatrix} \text{intrinsic angular} \\ \text{momentum of} \\ \text{protons } (\tfrac{1}{2}\hbar) \end{pmatrix} + \sum \begin{pmatrix} \text{intrinsic angular} \\ \text{momentum of} \\ \text{neutrons } (\tfrac{1}{2}\hbar) \end{pmatrix} + \sum \begin{pmatrix} \text{orbital angular momentum of} \\ \text{nucleons arising from their} \\ \text{motions in the nucleus } (m\hbar) \end{pmatrix}$$

The third term always gives $m\hbar$, where $m = 0, 1, 2, \ldots$. These three contributions to the nuclear spin of the nucleus can explain the stable nuclides that appear in nature. These are summarized in Table 26-3.

Table 26-3 Nuclear angular momenta of stable nuclides

Z	N	NUCLEAR SPIN
even	even	0
even	odd	$\frac{1}{2}, \frac{3}{2}, \frac{5}{2}, \frac{7}{2}$, etc.
odd	even	$\frac{1}{2}, \frac{3}{2}, \frac{5}{2}, \frac{7}{2}$, etc.
odd	odd	1, 2, 3

We recall that according to the Pauli exclusion principle, no two protons or neutrons can have the same set of quantum numbers. The protons will fill the available proton energy levels in pairs of antialigned proton spins and similarly for the neutrons. Finally, when the protons and neutrons complete closed shells, the total orbital angular momentum is zero and the total nuclear spin is zero. As an example, $^{60}_{28}Ni_{32}$ has a nuclear spin of zero.

The cases of Z even, N odd or Z odd, N even will have either a neutron or proton not paired with another nucleon. The nucleon with spin $\frac{1}{2}\hbar$ will combine with the orbital angular momentum of $m\hbar$. Since m is an integer, the resultant nuclear spin is $m\hbar \pm \frac{1}{2}\hbar = \frac{1}{2}\hbar, \frac{3}{2}\hbar, \ldots$. Two examples are $^{63}_{27}CO_{34}$ and $^{67}_{30}Zn_{37}$, which have spins of $\frac{3}{2}\hbar$ and $\frac{5}{2}\hbar$, respectively. The odd-odd nuclides have an unpaired proton and an unpaired neutron, and the total nuclear spin is $\hbar, 2\hbar, 3\hbar, \ldots$. The nuclide $^{2}_{1}H_{1}$ has a spin of $1\hbar$, and $^{10}_{5}B_{5}$ has a spin of $3\hbar$.

PROBLEMS

26-1 What is the threshold frequency of the γ ray that will initiate the nuclear photodisintegration

$^{9}_{4}Be\ (\gamma, p)\ ^{8}_{3}Li$

26-2 Two protons revolve in a circular orbit around the center of mass. If the combined orbital angular momentum is $\hbar$, show that the magnetic moment of the combination is 1 nuclear magneton.

26-3 A 6.5-MeV γ ray disintegrates a tritium nucleus into a neutron and a deuteron:

$^{3}_{1}H\ (\gamma, n)\ ^{2}_{1}D$

If the deuteron has a kinetic energy of 60 KeV, what is the kinetic energy of the neutron?

26-4 A free proton is placed in an external magnetic field B; because of the space quantization there are two possible orientations of its magnetic moment.

A photon whose energy is equal to the difference in energy between the two proton states produces a transition between one state and the other. Compute the proton resonance frequency if $B = 0.50$ Wb/m^2.

26-5 (a) Differentiate Equation (26-12), the Weizacker binding energy equation, with respect to Z and find the value of Z that makes the binding energy a maximum.

(b) If $A = 28$, find the value of Z that makes the binding energy a maximum.

26-6 Compute the binding energy per nucleon for $^{238}_{92}$U in joules, million electron volts, and atomic mass units.

26-7 What fraction of the negative or repulsive portion of the binding energy is attributed to Coulomb repulsion for $^{4}_{2}$He? For $^{40}_{20}$Ca? For $^{206}_{82}$Pb?

26-8 Compare the surface effect energies for $^{4}_{2}$He and $^{238}_{92}$U.

26-9 The diameter of an atom is of the order of 10^{-10} m. Assume that an electron is constrained within the dimensions of an atom (as opposed to a nucleus) and show that the energy of the electron will be several electron volts.

26-10 For the nuclides of $^{16}_{8}$O and $^{239}_{94}$Pu, compare the energies of volume effect, surface effect, and Coulomb repulsion.

RECOMMENDED READING

BARTLETT, J. H., "Nuclear Structure," *Nature* **130**, 165 (1932).

BETHE, Hans A., "What Holds the Nucleus Together?" *Sci. Am.*, September 1953.

FLOWERS, B. H., "The Nuclear Shell Model," *Progress in Nuclear Physics*, Academic, New York, 1952, Vol. 2, pp. 235-270.

MAYER, M. G., "On Closed Shells in Nuclei," *Phys. Rev.* **75**, 1894 (1949).

MAYER, M. G., "The Structure of the Nucleus," *Sci. Am.*, March 1951.

PAKE, George E., "Magnetic Resonance," *Sci. Am.*, August 1958.

PEIERLS, R. E., "Models of the Nucleus," *Sci. Am.*, January 1959.

WEISSKOPF, Victor F., and ROSENBAUM, E. P., "A Model of the Nucleus, *Sci. Am.*, December 1955.

27 The Neutron

Sir James Chadwick
(1891–)

Born in Manchester, England, Chadwick attended the universities of Manchester and Cambridge and the Charlottenburg Institution in Berlin, where he studied under Hans Geiger. Collaborating with Ernest Rutherford and C. D. Ellis, he wrote Radiations from Radioactive Substances *(1930), the most comprehensive book on nuclear phenomena then available. In 1932 he discovered that the radiations emitted from beryllium when bombarded by α particles are neutrons. For his work in identifying the neutron he was awarded the 1935 Nobel Prize in physics.*

27–1 DISCOVERY OF THE NEUTRON
27–2 PRODUCTION OF NEUTRONS
27–3 DETECTION OF NEUTRONS
27–4 NEUTRON CAPTURE

27-1 DISCOVERY OF THE NEUTRON

In 1930 the German physicists W. BOTHE and H. BECKER produced a very penetrating radiation by bombarding $^9_4$Be nuclei with 5.3-MeV α particles from a radioactive source. Since the radiation could readily penetrate several centimeters of lead and was not deflected by a magnetic or electrical field, it was assumed to be highly energetic γ rays.

In an attempt to measure absorption coefficients, IRENE CURIE-JOLIOT and her husband F. JOLIOT placed various absorbers such as Ag, Cu, and Pb between the radiation source and an ionization chamber. These absorbers had little or no effect. However, when absorbers rich in hydrogen (such as water, paraffin, and cellophane) were used, the ionization showed an increase in intensity. A thin aluminum absorber readily cut out most of the final radiation, and this led them to speculate that the γ rays were being scattered from protons much as in the Compton effect (Figure 27-1). They concluded that the protons were being knocked from the absorbers by the γ rays.

The experiment took place in the following two steps:

γ rays produced: $^4_2\text{He} + ^9_4\text{Be} \rightarrow ^{13}_6\text{C} + \gamma$

γ rays scattered: $h\nu_\gamma$ → Proton at rest → $h\nu_\gamma'$ Scattered photon, θ, φ, Recoil proton

The Joliots surmised that the α particles bombarded the $^9_4$Be to produce γ rays, which in turn scattered from a proton. The γ rays had a new frequency ν', and it was the recoil protons that were absorbed by the thin aluminum absorber.

The shift in wavelength for Compton scattering, from Equation (8-23), is

$$\Delta \lambda = \frac{h}{m_p c}(1 - \cos \theta)$$

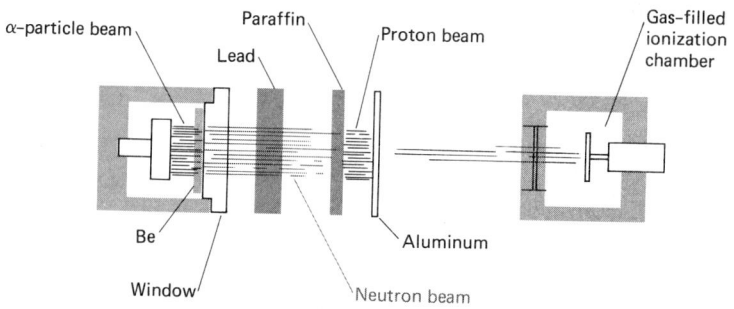

Figure 27-1 In the Curie-Joliot experiment, the unknown radiation beam readily penetrates lead, and a paraffin absorber increases the ionization chamber reading. A thin aluminum absorber readily cuts out most of the final radiation.

where m_p is the rest mass of the proton. This can also be written

$$\lambda' - \lambda = \frac{c}{v'} - \frac{c}{v} = \frac{h}{m_p c}(1 - \cos\theta)$$

or, after some algebraic manipulation,

$$m_p c^2 (h v - h v') = h v h v'(1 - \cos\theta)$$

where $E = hv$ is the energy of the incoming photon, $E = hv'$ is the energy of the scattered photon, and $K = hv - hv'$ is the energy transferred to the proton, and the proton rest energy is $m_p c^2 = 938$ MeV. This equation can now be written as

$$m_p c^2 K = E E'(1 - \cos\theta)$$

and in the case of a head-on collision ($\theta = 180°$), it is

$$m_p c^2 K = 2E(E - K)$$

The energy of the incoming photon must be

$$E = \tfrac{1}{2}(K + \sqrt{K^2 + 2m_p c^2 K}) = \tfrac{1}{2}K\left(1 + \sqrt{1 + \frac{2m_p c^2}{K}}\right) \qquad (27\text{-}1)$$

Only the positive value in front of the radical is chosen, since E must be positive. The recoil protons were measured to have a kinetic energy of 5.7 MeV. The energy of the incoming photon must then be around $E = 55$ MeV. This value is much too high to agree with any experimental evidence.

Now let us calculate the mass defect in the equation

$${}^4_2\text{He} + {}^9_4\text{Be} \rightarrow {}^{13}_6\text{C} + \gamma$$

to see how much energy is released in this reaction. The energy available is then found from

$$\begin{aligned}{}^4_2\text{He} &= 4.002604 \text{ amu} \\ {}^9_4\text{Be} &= \underline{9.012186 \text{ amu}} \\ & 13.014790 \text{ amu} \\ {}^{13}_6\text{C} &= \underline{13.003354 \text{ amu}} \\ & 0.011436 \text{ amu}\end{aligned}$$

where the energy released is 0.0114 amu × 931 MeV/amu = 10.6 MeV. But Compton-type scattering requires around 55 MeV of energy, so the conclusion must be that the unknown radiation produced in the Bothe–Becker experiment cannot be γ rays.

JAMES CHADWICK, working at the Cavendish Laboratory, suggested that the unknown radiation was not γ rays at all, but an uncharged particle (the neutron) of about the same size as a proton. Some 12 years earlier, Rutherford had suggested such a particle, but all attempts to observe it had been in vain. Chadwick proposed that the reaction should be

$$\boxed{{}^4_2\text{He} + {}^9_4\text{Be} \rightarrow {}^{12}_6\text{C} + {}^1_0 n} \qquad (27\text{–}2)$$

but since the mass of ${}^9_4\text{Be}$ was not known with precision, in order to make the computation of the neutron mass he used the reaction

$$ {}^4_2\text{He} + {}^{11}_5\text{B} \rightarrow {}^{14}_7\text{N} + {}^1_0 n \qquad (27\text{–}3)$$

since the masses of ${}^{11}_5\text{B}$ and ${}^{14}_7\text{N}$ were known (note that ${}^1_0 n$ is the symbolic representation of the neutron).

EXAMPLE 27–1: Calculate the energy available from the mass defect when the neutron is considered a part of the reaction.

SOLUTION: The equation is now

$$ {}^4_2\text{He} + {}^9_4\text{Be} \rightarrow {}^{12}_6\text{C} + {}^1_0 n$$

and the energy is found from the following calculations:

$$\begin{aligned}{}^4_2\text{He} &= 4.002604 \text{ amu} \\ {}^9_4\text{Be} &= \underline{9.012186 \text{ amu}} \\ \text{input mass} &= 13.014790 \text{ amu} \\ {}^{12}_6\text{C} &= 12.000000 \text{ amu} \\ {}^1_0 n &= \underline{1.008665 \text{ amu}} \\ \text{output mass} &= 13.008665 \text{ amu}\end{aligned}$$

The mass defect is

$$13.014790 \text{ amu}$$
$$-13.008665 \text{ amu}$$
$$\overline{0.006125 \text{ amu}}$$

and the energy released is 0.006125 amu × 931 MeV/amu = 5.70 MeV.

From Example 27–1, it is seen that Chadwick's explanation correctly explained the characteristics of the unknown radiation, and thus a new constituent of the atom was discovered and identified.

27–2 PRODUCTION OF NEUTRONS

Because the neutron is about the same size as a proton and has no charge, it has proved to be an important projectile in many experiments. We list several methods of producing neutrons for experimentation.

Neutron Source. Radium is a radioactive emitter of α particles, and when radium is mixed with beryllium, the α particles knock neutrons from the beryllium. Neutrons are emitted from this mixture according to the reaction in Equation (27–2),

$$^{4}_{2}\text{He} + ^{9}_{4}\text{Be} \rightarrow ^{12}_{6}\text{C} + ^{1}_{0}n$$

The neutron beam is not monoenergetic, since neutrons of different energies are produced.

Accelerated Charged Particles. When high-speed deuterons bombard a tritium target, neutrons are produced according to

$$^{2}_{1}\text{H} + ^{3}_{1}\text{H} \rightarrow ^{4}_{2}\text{He} + ^{1}_{0}n \tag{27-4}$$

This reaction has an advantage because the energy of the emitted neutron is known. Since a neutron is lighter than an α particle, it will carry off most of the energy available from the reaction in the form of kinetic energy. The "input" mass in this reaction is

$$^{2}_{1}\text{H} = 2.014102 \text{ amu}$$
$$^{3}_{1}\text{H} = 3.016049 \text{ amu}$$
$$^{2}_{1}\text{H} + ^{3}_{1}\text{H} = \overline{5.030151 \text{ amu}}$$

and the "output" mass is

$$^{9}_{4}\text{He} = 4.002604 \text{ amu}$$
$$^{1}_{0}n = 1.008665 \text{ amu}$$
$$^{4}_{2}\text{He} + ^{1}_{0}n = \overline{5.011269 \text{ amu}}$$

The mass defect between the "input" mass and the "output" mass is the source of the energy available from the reaction. The energy from the mass defect is

$$\begin{array}{r} 5.030151 \text{ amu} \\ -5.011269 \text{ amu} \\ \hline 0.018882 \text{ amu} \end{array}$$

The available energy is 0.0189 amu × 931 MeV/amu = 17.6 MeV. This represents the maximum energy carried away by neutrons. Since energy and momentum are conserved, the energy can be determined from the angle at which the neutrons are emitted with respect to the incident particles.

Photodisintegration. The interaction of γ rays with nuclei is also a means of producing neutrons. For example, the reaction

$$\gamma + {}^{9}_{4}\text{Be} \rightarrow {}^{8}_{4}\text{Be} + {}^{1}_{0}n \qquad (27\text{-}5)$$

shows a mass defect in a negative sense. The "output" mass is larger than the "input" mass, and energy is required from the outside to initiate the reaction. This is known as an *endoergic* reaction. The threshold energy or the minimum energy required of the γ ray to produce this reaction is 1.67 MeV.

Stripping Reaction. Very energetic neutrons can be produced when a high-energy deuteron strikes a target. The binding energy of the neutron to the deuteron is only about 2.2 MeV. When deuterons of energy of several hundred MeV strike a target, the neutron is easily separated and continues traveling in the forward direction with a kinetic energy approximately equal to one-half of the deuteron energy.

27-3 DETECTION OF NEUTRONS

Because the neutron has no charge, it cannot be deflected by an electrical or magnetic field nor can it be produced by ionization. Neutrons do not cause fluorescense, produce cloud chamber or emulsion tracks, or trigger Geiger counters. For these reasons, special techniques had to be developed to detect neutrons.

One way to detect neutrons is through the ionization produced by charged particles created in a nuclear reaction when neutrons are used as bullets. An ionization chamber can be made sensitive to neutrons by filling the chamber with boron trifluoride. The neutrons incident on the boron nucleus produce α particles according to the reaction

$${}^{1}_{0}n + {}^{10}_{5}\text{B} \rightarrow {}^{7}_{3}\text{Li} + {}^{4}_{2}\text{He}$$

It is the α particle that is now detected by the ionization chamber and that indirectly indicates the presence of neutrons.

Neutrons do not have an electrical charge, and a traveling neutron can get very close to and interact with a light particle such as a proton in an elastic head-on collision. The proton has practically the same mass as the neutron; and after a head-on collision, the neutron will come to rest and all of its kinetic energy will be transferred to the proton, which will move in the forward direction with an energy equal to the neutron energy. The proton will now produce ionization in an ionization chamber, indicating indirectly the presence of neutrons.

27–4 NEUTRON CAPTURE

A slow neutron is likely to interact with a nucleus and be absorbed by *neutron radiative capture*. As the name implies, the capture of the neutron by a nucleus in this case is signified by the emission of a γ ray,

$$_0^1 n + {}_{13}^{27}\text{Al} \rightarrow {}_{13}^{28}\text{Al} + \gamma \tag{27-6}$$

Thus, the presence of a neutron may be detected.

A neutron of energy about 1 MeV will most probably interact with a nucleus by the scattering process. As the neutron is scattered, it loses part of its energy by transfer to the nucleus (even for elastic collisions). The neutron will experience many scattering processes until its energy is decreased to about the order of kT, where k is the Boltzmann constant, $k = 1.38 \times 10^{-23}$ J/°K, and T is the absolute temperature. Neutrons with energy of the order of kT are called *thermal neutrons*; and at a room temperature of 27°C or 300°K, the energy of a thermal neutron is

$$kT = 1.38 \times 10^{-23} \times 300 = 4.14 \times 10^{-21} \text{ J} = 0.026 \text{ eV}$$

The cross section $\sigma(n, \gamma)$ for radiative neutron capture by Ag shown in Figure 27–2 decreases smoothly with increasing neutron speed or energy until a *resonance*

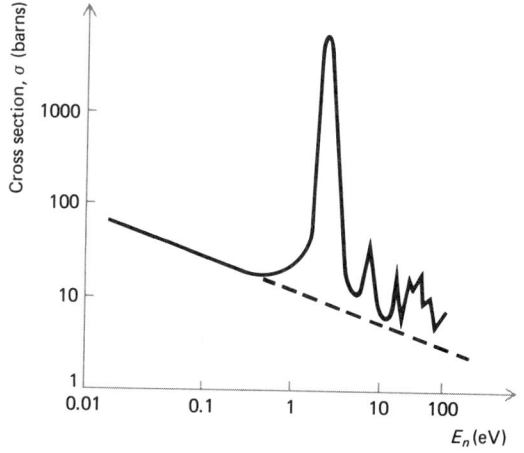

Figure 27–2 Neutron capture cross section σ for Ag versus energy in electron volts. (From R. Evans, *The Atomic Nucleus*, McGraw-Hill, New York, 1955. Used with permission.)

absorption appears. The cross section $\sigma(n, \gamma)$ is an indication of the *probability of neutron capture*. It is given by

$$\sigma(n, \gamma) = \frac{a}{v} \tag{27-7}$$

where a is a constant and v is the velocity of the neutron. It is easy to understand the $1/v$ law, since the probability of interaction is directly proportional to the time the neutron spends in the vicinity of a nucleus, and this time is inversely proportional to the speed of the neutron. For small energies, Figure 27–2 shows that the $1/v$ law is followed; but for energies greater than 1 eV, different *resonant energies* appear, which correspond to certain *nuclear energy levels*. Other elements show similar resonances in the cross section for capture. Because Cd-113 has a high cross section for neutron capture, cadmium bars are used in nuclear reactors to absorb neutrons and moderate the neutron flux.

PROBLEMS

27–1 How long will it take a 1-MeV neutron to cross a $^{235}_{92}\text{U}$ nucleus?

27–2 Compute the energy required to remove the least tightly bound neutron from a $^{12}_{6}\text{C}$ nucleus.

27–3 Compute the minimum photon energies required to produce photodisintegration of a $^{3}_{2}\text{He}$ nucleus at rest into (a) a proton and two neutrons, and (b) a deuteron and a neutron.

27–4 Compare the binding energy per nucleon with the energy necessary to remove the least tightly bound neutron from a nucleus of $^{16}_{8}\text{O}$.

27–5 A neutron with a kinetic energy of 5.7 MeV makes a head-on elastic collision with a $^{4}_{2}\text{He}$ nucleus at rest. What is the maximum recoil energy of the $^{4}_{2}\text{He}$ nucleus?

27–6 Nuclei of $^{7}_{3}\text{Li}$ are bombarded by 3-MeV protons. Calculate the energy of the neutrons emitted in the reaction, which come off at an angle of (a) 45°, and (b) 90° relative to the direction of the incoming protons.

27–7 Determine the energy of deuterons necessary to produce 25–MeV neutrons in the forward direction by bombarding lead with $^{3}_{1}\text{H}$ nuclei.

27–8 (a) Calculate the minimum kinetic energy of a neutron that can eject a 5-MeV proton from a $^{14}_{7}\text{N}$ initially at rest.
(b) What is the product atom?

27–9 Assume that a deuteron can be considered as a proton and a neutron with their centers separated by 2.5×10^{-15} m and a binding energy of 2.2 MeV.

Compare the energy associated with the nuclear forces with (a) the Coulomb energy of two protons at the same distance, and (b) the gravitational potential energy of the two nucleons at the same distance.

RECOMMENDED READING

BOTHE, W., and BECKER, H. Z., *Physik* **66,** 289 (1930).
  Written in German; no English translation available.
CHADWICK, J., "Possible Existence of a Neutron," *Nature* **129,** 312 (1932).
CHADWICK, J., and GOLDHABER, M., "A 'Nuclear Photo-effect': Disintegration of the Diplon by γ-Rays," *Nature* **134,** 237 (1934).
CURIE, Irene, *Compt. Rend.* **194,** 1412 (1931).
  Written in French; no English translation available.
HOWARD, R. E., *Nuclear Physics*, Wadsworth, Belmont, Calif., 1963, pp. 179–204.
HUGHES, D. J., *The Neutron Story*, Doubleday, Garden City, N.Y., 1959.
  A popular account of the discovery of the neutron—very interesting reading.
LAURITSEN, C. C., and CRANE, H. R., "Transmutation of Lithium by Deutrons and its Bearing on the Mass of the Neutron," *Phys. Rev.* **45,** 550 (1934).

28
Nuclear Reactions I

Robert Jemison Van de Graaff
(1901–1967)

A native of Tuscaloosa, Ala., Van de Graaff was awarded a Rhodes scholarship and received his Ph.D. in physics from Oxford University in 1928. During World War II, he was director of the Office of Scientific Research and Development's High Voltage Radiographic Project; at Massachusetts Institute of Technology, he conducted research and taught from 1931 to 1960. In 1933 Van de Graaff developed the high-voltage electrostatic generator that carries his name. The accelerator opened the way for the study of many types of nuclear reactions. Among his many honors are the Cresson medal and the Duddell medal.

28–1 NUCLEAR REACTIONS
28–2 Q VALUE OF A NUCLEAR REACTION
28–3 Q VALUE AND BINDING ENERGY

28-1 NUCLEAR REACTIONS

When a nucleus is bombarded with an energetic particle and some change occurs in the characteristics or identity of the nucleus, this event defines a *nuclear reaction*. The first experimental nuclear reactions were performed by Sir ERNEST RUTHERFORD in 1919. Using α particles with an energy of 7.68 MeV coming from Po-214, he bombarded N-14 and obtained O-17 and protons.

This reaction is

$$^4_2\text{He} + {}^{14}_7\text{N} \rightarrow {}^1_1\text{H} + {}^{17}_8\text{O} \tag{28-1}$$

Symbolically, this reaction, typical of many nuclear reactions, is written

$$\boxed{x + X \rightarrow y + Y} \tag{28-2}$$

where x corresponds to ^4_2He, the *bombarding particle*, called the *bullet*; X corresponds to $^{14}_7\text{N}$, the *target nucleus*; y corresponds to ^1_1H, the *product particle*; and Y corresponds to $^{17}_8\text{O}$, the *recoil nucleus* (see Figure 28-1). Frequently, Equation (28-2) is written in the shorthand notation

$$\boxed{X(x, y)\,Y}$$

with Equation (28-2) then written as

$$^{14}_7\text{N}\,(\alpha, p)\,{}^{17}_8\text{O}$$

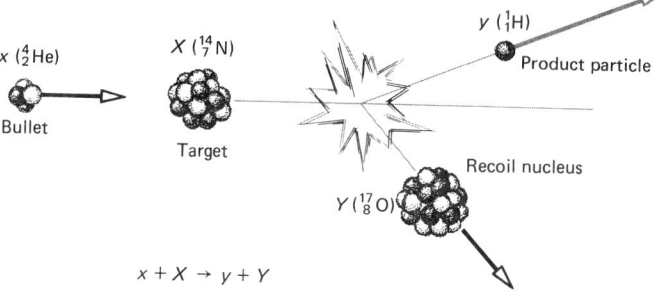

Figure 28-1 Schematics of a nuclear reaction.

This is called an (α, p) reaction, since the bombarding particle or bullet is an α particle and the product particle is a proton.

Until 1931, when ROBERT VAN DE GRAAFF constructed the first reliable high-voltage accelerator, the number of nuclear reactions that could be studied was limited to those initiated by α and β particles coming from naturally radioactive isotopes. Today many different particle accelerators, including Van de Graaff generators, linear accelerators, and cyclotrons, make available the study of a wide variety of nuclear reactions.

This chapter will be limited to the study of low-energy reactions in which the kinetic energy of the bullet particles is less than 10 MeV. The bullets described will be only light particles, including α and β particles, protons and neutrons, and γ rays and photons. The outgoing product particle will also be a light particle.

All nuclear reactions, high or low energy, are governed by the following principles:

1. *Conservation of electrical charge.* Electrical charge is conserved in nuclear reactions and

$$\sum Z_{\text{initial}} = \sum Z_{\text{final}}$$

the sum of proton numbers before a reaction is equal to the sum of the proton numbers after the reaction. In the example of Equation (28–1), note that the proton numbers are $2 + 7 = 1 + 8$.

2. *Conservation of the total number of nucleons.* The sum of the mass numbers before and after a reaction must be the same,

$$\sum A_i = \sum A_f$$

In Equation (28–1), the total number of nucleons is $4 + 14 = 1 + 17 = 18$.

3. *Mass-energy conservation.* For an isolated system, since mass and energy are interchangeable the total mass-energy of the system remains constant.
4. *Conservation of linear momentum.*
5. *Conservation of angular momentum.*

28–2 Q VALUE OF A NUCLEAR REACTION

In a reaction $X(x, y)Y$, x represents the bullet, X is the target, which is assumed to be at rest, y is the product particle, and Y is the recoil nucleus. The mass-energy balance from the conservation of mass-energy gives

$$\underbrace{m_x c^2 + K_x}_{\text{total energy of the bullet}} + \underbrace{M_X c^2}_{\text{rest energy of target}} = \underbrace{m_y c^2 + K_y}_{\text{total energy of product particle}} + \underbrace{M_Y c^2 + K_Y}_{\text{total energy of recoil nucleus}} \quad (28\text{–}3)$$

where $m_x c^2$, $M_X c^2$, $m_y c^2$, and $M_Y c^2$ are the rest energies of the bullet, target, product particle, and recoil nucleus, respectively. Since the reactions are low-energy reactions, the kinetic energies of the bullet K_x, the product particle K_y, and the recoil nucleus K_Y can be considered classically rather than relativistically. The kinetic energy is $K = (p^2/2m)$, where p is the linear momentum. Note that the target particle is assumed to be at rest and $K_X = 0$.

The *Q value* of the reaction is the energy available from the difference in mass of the bullet plus the target and the product particle plus the recoil nucleus. The Q value is then

$$Q = \underbrace{[(m_x + M_X) - (m_y + M_Y)]c^2}_{\text{input mass} \quad \text{output mass}} = \underbrace{K_y + K_Y}_{\text{kinetic energy of output particles}} - \underbrace{K_x}_{\text{kinetic energy of input particles}}$$

(28-4)

Note that if $m_x + M_X > m_y + M_Y$, then $K_y + K_Y > K_x$ and $Q > 0$.

When the input mass is larger than the output mass, $(m_x + M_X) > (m_y + M_Y)$, some mass is lost in the form of energy, which has been created at the expense of lost mass. The energy created is the difference between the output kinetic energy and the input kinetic energy. There has been a transformation of mass into energy according to the equation

$$(\Delta m)c^2 = \Delta E = Q \tag{28-5}$$

The mass loss or mass defect is Δm, and ΔE is the corresponding energy that has been created. Reactions in which $Q > 0$ are said to be *exoergic* because they release energy.

On the other hand, if $(m_x + M_X) < (m_y + M_Y)$, then $K_y + K_Y < K_x$. In this case, the output mass is larger than the input mass. Some mass has been created, apparently at the expense of the output kinetic energy. There has been a transformation of energy into mass according to the equation

$$\Delta m = \frac{\Delta E}{c^2}$$

Because $Q < 0$ and energy must be put in from some outside source, this type of reaction is said to be *endoergic*.

EXAMPLE 28-1: Compute the Q value of the reaction

$${}^{3}_{1}\text{H} \, (d, n) \, {}^{4}_{2}\text{He}$$

SOLUTION: This reaction is also written

$${}^{2}_{1}\text{H} + {}^{3}_{1}\text{H} \rightarrow {}^{1}_{0}n + {}^{4}_{2}\text{He}$$

In this reaction tritium target nuclei are bombarded with deuteron bullets, which results in a recoil nucleus of $^4_2$He and ^1_0n as product particles.

From the list of atomic masses in Table 6 of the Appendix we get the following masses:

$$\begin{aligned} \text{mass of } {}^2_1\text{H} = m_x &= 2.014102 \text{ amu} \\ \text{mass of } {}^3_1\text{H} = M_X &= \underline{3.016049 \text{ amu}} \\ m_x + M_X &= 5.030151 \text{ amu} \\ \\ \text{mass of } {}^1_0n = m_y &= 1.008665 \text{ amu} \\ \text{mass of } {}^4_2\text{He} = M_Y &= \underline{4.002603 \text{ amu}} \\ & \, 5.011268 \text{ amu} \end{aligned}$$

Then the mass defect is

$$\begin{aligned} (m_x + M_X) &= 5.030151 \text{ amu} \\ -(m_y + M_Y) &= \underline{5.011268 \text{ amu}} \\ & \, 0.018883 \text{ amu} \end{aligned}$$

and the Q value is

$$Q = 0.0189 \text{ amu} \times 931 \text{ MeV/amu} = 17.6 \text{ MeV}$$

Because Q is positive, the reaction is exoergic and 17.6 MeV of energy is released.

In previous examples, although we used nuclear masses, it was shown that the mass of electrons cancels out, and it has been convenient to use the table of atomic masses. In the following example, the mass of the electrons does not cancel out, and it must be included.

EXAMPLE 28–2: Calculate the Q value for the following nuclear decay:

$$^{30}_{15}\text{P} \rightarrow {}^{30}_{14}\text{Si} + {}^{0}_{-1}e + \nu$$

SOLUTION: In this nuclear decay, $^{30}_{15}$P decays into $^{30}_{14}$Si with the emission of an electron $^{0}_{-1}e$ and a particle ν called a neutrino.

The mass-energy balance is now

$$(M_P - 15m_e)c^2 = (M_{Si} - 14m_e)c^2 + m_e c^2 + K_{Si} + K_e + K_\nu$$

where M_P and M_{Si} are the respective atomic masses of $^{30}_{15}$P and $^{30}_{14}$Si. The Q value in this case is

$$Q = (M_P - M_{Si} - 2m_e)c^2 = K_{Si} + K_e + K_\nu$$

Since the neutrino has zero rest mass, it makes no contribution to the mass defect. Notice that two electron masses still remain in this equation, because an electron was one of the emitted particles.

The masses involved, then, are those of the initial particle,

$$\text{mass of } {}^{30}_{15}\text{P} = M_P = 29.97832 \text{ amu}$$

and the final particles,

$$\text{mass of } {}^{30}_{14}\text{Si} = M_{\text{Si}} = 29.973760 \text{ amu}$$
$$\text{mass of two electrons} = 2m_e = \underline{0.001100 \text{ amu}}$$
$$M_{\text{Si}} + 2m_e = 29.974860 \text{ amu}$$

The Q value is found from

$$M_P - (M_{\text{Si}} + 2m_e) = \begin{array}{r} 29.97832 \text{ amu} \\ \underline{-29.97486 \text{ amu}} \\ 0.00346 \text{ amu} \end{array}$$

and

$$Q = 0.00346 \text{ amu} \times 931 \text{ MeV/amu} = 3.22 \text{ MeV}$$

This reaction, then, is exoergic, with a release of 3.22 MeV of energy. For a radioactive decay, if it is found that $Q > 0$, then the decay process is possible.

28–3 Q VALUE AND BINDING ENERGY

If the Q value for a certain process is greater than zero, the process can occur spontaneously, while if $Q < 0$, the process is not possible unless some energy is provided from an external source.

Consider the nuclear reaction

$${}^{27}_{13}\text{Al} \, (\alpha, n) \, {}^{30}_{15}\text{P}$$

in which ${}^{27}_{13}\text{Al}$ nuclei are bombarded with α particles to produce neutrons and a recoil nucleus of ${}^{30}_{15}\text{P}$. This reaction can be written

$${}^{4}_{2}\text{He} + {}^{27}_{13}\text{Al} \rightarrow {}^{30}_{15}\text{P} + {}^{1}_{0}n + Q$$
$$x + X \rightarrow Y + y + Q$$

The Q value for this reaction is found from the mass defect calculations from the input masses,

$$\begin{array}{r} {}^{4}_{2}\text{He} = m_x = 4.00260 \text{ amu} \\ {}^{27}_{13}\text{Al} = M_X = \underline{26.98154 \text{ amu}} \\ m_x + M_X = 30.98414 \text{ amu} \end{array}$$

and the output masses,

$$\begin{array}{r} {}^{30}_{15}\text{P} = M_Y = 29.97832 \text{ amu} \\ {}^{1}_{0}n = m_y = \underline{1.00866 \text{ amu}} \\ m_y + M_Y = 30.98698 \text{ amu} \end{array}$$

The Q value is found from

$$(m_x + M_X) - (m_y + M_Y) = 30.98414 - 30.98698 = -0.00284 \text{ amu}$$

and

$$Q = -0.00284 \times 931 \text{ MeV/amu} = -2.64 \text{ MeV}$$

This reaction is endoergic and cannot take place spontaneously without some energy being added from the outside. In Chapter 29 it will be pointed out why even a threshold energy of 2.64 MeV is not enough to initiate this reaction.

Let us look at the same process now from the viewpoint of the binding energy. The binding energy equation for this process

$$^{4}_{2}\text{He} + ^{27}_{13}\text{Al} + \text{BE} = ^{30}_{15}\text{P} + ^{1}_{0}n$$

$$m_x + M_X + \text{BE} = M_Y + m_y$$

shows that the binding energy is

$$\text{BE} = (m_y + M_Y) - (m_x + M_X)$$

and evidently

$$\boxed{\text{BE} = -Q} \tag{28-6}$$

In this case, BE = $-(-2.64 \text{ MeV}) = 2.64 \text{ MeV}$; and we also conclude that if BE > 0, the process will not go spontaneously.

In summary, if $Q < 0$, then BE > 0, and the process will not occur spontaneously. If $Q > 0$, then BE < 0, and the process may occur spontaneously with a release of energy equal to Q.

PROBLEMS

28-1 Find the Q values for the following nuclear reactions, commonly called fusion reactions: (a) $^{2}_{1}\text{H}$ (d, n) $^{3}_{2}\text{He}$, (b) $^{2}_{1}\text{H}$ (d, p) $^{3}_{1}\text{H}$, and (c) $^{3}_{1}\text{H}$ (d, n) $^{4}_{2}\text{He}$.

28-2 Complete the following nuclear reactions by substituting the proper nuclide or particle for the question mark in each case.
(a) $^{10}_{5}\text{Bi}$ (?, α) $^{7}_{3}\text{Li}$
(b) $^{23}_{11}\text{Na}$ (d, ?) $^{24}_{12}\text{Mg}$
(c) $^{31}_{15}\text{P}$ (d, p) ?
(d) $^{59}_{27}\text{Co}$ (n, ?) $^{60}_{27}\text{Co}$

28-3 (a) Compute the energy released when $^{144}_{60}\text{Ne}$ decays into $^{140}_{58}\text{Ce}$ by emission of an α particle.
(b) Determine the energy carried away by the α particle and the kinetic energy of the recoil nucleus.

28-4 An α particle with a speed of 1.0×10^6 m/sec collides elastically with a nucleus of unknown mass. The scattering angle of the α particle and recoil nucleus are 60° and 30°, respectively. Find the mass of the nucleus.

28-5 Alpha particles are emitted when $^{214}_{84}$Po decays into $^{210}_{82}$Pb. If the kinetic energy K_α of the α particles and its range R in air are related by $R = 0.318 K_\alpha^{2/3}$, where K_α is in MeV and R is in meters, compute the range in air of the α particles emitted.

28-6 Describe the reactions that take place and compute the Q value for each of the following: $^{12}_{6}$C is bombarded by (a) 2-MeV protons, (b) 2-MeV deuterons, and (c) 4.2-MeV α particles. (Check the table of nuclides, Table 6 in the Appendix, for product particles.)

28-7 Compute the binding energy of $^{25}_{12}$Mg and determine whether it is stable or unstable.

28-8 (a) Determine the Q value for the reaction

$$^{14}_{7}N\ (\alpha, p)\ ^{17}_{8}O$$

(b) If the kinetic energy of the α particle is 7 MeV, calculate the approximate kinetic energy of the proton.

28-9 When a fast neutron is captured by a heavy nuclide, the following reactions are possible: (n, α), (n, p), and (n, γ). If the probability of the reaction is determined by the height of the potential barrier, write these equations according to their decreasing order of probability of occurrence.

28-10 A nuclear reaction takes place according to the scheme shown in Figure 28-2. Show that

$$Q = \left(1 + \frac{m_y}{M_Y}\right) K_y - \left(1 - \frac{m_x}{M_Y}\right) K_x - \frac{2\sqrt{m_x m_y K_x K_y}}{M_Y} \cos \theta_y$$

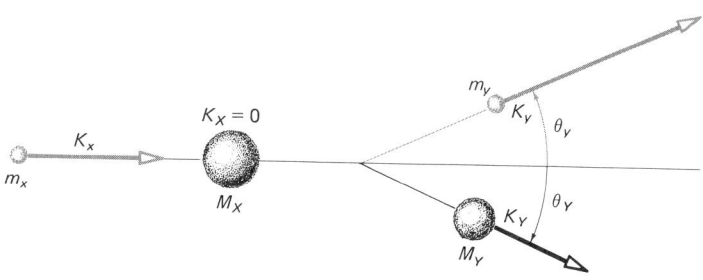

Figure 28-2

28-11 (a) Complete the following reaction:

$${}^{14}_{7}N\ (n, p)\ ?$$

(b) Calculate the Q value.
(c) Is the nuclide formed stable or unstable?

RECOMMENDED READING

Alonso, M., and Finn, E. J., *Fundamental University Physics*, Addison-Wesley, Reading, Mass., 1968, Vol. III, Chap. 8.

Arya, A. P., *Fundamentals of Nuclear Physics*, Allyn & Bacon, Boston, 1966, Chap. 4.

Bethe, H. A., and Morrison, P., *Elementary Nuclear Physics*, Wiley, New York, 1956, Chap. 20.

Hofstadter, R., "The Atomic Nucleus," *Sci. Am.*, July 1956.

Mayer, Maria G., "The Structure of the Nucleus," *Sci. Am.*, March 1951.

29
Nuclear Reactions II

Sir John Douglas Cockcroft
(1897–1967)

Born in Todmorden, England, and educated at the University of Manchester and St. John's College, Cockcroft was Jacksonian Professor of natural philosophy at Cambridge University (1939–1946). In 1932 he built the first high-energy particle accelerator that gave experimental evidence for Einstein's theory. Cockcroft was director of the Atomic Energy Establishment at Harwell, England, from 1946 to 1958. For their pioneer work on the transmutation of atomic nuclei by artificially accelerated atomic particles, he and E. T. S. Walton were awarded the 1951 Nobel Prize in physics.

29–1 KINETIC ENERGIES IN THE LABORATORY AND IN THE CENTER OF MASS FRAMES
29–2 THRESHOLD ENERGY OF AN ENDOERGIC REACTION
29–3 DERIVATION OF THE THRESHOLD EQUATION
29–4 CROSS-SECTION PROBABILITY

29-1 KINETIC ENERGIES IN THE LABORATORY AND IN THE CENTER OF MASS FRAMES

In analyzing collisions, it is often more convenient to introduce a coordinate system that moves along with the center of mass of all the particles involved. Also, looking at the collision problem from this different perspective, that is, as an observer at the center of mass, gives a better insight into what is occurring physically. Previously, we have viewed the collisions as pictured in Figure 29–1(a). In the center of mass frame, the observer views the collision occurring as shown in Figure 29–1(b).

In Figure 29–2, a system of particles $p_1, p_2, \ldots, p_n$ with its center of mass at O' is moving with a velocity $\mathbf{v}_0$ relative to a stationary laboratory reference frame. The reference frame with its center of mass at the origin O' is called the center of mass, or cm, frame. For simplicity, it is assumed that the laboratory frame and the cm frame are parallel and that the cm of the system of particles is moving with a velocity $\mathbf{v}_0 = v_0 \hat{\mathbf{i}}$.

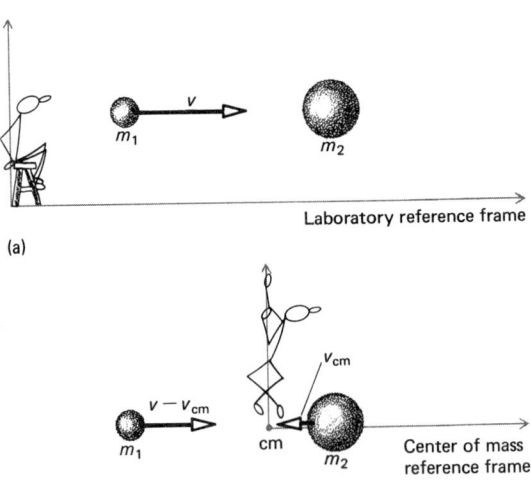

Figure 29–1 Generally, we observe events as a somewhat stationary observer in the laboratory. In the center of mass reference frame, we place ourselves, as the observer, at the center of mass of the two particles.

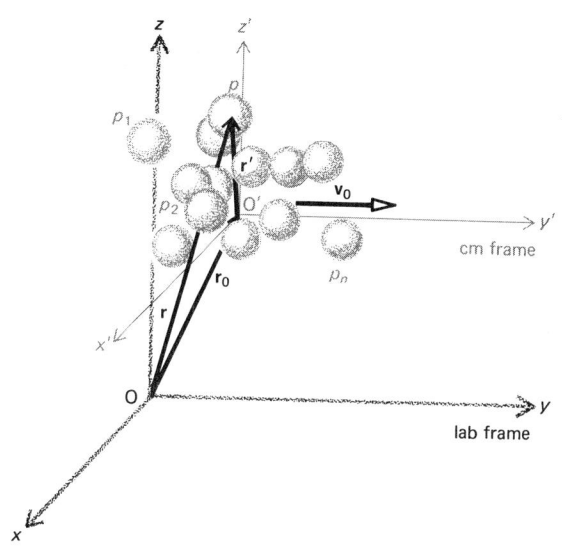

Figure 29-2 The center of mass of a group of particles $p_1, p_2, \ldots, p_n$ is moving with a velocity $\mathbf{v}_0$ relative to a fixed laboratory frame.

Consider a point $P(x, y, z)$ within the system of particles located by position vectors $\mathbf{r}$ and $\mathbf{r}'$ relative to the laboratory and cm frames, respectively. If $\mathbf{r}_0$ is the position vector of the cm relative to the laboratory frame, then

$$\mathbf{r} = \mathbf{r}_0 + \mathbf{r}' \tag{29-1}$$

A time derivative of r gives

$$\frac{d\mathbf{r}}{dt} = \frac{d\mathbf{r}_0}{dt} + \frac{d\mathbf{r}'}{dt}$$

where $d\mathbf{r}/dt = \mathbf{v}$, the velocity of point P relative to the laboratory frame; $d\mathbf{r}_0/dt = \mathbf{v}_0$, the velocity of the cm relative to the laboratory frame; and $d\mathbf{r}'/dt = \mathbf{v}'$, the velocity of point P relative to the cm frame. From these equations we see that it is possible to write

$$\mathbf{v} = \mathbf{v}_0 + \mathbf{v}' \tag{29-2}$$

Next, both sides of Equation (29-2) can be multiplied by the mass m of the particles at P to give

$$m\mathbf{v} = m\mathbf{v}_0 + m\mathbf{v}'$$

All of the particles within the system can be treated in this manner, and we can now write

$$\sum_i m_i \mathbf{v}_i = \sum_i m_i \mathbf{v}_0 + \sum_i m_i \mathbf{v}_i' \tag{29-3}$$

However,

$$\sum_i m_i \mathbf{v}_i' = \frac{d}{dt} \sum_i m_i \mathbf{r}_i' = 0$$

because the origin of the moving system is located at the cm, and by definition of the center of mass, $\sum_i m_i \mathbf{r}_i' = 0$. Equation (29-2) now becomes

$$\sum_i m_i \mathbf{v}_i = \left(\sum_i m_i\right) \mathbf{v}_0 = M\mathbf{v}_0 \qquad (29\text{-}4)$$

The total momentum of the system is equal to the linear momentum of the entire mass concentrated at the cm moving with a velocity $\mathbf{v}_0$ relative to the laboratory frame.

Equation (29-3) can now be squared and divided through by $\frac{1}{2}$,

$$\sum_i \tfrac{1}{2} m_i v_i^2 = \sum_i \tfrac{1}{2} m_i (\mathbf{v}_0 + \mathbf{v}_i')^2$$

or

$$\sum_i \tfrac{1}{2} m_i v_i^2 = \sum_i \tfrac{1}{2} m_i v_0^2 + \sum_i \tfrac{1}{2} m_i (v_i')^2 + \sum_i \tfrac{1}{2} m_i \mathbf{v}_0 \cdot \mathbf{v}_i'$$

which simplifies to

$$\sum_i \tfrac{1}{2} m_i v_i^2 = \tfrac{1}{2} M v_0^2 + \sum_i \tfrac{1}{2} m_i (v_i')^2 + \mathbf{v}_0 \cdot \sum_i m_i \mathbf{v}_i'$$

Since $\sum_i m_i \mathbf{r}_i' = 0$, $\sum_i m_i \mathbf{v}_i' = 0$, and

$$\underbrace{\sum_i \tfrac{1}{2} m_i v_i^2}_{\text{total KE in lab frame}} = \underbrace{\tfrac{1}{2} M v_0^2}_{\substack{\text{KE associated} \\ \text{with cm}}} + \underbrace{\sum_i \tfrac{1}{2} m_i (v_i')^2}_{\text{total KE in cm frame}} \qquad (29\text{-}5)$$

Thus, we see that the total kinetic energy relative to the lab frame, the energy we would ordinarily measure, is made up of two parts. One of these is the kinetic energy of the total mass moving with the speed at which the cm is moving, and the other portion of the energy is the kinetic energy of motion relative to the cm. This latter energy is the internal energy of the system; this will be seen to be energy that is available for a reaction to occur.

29-2 THRESHOLD ENERGY OF AN ENDOERGIC REACTION

Recall from the previous chapter that the Q value of the reaction shown schematically in Figure 29-3 is

$$Q = [(m + M_X) - (m + M_Y)]c^2 = K_y + K_Y - K_x$$

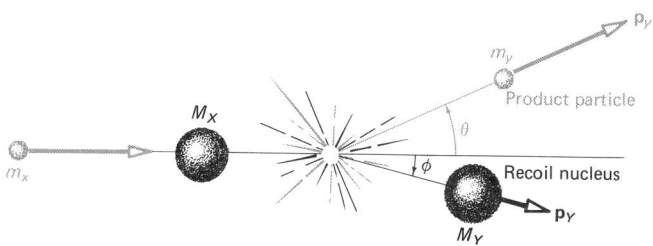

Figure 29-3 Schematic representation of a nuclear reaction $X(x, y) Y$.

For $Q > 0$, the reaction is exoergic; for $Q < 0$, it is endoergic. At first sight, the minimum energy required to initiate an endoergic reaction is

$$(K_x)_{\min} = -Q \qquad (29\text{-}6)$$

in which it has been assumed that $K_y = K_Y = 0$. This cannot be true, because this assumption leads to the conclusion that the linear momentum of the product particle p_y and the linear momentum of the recoil nucleus p_Y are each equal to zero. Before this reaction occurred, the bullet must have had a linear momentum of at least

$$p_x = \sqrt{2(K_x)_{\min} m_x} = \sqrt{-2Qm_x} \neq 0 \qquad (29\text{-}7)$$

Thus, if p_y and p_Y are zero, this would violate the principle of conservation of linear momentum.

In the next section, we shall see that the minimum threshold energy required for the reaction to go is

$$E_{\text{th}} = -Q \, \frac{m_x + M_X}{M_X} = (K_x)_{\min} \qquad (29\text{-}8)$$

which shows that evidently $E_{\text{th}} > (-Q)$. The difference

$$E_{\text{th}} - (-Q) = E_{\text{th}} + Q \qquad (29\text{-}9)$$

will appear as the energy carried away by the center of mass, while $-Q$ will be the minimum energy available for the reaction.

29-3 DERIVATION OF THE THRESHOLD EQUATION

From Figure 29-4, the x coordinate of the cm of the bullet-target system as seen from the laboratory frame is

$$x_{\text{cm}} = \frac{m_x x + M_X X}{m_x + M_X} \qquad (29\text{-}10)$$

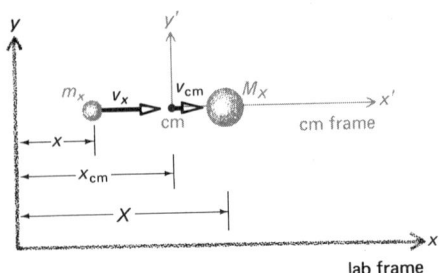

Figure 29-4 Collision of two particles in an endoergic reaction.

The velocity of the cm relative to the laboratory frame is then

$$v_{cm} = \left(\frac{m_x}{m_x + M_X}\right) v_x \qquad (29\text{-}11)$$

where $v_x = dx/dt$ is the speed of the bullet relative to the laboratory frame and $dX/dt = 0$ since the target is assumed to be at rest relative to the laboratory frame.

The total kinetic energy of the system before the collision is

$$K_x = \underbrace{\tfrac{1}{2}(m_x + M_X)v_{cm}^2}_{\text{KE carried away by cm, } K_{cm}} + \underbrace{\tfrac{1}{2}m_x(v_x')^2 + \tfrac{1}{2}M_X(v_X')^2}_{\text{KE available for reaction, } K_A} \qquad (29\text{-}12)$$

The first term $K_{cm} = \tfrac{1}{2}(m_x + M_X)v_{cm}^2$ is the kinetic energy associated with the cm, and $K_A = \tfrac{1}{2}m_x(v_x')^2 + \tfrac{1}{2}M_X(v_X')^2$ is the kinetic energy relative to the cm reference frame, because v_x' and v_X' are, respectively, the speeds of the bullet and target relative to the cm frame. This kinetic energy K_A is the kinetic energy available for the reaction.

From Equation (29-12), then

$$K_A = K_x - \tfrac{1}{2}(m_x + M_X)v_{cm}^2 \qquad (29\text{-}13)$$

or

> energy available for reaction = total input energy
> − energy carried away by the cm

Substituting Equation (29-11) into the energy Equation (29-13) gives

$$K_A = K_x - \tfrac{1}{2}(m_x + M_X)\frac{m_x^2 v_x^2}{(m_x + M_X)^2} = K_x - \tfrac{1}{2}\frac{m_x^2 v_x^2}{(m_x + M_X)} \qquad (29\text{-}14)$$

Since $K_x = \tfrac{1}{2}m_x v_x^2$, $m_x^2 v_x^2 = 2K_x m_x$, Equation (29-14) is then

$$K_A = K_x - \left(\frac{m_x}{m_x + M_X}\right) K_x = \left(\frac{M_X}{m_x + M_X}\right) K_x \qquad (29\text{-}15)$$

The minimum energy available for an endoergic reaction is

$$(K_A)_{min} = \left(\frac{M_X}{m_x + M_X}\right)(K_x)_{min} = -Q \tag{29-16}$$

and the minimum input energy for this reaction is

$$\boxed{(K_x)_{min} = E_{th} = \frac{-Q(m_x + M_X)}{M_X}} \tag{29-17}$$

From Equations (29-9) and (29-12), then, the energy carried away by the center of mass is

$$K_{cm} = E_{th} + Q \tag{29-18}$$

EXAMPLE 29-1: Find the threshold energy for the reaction

$$^{14}_{7}N\ (n,\ \alpha)\ ^{11}_{5}B$$

SOLUTION: The Q value for this reaction is found from

$$\begin{aligned}
^{1}_{0}n = m_x = &\quad 1.008665 \text{ amu} \\
^{14}_{7}N = M_X = &\quad 14.003074 \text{ amu} \\
\hline
m_x + M_X = &\quad 15.011739 \text{ amu}
\end{aligned}$$

and

$$\begin{aligned}
^{4}_{2}\alpha = m_y = &\quad 4.002603 \text{ amu} \\
^{11}_{5}B = M_Y = &\quad 11.009305 \text{ amu} \\
\hline
m_y + M_Y = &\quad 15.011908 \text{ amu} \\
(m_x + M_X) - (m_y + M_Y) = &\quad 15.011739 \text{ amu} \\
&\quad -15.011908 \text{ amu} \\
\hline
&\quad -\ 0.000169 \text{ amu}
\end{aligned}$$

or

$$Q = -0.00169 \text{ amu} \times 931 \text{ MeV/amu} = -0.16 \text{ MeV}$$

where $-Q = 0.16$ MeV is the energy for the reaction, and the minimum input energy is

$$(K_x)_{min} = E_{th} = \frac{-Q(m_x + M_X)}{M_X}$$

or

$$E_{th} = \frac{-(-0.16)(1 + 14)}{14} = 0.17 \text{ MeV}$$

From Equation (29-18), then, the energy carried away by the center of mass is

$$K_{cm} = E_{th} + Q = (0.17 - 0.16) \text{ MeV} = 0.01 \text{ MeV}$$

EXAMPLE 29-2: From Example 29-1, compute the speed of the center of mass before the reaction takes place.

SOLUTION: The energy carried away by the center of mass is

$$K_{cm} = 0.01 \text{ MeV} = \tfrac{1}{2}(m_x + M_X)v_{cm}^2$$

The speed of the cm is

$$v_{cm} = \sqrt{\frac{2K_{cm}}{m_x + M_X}} = \sqrt{\frac{2 \times 0.01 \text{ MeV} \times 1.60 \times 10^{-13} \text{ J/MeV}}{15 \times 1.67 \times 10^{-27} \text{ kg}}}$$

or

$$v_{cm} = 3.58 \times 10^5 \text{ m/sec}$$

29-4 CROSS-SECTION PROBABILITY

Even though the Q value of a nuclear reaction is found to be positive and hence exoergic, there is still not a 100% probability that a particular reaction will take place. The probability of occurrence for a particular reaction is measured by the cross section. (The cross section for Rutherford scattering of α particles was discussed in Chapter 13.)

The effective cross section for a nuclear reaction measures the target area surrounding the nucleus in which if a collision occurs a particular reaction will be produced. The reaction will take place if the bullet particle passes through this area; otherwise the reaction will not occur.

It is customary practice to associate a cross section with each particular type of nuclear interaction. A scattering cross section is used when dealing with nuclear scattering processes; an absorption cross section for nuclear absorption processes; and a fission cross section when studying nuclear collisions leading to the fission of the target nucleus.

To determine a reaction cross section, consider (Figure 29-5) a slab of target of thickness t and cross-sectional area A. If n is the number of target nuclei per unit volume, then the number of nuclei in the slab is nAt. Now, if σ is the cross section of each nucleus, or the area of interaction of each nucleus, the total exposed area for interactions will be $n\sigma At$. The probability for a nuclear reaction is

$$P = \frac{\text{total exposed area}}{\text{total area}} = \frac{n\sigma At}{A} = n\sigma t \qquad (29\text{-}19)$$

This probability is also equal to the ratio of the number of incident particles per second N that experience a reaction to the total number of impinging particles per

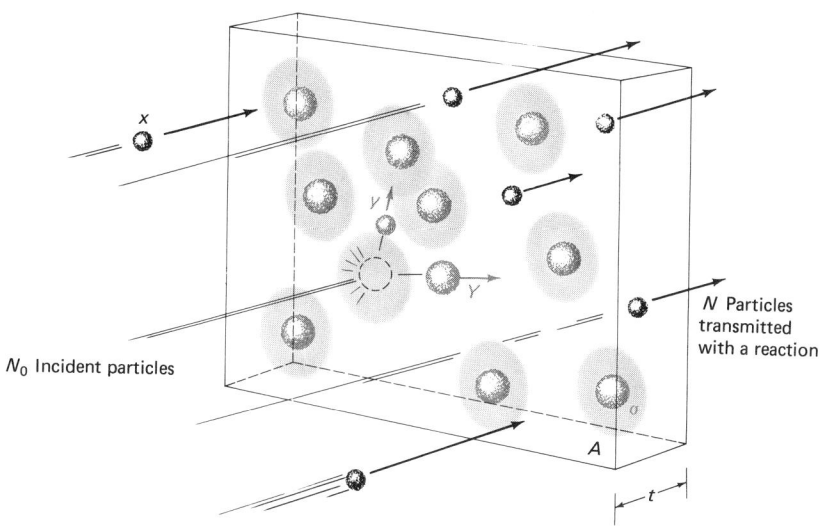

Figure 29-5 The probability for a nuclear reaction is given by the fraction $f = N/N_0 = n\sigma t$.

second N_0. The probability of a reaction occurring is then

$$P = \frac{N}{N_0} = n\sigma t \qquad (29\text{-}20)$$

and it is seen to be directly proportional to σ.

If the reaction is endoergic, the cross section is zero if the energy of the bullets is less than the threshold energy.

EXAMPLE 29-3: A thin sheet of Co-59, 0.02 cm thick, is irradiated with a neutron flux of 10^{16} neutrons/m²-sec. If the radiative capture reaction that takes place is

$$^{59}_{27}\text{Co}\ (n, \gamma)\ ^{60}_{27}\text{Co}$$

how many nuclei of Co-60 will be produced at the end of a 2-h irradiation period?

SOLUTION: The density of Co-59 is 8.9×10^3 kg/m³, and the cross section for neutron capture by Co-59 is 30 barns.

The number of target nuclei per unit volume is

$$n = \frac{\rho N_A}{M} = \frac{(8.9 \times 10^3 \text{ kg/m}^3)(6.02 \times 10^{26} \text{ nuclei/kg-mole})}{59 \text{ kg/kg-mole}}$$

$$= 9.1 \times 10^{28} \text{ nuclei/m}^3$$

The number of neutrons per square meter per second that will experience an interaction is

$$N = N_0 n \sigma t = 10^{16} \frac{\text{neutrons}}{\text{m}^2\text{-sec}} \times 9.1 \times 10^{28} \frac{\text{nuclei}}{\text{m}^3}$$
$$\times 30 \times 10^{-28} \text{ m}^2/\text{nuclei} \times 2 \times 10^{-4} \text{ m}$$
$$= 5.5 \times 10^{14} \text{ neutrons/m}^2\text{-sec}$$

This is also the number of Co-60 atoms per square meter of target material that will be produced. At the end of a 2-h period, there will be

$$5.5 \times 10^{14} \frac{\text{neutrons}}{\text{m}^2\text{-sec}} (3600 \text{ sec}) = 2 \times 10^{18} \text{ nuclei/m}^2$$

or

$$2 \times 10^{18} \frac{\text{nuclei}}{\text{m}^2} \times 60 \frac{\text{amu}}{\text{nuclei}} \times 1.6 \times 10^{-27} \frac{\text{kg}}{\text{amu}}$$
$$= 1.92 \times 10^{-7} \frac{\text{kg}}{\text{m}^2} = 0.192 \text{ mg/m}^2$$

How long would it take to produce 1 lb of Co-60 if the target area were 5 cm^2?

PROBLEMS

29-1 Nuclides of $^{32}_{16}$S are bombarded by α particles, and protons are emitted. Compute (a) the Q value of this reaction, (b) the threshold energy, (c) the kinetic energy carried away by the cm, and (d) the speed of the cm before the reaction takes place.

29-2 Given the reaction

$$^{7}_{3}\text{Li} (p, n) \, ^{7}_{4}\text{Be}$$

determine (a) the threshold energy, and (b) the neutron–proton mass difference using the atomic masses of Li-7 and Be-7.

29-3 An α particle of energy 7.67 MeV collides elastically with a proton. The kinetic energy of the α particle after the collision is 2.56 MeV, and it scatters at an angle of 12° from the initial direction. Find the momentum and scattering angle of the proton.

29-4 An incident α particle with a kinetic energy of 7.68 MeV initiates the reaction

$$^{14}_{7}\text{N} (\alpha, p) \, ^{17}_{8}\text{O}$$

If the N-14 is initially at rest, calculate the kinetic energy of the proton. For simplicity, assume that the proton and O-17 nucleus move off in the same direction as the incident α particle path.

29-5 (a) Show that the reaction

$$^{16}_{8}O\ (n, \alpha)\ ^{13}_{6}C$$

is endoergic.

(b) Compute the kinetic energy of the neutron necessary to initiate this reaction. Make this computation relative to the laboratory frame.

29-6 Determine which of the following nuclides can be α particle emitters: $^{5}_{3}Li$, $^{8}_{3}Li$, $^{12}_{5}B$, and $^{5}_{2}He$.

29-7 A proton of 1.0 MeV initiates the reaction

$$^{7}_{3}Li\ (p, \alpha)\ ^{4}_{2}He$$

(a) How much energy is released by the reaction?
(b) Determine the energy carried away by the cm.

29-8 A 4.0-MeV γ ray disintegrates a deuteron according to the reaction

$$d\ (\gamma, p)\ n$$

If the proton moves off at right angles to the original direction of the γ ray, find (a) the kinetic energy of the neutron, (b) the direction of the neutron motion, and (c) the kinetic energy of the proton.

29-9 A particle of mass m_x has a kinetic energy of K_x in the laboratory frame. When the particle is projected against a nucleus of mass M_X at rest in the same laboratory frame, the following nuclear reaction takes place:

$$M_X\ (m_x, m_y)\ M_Y$$

Prove that (a) the total kinetic energy in the cm system is

$$K_{cm} = \left(\frac{m_x}{m_x + M_X}\right) K_x$$

and (b) the energy available for the reaction is

$$K_A = \left(\frac{M_X}{m_x + M_X}\right) K_x$$

RECOMMENDED READING

ALONSO, M., and FINN, E. J., *Fundamental University Physics*, Addison-Wesley, Reading, Mass., 1968, Vol. III, Chap. 8.

BETHE, H. A., and MORRISON, P., *Elementary Nuclear Physics*, Wiley, New York, 1956, Chap. 20.

GRIFFY, T., "Resource Letter NR-1 on Nuclear Reactions," *Am. J. Phys.* **35**, 297 (1967).

KAPLAN, I., *Nuclear Physics*, Addison-Wesley, Reading, Mass., 1955, Chap. 16.

30 Radioactivity I

Antoine Henry Becquerel
(1852–1908)

A native of Paris and educated at the École Polytechnique, Becquerel was appointed professor there in 1895. He later succeeded to his father's chair at the Musée d'Histoire Naturelle. In 1896 he discovered spontaneous radioactivity in uranium salts. He later researched various aspects of light, including polarization and absorption by crystals. For his study of the phenomenon of radioactivity, he, with his associates Pierre and Marie Curie, was awarded the Nobel Prize in 1903.

30–1 RADIOACTIVITY
30–2 DISINTEGRATION CONSTANT
30–3 HALF-LIFE AND MEAN LIFE
30–4 CURVE OF GROWTH
30–5 RADIOACTIVE SERIES
30–6 DATING BY RADIOACTIVE DECAY

30–1 RADIOACTIVITY

In 1896 the French physicist HENRI BECQUEREL placed a piece of uranium ore on a package of undeveloped photographic plates which were wrapped in black paper for protection. On developing the photographic plate, he was amazed to find that an unknown kind of radiation had come from the ore and left an image on the plate in the shape of the ore.

By 1898 MARIE and PIERRE CURIE had succeeded in isolating a portion of 1 g of active substance from about a ton of pitchblende. The new substance was named *polonium* in honor of Mme Curie's native Poland. The Curies were awarded the Nobel Prize in physics jointly with Becquerel in 1903 for this discovery. In 1911 Mme Curie was awarded the Nobel Prize in chemistry for isolating the naturally radioactive element radium. English physicists Ernest Rutherford and Frederic Soddy demonstrated that in all radioactive processes, a transmutation of elements occurs.

Rutherford discovered that a magnetic field separated the radiation into α particles, which are helium nuclei, and β particles, which are electrons. Later Paul Willard found a still different type of radiation, γ rays, which are a very energetic form of electromagnetic waves. Any nuclide that changes its structure by giving off γ rays or some nuclear particle like α, β^-, or β^+ particles is called a *radioactive* nucleus. There are 272 stable nuclei of the elements found in nature, and all of the others, known as *radioisotopes*, are radioactive in some form.

The following processes are representative of the disintegration of the natural radioactive nuclides:

$$\alpha \text{ decay:} \quad {}_Z^A M \to {}_{Z-2}^{A-4} M + {}_2^4 He$$

$$\beta^- \text{ decay:} \quad {}_Z^A M \to {}_{Z+1}^{A} M + {}_{-1}\beta^0 + \bar{\nu}$$

$$\beta^+ \text{ decay:} \quad {}_Z^A M \to {}_{Z-1}^{A} M + {}_{+1}\beta^0 + \nu \qquad (30\text{--}1)$$

$$\beta^- \text{ capture:} \quad {}_Z^A M + {}_{-1}\beta^0 \to {}_{Z-1}^{A} M + \nu$$

$$\gamma \text{ decay:} \quad ({}_Z^A M)^* \to {}_Z^A M + \gamma$$

In these processes ν and $\bar{\nu}$ represent the neutrino and the antineutrino, which will be discussed shortly, and $({}_Z^A M)^*$ represents an atom in an excited energy state. The atom returns to the ground state ${}_Z^A M$ by the emission of a γ ray.

30-2 DISINTEGRATION CONSTANT

If N represents the number of atoms present at a given time and dN represents the number of disintegrations during a time interval dt, then

$$\text{probability of disintegration} = -\frac{dN}{N} \tag{30-2}$$

where the minus sign indicates that dN is always negative. The probability of disintegration per unit time λ is

$$\lambda = -\frac{1}{N}\frac{dN}{dt} = \text{constant} \tag{30-3}$$

where λ is identified as the *disintegration constant*. The *activity* of a sample is defined by $A = |(dN/dt)| = \lambda N$ disintegrations/sec and represents the rate at which disintegrations of nuclei occur. The unit for measuring activity is the curie (Ci), which represents 3.70×10^{10} disintegrations/sec, or the millicurie (mCi), which is 3.70×10^{7} disintegrations/sec.

Equation (30-3) can be rewritten in a slightly different form, and if there are N_0 atoms at $t = 0$, then

$$\int_{N_0}^{N} \frac{dN}{N} = \int_{t=0}^{t} -\lambda\, dt$$

or

$$\ln \frac{N}{N_0} = -\lambda t$$

which in exponential form is

$$N = N_0 e^{-\lambda t} \tag{30-4}$$

When Equation (30-4) is multiplied through by λ, it becomes

$$\lambda N = \lambda N_0 e^{-\lambda t} \tag{30-5}$$

where $\lambda N_0 = A_0$ is the initial activity, and $\lambda N = A$ is the activity at some time t. This gives for the activity

$$A = A_0 e^{-\lambda t} \tag{30-6}$$

or, in logarithmic form,

$$\ln A = \ln A_0 - \lambda t \tag{30-7}$$

As a typical example, the exponential decay of activity of P-32 is shown in Figure 30–1. If the natural logarithm of the activity is plotted against time as in Figure 30–2, according to Equation 30–7 it will produce a straight line that has a slope equal to $-\lambda$. The disintegration constant λ for P-32 from the slope of the line in Figure 30–2 is

$$\lambda = \frac{10.8 - 7.0}{78 - 0} = 0.049 \text{ days}^{-1}$$

30–3 HALF-LIFE AND MEAN LIFE

The *half-life* $T_{1/2}$ of a radioisotope is defined as the elapsed time in which the number of atoms decays to one-half the initial number, or the time in which the activity drops to one-half the initial activity. When $t = T_{1/2}$, then, the number of atoms of a given kind present is $N = \frac{1}{2}N_0$ and, from Equation (30–4),

$$\tfrac{1}{2}N_0 = N_0 e^{-\lambda T_{1/2}} \quad \text{or} \quad \tfrac{1}{2} = e^{-\lambda T_{1/2}}$$

Taking the natural logarithm of both sides gives

$$-\ln 2 = -\lambda T_{1/2}$$

or the half-life is

$$T_{1/2} = \frac{\ln 2}{\lambda} = \frac{0.693}{\lambda} \tag{30-8}$$

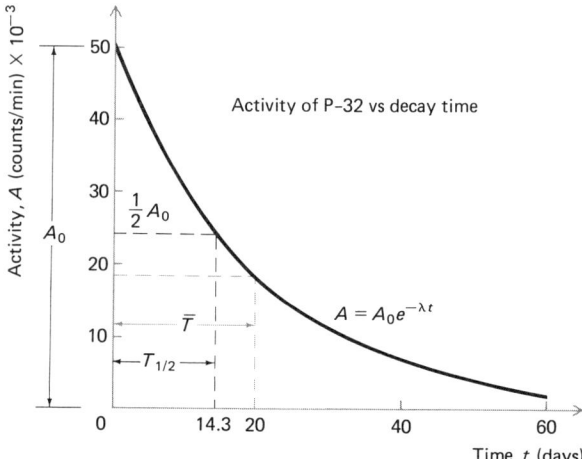

Figure 30–1 The curve shows the exponential decay of P-32, which has a half-life of 14.3 days. The sample originally had an activity of 50,000 disintegrations/min.

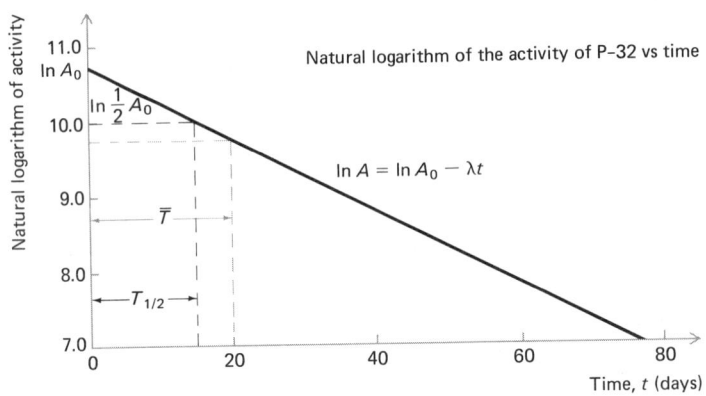

Figure 30-2 When the natural logarithm of Equation (30-6) is taken, it gives $\ln A = \ln A_0 - \lambda t$. When $\ln A$ is plotted against t, it gives a straight line with a slope equal to $-\lambda$.

From Figure 30-1, showing the decay in activity of P-32, the activity has dropped to one-half the original value when it is at 25,000 counts/min. The corresponding time then shows the half-life for P-32 to be 14.3 days. By using this value in Equation (30-8), we can find the value of the disintegration constant λ.

The *mean life* or *average life* of a radioactive nucleus is the average lifetime of all the atoms in a given sample. Therefore, if there are dN_1 atoms with a lifetime t_1, dN_2 atoms with a lifetime t_2, and so on, then the mean life $\overline{T}$ is defined as

$$\overline{T} = \frac{t_1\, dN_1 + t_2\, dN_2 + \cdots}{dN_1 + dN_2 + \cdots} \tag{30-9}$$

or stated in integral calculus,

$$\overline{T} = \frac{\int_{t=0}^{\infty} t\, dN}{\int_{N_0}^{0} dN} = -\frac{\int_{t=0}^{\infty} t\, dN}{N_0} \tag{30-10}$$

However,

$$dN = d(N_0 e^{-\lambda t}) = -\lambda N_0 e^{-\lambda t}\, dt$$

Substitution into Equation (30-10) now gives

$$\overline{T} = -\frac{\int_0^\infty -\lambda N_0 t e^{-\lambda t}\, dt}{N_0} = \lambda \int_0^\infty t e^{-\lambda t}\, dt \tag{30-11}$$

Then from integration by parts of Equation (30-11),

$$\overline{T} = \lambda\left(-\frac{t}{\lambda} e^{-\lambda t} + \frac{1}{\lambda}\int_0^\infty e^{-\lambda t}\, dt\right)$$

$$= \lambda\left(-\frac{t}{\lambda} e^{-\lambda t} - \frac{1}{\lambda^2} e^{-\lambda t}\right)_0^\infty$$

or finally,

$$\boxed{T = \frac{1}{\lambda}} \qquad (30\text{-}12)$$

The mean life and the half-life are related by

$$\overline{T} = \frac{T_{1/2}}{\ln 2} \quad \text{or} \quad T_{1/2} = 0.693 \qquad (30\text{-}13)$$

30-4 CURVE OF GROWTH

When a decaying nucleus, called the *parent nucleus P*, disintegrates there is a *decay* of this parent nucleus; and at the same time there is a *growth* of a product nucleus, called the *daughter nucleus D*. Schematically, this is

parent $P \to$ daughter D

At time $t = 0$, the number of parent nuclei is N_0 while the number of daughter nuclei is zero. At some time t, the number of parent nuclei is N while the number of daughter nuclei is $N' = N_0 - N$.

The equation describing the decay of the parent P is

$$N = N_0 e^{-\lambda t}$$

while that describing the growth of the daughter is

$$N' = N_0 - N = N_0(1 - e^{-\lambda t}) \qquad (30\text{-}14)$$

The exponential decay of P-32 is shown in Figure 30–3, where the decay scheme is

$$(^{32}_{15}P)^* \to {}^{32}_{16}S + {}^{0}_{-1}\beta + \bar{\nu}$$

Notice that at the half-life time the daughter and parent nuclei are equal in number.

Up to this point, the growth of the daughter has been described assuming that it is a stable nuclide. If the daughter nuclei are unstable and decay also, the number N' of the daughter nuclei may be at a given instant increasing or decreasing. The rate at which the number of daughter nuclei is changing is

$$\frac{dN'}{dt} = \lambda N - \lambda' N' \qquad (30\text{-}15)$$

where λ and λ' are the disintegration constants of the parent and daughter nuclei. The term λN is the increase in daughter nuclei per unit time due to the disintegration of the parent, and $\lambda' N'$ is the decrease of daughter nuclei due to their own decay.

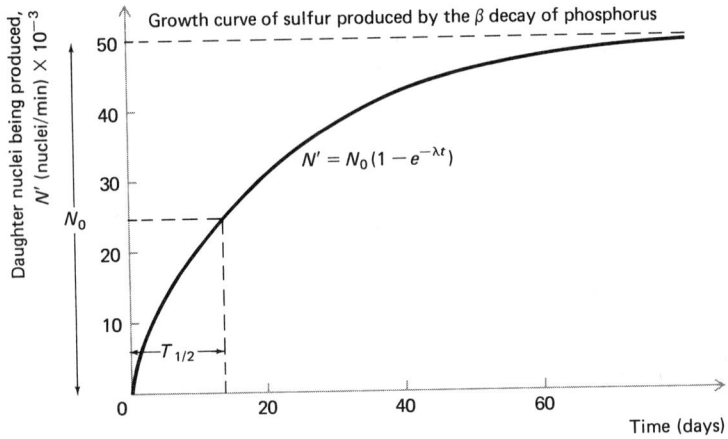

Figure 30–3 As the parent nuclei of P-32 decay by β emission, there is a growth of the stable daughter nuclei S-32.

Equation (30–15) can now be written

$$\frac{dN'}{dt} + \lambda'N' = \lambda N$$
$$= \lambda N_0 e^{-\lambda t} \tag{30–16}$$

If it is assumed that at $t = 0$, $N' = 0$, then the solution of Equation (30–16) is

$$\boxed{N' = N_0 \left(\frac{\lambda}{\lambda' - \lambda}\right)(e^{-\lambda t} - e^{-\lambda' t})} \tag{30–17}$$

It is left to the student to prove that Equation (30–17) is the proper solution of Equation (30–16).

Observe that Equation (30–17) now shows that when $t = 0$, $N' = 0$ and when $t \to \infty$, $N' \to 0$. Figure 30–4 shows a curve of a daughter nucleus that is unstable and decays.

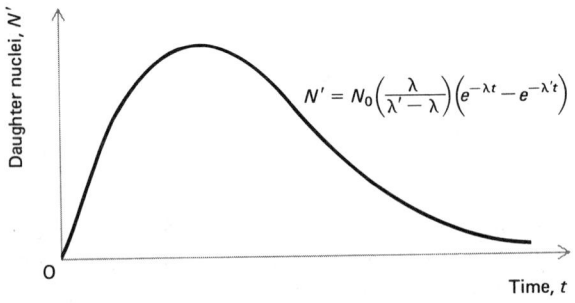

Figure 30–4 Growth and decay of an unstable daughter nucleus.

When the parent nucleus has a very long half-life compared with the daughter nucleus, for smaller values of t, $\lambda' - \lambda \cong \lambda'$ and $e^{-\lambda t} \to 1$ since λ is very small compared to λ'. Equation (30–17) then becomes

$$N' = N_0 \frac{\lambda}{\lambda'} (1 - e^{-\lambda' t}) \qquad (30\text{–}18)$$

For large values to t and $\lambda' \gg \lambda$, $e^{-\lambda' t} \to 0$ and $\lambda' - \lambda \cong \lambda'$, and Equation (30–17) is then

$$N' = N_0 \frac{\lambda}{\lambda'} e^{-\lambda t} \qquad (30\text{–}19)$$

Since the number of parent nuclei at a given time is $N = N_0 e^{-\lambda t}$, Equation (30–19) gives

$$N' = N \frac{\lambda}{\lambda'}$$

or

$$\boxed{N\lambda = N'\lambda'} \qquad (30\text{–}20)$$

This condition is called *secular equilibrium*, and it simply states that after a long period of time the rate of growth of the daughter is equal to its rate of decay so that N' remains practically constant. This condition can also be obtained directly from Equation (30–15) by making $dN'/dt = 0$.

30–5 RADIOACTIVE SERIES

There are only 14 radioactive nuclides found in nature whose half-life is of the same order or larger in magnitude than the age of the universe, which is estimated as 4.5×10^9 yr. These are listed in Table 30–1.

Table 30–1 Natural radioactive nuclides with half-lives of the same order or larger than the age of the universe (4.5×10^9 yr)

NUCLIDE	DECAY MODE	HALF-LIFE (yr)	
K-40	β^-, e capture	1.2×10^9	decay into stable daughter
V-50	e capture	4.0×10^{14}	decay into stable daughter
Rb-87	β^-	6.2×10^{10}	decay into stable daughter
In-115	β^-	6.0×10^{14}	decay into stable daughter
La-138	β^-, e capture	1.0×10^{11}	decay into stable daughter
Ce-142	α	5.0×10^{15}	decay into stable daughter
Nb-144	α	3.0×10^{15}	decay into stable daughter
Sm-147	α	1.2×10^{11}	decay into stable daughter
Lu-176	β^-	5.0×10^{10}	decay into stable daughter
Re-187	β^-	4.0×10^{12}	decay into stable daughter
Pt-192	α	1.0×10^{15}	decay into stable daughter
Th-232	α	1.4×10^{10}	ten active generations
U-235	α	7.1×10^9	ten active generations
U-238	α	4.5×10^9	ten active generations

Table 30-2 The four radioactive series

NAME	STARTING NUCLEUS	MASS NUMBER	n INITIAL	n FINAL	HALF-LIFE STABLE NUCLEUS (yr)	FINAL STABLE NUCLEUS
thorium	$^{232}_{90}$Th	$4n$	58	52	1.39×10^{10}	$^{208}_{82}$Pb
neptunium	$^{237}_{93}$Np	$4n + 1$	59	52	2.20×10^{6}	$^{209}_{83}$Bi
uranium-radium	$^{238}_{92}$U	$4n + 2$	59	51	4.51×10^{9}	$^{206}_{82}$Pb
actinium	$^{235}_{92}$U	$4n + 3$	58	51	7.15×10^{9}	$^{207}_{82}$Pb

The first 11 nuclides in Table 30-1 decay into stable daughters, but the last three, Th-232, U-235, and U-238, decay into daughters that are in turn radioactive and decay into other radioactive daughters through several generations until a stable nuclide is reached. The final stable nuclides into which these three naturally radioactive series decay are three isotopes of lead, Pb-208, Pb-207, and Pb-206, respectively. The four possible radioactive series are listed in Table 30-2.

Because of the short half-life of neptunium, even traces of this element in nature are undetectable by ordinary methods. With the discovery of *transuranic elements* (elements of atomic number greater than 92), physicists were able to

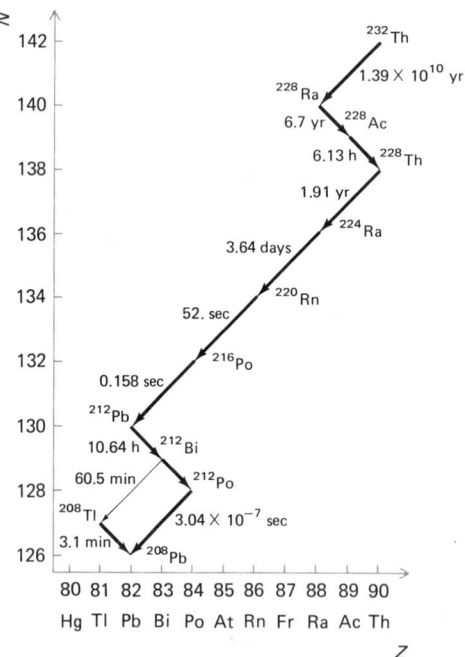

Figure 30-5 The thorium series. The mass number of each element in the series is given by $4n$, where n is an integer.

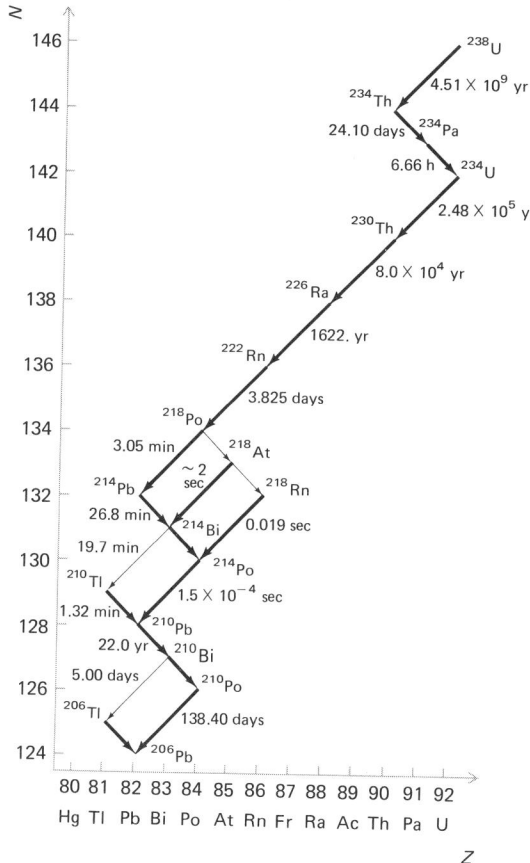

Figure 30–6   The uranium–radium series. The mass number of each element in the series is given by $4n + 2$, where n is a positive integer.

produce new isotopes and it was then possible to trace the neptunium series, which actually has its origin with americium and plutonium.

The mass number of each member of a given series found in nature was found to fit the equations $4n$, $4n + 2$, and $4n + 3$, where n is an integer. The fact that $4n + 1$ was missing led physicists to speculate about a missing series, which eventually showed up as the neptunium series. Figures 30–5, 30–6, and 30–7 show the transmutation for the series $4n$, $4n + 2$, and $4n + 3$.

30–6 DATING BY RADIOACTIVE DECAY

The decay of radioactive elements is independent of the physical and chemical conditions imposed on them; and although the decay of an individual particle from a given nucleus is a random process, the gross decay of the many nuclei in a given

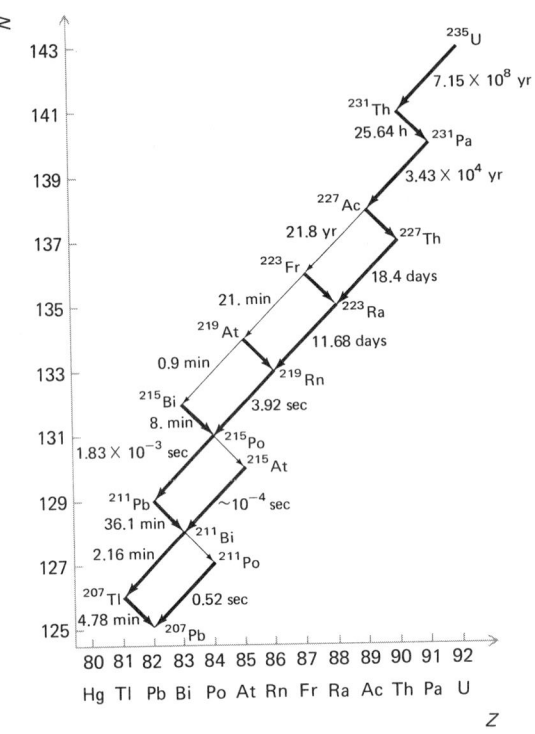

Figure 30-7 The uranium–actinium series. The mass number of each element in the series is given by $4n + 3$, where n is a positive integer.

sample provides a very convenient way of measuring times. In 1913 Joly and Rutherford suggested that if igneous rock, formed as a result of a prehistoric volcanic eruption, contained a small amount of uranium it would steadily decay, leaving less uranium and depositing more stable Pb-206 according to Figure 30-6. By measuring the ratio of uranium to lead in rock samples, a rather exact time can be determined for the origin of the geological deposits. Uranium dating measures times of the order of millions of years.

Another similar method of dating is the decay of the isotope K-40 into Ca-40 or A-40 by positron emission,

$$^{40}_{19}K \rightarrow {}^{40}_{18}A + \beta^+ + \nu$$

with a half-life of 1.3×10^9 yr. Because it is more readily measured, A-40 is most often used. A geological sample is first heated to a temperature of 400°C for 12–48 h to boil off any A-40 that may have been picked up from the surrounding air. (For the sealed-off moon rocks, this would be minimized.) Then the sample is heated until it is melted, and the A-40 within the sample is boiled off and sent through a mass spectrometer. By measuring the ratio of potassium to argon within the sample, the date of formation of the sample can be established.

Since all plants use carbon dioxide from the atmosphere for growth, a portion of the carbon in plants is radioactive C-14 and the plants are slightly radioactive. When a plant dies, no additional C-14 is taken in, and that within the plant body begins to decay without being replaced. Measurement of the relative amounts of C-14 and C-12 in an organic archeological sample provides a sensitive method of dating. Some interesting archeological events dated by W. F. Libby by radiocarbon dating are listed in Table 30-3.

Another element that is continuously produced in the atmosphere by cosmic-ray bombardment is tritium, ^3_1H. It is formed by neutron collision with nitrogen according to

$$^{14}_{7}N + ^{1}_{0}n \rightarrow ^{12}_{6}C + ^{3}_{1}H$$

Tritium, the heavy isotope of hydrogen, decays with a half-life of 12.4 yr by β decay into the stable helium isotope ^3_2He:

$$^{3}_{1}H \rightarrow ^{3}_{2}He + \beta^- + \bar{\nu}$$

A small percentage of water contains ^3_1H instead of ^1_1H, and by measuring the relative amounts of ^3_1H and ^1_1H in samples of water at various locations, some indication is given as to how long ago this water was in the form of rain.

Table 30-3 Some typical samples dated by radiocarbon dating

SAMPLE	KNOWN AGE (yr)	C-14 AGE (yr)
1. Charcoal from Lascaux cave (well known for wall paintings)		15,516 ± 900
2. Slab of wood from roof beam of tomb of Vizier Hemaka	4700 – 5100	4,883 ± 20
3. Cypress beam from tomb of Pharaoh Sneferu	4575 ± 75	4,802 ± 210
4. Deer antler from Indian site in Butler County, Kentucky		4,333 ± 450
5. Wood from the deck of the funeral ship from tomb of Pharaoh Sesostris III	3750	3,621 ± 180
6. Bushman painting, South Africa		3,368 ± 200
7. Wood from giant redwood tree known as the Centennial Stump, felled in 1874	2928 ± 51	2,710 ± 130
8. Dead Sea scrolls	2054 ± 100	1,917 ± 200

NOTE: Dates are taken from W. F. Libby, *Radiocarbon Dating*, University of Chicago Press, Chicago, 1965. All of the dates were obtained prior to the fall of 1954. Radiocarbon dates are now published in *Radiocarbon*, a supplement to the *American Journal of Science*.

PROBLEMS

30–1 The activity of 1 g of $^{226}_{88}$Ra is 1 Ci. From this find the mean life and the half-life of radium.

30–2 Determine an expression for the "third-life" of a radioactive element, that is, the time required for the element to decay to one-third of its original amount.

30–3 Uranium ($^{238}_{92}$U) is an α-particle emitter, and it is found that 1 g will emit 12,200 α particles/sec. With this information, determine the half-life of uranium.

30–4 The disintegration constant of $^{210}_{82}$Pb is 10^{-9} sec^{-1}. Compute the disintegration time for two-fifths of a given sample of Pb-210.

30–5 The half-life of $^{238}_{92}$U for α-particle disintegration is 4.5×10^9 yr. Compute the activity of 1 g of uranium in disintegrations per second.

30–6 The half-life of $^{210}_{83}$Bi is 5 days. (a) Compute the disintegration constant in h^{-1}. (b) What fraction of a given sample will remain after 15 days?

30–7 One-billionth of a kg-mole of $^{226}_{88}$Ra is in equilibrium with its products $^{222}_{86}$Rn and $^{218}_{84}$Po. Determine how many atoms of each element are present.

30–8 Show from Equation (30–17) that if N_0 is the number of parent atoms initially and λ and λ' are the disintegration constants of parent and daughter, respectively, the time at which the number of daughter atoms is a maximum is given by

$$t_m = \frac{\ln(\lambda'/\lambda)}{\lambda' - \lambda}$$

30–9 A radioactive element disintegrates for an interval of time equal to the average life. What fraction of the original amount has disintegrated in this time?

30–10 What mass of each of the following elements will have an activity of 1 mCi: (a) U-238, (b) Th-232?

30–11 A sample of $^{223}_{88}$Ra is found initially to have an activity of 3.0×10^4 disintegrations/min. From the detector readings of the activity of this sample shown on p. 397, plot a graph and determine the half-life and the mean life of this sample.

ELAPSED TIME (days)	ACTIVITY OF $^{223}_{88}$Ra (disintegrations/min)
0	30×10^4
10	17×10^4
20	9.0×10^4
30	5.0×10^4
40	3.0×10^4
50	1.5×10^4
60	1.0×10^4
80	0.3×10^4

30–12 Strontium-90 ($^{90}_{38}$Sr) is a β^- emitter with a half-life of 28 yr. How long will it take for 90% of the original amount of a given sample to disintegrate?

30–13 In an ore sample both uranium and lead (Pb-206) are found. If the sample contains 0.85 g of Pb-206 for each gram of U-238, determine the age of the ore sample.

30–14 A radioactive sample of material emits 10×10^{-6} W of radiation at some given time and 1×10^{-6} W after 5 h. What is the half-life of the sample?

30–15 The radioactive isotope $^{228}_{89}$Ac decays into $^{228}_{90}$Th with a half-life of 6.13 h. The $^{228}_{90}$Th decays into Ra-224 with a half-life of 1.91 yr. Make a rough sketch of the number of nuclei of Th-228 present versus time. Assume that at $t = 0$ there are N_0 nuclei of Ac-228 and none of Th-228.

30–16 The radioisotope $^{212}_{83}$Bi, with a half-life of 61 min, emits both α and β particles. Of the disintegrations, 34% are α particles and the remaining 66% are β particles. In a 1.0-μg sample of Bi-212, determine (a) the initial activities for α particles and for β particles, and (b) the respective α and β activities after 3.0 h.

30–17 The abundance of $^{234}_{92}$U in natural uranium is 0.0058%. If the half-life of $^{238}_{92}$U is 4.51×10^9 yr, what is the half-life of U-234?

30–18 The half-life of $^{210}_{84}$Po is 138.4 days. Calculate the activity in millicuries after 1 yr if the initial amount was 1.0 μg.

RECOMMENDED READING

BOTHE, W., and BECKER, H., *Z. Physik* **66**, 289 (1930).

COWAN, C., ATLURI, C. R., and LIBBY, W. F., "Possible Anti-Matter Content of the Tunguska Meteor of 1908," *Nature* **206**, 861–865 (1965).
A feasible explanation of the mystery of the Tunguska meteor is based on results from C-14 dating.

CURIE, I., and JOLIOT, F., *Compt. Rend.* **194,** 273 (1932).

DALRYMPLE, C. Brent, and LANPHERE, Marvin A., *Potassium-Argon Dating: Principles, Techniques, and Applications to Geochronology,* Freeman, San Francisco, 1969.

GAMOW, George, *The Atom and Its Nucleus,* Prentice-Hall, Englewood Cliffs, N.J., 1961, Chap. 6.

KAPLAN, I., *Nuclear Physics,* Addison-Wesley, Reading, Mass., 1962.

LIBBY, Willard F., *Radiocarbon Dating,* University of Chicago Press, Chicago, 1965.
Interesting account of radiocarbon dating along with many intriguing events dated by this method.

MANN, W., and GARFINKEL, S., *Radioactivity and Its Measurement,* Van Nostrand Reinhold, New York, 1966.

31
Radioactivity II

Marie Sklodowska Curie
(1867–1934)

Born in Warsaw, Mme Curie studied at the Sorbonne where, in 1906, she succeeded her husband, Pierre, becoming the first woman appointed titular professor there. She did extensive research on the radiation phenomena—she studied and measured radiations given off by uranium, tested all elements for radioactivity, discovered and isolated polonium and radium. She received the Nobel Prize for physics (with Pierre) in 1903, and for chemistry in 1911. She died of leukemia caused by overexposure to radioactive substances.

31–1   ALPHA DECAY
31–2   POSITRON DECAY
31–3   ELECTRON DECAY
31–4   ELECTRON CAPTURE
31–5   GAMMA DECAY
31–6   RADIOLOGICAL HEALTH HAZARDS

31–1 ALPHA DECAY

Most of the heavy nuclides ($Z \geq 84$, $A \geq 208$) decay by α-particle emission according to the pattern

$$^{A}_{Z}P \rightarrow {}^{A-4}_{Z-2}D + {}^{4}_{2}He \tag{31-1}$$

As a specific example, consider the following decay:

$$^{226}_{91}Pa \rightarrow {}^{222}_{89}Ac + {}^{4}_{2}He$$

The Q value for this process is found from an input mass of

$$\text{Pa-226} = m_p = 226.0280 \text{ amu}$$

and an output mass of

$$\begin{array}{rl} \text{Ac-222} = m_d = & 222.0178 \text{ amu} \\ \text{He-4} = m_\alpha = & \underline{4.0026 \text{ amu}} \\ & 226.0204 \text{ amu} \end{array}$$

The Q value is then

$$m_p - (m_d + m_\alpha) = \begin{array}{r} 226.0280 \text{ amu} \\ -226.0204 \text{ amu} \\ \hline 0.0076 \text{ amu} \end{array}$$

$$Q = 0.0076 \text{ amu} \times 931 \text{ MeV/amu} = 7.07 \text{ MeV}$$

Since $Q > 0$, there is energy released in this reaction when it takes place.

Let us analyze α decay in more detail. In terms of the mass-energy conservation principle, Equation (31–1) is written as

$$m_p c^2 = m_d c^2 + m_\alpha c^2 + K_d + K_\alpha \tag{31-2}$$

where K_d and K_α are the kinetic energies of the daughter and α particles, respectively. The Q value is

$$Q = K_d + K_\alpha = [m_p - (m_d + m_\alpha)]c^2 > 0 \tag{31-3}$$

The kinetic energy of the parent has not been included since it is assumed to be at rest.

Conservation of linear momentum requires that

$$m_\alpha v_\alpha = m_d v_d \tag{31-4}$$

from which we can write

$$\frac{m_\alpha^2 v_\alpha^2}{2} = \frac{m_d^2 v_d^2}{2}$$

or, in terms of kinetic energy,

$$m_\alpha K_\alpha = m_d K_d \tag{31-5}$$

In terms of atomic mass units, the masses are

$$m_\alpha = 4$$
$$m_d = A - 4$$

and Equation (31-5) is now written as

$$4K_\alpha = (A - 4)K_d$$

or

$$\boxed{K_d = \left(\frac{4}{A-4}\right) K_\alpha} \tag{31-6}$$

The Q value from Equation (31-3) is then

$$Q = \left(\frac{4}{A-4}\right) K_\alpha + K_\alpha = \left(\frac{A}{A-4}\right) K_\alpha$$

and finally,

$$\boxed{K_\alpha = \left(\frac{A-4}{A}\right) Q} \tag{31-7}$$

which is the kinetic energy of the α particle in terms of the Q value. Equation (31-7) shows that the beam of α particles is monoenergetic, as shown in Figure 31-1. A combination of Equations (31-6) and (31-7) gives the kinetic energy of the daughter as

$$\boxed{K_d = \frac{4}{A} Q} \tag{31-8}$$

For very heavy nuclides, A is large, $A - 4 \cong A$, and the kinetic energies of the α particle and the daughter become

$$K_\alpha \cong Q \quad \text{and} \quad K_d \cong 0$$

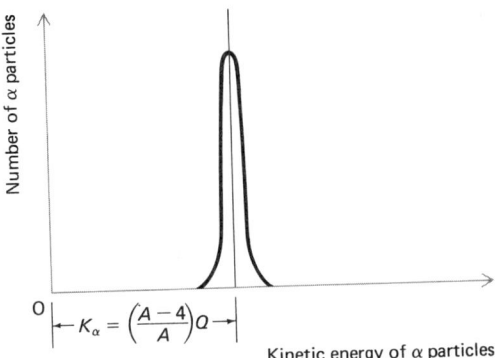

Figure 31-1 Energy spectrum of α particles in an α-decay process is practically monoenergetic.

that is, the α particle carries off practically all of the energy available in the form of kinetic energy.

Some of the first radioactive isotopes that were isolated, such as Ra-226, Em-222, and Po-218, are α-particle emitters. The polonium isotope Po-214 emits very energetic α particles (7.7 MeV). Because α particles are large, containing four nucleons, and because they have two ($+2e$) charges, their range of travel is short. Most α particles are stopped by several centimeters of air or by a sheet of paper. The range of α particles emitted from uranium is 2.7 cm in air, while from radium the range is 3.3 cm. In the uranium series, the nuclide with the shortest half-life (162 μsec) emits the most energetic α particles, with a range of 6.9 cm in air.

There is a relation between the half-life of α emitters and the kinetic energy (or range) of the α particles emitted. It was first expressed in empirical form by the Geiger-Nutall law:

$$\ln T_{1/2} = A \ln K + B \qquad (31-9)$$

where the constant A is very nearly the same for the three natural radioactive series and B is a different constant for each series.

This once empirical equation was put on a sound theoretical basis by GEORGE GAMOW in his quantum mechanical explanation of the tunnel effect.

31-2 POSITRON DECAY

Positron or β^+ decay is schematically represented by

$$_Z^A P \rightarrow _{Z-1}^A D + _1\beta^0 + \nu \qquad (31-10)$$

where P and D are the parent and daughter nuclei, $_1\beta^0$ is the positron, and ν is a *neutrino*. The neutrino is an elusive particle with no charge and zero rest mass.

It was predicted on theoretical grounds by W. PAULI, and found experimentally by F. REINES and C. L. COWAN in 1956.*

In terms of the mass-energy conservation principle, positron decay is

$$(m_p - Zm_e^-)c^2 = [m_d - (Z-1)m_e^-]c^2 + m_e^+c^2 \\ + K_d + K_\beta + K_\nu \quad (31\text{-}11)$$

where m_p, m_d, and $m_e^\pm$ are the atomic rest masses of the parent, daughter, and positron or electron, and K_d, K_β, and K_ν are the kinetic energies of the daughter, positron, and neutrino, respectively.

The Q value of the process is

$$Q = K_d + K_\beta + K_\nu = (m_p - m_d - 2m_e)c^2 \quad (31\text{-}12)$$

and for positron decay to be feasible,

$$\boxed{m_p > m_d + 2m_e}$$

which means that $Q > 0$.

The isotope $^{107}_{49}\text{In}$ is a positron emitter according to the scheme

$$^{107}_{49}\text{In} \rightarrow {}^{107}_{48}\text{Cd} + {}_{+1}\beta^0 + \nu$$

The input mass is

In-107 $= m_p = 106.9182$ amu

and the output mass is

$$\begin{aligned} \text{Cd-107} = m_d &= 106.9065 \text{ amu} \\ 2m_e = 2(0.00055) &= \underline{0.0011 \text{ amu}} \\ m_d + 2m_e &= 106.9076 \text{ amu} \end{aligned}$$

The mass difference is then

$$\begin{aligned} m_p - (m_d + 2m_e) = &106.9182 \text{ amu} \\ -&\underline{106.9076 \text{ amu}} \\ &0.0106 \text{ amu} \end{aligned}$$

and the Q value in MeV is

$$Q = 0.0106 \text{ amu} \times 931 \text{ MeV/amu} = 9.9 \text{ MeV}$$

The energy spectrum of positrons is a continuous distribution with a definite cutoff as shown in Figure 31–2. If there are no neutrinos emitted, the beam of

---

* A discussion of the historical development of the neutrino concept may be found in F. Reines and C. L. Cowan, "Neutrino Physics," *Phys. Today*, August 1957, p. 12.

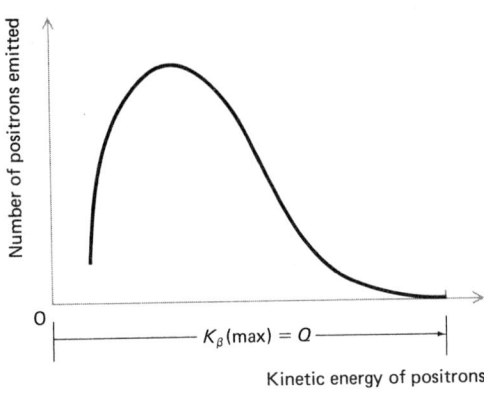

Figure 31–2 Energy spectrum of positrons in β^+ decay.

positrons must be monoenergetic, with only one linear momentum for the positrons

$$p_\beta = p_d$$

which must equal the momentum of the daughter since the parent is essentially at rest.

To explain the continuous distribution of energies a third particle, the neutrino, is needed. Since the parent is assumed to be at rest, the conservation of linear momentum requires that

$$\boxed{\mathbf{p}_d + \mathbf{p}_\beta + \mathbf{p}_\nu = 0} \qquad (31\text{--}13)$$

The inclusion of the neutrino in the picture provides an explanation for the continuous energy spectrum of positrons because $\mathbf{p}_d$, $\mathbf{p}_\beta$, and $\mathbf{p}_\nu$ can have many combinations of values and still add vectorially to zero. The sharp cutoff point K_β (max) $= Q$ in Figure 31–2 corresponds to the case in which $K_d = K_\nu = 0$, with the positrons carrying away all of the energy available from the reaction.

Note that in Equation (31–10) the proton number of the daughter is $Z - 1$ while the parent is Z, which indicates that the daughter has one less proton. At the same time the mass number of the daughter is A, the same as the parent. The neutron number of the daughter is $N' = A - (Z - 1) = (A - Z) + 1$, which is one more neutron than the parent has. Positron decay may be interpreted as a process of transformation of a proton into a neutron accompanied by emission of a positron and a neutrino. This is a transformation of the form

$$\boxed{{}_1^1p \rightarrow {}_0^1n + {}_{+1}\beta^0 + \nu} \qquad (31\text{--}14)$$

31-3 ELECTRON DECAY

Electron or β^- decay is represented by

$$_Z^A P \rightarrow \, _{Z+1}^A D + \, _{-1}\beta^0 + \bar{\nu} \tag{31-15}$$

where $_{-1}\beta^0$ is an electron and $\bar{\nu}$ is an antineutrino.* The *antineutrino*, like the neutrino, has no charge or mass. It is the antiparticle of the neutrino; the subtle difference between the two particles will be examined in Chapter 37. The mass-energy conservation relation of this process is

$$(m_p - Zm_e)c^2 = [m_d - (Z+1)m_e]c^2 + m_e c^2 + K_d + K_\beta + K_{\bar{\nu}} \tag{31-16}$$

where $K_{\bar{\nu}}$ is the kinetic energy of the antineutrino.

The Q value for electron decay is

$$\boxed{Q = K_d + K_\beta + K_{\bar{\nu}} = (m_p - m_d)c^2} \tag{31-17}$$

and when $m_p > m_d$, $Q > 0$ and the decay can take place.

Let us analyze the process

$$_6^{15}C \rightarrow \, _7^{15}N + \, _{-1}\beta^0 + \bar{\nu}$$

to see if C-15 is an electron emitter. The mass difference is found from

$$\text{C-15} = m_p = 15.01060 \text{ amu}$$
$$\underline{\text{N-15} = m_d = 15.00011 \text{ amu}}$$
$$m_p - m_d = 0.01049 \text{ amu}$$

and the Q value is

$$Q = 0.0105 \text{ amu} \times 931 \text{ MeV/amu} = 9.8 \text{ MeV}$$

Since Q is positive, the process is possible.

The energy spectrum is a continuous spectrum with a sharp cutoff point, just as in the case of the energy spectrum of positrons shown previously in Figure 31-2. The emitted electrons would be monoenergetic with only one possible value of linear momentum if it were not for the fact that the energy is shared with the antineutrino. The cutoff energy, which is the maximum energy of the emitted electrons, corresponds to the case where $K_\nu = 0$ and $K_\beta = Q$.

Finally, Equation (31-15) shows that the daughter nucleus has $Z + 1$ protons, or one more proton than the parent and one less neutron. This implies a decay of the neutron according to the scheme

$$\boxed{_0^1 n \rightarrow \, _1^1 p + \, _{-1}\beta^0 + \bar{\nu}} \tag{31-18}$$

---

* The first attempt to detect the antineutrino was made at the Hanford Engineering Works in 1953. See F. Reines and C. L. Cowan, "Detection of the Free Neutrino," *Phys. Rev.* **92**, 830 (1953).

31-4 ELECTRON CAPTURE

In some instances, the nucleus can absorb one of the orbital electrons of the atom. Since the electrons most likely to be captured are those closest to the nucleus, it is the electrons in the K shell that are absorbed most often. For this reason, electron capture is commonly called K *capture*. Captures from the L and M shells are also possible, but much less probable.

When K capture occurs, the daughter has a proton number of one less than the parent atom. The remaining electrons rearrange themselves to correspond to the new structure of the atom, and x rays characteristic of the new atomic system are emitted during the process.

The scheme for K capture is

$$_{-1}\beta^0 + {}_Z^A P \rightarrow {}_{Z-1}^A D + \nu \tag{31-19}$$

According to the mass-energy principle, the process is written

$$m_e c^2 + (m_p - Zm_e)c^2 = [m_d - (Z-1)m_e]c^2 + K_d + K_\nu \tag{31-20}$$

with a Q value of

$$Q = K_d + K_\nu = (m_p - m_d)c^2 \tag{31-21}$$

where K_d and K_ν are kinetic energies of the daughter and the neutrino, respectively. For the process to be possible it is necessary that $Q > 0$ or $m_p > m_d$.

The transformation of V-49 into Ti-49 is a classical example of K capture:

$$_{-1}\beta^0 + {}_{23}^{49}V \rightarrow {}_{22}^{49}Ti + \nu \tag{31-22}$$

To show that the process is possible, let us compute the Q value of the reaction. The mass difference is

$$\begin{aligned} \text{V-49} = m_p &= 48.948523 \text{ amu} \\ \text{Ti-49} = m_d &= 48.947867 \text{ amu} \\ \hline m_p - m_d &= 0.000656 \text{ amu} \end{aligned}$$

which represents a Q value of

$$Q = 0.000656 \text{ amu} \times 931 \text{ MeV/amu} = 0.61 \text{ MeV}$$

Since the number of protons in the daughter nucleus is one less than in the parent and the number of neutrons is one more, K capture can be interpreted as

$$_{-1}\beta^0 + {}_1^1 p \rightarrow {}_0^1 n + \nu \tag{31-23}$$

There are only two particles that have been produced, and according to the principle of conservation of linear momentum, they should move off in opposite directions with equal but oppositely directed momenta. This assures us that the neutrinos as well as the daughter nuclides move with definite kinetic energies. The energy spectra of the daughter and the neutrino are monoenergetic.

31-5 GAMMA DECAY

Stable nuclides are usually in the state of least energy or ground state, but they can be excited by particle or photon bombardment. One way an excited nucleus can return to the ground state is by the emission of γ rays. Gamma-ray decay is represented schematically by

$$(^{A}_{Z}M)^* \rightarrow {}^{A}_{Z}M + \gamma \tag{31-24}$$

The star (*) indicates an excited nucleus, and both the daughter and the parent have the same structure of nuclear particles. It is the conservation of linear momentum and the mass-energy conservational principle that govern γ-ray decay. If E^* is the energy associated with the excited state and E is the energy of the ground state, then the γ rays have an energy

$$h\nu = E^* - E \tag{31-25}$$

where ν is the frequency of the emitted γ ray.

If the decaying nucleus is initially at rest, the conservation of linear momentum requires that it must recoil with a momentum p_d equal and oppositely directed to the momentum of the emitted photon, or

$$p_d = \frac{h}{\lambda} = \frac{h\nu}{c} \tag{31-26}$$

where λ and ν are the wavelength and frequency, respectively, of the emitted γ-ray photon.

The nature of absorption of γ rays is different from that of charged particles such as α or β rays. Alpha and beta particles lose their energy by inelastic collisions and slow down until they eventually come to rest. Gamma rays do not "slow down," but as they pass through matter their intensity is reduced according to

$$\boxed{I = I_0 e^{-\mu x}} \tag{31-27}$$

where μ is the absorption coefficient and x is the thickness of the absorbing material. Gamma rays do not have a definite range, but are similar to x rays in that the intensity is reduced through one or a combination of the following processes: photoelectric effect, Compton scattering, or pair production.

31-6 RADIOLOGICAL HEALTH HAZARDS

The fact that nuclear radiation such as γ rays, α particles, β particles, and neutrons can cause damage to the human body is well known. Within one year after the discovery of the x rays by Roentgen, a manufacturer of x ray tubes noticed x ray

Table 31-1 Summary of effects of acute whole-body ionizing radiation doses

EFFECT \ RANGE	100–200 rems	200–600 rems	600–1000 rems	1000–5000 rems	OVER 5000 rems
incidence of vomiting	100 rems: 5% 200 rems: 50%	300 rems: 100%	100%	100%	100%
delay time	3 h	2 h	1 h	$\tfrac{1}{2}$ h	$\tfrac{1}{4}$ h
leading organ	hematopoietic tissue	hematopoietic tissue	hematopoietic tissue	gastrointestinal	central nervous
characteristic signs	moderate leukopenia	severe leukopenia; purpora; hemorrhage; infection; epilation above 300 rems	hematopoietic tissue	diarrhea; fever	convulsions; tremor; ataxia; lethargy
critical period post-exposure	—	4–6 weeks	4–6 weeks	5–14 days	1–48 h
therapy	reassurance; hematologic surveillance	blood transfusions; antibiotics	possible bone marrow transplant	maintenance of electrolyte balance	sedatives
prognosis	excellent	good	guarded	hopeless	hopeless
convalescent period	several weeks	1–12 months	long	—	—
incidence of death	none	0–80% (variable)	80–100% (variable)	90–100%	90–100%
death occurs within	—	2 months	2 months	2 weeks	2 days
cause of death	—	hemorrhage; infection	hemorrhage; infection	circulatory collapse	respiratory failure; brain endema

SOURCE: Taken from class notes compiled by Prof. D. L. Hathway, U.S. Naval Academy.

damage to his hands and sought medical treatment. Becquerel received a radiation burn on his chest from carrying a vial containing radium in his vest pocket.

The harmful effects of nuclear radiations appear to be due to the ionization or excitation of atoms in living cells by the Compton effect, bremsstrahlung, photoelectric effect, and so on. Some of the cell constituents are altered or destroyed by ionization, and some of the products formed may act as poisons. Examples of damage are the breaking up of chromosomes, swelling of the nucleus of a cell or of the entire cell, changes in the permeability of cell membranes, and destruction of cells.

Frequently, normal cell replacement is inhibited due to the inability of cells to reproduce after being subjected to ionizing radiations. The most sensitive cells are those in the bone marrow, lymph glands, lining of the mouth and intestines, reproductive organs, hair follicles, and skin. Liver and kidney tissues are moderately sensitive, while nerve, brain, and muscle tissues are least sensitive.

Since various radiations are absorbed at different rates, a unit of absorbed dose of radiation called the *rad* is used to standardize measurements of exposure. One rad is defined as the absorbed dose of any nuclear radiation that results in the absorption of 0.01 J of energy per kilogram of absorbing material. Since the absorbed dose required to produce a certain effect may be different for various types of radiation, this difference is expressed by the relative biological effectiveness (RBE) of the particular nuclear radiation. The RBE is defined as the ratio of the absorbed dose in γ rays that will produce the same effect. Finally, a dose unit for biological effect is the *rem* (roentgen equivalent man), which is the dose in rads multiplied by the RBE.

The RBE for x rays or γ rays is approximately one, as is the RBE for β particles. The RBE for α particles from radioactive sources varies from 10 to 20, depending on the energy of the α particle and the tissue being irradiated. A dose of α radiation may be 10 to 20 times more damaging than the same dose of β radiation. For neutrons the RBE varies from 1 to 10, depending on the neutron energy and the tissue.

Table 31-1 compares the effects of various doses of ionizing radiation.

PROBLEMS

31-1 The nucleus $^{222}_{86}$Rn emits an α particle with a speed of 1.6×10^7 m/sec. Compute the momentum and kinetic energy of (a) the α particle, and (b) the recoil nucleus. (c) What is the speed of the recoil nucleus?

31-2 Find the separation energy of (a) the least tightly bound neutron in O-16, and (b) the least tightly bound proton in the same nucleus. (c) Determine the binding energy of O-16 and compare this with the separation energies in (a) and (b).

31-3 The nuclide $^{60}_{27}$Co experiences β^- decay, and $^{60}_{28}$Ni in an excited state is produced. What is the recoil energy of Ni-60 when a 1.30-MeV γ ray is emitted?

31-4 Show whether $^9_5$B is an α emitter or not.

31-5 What are the possible decay modes of $^{126}_{53}$I?

31-6 The nuclide $^{227}_{92}$U is an α emitter. (a) Into what nuclide will it decay? (b) Determine the kinetic energy and momentum of the α particle. (c) What is the recoil energy of the daughter nucleus?

31-7 Radium-226 emits α particles of 4.78 MeV and 4.60 MeV. Show the schematics of the decays in an energy level diagram and determine the energies of the γ rays emitted after the α decay.

31-8 (a) Determine the Q value of the β^+ decay of $^{13}_7$N.
(b) What is the maximum energy of the emitted positron?

31-9 Analyze the stability of the following nuclides: $^5_3$Li, $^8_3$Li, and $^8_5$B.

31-10 The nuclide $^{219}_{86}$Rn emits four groups of α particles of energies 6.817, 6.551, 6.423, and 6.211 MeV. It also emits γ rays of energies 0.128, 0.210, 0.267, 0.340, 0.392, and 0.605 MeV. Construct and label an energy level diagram for these processes.

31-11 Prove that $^{80}_{35}$Br may decay by β^+, β^-, or electron capture by computing the Q value for each process.

31-12 The range and energy of an α particle in air are related by $R = 0.318E^{3/2}$, where E is in million electron volts and R is in meters. If Po-212 emits α particles with ranges 11.2, 11.0, 9.58, and 8.50, what are the corresponding energies?

31-13 (a) Prove that $^7_4$Be decays by electron capture.
(b) Determine the momentum and energy of the neutrino and the daughter nucleus $^7_3$Li.

31-14 What will be the resulting nucleus when $^{238}_{92}$U experiences successively eight α decays and six β decays?

31-15 Compute the thickness of lead required to reduce the intensity of γ rays from a given radioisotope to one-tenth of the original intensity if the linear coefficient of absorption is 46 m^{-1}.

RECOMMENDED READING

COWAN, C. L., REINES, F., and HARRISON, F. B., "Upper Limit on the Neutrino Magnetic Moment," *Phys. Rev.* **96**, 5, 1294 (1954).

CURIE, P., and CURIE, M., "Sur une substance nouvelle radio-active, contenue dans la pechblende," *Compt. Rend.* **127**, 175 (1898).

CURIE, P., CURIE, M., and BEMONT, G., "Sur une nouvelle substance fortement radioactive, contenue dans la pechblende," *Compt. Rend.* **127,** 1215 (1898).

MANN, W., and GARFINKEL, S., *Radioactivity and Its Measurement*, Van Nostrand Reinhold, New York, 1966.

MORRISON, Philip, "The Neutrino," *Sci. Am.*, January 1956.

PERLMAN, I., "Alpha Radioactivity and the Stability of Heavy Nuclei," *Nucleonics* **7,** No. 2, 3 (1949).

ROWLAND, S., "Methods for Measuring Very Long and Very Short Half-Lives," *Nucleonics* **3,** No. 3, 2 (1945).

RUTHERFORD, E., *Phil. Mag.* **47,** 109 (1899).

RUTHERFORD, E., *Nature* **65,** 366 (1903).

RUTHERFORD, E., *Phil. Mag.* **12,** 348 (1906).

RUTHERFORD, E., *Radioactivity*, 2nd ed., Cambridge University Press, London, 1909.

32
Fission and Fusion

Enrico Fermi
(1901–1954)

Fermi was born in Rome and received his doctorate from the University of Pisa in 1922. He was the first to bring about nuclear transformation of heavy elements by neutron bombardment, and for his work with artificial radioisotopes and nuclear reactions, he was awarded the 1938 Nobel Prize in physics. In 1939 he came to the United States, where he became a citizen in 1945. At the University of Chicago in 1942, he constructed the first self-sustaining, nuclear fission chain reaction. The artificial element fermium was named in his honor in 1955.

32–1 FISSION
32–2 FUSION
32–3 NUCLEAR REACTORS

32–1 FISSION

In experiments performed by ENRICO FERMI, EMILIO SEGRÈ, and co-workers in 1934, the bombardment of uranium by neutrons produced several β-ray activities with different half-lives. Since uranium decays by α-particle emission with a very long half-life, it was assumed that transuranic elements ($Z > 92$) were being formed. However, OTTO HAHN and FRITZ STRASSMAN, at the Kaiser Wilhelm Institute in Berlin in 1938, showed by very careful chemical analysis that one of the radioactive elements produced when uranium is bombarded by neutrons is an isotope of barium, $^{141}_{56}$Ba. In 1939 OTTO FRISCH and LISA MEITNER suggested that the uranium was undergoing a *nuclear fission* process, in which the nucleus was being fragmented into particles of comparable size, $^{141}_{56}$Ba and $^{92}_{36}$Kr.

It was the discovery that neutrons can be slowed down by elastic collision with nuclei that led Fermi and his collaborators to experiments in which neutrons produced nuclear reactions. As previously discussed, the slow neutrons have a high cross section for capture (about 550 barns for thermal neutrons incident on $^{235}_{92}$U) because they spend a greater length of time in the field of the nucleus. Slow neutrons are most often produced by moderation of the neutron beam by materials rich in hydrogen, such as water or paraffin.

The schematic equation for the fission process is

$$^{235}_{92}\text{U} + n \rightarrow \, ^{236}_{92}\text{U}^* \rightarrow X + Y + \text{neutrons} \tag{32-1}$$

where n is a slow neutron, $^{236}_{92}$U* is a highly unstable isotope, and X and Y are the fission fragments. The fragments are not uniquely determined, because there are various combinations of fragments possible and a number of neutrons are given off. Typical of these fission reactions are

$$^{1}_{0}n + \, ^{235}_{92}\text{U} \rightarrow \, ^{236}_{92}\text{U}^* \rightarrow \, ^{144}_{56}\text{Ba} + \, ^{89}_{36}\text{Kr} + 3\,^{1}_{0}n + Q \tag{32-2}$$

and

$$^{1}_{0}n + \, ^{235}_{92}\text{U} \rightarrow \, ^{236}_{92}\text{U}^* \rightarrow \, ^{140}_{54}\text{Xe} + \, ^{94}_{38}\text{Sr} + 2\,^{1}_{0}n + Q \tag{32-3}$$

where Q is the energy released in the reaction. The β-ray activity observed by Fermi and his collaborators came from the radioactive decay of fragments such as

$$^{140}_{54}Xe \xrightarrow{\beta^-} {}^{140}_{55}Cs \xrightarrow{\beta^-} {}^{140}_{56}Ba \xrightarrow{\beta^-} {}^{140}_{57}La \xrightarrow{\beta^-} {}^{140}_{58}Ce \qquad (32\text{--}4)$$

and

$$^{94}_{38}Sr \xrightarrow{\beta^-} {}^{94}_{39}Y \xrightarrow{\beta^-} {}^{94}_{40}Zr$$

from the fission reaction in Equation (32–2).

As the neutron is captured by the $^{235}_{92}U$ nucleus, the energy is shared among the degrees of freedom of the particles within the nucleus. Then, like a liquid drop, the nucleus undergoes deformations until Coulomb repulsion distorts the nucleus and its fragments into particles as shown in Figure 32–1. The distribution of fission fragments produced by slow neutrons captured by uranium is shown in Figure 32–2.

When compared with the stability curve in Figure 26–2, all of the fragments fall to the left of the curve. Because these fragments have an excess number of neutrons, they are unstable. These excited fragments give off a number of *prompt neutrons* within about 10^{-13} sec. In addition to these prompt neutrons, a small number of *delayed neutrons* with half-lives of the order of seconds are also given off. These few delayed neutrons will be seen to be important in controlling the rate of reactions in a reactor.

The fission process releases a rather large amount of energy, and a further smaller amount of energy is released in the radioactive decay of fragments. This

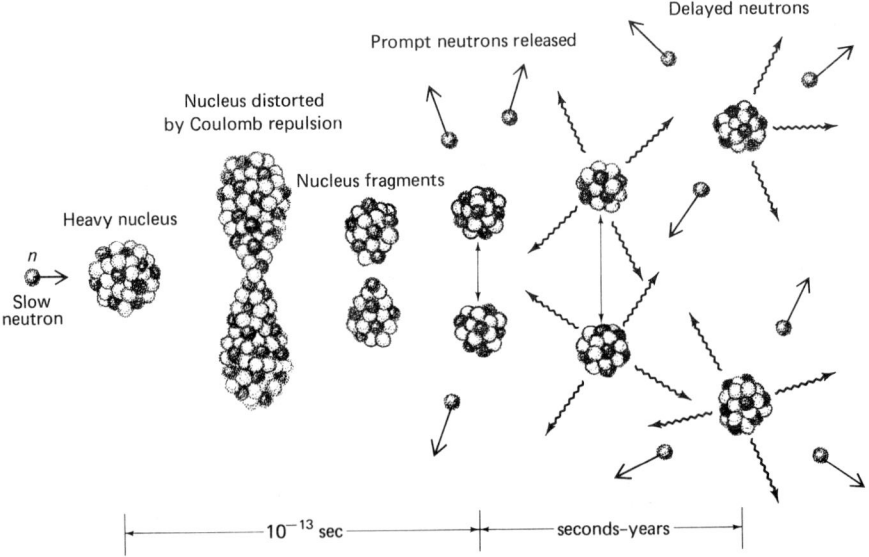

Figure 32–1 Nuclear fission of a heavy nucleus by a slow neutron.

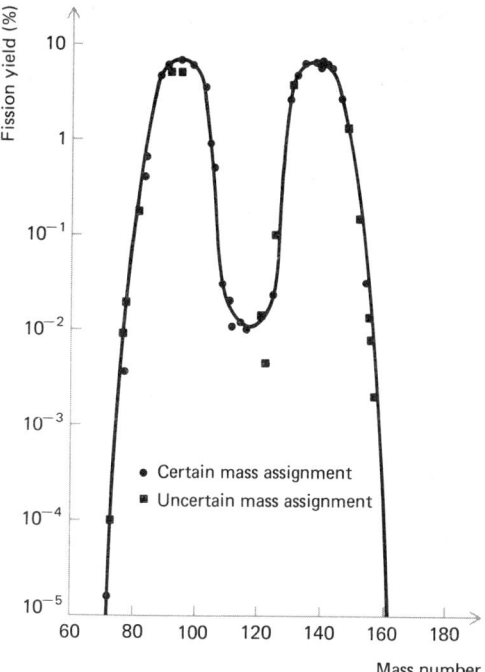

Figure 32–2 Distribution of fission fragments from U-235 bombarded by slow neutrons. [From *Rev. Mod. Phys.* **18**, 539 (1946).]

energy released appears as kinetic energy of the fragments and as kinetic energy of the electrons, photons, neutrinos, and neutrons released.

As a rough calculation of the energy released in nuclear fission, consider the fission reaction in Equations (32–3) and (32–4). The unstable nucleus $^{236}_{92}U^*$ splits and produces as the end products $^{140}_{58}Ce$, $^{94}_{40}Zr$, six β particles, and two neutrons. The isotopic masses before and after fission are as follows:

BEFORE FISSION	AFTER FISSION
$^{235}_{92}U$ = 235.0439 amu	$^{140}_{58}Ce$ = 139.9054 amu
$^{1}_{0}n$ = 1.0087 amu	$^{94}_{40}Zr$ = 93.9036 amu
236.0526 amu	$2^{1}_{0}n$ = 2.0173 amu
	$6\beta^-$ = 0.0033 amu
	235.8296 amu

The difference in masses is 0.223 amu, which represents a very large Q value of 0.223 amu × 931 MeV/amu = 208 MeV. The energy released in an α-particle disintegration is only around 5 MeV, and the energy released in a chemical combustion process is of the order of 4 eV!

Because low-energy neutrons can initiate fission in $^{235}_{92}U$, the neutrons given off in a fission process can be used to initiate additional fissions, which then release additional energy. In an "atomic pile" constructed by Fermi at the University of Chicago, the first controlled self-sustaining chain reaction was put into operation on December 2, 1942.

Fission can be initiated by bombardment of heavy nuclei by particles other than neutrons, including protons, α particles, and γ rays. Also, nuclear fission can be accomplished in lighter elements. As a typical example, 60-MeV protons from a cyclotron were used to produce the following fission reaction:

$$^{63}_{29}Cu + ^{1}_{1}p \rightarrow ^{38}_{17}Cl + ^{25}_{13}Al + ^{1}_{0}n \tag{32-5}$$

32-2 FUSION

The fact that the sun did not seem to cool down over a period of many years, despite the tremendous quantities of energy it produced, was a puzzle until the discovery of nuclear reactions. In 1938 HANS BETHE suggested that a nuclear reaction in which two nuclei came together to form a single heavier nucleus was responsible for the large quantities of energy released by the sun and other stars. These reactions were known as nuclear *fusion*.

The binding energy curve in Figure 26–6 shows that a combination of two light nuclei on the steep portion of the curve forms a heavier nucleus, which has a larger binding energy per nucleon than either of the lighter nuclei. Because there is a greater binding energy, there is a decrease in nuclear mass, which results in a positive Q and a release in energy.

At the high temperatures and pressures found in the interior of stars, molecules dissociate into atoms and the atoms are highly ionized to form a very hot and very active *plasma*. Under these conditions Coulomb barriers are overcome, and the atoms enter into nuclear fusion reactions and liberate large quantities of energy. Since hydrogen is the most abundant element, Bethe suggested the following carbon cycle* as one of the most important nuclear reactions for release of energy by fusion:

(a) $^{1}_{1}H + ^{12}_{6}C \rightarrow ^{13}_{7}N + \gamma$

(b) $^{13}_{7}N \rightarrow ^{13}_{6}C + ^{0}_{+1}e + \nu$

(c) $^{1}_{1}H + ^{13}_{6}C \rightarrow ^{14}_{7}N + \gamma$

(d) $^{1}_{1}H + ^{14}_{7}N \rightarrow ^{15}_{8}O + \gamma$ \hfill (32–6)

(e) $^{15}_{8}O \rightarrow ^{15}_{7}N + ^{0}_{+1}e + \nu$

(f) $^{1}_{1}H + ^{15}_{7}N \rightarrow ^{12}_{6}C + ^{4}_{2}He$

By checking the nuclei on both sides of the equations, we see that C-12 acts like a catalyst in that it was part of the initial reaction and a product in the final reaction.

* See Hans Bethe, "Carbon Cycle," *Phys. Rev.* **55**, 434 (1939).

Notice also that four hydrogen nuclei (protons) are consumed and three γ rays, two neutrinos, two positrons, and a helium nucleus are formed. The reaction cycle is essentially the reaction.

$$4{}_1^1H \to {}_2^4He + 2{}_{+1}^0e + 2\nu + Q$$

in which the mass difference is

$$\begin{array}{r} 4{}_1^1H = 4.031300 \text{ amu} \quad {}_2^4He = 4.002603 \text{ amu} \\ 2{}_{+1}^0e = 0.001098 \text{ amu} \\ \hline 4.003701 \text{ amu} \end{array}$$

The mass difference $4.031300 - 4.003701 = 0.0276$ amu represents a Q value of 0.0267 amu $\times$ 931 MeV/amu $= 25.7$ MeV, which is the energy available for the γ rays, neutrinos, and product particles from the reaction.

Another important fusion cycle involves protons. The proton–proton cycle is

(a) $\quad {}_1^1H + {}_1^1H \to {}_1^2H + {}_{+1}^0e + \nu$
(b) $\quad {}_1^2H + {}_1^1H \to {}_2^3He + \gamma$ \hfill (32-7)
(c) $\quad {}_2^3He + {}_2^3He \to {}_2^4He + 2{}_1^1H$

In this cycle of reactions, fusion reactions (a) and (b) must occur twice to produce the He-3 isotope in (c). The net result of these reactions is that four hydrogen nuclei (protons) are combined to produce an α particle, two positrons, and two γ rays. The energy released in this cycle is 26.2 MeV. Other typical fusion reactions are shown in Figure 32–3.

The use of nuclear fusion reactions for a controlled source of energy is feasible, but it presents some rather severe engineering problems. Not the least of these is the design of a "container" in which a very hot plasma can be contained under high pressure to initiate a fusion reaction. Since almost any container in the ordinary sense of the word would melt in the presence of a plasma, attempts are being made to contain and control plasmas trapped in a specially shaped magnetic field. By increasing the field and changing the shape of the field, it is hoped that the plasma in this "magnetic bottle" can be raised to the required temperature and pressure for fusion reactions.

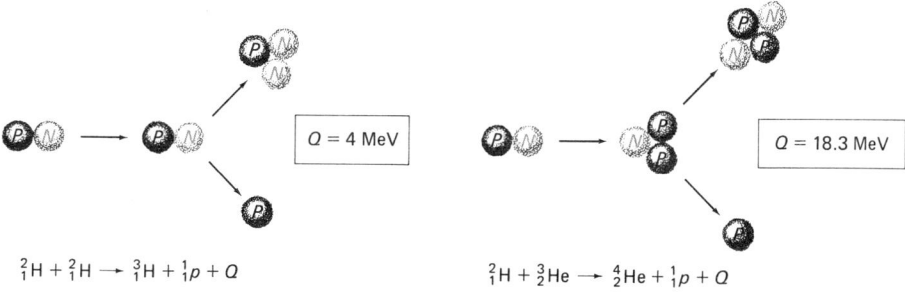

Figure 32–3 Typical examples of fusion reactions.

Another interesting development has been the use of a very-high-powered (peak power of 50 GW) neodynium-glass pulsed laser to produce a plasma temperature of 7–10 million °K, in which fusion neutrons are produced from a target of solid deuterium.

32–3 NUCLEAR REACTORS

Notice in Equations (32–2) and (32–3) that once fission occurs, neutrons are produced which can start other fission reactions, which in turn produce more neutrons for more fission reactions, and so on for a chain reaction. The purpose of the *nuclear reactor* is to initiate nuclear fission reactions, to control these reactions, and to extract the energy produced.

The control of neutrons is the key to the function of the reactor. For each thermal neutron absorbed in U-235 fission, an average of 2.5 neutrons are released. Over 99% of these are prompt neutrons emitted within $\sim 10^{-13}$ sec; and they have energies of 1 or 3 MeV and speeds of 10,000 m/sec, far above the velocity of thermal neutrons. Only a few percent of these neutrons are effective in initiating more fission, because competing processes render them ineffective. Some of the neutrons are simply scattered elastically, some escape the region of interaction, and some are absorbed into nuclei without causing fission.

A substance known as a *moderator* must be used to control the neutron flux. Since protons have a mass about equal to the mass of a neutron, it would seem that hydrogen would be efficient in slowing down neutrons by exchange of energy in elastic collisions. However, hydrogen has a high cross section for the reaction ^1_1H (n, γ) ^2_1H, which causes the neutrons to be removed from circulation. A good moderator should contain a light element that can slow down neutrons by an exchange of energy, but it must also have a small cross section for other types of reactions that will absorb and use up available neutrons.

For a given moderator material there is a neutron cross section for scattering-type collisions Σ_s and a cross section for absorption of neutrons Σ_a. Another useful parameter is ξ, the average of the natural logarithm ratio of the energy lost in one collision.

$$\xi = \left(-\ln \frac{E}{E_0}\right)_{av} \quad (32\text{–}8)$$

where E and E_0 represent the energies before and after the collision.

For values of atomic mass $A > 10$, this can be approximated by

$$\xi \cong \frac{2}{A + \frac{2}{3}} \quad (32\text{–}9)$$

A measure of the capacity for the nuclei in a given 1 cm³ of material to slow down neutrons is determined from the effectiveness of a moderator as measured by a parameter known as the moderating ratio, which is given by

$$\boxed{\text{moderating ratio} = \xi \frac{\Sigma_s}{\Sigma_a}} \qquad (32\text{--}10)$$

Table 32–1 compares the effectiveness of various materials as moderators.

EXAMPLE 32–1: How many collisions must a 5-MeV neutron undergo with Pb-206 nuclei before it is thermalized?

SOLUTION: For Pb, Equation (32–9) shows that

$$\xi = \frac{2}{A + \frac{2}{3}} = \frac{2}{206 + \frac{2}{3}} = 0.0096$$

Since a thermal neutron has a kinetic energy of $\frac{1}{25}$ eV, the ratio indicating the loss in energy is

$$\frac{5 \times 10^6 \text{ eV}}{\frac{1}{25} \text{ eV}} = 1.25 \times 10^8$$

and

$$\ln(1.25 \times 10^8) = 18.64$$

Since ξ is the average of the logarithm of the ratio of the energy lost in one collision, the average total number of collisions is

$$\frac{18.644}{0.0096} = 1940 \text{ collisions}$$

Table 32–1 Moderating ratios

MATERIAL	MASS NUMBER	ξ	COLLISIONS TO THERMALIZE	MODERATING RATIO
hydrogen, H	1	1.00	18	66
water, H$_2$O	2[a]	0.93	19	67
heavy water, D$_2$O	4[a]	0.51	35	21,000
helium (atm. pressure), He	4	0.53	43	94
beryllium, Be	9	0.21	86	160
carbon, C	12	0.16	114	170
uranium, U	238	0.0084	2172	0.009

[a] Only the hydrogen and deuterium are effective in moderating the neutrons.

Since pure uranium metal contains 99.3% $^{238}_{92}U$, 0.7% $^{235}_{92}U$, and a trace of 0.006% $^{234}_{92}U$, some fast neutrons are absorbed by $^{238}_{92}U$ and some are absorbed by fission products. When the number of neutrons lost or absorbed by means other than fission reactions is greater than the number of neutrons being produced by fission, the self-sustaining chain reaction dies down and stops, and the reactor is said to be *subcritical*. If the number of neutrons available for fission increases with each fission, the power output rises and the reactor is *supercritical*. When the reactions are such that one fission produces one neutron, which initiates a second fission the reactor is *critical* and the chain reactions are self-sustaining and controlled.

To control the criticality of a reactor, *control rods*, usually containing cadmium or boron, which readily absorb neutrons, are inserted into the reactor. When the control rods are inserted all the way into the reactor, so many neutrons are absorbed that the reactor *shuts down*. The control rods are then withdrawn until enough neutrons are present for the reactor to start toward supercritical, and then they are reinserted until it is critical and the reactor operates at a fixed power level.

Most neutrons produced in fission are prompt neutrons, and they are thermalized very quickly. Mechanical control of the reactor would be a difficult problem if it were not for the small percentage of delayed neutrons given off by the unstable fragments resulting from fission.

An early model pile reactor is shown in Figure 32–4. It contains about 52 tons

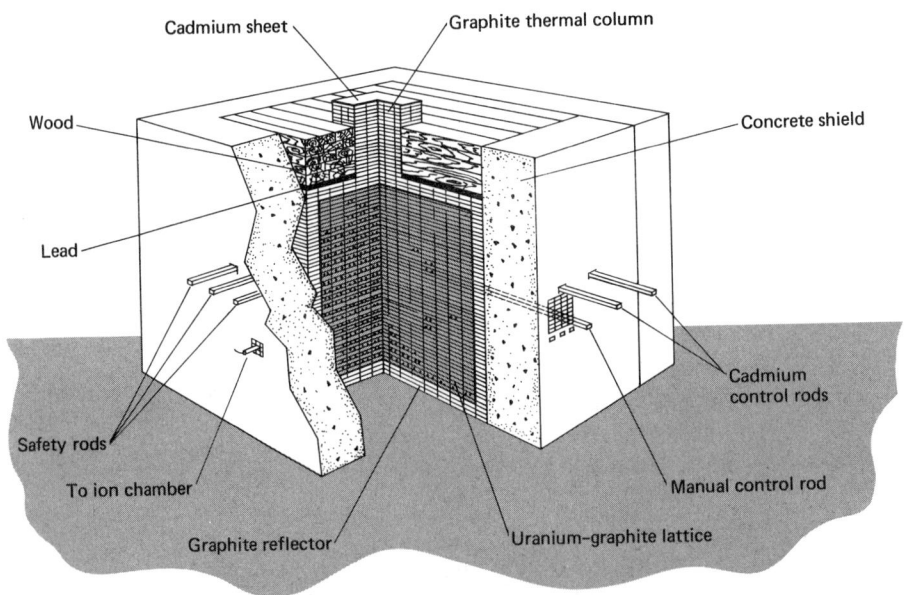

Figure 32–4 Chicago pile reactor located at the Argonne National Laboratory. This early type of experimental reactor consisted of a lattice composed of uranium interspersed with a graphite moderator.

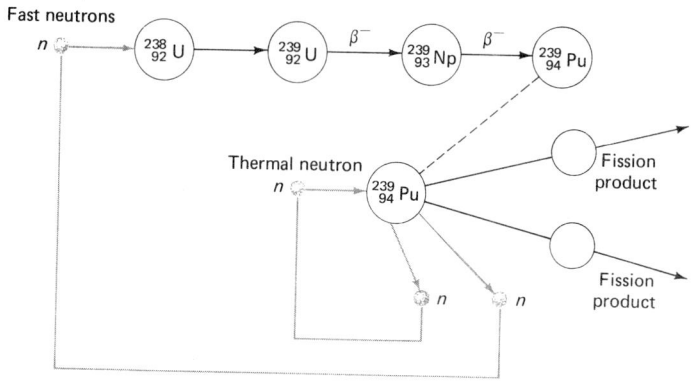

Figure 32–5 Fast neutrons can be used to produce a thermal fissionable isotope from the more common uranium isotope $^{238}_{92}U$ for use in a breeder reactor.

of uranium in the form of lumps interspersed with about 472 tons of graphite, which acts as the moderating substance. There are five control rods made of 17-ft-long brass strips covered with cadmium. As an automatic safety feature, there are three rods that have weights on them to pull the rods into the reactor in case of an electrical power failure.

Because $^{235}_{92}U$ is the only naturally occurring isotope that undergoes fission by thermal neutrons, within a few decades reactors depending on this fuel may be uneconomical to operate. However, *breeder reactors* use fast neutrons, which are captured by $^{236}_{92}U$, which by consecutive β^- decays decay into $^{239}_{94}Pu$, which is fissionable by thermal neutrons. The scheme for the breeder reactors is illustrated in Figure 32–5. Notice that two neutrons in excess of those captured and lost by other processes must be produced to maintain the reaction. These reactors are designed so that they make available more fuel than they consume. Within a few years, a breeder reactor can make available double the original quantity of fuel.

PROBLEMS

32–1 A thermal neutron initiates fission of $^{235}_{92}U$, which splits according to the following fission chain:

$$^1_0n + {}^{235}_{92}U \longrightarrow {}^{98}_{40}Zr + {}^{136}_{52}Te + 2{}^1_0n$$

$$^{92}_{40}Z \xrightarrow{\beta^-} {}^{98}_{41}Cb \xrightarrow{\beta^-} {}^{98}_{42}Mo$$

and

$$^{136}_{52}Te \xrightarrow{\beta^-} {}^{136}_{53}I \xrightarrow{\beta^-} {}^{136}_{59}Xe$$

From the difference in nuclear masses involved, determine the energy released in this fission process.

32–2 From Figure 26–6, estimate the energy released when a nucleus of mass number $A = 240$ splits into three equal fragments.

32–3 At what rate will $^{235}_{92}U$ metal be consumed by a reactor operating at an output of 100 kW?

32–4 It is thought that some stars produce energy by the fusion of three α particles to form a $^{12}_{6}C$ nucleus. What is the energy released in such a fusion?

32–5 When coal is burned, 7500 kcal/kg of heat are released. How many tons of coal have the fuel value equivalent to 1 kg of $^{235}_{92}U$?

32–6 From the binding energy curve in Figure 26–6, what is the Q value for fusion of (a) $^{1}_{1}H + ^{2}_{1}H$, (b) $^{14}_{7}N + ^{60}_{27}Co$, and (c) $^{60}_{27}Co + ^{200}_{80}Hg$.

32–7 For a 2.0-MeV neutron to be slowed to an energy of $\frac{1}{25}$ eV on the average, how many collisions would be required if the neutron moderator were (a) carbon, (b) iron, or (c) gold?

32–8 If the absorption cross section for cadmium is 2450 and the scattering cross section is 7, what is the moderating ratio?

32–9 What is the moderating ratio for nickel if it has an absorption cross section of 4.8 and a scattering cross section of 17.5? How many collisions will be required to thermalize a 1.0-MeV neutron?

32–10 What is the value of ξ and the moderating ratio for beryllium carbide, Be_2C.

32–11 Determine the Q value for the fission reaction

$$^{63}_{29}Cu + ^{1}_{1}H \rightarrow ^{24}_{11}Na + ^{39}_{19}K + ^{1}_{0}n$$

32–12 If an atomic submarine consumes 6×10^4 kW of energy continuously, what is the mass of $^{235}_{92}U$ used per day in operating the submarine?

32–13 Show all of the steps in the proton–proton cycle in Equation (32–7), and calculate the energy released in each step.

RECOMMENDED READING

BUMP, T. R., "A Third Generation of Breeder Reactors," *Sci. Am.*, May 1967.

LEACHMAN, R. P., "Nuclear Fission," *Sci. Am.*, August 1965.

POST, Richard F., "Fusion Power," *Sci. Am.*, December 1957.

SEABORG, Glenn T., and BLOOM, Justin L., "Fast Breeder Reactor," *Sci. Am.*, November 1970.

SEGRÈ, EMILIO, *Enrico Fermi, Physicist*, University of Chicago Press, Chicago, 1970.

WEINBERG, Alvin M., "Power Reactors," *Sci. Am.*, December 1954.

33
Particle Detectors

Johannes (Hans) Wilhelm Geiger
(1882–1945)

Geiger was born in Neustadt, Germany, and studied at the universities of Munich and Erlangen. At Cavendish Laboratory, he worked with Ernest Rutherford and E. Marsden (1906–1912), where they confirmed the fact that atoms consist of a small nucleus surrounded by electrons rather than a sphere of distributed positive and negative charges. Geiger also determined the radioactive period of radium and, in 1908, perfected the counter for β or cosmic-ray particles (Geiger–Müller counter). He is the recipient of the Duddell medal and the Hughes medal.

33–1 PROPERTIES OF PARTICLES
33–2 NUCLEAR EMULSIONS
33–3 TRACK CHAMBERS
33–4 ELECTRONIC DETECTORS

33–1 PROPERTIES OF PARTICLES

As the nineteenth century drew to a close, scientists pictured the material universe as being constructed in terms of atoms that were unique and distinct from the elements. These different atoms were considered to account for all of the chemical reactions and for the vast complexity of nature, although the reasons for differences between atoms was not well understood at that time. As we have seen, an important step toward this understanding occurred in 1897, when Thomson discovered a subatomic unit of the atom—the electron. From the pioneering work of Thomson and Millikan, two new concepts were established:

1. Distinct and unique particles exist that are smaller than chemical atoms and that are constituents of these atoms.
2. Definite properties, such as charge and mass, can be assigned to these particles to identify them.

From this rudimentary beginning, a long list of elementary particles has evolved, each of which is described by a list of numbers giving the observable characteristics of that particle. These characteristics have been compiled from a variety of experiments, ranging from simple one-man experiments to giant particle accelerators to astronomical observations of the cosmos. The list for each particle begins with the rest mass of the particle and then its electrical charge, the value of an intrinsic angular momentum associated with the particle, the values of other sorts of charge that determine how the particles will interact with other particles, and so on. These numbers and algebraic signs are the quantum numbers of the elementary particles and constitute the true description of each particle.

Since the quantum numbers are the physically observable quantities, a certain physical reality is ascribed to these numbers, and each is treated as having a definite physical consequence when the particle interacts with other particles. It should be kept in mind, however, that the raw data of every physical observation are just intervals in space and in time. The underlying concepts of space and time enter into any discipline of physics. Let us now investigate a variety of methods of detecting particles.

33–2 NUCLEAR EMULSIONS

To measure lengths and times associated with successive positions of a particle so that its trajectory may be determined, it is necessary to locate the particle accurately with some measuring device. The particles may be of the order of 10^{-15} m in diameter and quite unobservable by any ordinary means, including the most powerful microscopes. To overcome this limitation, the particle must be made to affect a much larger region than its own volume in such a way that changes in the larger region can be observed.

An electron with dimensions of perhaps 10^{-10} m may be made to pass through an undeveloped photographic plate. In doing so, its path will take it through many grains of silver halide, each of which has dimensions of the order of 10^{-6} m. Excitation of tiny regions in each grain produced by the passage of the electron will then result in the blackening of that entire grain when the plate is developed. An amplification of about 10,000 has taken place in the size of the region being dealt with. Now, although the electron itself is far too small to be observable, the blackened silver grain is easily observed under a microscope. A string of such grains then defines the trajectory of the electron in passing through the plate. Photographic emulsions are now in use that can respond in this way to the passage in any of the known charged particles. It is the electrical charge on the particle that excites the grains in passing through it.

The sensitive emulsions in the majority of films are extremely thin. The first satisfactory emulsions, containing about 80% AgBr (silver bromide), were prepared by C. F. POWELL and his co-workers at the University of Bristol, England, in the period 1935–1940. Since the emulsions are very thin (about 1 mm thick), they can be stacked in layers as shown in Figure 33–1. In the figure, an incoming beam of high-energy particles *AB* produces an event at *C*, and the tracks *CD* and *CF* of two particles that have been produced can be studied by microscope observation. Measurements are made of the ranges, track densities, and direction of travel of the particles born at *C*. The years following World War II saw the production of emulsions highly sensitive to all kinds of charged particles and the use of large blocks of emulsions many inches thick, like the one shown in Figure 33–1. The complex paths produced by high-energy particles are then captured in detail in the thick block of emulsions. The density and range

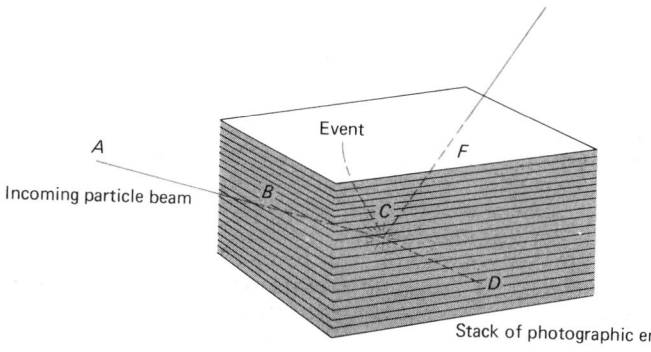

Figure 33–1 Nuclear photographic emulsions are stacked to provide a thick target for the detection of high-energy particles.

of the tracks depend on the energy and charge of the particle. As an example, in an average emulsion a 55-MeV α particle can travel about 1 mm, while the same range can be produced by an electron of only 0.7 MeV. The differences in track densities due to different energies and charges on particles are illustrated in Figure 33–2.

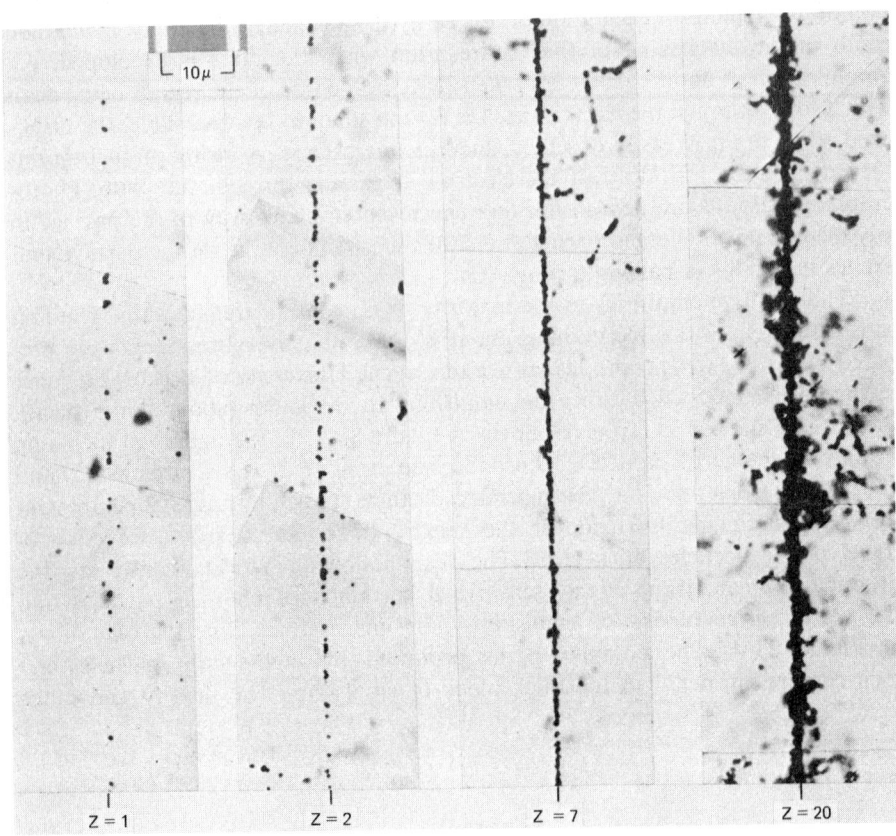

Figure 33–2 Photomicrograph of primary cosmic-ray nuclei registered in the sensitive photographic layer of a nuclear research emulsion suspended from a balloon near the top of the atmosphere. From left to right, the tracks are those of hydrogen, helium, nitrogen, and calcium nuclei traveling at speeds close to that of light. The blackening of the developed silver grains which constitutes each track is made possible by the ionization produced by the cosmic-ray nuclei along their paths. The letter Z denotes the atomic number, that is, the number of constituent protons in each nucleus. (Courtesy of Nathan Seeman, Laboratory for Cosmic Ray Physics, Naval Research Laboratory.)

33–3 TRACK CHAMBERS

Cloud Chambers. The first observations of particle tracks were made in 1911 by C. T. R. WILSON, who studied the condensation of liquids from supersaturated vapors onto electrically charged gas ions. This spectacular development was achieved by constructing a glass chamber which contained air and some condensable, volatile liquid such as alcohol. The chamber was placed under a moderate compression and allowed to come to thermal equilibrium with its surroundings. Supersaturation was then produced by a sudden adiabatic expansion of the gas when a piston in the bottom of the chamber was quickly released. Electrically charged ions present under these conditions act as condensation nuclei for the vapor, and droplets of fog grow on them. If these ions have been produced by the passage of a charged particle through the gas, a track is formed in the chamber. By a suitable arrangement of flash lamps, mirrors, and a camera, stereoscopic photographs can be made of the tracks.

These chambers, designed by Wilson, are called *cloud chambers*. Wilson described the origin of his concept of the cloud chamber in the following terms:

"In September 1894 I spent a few weeks in the observatory which then existed on the summit of Ben Nevis, the highest of the Scottish hills. The wonderful optical phenomena shown when the sun shone on the clouds surrounding the hill top and specially the coloured rings surrounding the sun or surrounding the shadow cast by the hilltop or observer on mist or cloud, greatly excited my interest and made me to imitate them in the laboratory.

"At the beginning of 1895 I made some experiments for this purpose—making clouds by expansion of moist air after the manner of Coulier and Aitken. Almost immediately I came across something which promised to be of more interest than the optical phenomena which I had intended to study."*

A diagram of an early cloud chamber is shown in Figure 33–3.

Bubble Chambers. The first track chamber for elementary particles was able to show the trajectories of β and α particles and of the fast electrons produced in the gas by γ rays. The technique, modified and developed in many ingenious ways, is still one of the prime tools of particle detection. One of the modifications consists of the production of tracks in a gas that does not need to undergo a preliminary expansion and so is continuously sensitive to the passage of swift particles. In these chambers, a large temperature gradient is set up in a gas-alcohol mixture so that in a given region the gas is supersaturated at all times and is continuously renewed by diffusion of fresh vapor from other parts of the chamber.

The most significant development in the observation of elementary particles, the *bubble chamber*, invented by D. A. GLASER in 1952, might be considered as having evolved from the cloud chamber. The necessary low density of the gas in

---

* C. T. R. Wilson, Nobel Lecture, December 12, 1927.

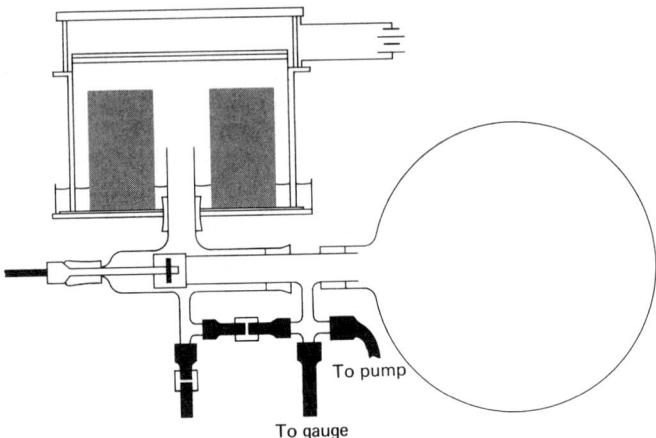

Figure 33-3 The C. T. R. Wilson cloud chamber.

the cloud chambers means that most particles escape from the chamber before depositing sufficient detail in the form of systems of tracks, and the probability that a given particle will produce an interesting event is low. The substitution of a superheated liquid for the supersaturated gas means an increase of more than 1000-fold in the density of the chamber. This increased target density in the energy-absorbing region makes it possible to investigate high-energy particle interactions in great detail.

The bubble chamber operates by having a volatile liquid under compression and in thermal equilibrium with its surroundings. Just prior to the passage of, say, a beam of fast protons from an accelerator through the chamber, a small, adiabatic expansion is allowed to take place. This expansion causes the liquid to become superheated to the point of boiling. Ions then produced by the passage of charged particles supply discontinuities in the liquid, and small bubbles are formed and grow along the trail of the discontinuities. When the bubbles have grown sufficiently large, a flash lamp is triggered, and stereoscopic photographs of the chamber are taken. Reproductions of these photographs permit many accurate measurements to be made on the events produced in the chamber.

The chambers produced to date extend in size up to 5 m in diameter, and much larger ones are under construction. The aim is to construct chambers that are large enough when used with any particular accelerator of elementary particles to contain all of the detailed tracks produced in the beam of that accelerator. As larger and larger accelerators are constructed, bubble chambers become correspondingly larger. The 80-in. liquid hydrogen bubble chamber at the Brookhaven National Laboratory is shown in Figure 33-4. The stainless steel chamber contains 900 liters of liquid hydrogen at a temperature of $-259\,°C$. The magnet, which requires 4 million watts of electrical power, provides a uniform magnetic field throughout

CHAPTER 33: PARTICLE DETECTORS • **429**

Figure 33–4   The 80-in. liquid hydrogen bubble chamber at the Brookhaven National Laboratory. In the lower portion of the picture, a technician is seen removing one of three automatic cameras that photograph the particle tracks. (Courtesy of Brookhaven National Laboratory.)

the chamber of 2.0 Wb/m². A schematic cross section is illustrated in Figure 33-5. Finally, Figure 33-6 shows several events that have been photographed in a bubble chamber.

Spark Chambers. A relatively new approach to the detection of particle events is the *spark chamber*. Essentially, a spark chamber is a set of conducting plates, usually aluminum, alternately connected to a source of high dc voltage (about 20 kV). Figure 33-7 shows a schematic cross section of a typical spark chamber. Sudden application of very high voltages to alternate plates, while the others are left at ground potential, results in very high electrical fields across the gaps. Electrical breakdown then occurs along the trails of ions, so the trajectory of a given particle through the system is marked by a series of sparks. The spark trails are photographed stereoscopically and reprojected for measurement in a similar fashion to those used in the bubble chamber and the cloud chamber.

Again, many variations of the spark chamber have been used, from very small ones that record the positions of the sparks electrically rather than optically and that are flown in rockets and balloons, to extremely large spark chambers weighing hundreds of tons. These latter chambers have been used to investigate the interactions of neutrinos, where massive targets must be supplied to offset the extremely low probability that a neutrino will produce an interaction in the equipment.

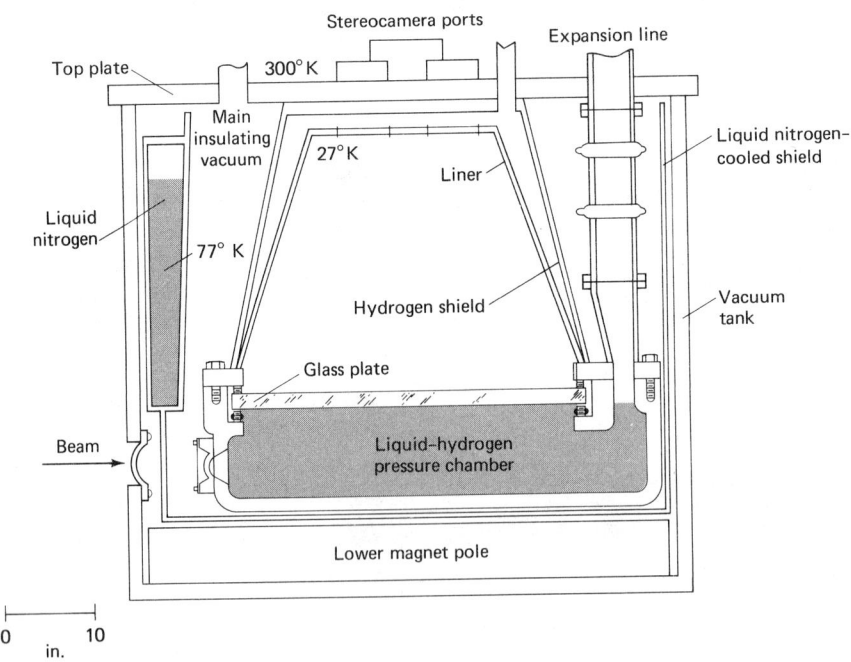

Figure 33-5 Schematic diagram of a typical bubble chamber.

CHAPTER 33: PARTICLE DETECTORS · **431**

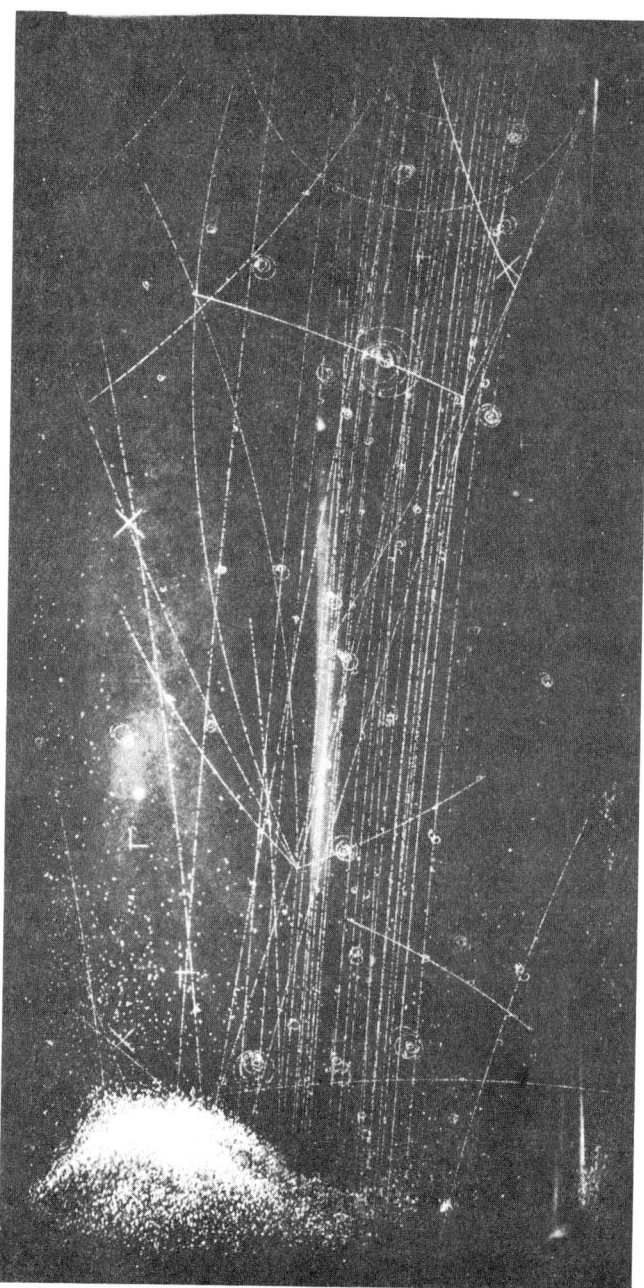

Figure 33-6 Photograph of proton–antiproton annihilation taken in a 20-in. bubble chamber. Most of the small spiral paths in the picture are made by electrons. (Courtesy of Brookhaven National Laboratory.)

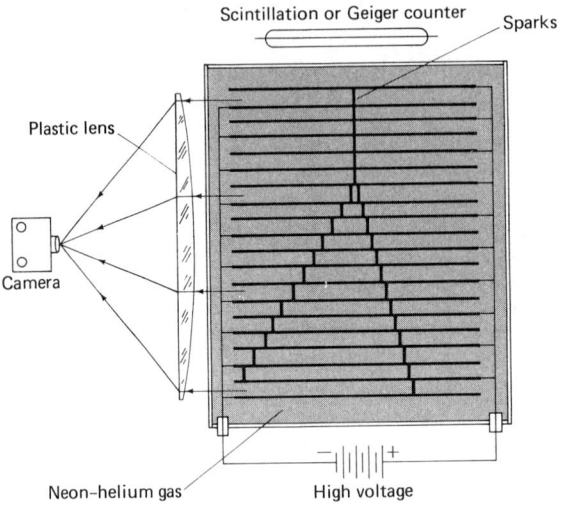

Figure 33-7   A cross-sectional view of a typical spark chamber.

Figure 33-8   Spark-chamber cosmic ray "telescope." A 1-m^3 spark chamber with horizontal plates is shown at the left of the picture. On the right is a large mirror from which one face of the chamber is seen by the camera in the center. Another mirror (out of view on the left) projects a second face into the camera so that a stereoscopic view of cosmic-ray tracks is obtained on film. (Courtesy of Prof. C. L. Cowan.)

As in the cloud chamber, the gaseous ions take some tens of seconds to recombine, so the "memory" of the chamber is quite ample for electronic circuits to be activated and to trigger the chamber when a particular kind of event has probably occurred in the chamber. For this reason, other sorts of particle detectors are often used to trigger the system when such an interesting event has been detected. The lifetimes of the discontinuities in the bubble chamber are much shorter, so this chamber must be triggered a short time prior to the passage of particles of interest. This restricts the use of bubble chambers to beams from accelerators that are cyclic and are accurately timed.

A photograph of a spark chamber at the Catholic University of America is shown in Figure 33–8. This chamber, which is used as a cosmic-ray detector, consists of 36 aluminum plates sandwiched with scintillating plastic plates. Figure 33–9 shows tracks obtained from this spark chamber.

33-4 ELECTRONIC DETECTORS

Ionization Chambers. An extensive class of particle detectors exists that depends on the collection of gaseous ions produced by the passage of a charged particle and the electrical signal that this generates in an external circuit. In some ways the simplest of these detectors is the *ionization chamber*. This chamber contains a suitable gas in which ions have a rather long lifetime and several electrodes maintained at a moderate voltage for the collections of the ions. The gas may be under some pressure in order to enhance the sensitivity of the device by providing more target gas for the impinging elementary particles.

The energy deposition is proportional to the number of ions produced in the gas, and since the energy deposition is a measure of the charge and velocity of the particle, these are the quantities that are measured with ionization chambers. The collection of charged gas ions on electrodes of low capacitance generates voltage that, upon amplification, constitutes the output signal of the chamber. Since the chamber integrates the total signal from all the charged particles passing through it, and since the electrical circuit may be designed for a wide dynamic range, the ionization chamber is often used to detect and measure the simultaneous passage of a large number of particles. Such particle bursts are found in the beams of accelerators, for instance.

In its simplest form, the ionization chamber may consist of a gas-filled container as pictured in Figure 33–10. The two electrodes of this chamber are the outer metal cylinder connected to the terminal of a dc power supply and the central electrode, which is a straight wire connected in series to a resistor to the positive side of the power supply (about 200 V). A thin mica window allows photons or charged particles to enter the chamber. Upon entering the chamber, the charged particles or photons ionize some of the gas inside the cylinder. The ions produced are then attracted toward the electrodes because of the action of the electrical field

434 • PART FOUR: THE NUCLEUS

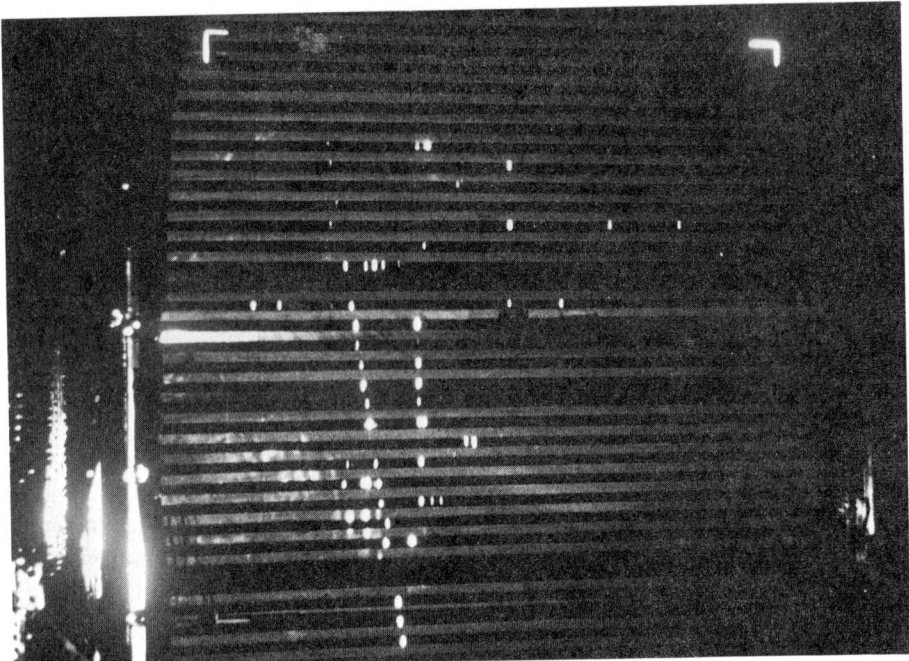

Figure 33-9 A cosmic ray interacts in the spark chamber. The long track of a cosmic ray particle, which comes down and to the right, shows an interaction high in the chamber with the production of a secondary particle and gamma rays, as evidenced by the track on the left and a scattering of single sparks below it. (Courtesy of Prof. C. L. Cowan.)

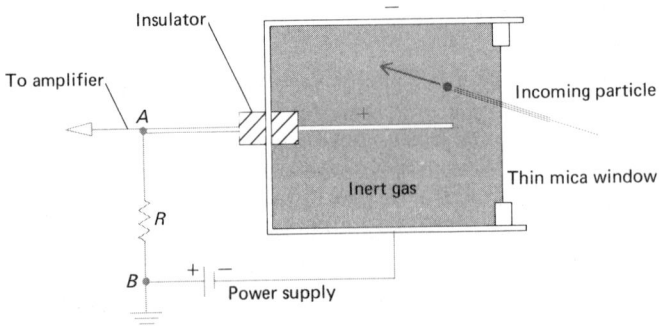

Figure 33-10 Typical ionization chamber.

existing between the central wire and the outer cylinder. A voltage pulse then appears between A and B, and this pulse is then amplified and registered.

A graph showing the magnitude of the voltage pulse versus the applied voltage between the electrodes appears in Figure 33-11. The region AB, from a few volts to about 200 V, corresponds to a typical ionization chamber. In this region, all of the ions produced by the incident particles or photons are collected by the electrodes. Since the pulse height in region AB is practically independent of the applied voltage, the chamber does not measure the energy of the incoming particles. The energy required to make an ion pair is about 32 eV; but because the ionization produced by a single charged particle is very small, the chamber detects bursts of particles instead of individual particles. It can, however, distinguish between bursts of α particles and bursts of β particles. Because x rays and γ rays readily ionize gases, they are easily detected by the ionization chamber.

Proportional Counters. A second type of gas detector, derived in a sense from the ionization counter, is the *proportional counter*. In this device, a voltage of 250–800 V is applied to the collection electrodes so that the original gas ions produced by the particle are accelerated to sufficient energy to generate other ions by collision with neutral molecules of the gas. The total charge collected for each particle passing through the chamber is thus larger than the original charge produced and is proportional to it. Such counters may be accurately calibrated to yield distinctive voltage pulses characteristic of different particles, or they may be set to ignore some types of particles completely. Alpha particles, for instance, are readily distinguished from electrons or protons by the larger voltage pulses they produce, stemming from their greater electrical charge. As shown in Figure 33-11,

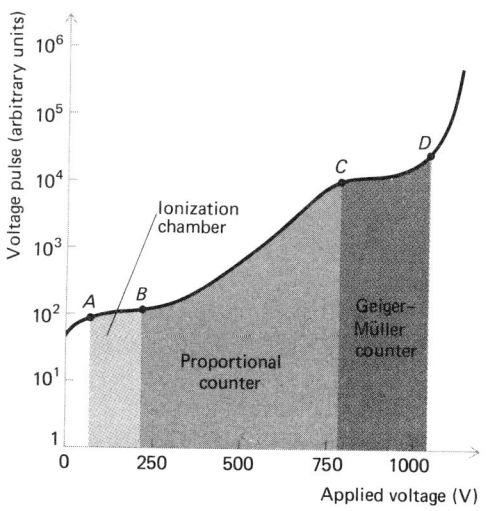

Figure 33-11 Graph of the voltage pulse produced by a charged particle or ionizing radiation versus the applied voltage for a gas-filled detector.

the region *BC* corresponds to the proportional counter. The pulse height is practically proportional to the energy of the incoming particle. It counts not only the incoming particles, but also the energy of the particles.

Geiger–Müller Counters. If the electrodes of a gas counter are shaped in such a way that there is a high field near one of them when a moderately high voltage is applied to it, then the gas amplification of an ionic charge in the region of high voltage enters an avalanche condition where essentially all of the gas present in the local volume of the electrode is ionized. This initiates a very large pulse of voltage on the electrode compared to the types of counters discussed previously. The pulse height is independent of the amount of ionization originally produced by the particle and may be triggered by only a single ion entering the vicinity of the high-voltage electrode. This counter, developed by GEIGER and MÜLLER (1913), is very simple to construct and is extremely sensitive to the passage of charged particles. The fine wire electrode at the center of the cylinder is held at a voltage of near 1000 V and is connected to a circuit of low capacitance and high impedance. The counter is filled with an inert gas to a pressure of a few centimeters of mercury and contains a trace of an organic vapor that quenches the initial discharge soon after it is initiated.

The region *CD* on the graph in Figure 33-11 between 800 and 1000 V is the operational region for the Geiger–Müller counter. The pulse height is practically constant and follows a plateau on the curve. The counter does not distinguish between the types of particles, nor does it measure their energies. It is usually designed to count β particles and x rays and γ rays. For applied voltages much larger than 1000 V, a continuous discharge takes place with the breakdown of the gas within the chamber.

Scintillation Detectors. There are substances that emit a flash of light following the deposition of energy in them by the passage of a fast charged particle. These substances are known as *scintillators* and may be liquid, crystalline, or plastic. The zinc sulphide screens used by Rutherford and his contemporaries to explore the scattering of α particles are scintillators.

Scintillators are all characterized by the possession of atomic or molecular "optical" levels that are excited by the Coulomb fields of passing charged particles. The optical levels are usually supplied by a trace impurity of a special sort dissolved in an otherwise highly purified medium. Among the crystalline, inorganic scintillators, sodium iodide crystals containing a trace of thallium are very common. Because of the high atomic number of the iodine, they are quite sensitive to γ radiation, which ejects fast photoelectrons from the iodine atom. Crystalline napthalene and crystalline anthracene are representative of the organic crystals used. Highly purified toluene, triethylbenzene, and mineral oil are some of the solutes used in liquid scintillator systems. These are then activated by dissolving trace amounts of oxazoles and often some napthalene or anthracene.

Very large volume detectors may be built with liquid scintillator tanks, so large target areas may be supplied to detect very weak signals. The natural radioactivity of potassium in the human body has been detected by building large liquid systems that surround the human subject during counting. The extremely weak signals produced by neutrinos coming from a nuclear reactor were first detected in a large liquid scintillation system used to demonstrate the existence of this particle.*

The amount of light emitted by the scintillator when a particle passes through it is very small and is usually far below the level that may be detected by the human eye or a photographic plate. For this reason, the light is collected by reflection from the interior of the liquid tanks and allowed to fall on the faces of highly sensitive phototubes. These are usually of the photomultiplier type. In this tube, weak light pulses fall onto a sensitive photocathode, which then emits a few electrons into the interior of the tube. These electrons are then accelerated in a high electrical field and fall onto a second electrode, where they "splash" out more electrons. This larger group is then accelerated in turn to fall onto a third electrode to "splash" out yet more electrons. This cascade sequence is made to proceed through as many as 12 or 14 stages until the original charge produced by the light is multiplied by 10^9 or more. The resulting current from the tube produces a voltage pulse that is easily measured and that can be made proportional to the amount of light collected and to the energy deposited by the charged particle in the scintillator.

The length of the pulse in time can be made to approach the time for the collection of the light itself from the scintillator, of the order of 10^{-9} sec, so time intervals approaching this in accuracy can be measured. When used in conjunction with, say, a bubble chamber, these pulses may be used to sort out a particular kind of event from a large background of many other kinds and then to trigger the bubble chamber when the desired event occurs. Systems of various detectors are thus built up to work together in the measurement of characteristics of elementary particles.

Figure 33–12 is a scintillation detector used in conjunction with a photomultiplier tube. Incoming charged particles or γ rays produce a flash at A. Then, by the photoelectric effect, an electron at point B in the photocathode is released. More electrons are produced by secondary emission, until the effect of the single electron has been multiplied many times. The accelerating potential between two consecutive electrodes is about 100 V. The amplified signal is finally collected at the plate P and detected by a sensitive microammeter in series with a resistor. The voltage pulse is then monitored by a recorder.

Čerenkov Detectors. It is commonly recognized that the highest velocity that can be obtained in nature is that of light traveling in a vacuum. Particles of high energy travel almost this fast if their kinetic energy equals or exceeds their rest mass energy. Such relativistic particles retain their velocity almost equal to that of

---

* See F. Reines and C. L. Cowan, "Neutrino Physics," *Phys. Today*, August 1957, pp. 12–18.

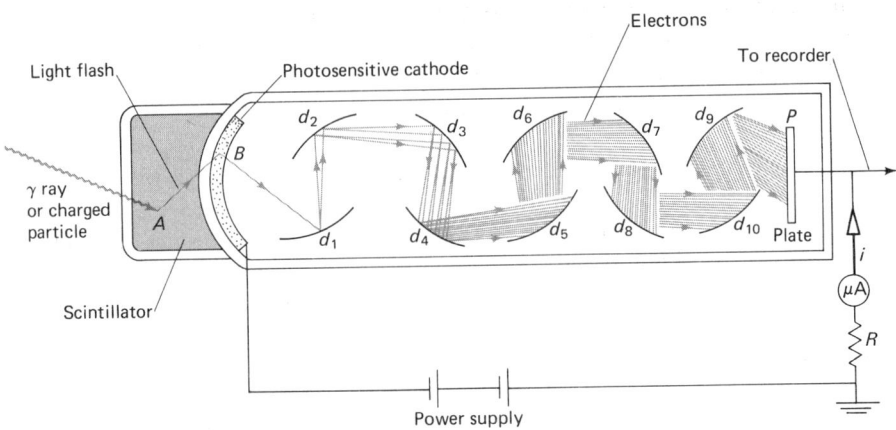

Figure 33-12 Scintillation detector used with a photomultiplier tube. The curved surfaces $d_1, d_2, \ldots$, where secondary emission of electrons takes place, are commonly called *dynodes*. Between every two consecutive dynodes there is always an accelerating potential of about 100 V.

light in a vacuum even when they penetrate matter, until they lose sufficient energy to become nonrelativistic. Light itself, on the other hand, slows down immediately upon entering a material of refractive index greater than unity. If c is the velocity of light in a vacuum, and n is the index of refraction of the medium, then the velocity of light in the medium is

$$v = \frac{c}{n}$$

The index of refraction for, say, Plexiglas, is about 1.50, so that light travels in Plexiglas with a velocity of about $0.66c$.

It may happen then that a relativistic charged particle is traveling in a transparent medium faster than light travels in the medium. In this case, a sort of electromagnetic "bow" wave is emitted by the particle in the form of visible light. This light, discovered by P. A. ČERENKOV in 1934, can then be collected and made to fall onto a photomultiplier tube as is done with scintillation light. Čerenkov emission thus has a threshold for a particle in terms of the velocity of that particle (it must be greater than c/n). This effect can be used to sort out highly energetic particles from a background of slower, less energetic particles that might confuse other kinds of detectors. The light is emitted in a narrow cone about the direction of travel of the particle, so optics may be used to choose this light from an ambient background that is traveling in many directions. The direction of approach of the particle may thus be defined by using Čerenkov counters. (For the development of this effect, P. A. Čerenkov, I. M. Frank, and I. M. Tamm (U.S.S.R.) were awarded the Nobel Prize in physics in 1958.) For this reason, they are commonly employed as parts of a trigger system for bubble chambers, cloud chambers, and spark chambers.

Since the index of refraction for *all* substances is greater than unity, even gases serve as detecting media for the Čerenkov effect. The index of refraction of a gas is dependent on the pressure of the gas, so detectors with variable thresholds may be constructed by controlling the gas pressure in them. The lower the pressure, the closer n is to unity, and the higher the velocity threshold for the particle becomes. An interesting application of the Čerenkov effect in gas is made in detecting the light produced by cosmic rays as they penetrate the atmosphere of the earth. In detecting this light (on dark, moonless nights), photomultipliers are mounted in large searchlight reflectors in place of the carbon arc usually used there, and the reflectors are pointed toward the sky. If several reflectors are used and are spaced some distance apart, they may be focused onto a small region of the sky and operated in coincidence to eliminate unwanted light signals. They then detect cosmic-ray showers produced in that part of the atmosphere. The magnitude and duration of the light flashes are used to determine the size of the shower, while the different times of arrival of the Čerenkov light front at the various detectors yields the direction at which the shower approaches the earth.

PROBLEMS

33-1 Sometimes the range of charged particles in a given material is expressed by R_m (mg/cm²), meaning that a column of material 1 cm² in cross section and containing R_m mg will be able to stop an α particle. Show that R_L, the range in centimeters of the particle in that medium, is given by

$$R_L = \frac{R_m}{\rho} \times 10^{-3}$$

where ρ is the density of the medium in grams per cubic centimeter.

33-2 If the range of α particles from Bi-214 in Cd-112 is $R_L = 2.42 \times 10^{-3}$ cm and the density of Cd is 8.65 g/cm³, find the range R_m of the α particles in that medium.

33-3 The range constant of charged particles in a given medium is defined by

$$K = \frac{R_m}{A^{1/2}}$$

where A is the mass number of the medium. An α particle from Bi-214 has an energy of 5.5 MeV and is traveling through a thin foil of Au-197, which has a density of $\rho = 19.3$ g/cm³. If it is known that for Au, $K = 1.93$ mg/cm², find approximately (a) R_L for the α particle in centimeters, and (b) the average retarding force acting on the α particles.

33-4 A charged particle of charge Ze is traveling with a constant speed v along the x axis as shown in Figure 33-13. An electron is located at point A at a distance a from the x axis. The length of the segment BC is L, and its midpoint is at M. (a) Show that the linear momentum given to the electron along the y direction is approximately given by

$$\Delta p_y = \int_{t_1}^{t_2} F \, dt = \int_{x_1}^{x_2} \frac{KaZe^2}{(a^2 + x^2)^{3/2}} \frac{dx}{v}$$

(b) Evaluate Δp_y for the interval $x_1 = 0.5L$ to $x_2 = 1.5L$.

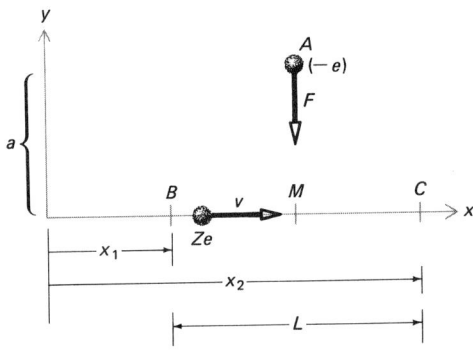

Figure 33-13

33-5 A particle passing through a medium loses energy primarily through excitation and ionization of the medium and from Coulomb interaction with atomic electrons. The specific energy loss in this process is given by

$$\frac{dE}{dx} = -\frac{4\pi e^4 z^2}{(4\pi\varepsilon_0)^2 mv^2} N_e Z \ln \frac{2mv^2}{I}$$

where z and Z are the proton numbers of the particle and the medium, respectively, v is the particle speed, N_e is the electron density of the medium (electrons/m³), I is the mean ionization potential, and m and e are the mass and charge of the electron. (a) An α particle traveling in helium gas, which has an ionization potential of 25 eV, has an energy of 4.0 MeV. Compute approximately, in MeV per centimeters, the specific energy loss. (b) What will be the total energy lost by the α particle after traveling a distance of 3 cm?

33-6 Compute the average current produced in an ionization chamber if 400 α particles are entering the chamber during a 1.5-min interval. If the energy of the α particles is 4.2 MeV, what is the value of the resistor in Figure 33-10 that will give a voltage pulse of 0.01 V across the resistor?

33-7 An elementary particle travels 1.0 m in a bubble chamber before it decays. If its speed is 0.98c, find (a) its real lifetime, and (b) the lifetime in the laboratory frame.

RECOMMENDED READING

ČERENKOV, P. A., *Phys. Rev.* **52,** 378 (1937).
GLASER, D. A., *Phys. Rev.* **91,** 762 (1963).
HOWARD, R. A., *Nuclear Physics*, Wadsworth, Belmont, Calif., 1963, Chap. 4.
POWELL, C. F., and OCCHIALINI, G. P. S., *Nuclear Physics in Photographs*, Clarendon Press, Oxford, 1947.
WILSON, C. T. R., *Proc. Roy. Soc.* **A85,** 285 (1911).
WILSON, C. T. R., *Proc. Roy. Soc.* **87,** 277 (1912).
WILSON, C. T. R., *Proc. Roy. Soc.* **104,** 1 (1923).
YANG, C. E., *Elementary Particles*, Princeton University Press, Princeton, N.J., 1961.

34
Particle Accelerators

Ernest Orlando Lawrence
(1901–1958)

Born in Canton, S. Dak., Lawrence received his Ph.D. from Yale University in 1925; he was professor at the University of California from 1927 until his death. His extensive research in nuclear physics included the invention of the cyclotron (1931), with which he produced artificial radioactivity, investigations into the structure of the atom, and applications of radioisotopes in medicine and biology. For his work in physics, Lawrence received the Cresson medal, the Hughes medal, the Duddell medal, and the Nobel Prize in 1939.

34–1 ACCELERATORS
34–2 COCKCROFT–WALTON GENERATOR
34–3 VAN DE GRAAFF GENERATOR
34–4 CYCLOTRON
34–5 BETATRON
34–6 LINEAR ACCELERATOR

34–1 ACCELERATORS

The earliest sources of charged particles and γ rays, as we have seen, were provided by the decay of naturally radioactive elements such as radium and thorium. These elements eject α particles with energy ranges up to 5 or 6 MeV, β particles (fast electrons) with comparable energies, and γ rays with energies up to about 3 MeV. These naturally occurring particles were studied first to determine their own nature and later were used as projectiles to bombard matter of various sorts to investigate how they interact with the elements. It was this latter type of investigation that led Rutherford to the discovery of the atomic nucleus.

With the knowledge of the atomic nucleus and the realization that some of the electrons surrounding it are rather easily removed to yield a charged ion, it became evident that if a very large potential drop existed between electrodes in some sort of electrical machine, these ions might be accelerated, to furnish a variety of projectiles for study. Quite a number of high-voltage machines were constructed in an attempt to do this, most being adaptations of the high-voltage static machines common to the physics laboratory of that era. In general, however, they were unstable in voltage regulation and could handle only very small ion currents between their electrodes.

All of the machines that can produce a high dc voltage fall under the general denomination of *accelerators*. In this chapter, a general survey of the most common accelerators will be made.

34–2 COCKCROFT–WALTON GENERATOR

In 1932, following the suggestion of Sir Ernest Rutherford, J. D. Cockcroft and E. T. Walton constructed a dc particle accelerator with which they were able to produce a nuclear reaction. An early version of the Cockcroft–Walton accelerator shown in Figure 34–1 uses a stream of electrons to ionize hydrogen gas and produce protons. These protons are then accelerated through a potential difference of 0.15 MeV and allowed to fall on a target of thin lithium foil. The protons eject α particles with an energy of 8.5 MeV according to the reaction

$$^{1}_{1}\text{H} + ^{7}_{3}\text{Li} \rightarrow ^{4}_{2}\text{He} + ^{4}_{2}\text{He} \qquad (34\text{–}1)$$

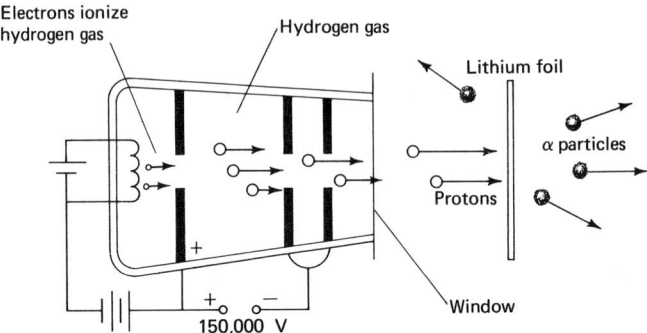

Figure 34–1   Cockcroft–Walton generator accelerates protons from ionized hydrogen gas.

The equation of mass-energy conservation from Equation (34–1) is

$$m_p c^2 + K_p + m_{Li} c^2 = 2m_\alpha c^2 + 2K_\alpha \tag{34-2}$$

where $K_p = 0.15$ MeV. The Q value for this reaction is

$$Q = (m_p + m_{Li})c^2 - 2m_\alpha c^2 = 2K_\alpha - K_p$$

where the mass difference is

$$\begin{aligned} m_p &= 1.007825 \text{ amu} \\ m_{Li} &= 7.016005 \text{ amu} \\ &\, 8.023830 \text{ amu} \\ 2m_\alpha &= 8.005208 \text{ amu} \\ \text{mass difference} &= 0.018622 \text{ amu} \end{aligned}$$

The Q value in million electron volts is 0.0186 amu × 931 MeV/amu = 17.35 MeV, with the net energy released being

$$Q - K_p = 17.35 - 0.15 = 17.2 \text{ MeV}$$

Each of the two α particles acquires an energy of 8.6 MeV, which agrees well with the experimental results of Cockcroft and Walton.

Almost immediately, Cockcroft and Walton improved this first version of their accelerator by using high-voltage transformers to charge a set of capacitors in parallel, and a set of rectifiers to discharge them in series. In the first such accelerator, the voltage applied to an evacuated tube produced protons of 0.8 MeV. A schematic diagram of an improved accelerator is shown in Figure 34–2.*

---

* See, for example, J. D. Cockcroft and E. T. S. Walton, *Proc. Roy. Soc.* **A137**, 229 (1933).

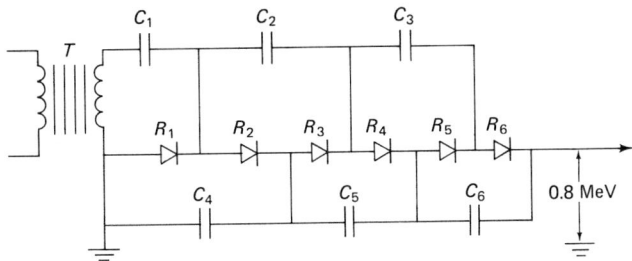

Figure 34–2 Improved Cockcroft–Walton accelerator. A high-voltage step-up transformer T with capacitors $C_1, \ldots, C_6$ and rectifiers $R_1, \ldots, R_6$.

34–3 VAN DE GRAAFF GENERATOR

Around the same time that the Cockcroft–Walton accelerator was developed, ROBERT VAN DE GRAAFF at Princeton University developed a simple but effective accelerator. The Van de Graaff generator shown in Figure 34–3 utilizes a belt running between two pulleys to transfer electrical charge from a high-voltage source that, through a set of fine points, "sprays" charge onto the pulley. A second set of points within the dome removes the charge, which flows to the exterior of the dome and raises it to a high potential.

An ion source located in the dome provides charged particles, which are then accelerated down an evacuated tube to strike a target at ground potential. The

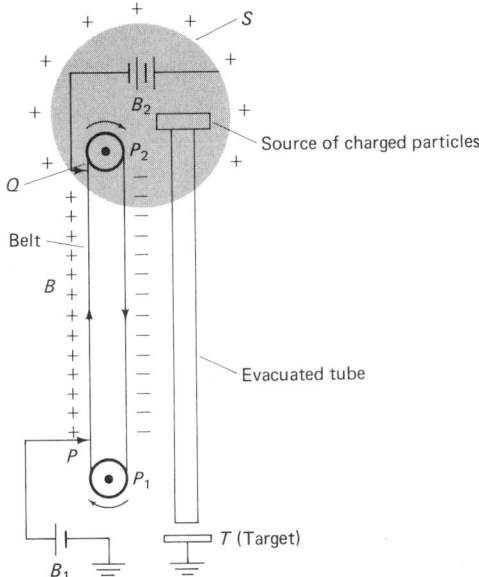

Figure 34–3 Cross-section of a Van de Graaff generator.

first voltage providing particles in excess of 1 MeV were attained in Washington, D.C., using this machine. The Van de Graaff accelerator is capable of producing very stable beams carrying charged particles in the hundred microampere range (about 10^{21} particles/sec) and with energy up to about 20 MeV. The Van de Graaff is the most commonly used accelerator, coming in a variety of sizes and energy ranges. Figure 34–4 is a photograph of a 5-MeV Van de Graaff generator installed at the Massachusetts Institute of Technology.

34–4 CYCLOTRON

Almost simultaneously with these first machines, yet another type, the cyclotron, was successfully tested by E. O. LAWRENCE and M. S. LIVINGSTON at the University of California at Berkeley. The essential part of the cyclotron is two hollow metal chambers, shown in Figure 34–5, known as "dees." The dees are separated along their diameter and connected to a radio-frequency supply at a frequency of about 10^6 Hz. The dees are placed in an evacuated chamber between the poles of a powerful magnet, which provides a magnetic field of up to several tesla.

Figure 34–4 Double-exposure photograph of the 5.5-MeV Van de Graaff ion accelerator showing the high-voltage terminal suspended in cantilever by an insulating structure covered with circular aluminum rings that define equipotential planes. The tank is mounted on rail-guided wheels so that it may be rolled away to expose the high-voltage structure. (Courtesy of Nuclear Sciences Division, Naval Research Laboratory.)

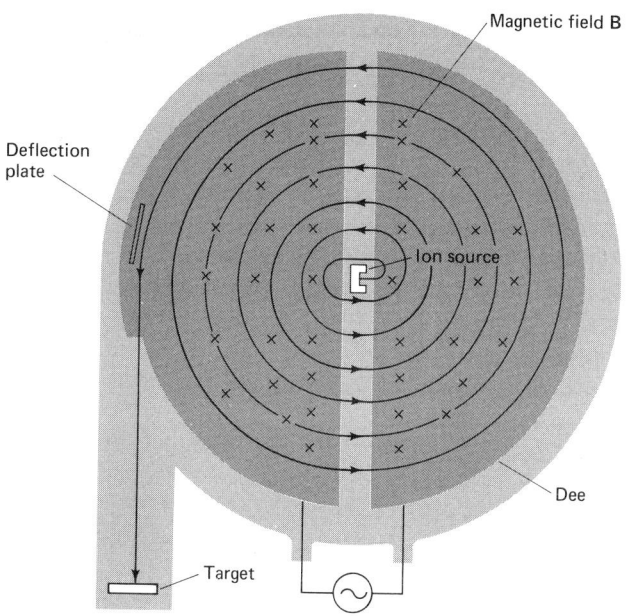

Figure 34–5 Positive ions are repeatedly accelerated between the dees and follow a circular path perpendicular to the magnetic field within the cyclotron.

A charged particle leaves a source S between the dees with a velocity v_0. Under the action of the magnetic field **B**, the force acting on the particles is

$$F = qv_0 \times \mathbf{B} \qquad (34\text{-}3)$$

or, since **B** is perpendicular to the force,

$$F = qv_0 B$$

From Newton's second law, then,

$$F = qv_0 B = \frac{mv_0^2}{r_0}$$

where m is the particle mass.

The particle describes a semicircular path of radius

$$r_0 = \frac{mv_0}{qB}$$

As the particle leaves the dee, the electrical field produced by the radio-frequency source changes direction. The field is synchronized in such a way that just as the particle leaves one dee, it is given a "kick" and accelerated into the other dee, where, because it now has a greater speed, it travels a path of radius

$$r = \frac{mv}{qB} \tag{34-4}$$

The trajectory of the particle inside the dees continues in a sort of increasing spiral until it reaches the maximum radius, and the particle is then diverted from the cyclotron toward a target. The particles emerge from the cyclotron with an energy equivalent to a fall through a potential much greater than that used in the radio-frequency field to drive them.

The frequency of revolution of a particle is given by

$$\boxed{f = \frac{v}{2\pi r} = \frac{vqB}{2\pi mv} = \left(\frac{q}{m}\right)\frac{B}{2\pi}} \tag{34-5}$$

which is independent of the radius r and the speed v. The maximum kinetic energy of the particles when they leave the cyclotron is

$$\boxed{K = \tfrac{1}{2}mv_{max}^2 = \frac{1}{2}\frac{q^2}{m^2}B^2 r_{max}^2} \tag{34-6}$$

EXAMPLE 34-1: What are the design parameters for a cyclotron that is to accelerate α particles to a maximum energy of 20 MeV? The dees are to have a radius $r = 0.50$ m.

SOLUTION: The magnetic field, from Equation (34-6), is

$$B = \left(\frac{2mK}{q^2 r^2}\right)^{1/2}$$

and for an α particle the calculation becomes

$$B = \left[\frac{2(4 \times 1.67 \times 10^{-27} \text{ kg})(20 \text{ MeV} \times 1.6 \times 10^{-13} \text{ J/MeV})}{(2 \times 1.6 \times 10^{-19} \text{ C})^2 (0.50 \text{ m})^2}\right]^{1/2}$$

$$= 1.3 T$$

The radio-frequency field must be driven at

$$f = \left(\frac{q}{m}\right)\left(\frac{B}{2\pi}\right) = \left(\frac{2 \times 1.6 \times 10^{-19} \text{ C}}{4 \times 1.67 \times 10^{-27} \text{ kg}}\right)\left(\frac{1.3 T}{2 \times 3.14}\right)$$

$$= 9.9 \times 10^6 \text{ Hz} = 9.9 \text{ MHz}$$

CHAPTER 34: PARTICLE ACCELERATORS · **449**

The first cyclotron ever built, a 4.5-in. brass vacuum chamber constructed by Lawrence in 1930, is shown in Figure 34–6. The cyclotron has become the inspiration for a wide family of accelerators of different detail but all of which employ the timed or phased interaction between a high-frequency electromagnetic field and a beam of ions. Such machines now produce beams of protons in the tens of billions of electron volts (1 GeV = 10^9 eV). The newest cyclotron under design, to be installed in Illinois, will deliver proton beams with an energy of about 200 GeV. It is to be noted that the rest energy of a proton as given by the relation $E = mc^2$ is 935 MeV, or just under 1 GeV.

Figure 34–6   The first cyclotron, constructed by E. O. Lawrence and his graduate student, M. Stanley Livingston. The $4\frac{1}{2}$-in.-diameter chamber was placed between the poles of a 4-in.-diameter magnet which produced a magnetic field of 1.3 Wb/m². (Courtesy of Lawrence Radiation Laboratory.)

Figure 34-7 Professor Lawrence at the controls of his 27-in. cyclotron as it looked in 1934. (Courtesy of Lawrence Radiation Laboratory.)

34-5 BETATRON

The American physicist D. W. Kerst at the University of Illinois in 1941 designed a machine, the betatron, specifically for accelerating electrons. The betatron is basically different from the cyclotron in that it uses an oscillating magnetic field, called an *induction field*, to drive the electrons in a circular orbit.

Cross-sectional views of the betatron are shown in Figures 34–8(a) and 34–8(b). A vacuum glass, doughnut-shaped chamber containing the source of electrons is mounted between the poles of a powerful electromagnet.

Most betatrons operate conveniently from a 60-Hz ac source. As the magnetic field varies in time, there is induced an electromotive force

$$\mathscr{E} = \frac{d\Phi}{dt}$$

where $\Phi = \pi r^2 B_{av}$ is the magnetic flux enclosed by the orbiting electrons and B_{av}

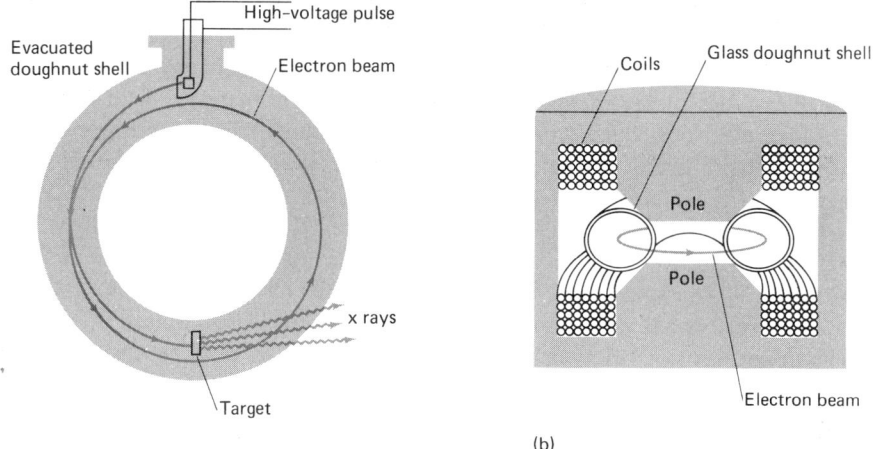

Figure 34–8 (a) A high-energy beam of electrons from a betatron produces x rays upon striking a metal target. (b) Cross section of betatron showing the arrangements of the magnetic coils and the evacuated, doughnut-shaped shell in which the electrons are accelerated.

is the average magnetic field inside the orbit. The resulting electrical field along the orbit is

$$E = -\frac{\mathscr{E}}{2\pi r} = \frac{1}{2\pi r}\frac{d(\pi r^2 B_{av})}{dt} = \frac{r}{2}\frac{dB_{av}}{dt} \tag{34-7}$$

Like Equation (34–4), the linear momentum is

$$mv = qBr$$

and at a constant radius r, a change in the magnetic field dB produces the following change in momentum:

$$d(mv) = qr\, dB \tag{34-8}$$

This change in momentum corresponds to an impulse

$$d(mv) = F\, dt = qE\, dt$$

or

$$d(mv) = q\frac{r}{2}\frac{dB_{av}}{dt}dt = q\frac{r}{2}dB_{av}$$

Comparison with Equation (34–8) then gives

$$dB_{av} = 2\, dB$$

or

$$\boxed{B_{av} = 2B} \tag{34-9}$$

For a stable orbit, the magnetic field must be shaped such that the average magnetic field over the space enclosed by the orbit is twice the magnetic field at the orbit.

In Figure 34–8(b), coils induce a strong magnetic field between the poles of a magnet. The poles of the magnet are shaped so that the magnetic field not only energizes the electrons, but keeps them in a stable orbit. After the electrons have completed enough orbits to achieve the desired energy, the field is adjusted until the electrons strike a target where very energetic photons are emitted. The parameters of the General Electric 100-MeV betatron are typical. This instrument has a pole face diameter of about 2 m and a magnet that weighs 130 tons. To achieve an energy of 100 MeV, the electrons pick up about 420 eV each turn, for a total of 2.4×10^5 turns, which is equivalent to a linear distance of 790 miles.

The upper limit of betatron performance is limited by two factors. Once the electrons achieve speeds close to the speed of light, relativistic effects become important and further acceleration of the particles becomes slight. Also, since they are charged particles traveling in a circular path, they are being accelerated continually, and energy is lost by radiation from a charged accelerated particle.

The synchrotron is an adaptation of the betatron that compensates for the relativistic effects. Electrons can reach energies of 300 MeV, where they have a mass something like 600 times their rest mass.

34–6 LINEAR ACCELERATOR

The concept of linear accelerators (linacs) was suggested around 1930, and various heavy-ion linacs were built, but these were not capable of producing significant nuclear disintegrations. At the end of World War II, LUIS ALVAREZ and his co-workers built a successful linear accelerator which produced 32-MeV protons.

Figure 34–9 shows schematically how the linear accelerator works. Protons accelerated by a Van de Graaff generator are introduced into the linear accelerator at an initial energy of 4 MeV. A series of hollow metallic cylinders is connected alternately to a radio-frequency generator, which boosts the electrons along the length of the accelerator. The particles pick up greater speeds as they are accelerated, so each tube through which it passes must be correspondingly longer. The Stanford 2-mile linear accelerator (SLAC) accelerates electrons to energies up to 20 GeV.

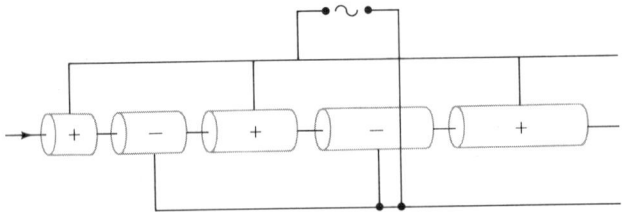

Figure 34–9 Very simplified version of the Luis Alvarez linac.

Table 34–1 Some important particle accelerators

DATE	ACCELERATOR	PARTICLE	ENERGY
1932	Cockcroft–Walton generator, Cavendish Laboratory, England	proton	0.5 MeV
	Lawrence–Livingston cyclotron, University of California, Berkeley	proton	1.2 MeV
1942	Kerst betatron, General Electric Company	electron	20 MeV
1950	Stanford mark II linac, Stanford University	electron	35 MeV
	Alvarez proton linac, University of California, Berkeley	proton	32 MeV
1952	Brookhaven National Laboratory proton synchrotron (cosmotron)	proton	2.2 GeV
1957	Proton synchrotron (synchrophasotron), Dubna, U.S.S.R.	proton	10 GeV
1959	CERN alternating-gradient proton synchrotron, Switzerland	proton	28 GeV
1966	Stanford 2-mile linear accelerator (SLAC), Stanford University	electron	20 GeV
1967	Cornell alternating-gradient electron synchrotron, Cornell University	electron	10 GeV
	Alternating-gradient proton synchrotron, Serpukhov, U.S.S.R.	proton	76 GeV
1971	National Accelerator Laboratory proton synchrotron, Weston, Ill.	proton	500 GeV

SOURCE: Most of the data for this table is taken from M. Stanley Livingston, *Particle Accelerators: A Brief History*, Harvard University Press, Cambridge, Mass, 1969.

The development of accelerators of all types is indicated in Table 34–1. The National Accelerator Laboratory at Weston, Ill., is completing an accelerator to produce 500-GeV protons. A Cockcroft–Walton generator sends 750-KeV protons into a linac accelerator, which boosts them to 200 MeV. They are then sent into an injection synchrotron, where they are given an energy of 8 GeV before being sent into the main synchrotron and boosted up to energies of 500 GeV. Other accelerators now in the design stage will boost particles to energies greater than 1000 GeV.

PROBLEMS

34–1 What is the velocity of a proton accelerated (a) to 0.15 MeV (Cockcroft–Walton accelerator)? (b) To 1.0 MeV (Van de Graaff generator)? (c) To 20 MeV (cyclotron)? (d) To 200 BeV (accelerator under construction)?

34–2 Sketch a feasible design to orbit satellite magnets about the earth to capture very-high-energy cosmic rays to be used as accelerating particles.

34–3 A Van de Graaff generator accelerates protons through a potential difference of 2×10^6 V. The running belt delivers positive charge to the spherical conductor at the rate of 3.0×10^{-3} A. If 10^4 protons/m²/sec reach the target, compute (a) the power expended to run the belt, (b) the speed of the protons when they reach the target, and (c) the current delivered to a target 20 cm² in area.

34-4 A cyclotron with a radius of 1.0 m has a magnetic field of $2.0T$. For deuterons accelerated in this device, determine (a) the frequency of revolution, and (b) the maximum energy imparted to the deuterons. (c) Through what potential difference would they have had to travel to attain these same energies?

34-5 In a betatron of diameter 1.52 m, the accelerating potential is 500 V for an electron in a stable orbit. If the electrons orbit at the rate of 5×10^5 rev/sec, what are (a) the linear speed of the electrons, (b) the final energy of the electrons in MeV, and (c) the linear momentum? (Are the electrons relativistic?)

34-6 In a betatron, the diameter of a stable orbit is 2.0 m and the accelerating potential per orbit is 500 V. If electrons are injected at 50 KeV and have a final energy of 50 MeV, calculate (a) the total number of revolutions of the electron, and (b) the maximum relativistic linear momentum.

34-7 A cyclotron has an accelerating potential between the dees of 100 kV, and the magnetic field is $1.5T$. If protons are accelerated to a maximum energy of 20 MeV, compute (a) the revolutions made by the proton to achieve this energy, (b) the frequency of revolution of the proton, (c) the maximum speed of the proton, and (d) the maximum radius of the orbit.

34-8 The linear accelerator located in Illinois is designed to accelerate protons to 200 BeV. (a) What is the de Broglie wavelength of these protons? (b) How does this wavelength compare to the dimensions of a nuclear structure?

34-9 A charged particle accelerated in a circular orbit loses energy according to the following equation (developed by Schwinger):

$$\text{energy loss per rev} = 12\pi \frac{e^2}{r} \left(\frac{E}{m_0 c^2} \right)^4 \times 10^9 \text{ J}$$

Determine the energy loss for (a) a 50-MeV electron in an orbit with a radius 2 m, and (b) a 50-MeV proton in the same orbit.

RECOMMENDED READING

DAVIS, Nuel Pharr, *Lawrence and Oppenheimer*, Simon & Schuster, New York, 1968.
  A biographical account of two eminent physicists involved in the development of accelerators and the atomic bomb.

HALLIDAY, David, and RESNICK, Robert, *Physics for Students of Science and Engineering*, Wiley, New York, 1960.
  The mechanics of the betatron are simply and well presented in Chapter 35.

HOWARD, R. A., *Nuclear Physics*, Wadsworth, Belmont, Calif., 1963, pp. 433–462.

Lawrence Radiation Laboratory Catalogue, Berkeley, Calif., 1969.

MARGENAU, H., and BERGAMINI, D., *The Scientist*, Life Science Library, Time Inc., New York, 1964.
  An excellent book intended to motivate students toward science. The illustrations are superb and should be part of the library of every physicist or physics student.

ROSE, Peter H., and WITTKOWER, Andrew B., "Tandem Van de Graaff Accelerators," *Sci. Am.*, August 1970.

SLOANE, D. H., and LAWRENCE, E. O., "The Production of High Speed Ions Without the Use of High Voltages," *Phys. Rev.* **38,** 2021 (1931).

WHITE, H. E., *Introduction to Atomic and Nuclear Physics*, Van Nostrand Reinhold, New York, 1964, Chap. 28.

WILSON, R. R., and LITTAUER, R., *Accelerators*, Doubleday, Garden City, N.Y., 1960.
  The whole book is fascinating and very well written.

35
Solid State I

John Bardeen
(1908–)
A native of Madison, Wis., Bardeen received his Ph.D. from Princeton University in 1936. He was a physicist at the Naval Ordnance Laboratory in Washington, D.C., and at the Bell Telephone Laboratories. His research in the theory of solid state and low-temperature physics, particularly semiconductors and superconductivity, is now universally recognized as providing a correct account of these phenomena. In 1956 Bardeen and his associates, William Shockley and Walter Brattain, were awarded the Nobel Prize in physics for invention of the transistor. In 1972 he shared the Nobel Prize in physics with Leon Cooper and John Schrieffer for their fundamental work on superconductivity. He is the first person to receive two Nobel prizes in the same field.

35–1 CRYSTALS
35–2 METALS
35–3 BAND THEORY

35–1 CRYSTALS

The attractive forces between the particles composing solids are so strong that the particles are not free to move about. However, the particles are far from motionless, because each particle oscillates back and forth continuously under the influence of elastic forces between neighboring particles. Although the structure of some solids, such as glass and rubber, shows no regularity in the arrangement of particles, most solids are composed of particles arranged in a very ordered fashion. The very regular geometric character of crystals and the manner in which crystals grow suggests that the atoms within them are very ordered and symmetrically arranged.

In 1912 FRIEDRICH, KNIPPING, and LAUE, by x-ray diffraction, showed that crystals are composed of atoms in a systematic arrangement. LAWRENCE BRAGG in 1913 made determinations of the crystal structures of KCl, NaCl, KBr, and KI, also by using x-ray diffraction. These measurements gave physical meaning to the structure of solids on the atomic scale.

The structure and rigidity of solids come from various forms of electrostatic forces that link the composite atoms together. The simplest crystalline structure, the NaCl crystal, which is shown in Figure 35–1, has a sodium atom and a chlorine

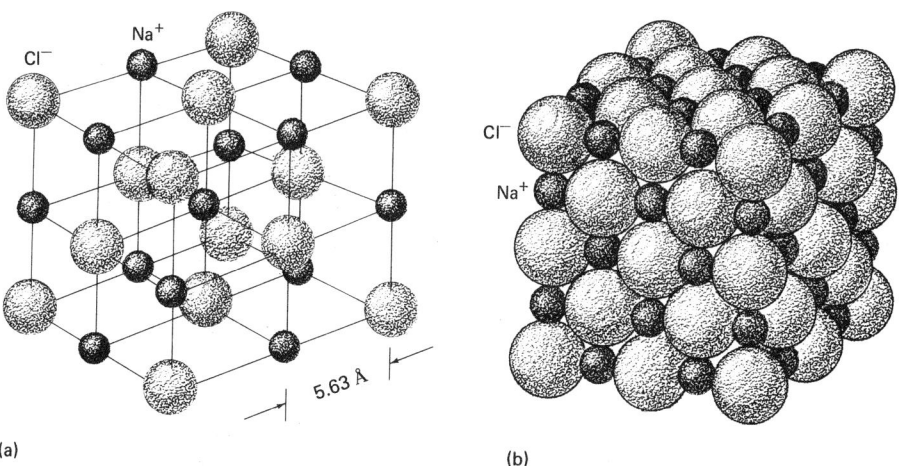

Figure 35–1 (a) Simple cubic structure of NaCl crystal. (b) The ions of sodium and chlorine occupy the sites shown in (a).

atom located in a cubical geometry with a different kind of atom at each corner. Figures 35–2(a) and 35–2(b) show neutral sodium with a single electron in its outer shell, and neutral chlorine with an outer shell filled except for one electron. The ions of these atoms are oppositely charged, and the electrostatic attraction between ions forms the simplest chemical bond, the ionic bond. This sort of bond is illustrated in Figure 35–3(a).

Another strong bond is the covalent bond between two atoms. A typical example of the covalent bond is binding of carbon atoms to make up the diamond crystal. Each carbon atom shares a pair of electrons in its outer shell with other neighboring carbon atoms to produce the covalent bond. This is illustrated schematically in Figure 35–3(b).

In metallic crystals, the metallic bond arises when all of the atoms share all of the valence electrons. In metals, the electrons are so loosely bound that they are free to move through the crystal lattice as shown in Figure 35–3(c). However, when the charge distribution of an electrically neutral atom is displaced, a weak interaction arises between neighboring atoms. This is a molecular bond known as the van der Waals bond [see Figure 35–3(d)], and solid H_2 is a typical example.

Because the hydrogen is small and its charge can be readily displaced, it is frequently attracted by two atoms to form a weak bond known as the hydrogen bond. [See Figure 35–3(e)]. It is the hydrogen bond that is largely responsible for the interaction of H_2O molecules that gives rise to the unique physical properties of water and ice. This type of bonding is shown in Figure 35–4 for one form of ice crystal.

Although for most crystals one particular chemical bond will predominate as the binding force, in reality a combination of different types of bonds will exist.

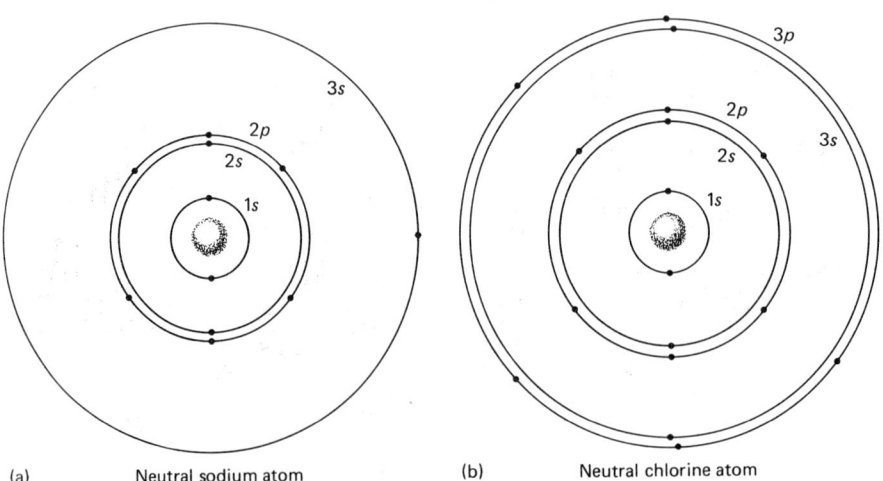

Figure 35–2 (a) A singly charged ion of sodium has an electron configuration of $Na^+: 1s^2 2s^2 2p^6$. (b) A singly charged ion of chlorine has an electron configuration of $Cl^-: 1s^2 2s^2 2p^6 3s^2 3p^6$.

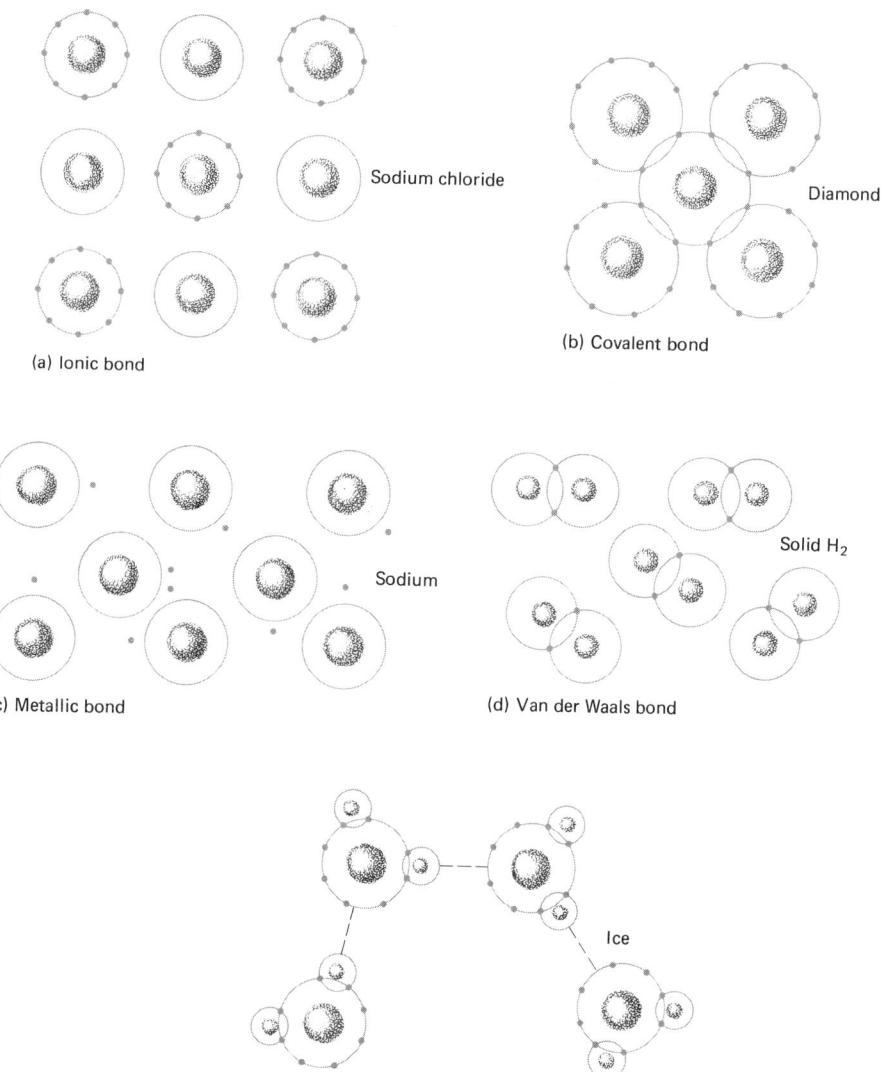

Figure 35–3 Schematic representations of the types of bonds that hold solids together. (a) Ionic bond. (b) Covalent bond. (c) Metallic bond. (d) Van der Waals bond. (e) Hydrogen bond.

460 · PART FOUR: THE NUCLEUS

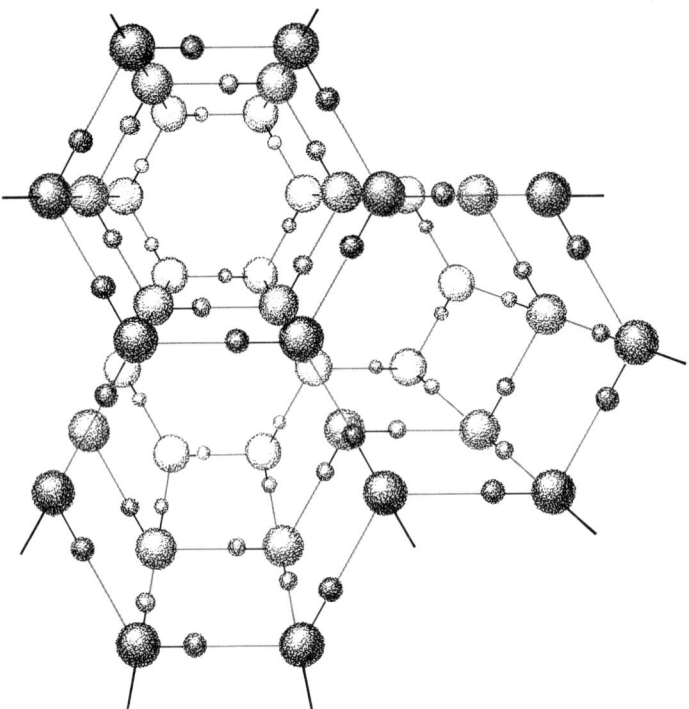

Figure 35–4 Ice crystal. Some of the unique physical properties of ice are due to the hydrogen bond.

35–2 METALS

About four-fifths of all the chemical elements can be classified as metals. Since the metals contain only 1, 2, or 3 valence electrons, they have a tendency to lose readily those electrons that are the least tightly bound. On the basis of this fact and experimental evidence that the mobile electrons are the current carriers in metals, DRUDE and LORENTZ in the early 1900s proposed a classical free electron gas model. The valence electrons were assumed to be completely free within the metal and thus they would behave much as in a classical gas.

Although this is an oversimplification, the theory was able to describe correctly some of the important physical properties of matter. Electrons in a metal are much more densely packed together than are molecules in an ideal gas. This requires that some attention be given to the Coulomb interaction with other electrons, and it implies a very high pressure within the electron gas. Both of these effects cause the observed results to deviate from classical predictions.

The most important drawback, however, is the fact that the electron itself is just not describable in classical terms. As a first approximation in applying quantum theory to the free electrons in a metal, it will be assumed that the electrons move in only one dimension and are bounded by an infinite potential at the edge of the metal. There is experimental support for these assumptions, because few electrons are scattered from other free electrons, and at very low temperatures a pure metal may have a mean free path for electrons of the order of 1 cm. Also, even at room temperature, electrons meet a high potential at the surface of the metal because, although they are relatively free to move within the metal, they do not leave the metal.

Thus, the wave function $\psi_n(x)$ of a free electron within the metal can be treated much like a particle in a box, as described in Chapter 15. The effect of the ions forming the crystal lattice is small, and the average potential energy is assumed to be zero. The Schrödinger equation for the free electron can be written as

$$-\frac{\hbar^2}{2m}\frac{d^2\psi_n(x)}{dx^2} = E_n\psi_n(x) \tag{35-1}$$

or

$$\frac{d^2\psi_n(x)}{dx^2} - \frac{2m}{\hbar^2}E_n\psi_n(x) = 0$$

Because the electron is contained within the metal, it is required that the wave function evaluated at the boundary of the metal be $\psi_n(0) = 0$ and $\psi_n(L) = 0$. This second-order differential equation has a normalized solution

$$\boxed{\psi_n(x) = \left(\frac{2}{L}\right)^{1/2} \sin\left(\frac{n\pi}{L}\right)x} \tag{35-2}$$

When the solution is substituted into Equation (35–1), the energy eigenvalues are found to be

$$\boxed{E_n = \frac{\hbar^2}{2m}\left(\frac{n\pi}{L}\right)^2 = n^2\left(\frac{\hbar^2\pi^2}{2mL^2}\right)} \tag{35-3}$$

The Pauli exclusion principle applies to solids as well as atoms or molecules. The exclusion principle applied to the free electrons means that no two electrons will have the same set of quantum numbers n and m_s. Since the electrons may have a spin of $m_s = +\frac{1}{2}$ or $m_s = -\frac{1}{2}$, there will be two unique states for each quantum number n. Let us recall from Chapter 24 that particles that are subject to the Pauli exclusion principle also obey Fermi–Dirac statistics.

For a total of N electrons, the states $n = 0, 1, 2, 3, \ldots, \frac{1}{2}N$ (assuming N to be even) will be occupied by two electrons each. The electrons in this one-dimensional model have a range of energies from zero when $n = 0$ to some maximum value when $n = n_F$. The energy associated with the highest filled energy level n_f is called the *Fermi energy*.

For a three-dimensional model, where the metal is a cube of side L, the Schrödinger equation (35-1) for a wave function $\psi_n(x, y, z)$ is written

$$\frac{\partial^2 \psi_n}{\partial x^2} + \frac{\partial^2 \psi_n}{\partial y^2} + \frac{\partial^2 \psi_n}{\partial z^2} + \frac{2m}{\hbar^2} E \psi_n = 0 \qquad (35\text{-}4)$$

It is now required that the wave function be zero at each boundary, and this gives a wave function

$$\psi_n(x, y, z) = \left(\frac{8}{L^3}\right)^{1/2} \sin \frac{n_x \pi x}{L} \sin \frac{n_y \pi y}{L} \sin \frac{n_z \pi z}{L} \qquad (35\text{-}5)$$

The energy eigenvalues are

$$E = (n_x^2 + n_y^2 + n_z^2) \left(\frac{\hbar^2 \pi^2}{2mL^2}\right) \qquad (35\text{-}6)$$

Just as in the one-dimensional case, the Pauli exclusion principle imposes the restriction that no more than two electrons can exist in a given state. If N is the total number of electrons, $\tfrac{1}{2}N$ is the number of sets of positive integers n_x, n_y, and n_z.

If $n_{max} = \sqrt{n_x^2 + n_y^2 + n_z^2}$ is pictured as the radius vector of a spherical surface, the total number of sets of integers is given by the volume of the sphere. However, since only positive values of the integers are used, only the positive octant of the sphere is considered (see Figure 35-5). The number of sets of positive integers is

$$\tfrac{1}{2}N = \tfrac{1}{8}(\tfrac{4}{3}\pi n_{max}^3)$$

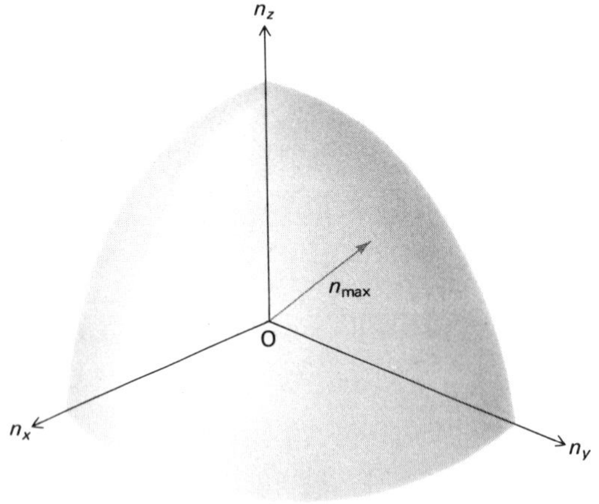

Figure 35-5 The number of sets of positive integers n_x, n_y, and n_z is given by one-eighth the volume of a sphere whose volume is given by $\tfrac{4}{3}\pi n_{max}^3$.

and because there are two spin orientations allowed for each state, the total number of distinct quantum states (n, m_s) is

$$N = \tfrac{1}{8}(\tfrac{4}{3}\pi n_{max}^3)2 = \frac{\pi n_{max}^3}{3} \tag{35-7}$$

now

$$n_x^2 + n_y^2 + n_z^2 = n_{max}^3 = \left(\frac{3N}{\pi}\right)^{2/3}$$

and the maximum kinetic energy, the Fermi energy, of any electron is

$$E_F = \left(\frac{3}{\pi}\right)^{2/3} \frac{\hbar^2 \pi^2}{2m} \left(\frac{N}{L^3}\right)^{2/3} = \frac{\hbar^2 \pi^2}{2m} \left(\frac{3N}{\pi V}\right)^{2/3} \tag{35-8}$$

where $L^3 = V$.

In the same manner that Equation (35–8) was approached, we find that the total number of electrons with energy less than E is

$$N = \left(E \frac{2m}{\hbar^2 \pi^2} L^2\right)^{3/2} \frac{\pi}{3} \tag{35-9}$$

The density of electrons with respect to energy can be found by differentiating Equation (35–9),

$$\frac{dN}{dE} = \frac{V}{2\pi^2}\left(\frac{2m}{\hbar^2}\right)^{3/2} E^{1/2}$$

The number of electrons with energy between E and $E + dE$ is

$$N(E + dE) - N(E) = \frac{dN}{dE} dE = P(E)\, dE$$

$$P(E)\, dE = \frac{V}{2\pi^2}\left(\frac{2m}{\hbar^2}\right)^{3/2} E^{1/2}\, dE \tag{35-10}$$

In Figure 35–6, the electron density per unit density $[(1/V)(dN/dE)]$ has been plotted versus E.

EXAMPLE 35–1: Calculate the kinetic energy of a free electron in the highest-energy state in sodium metal.

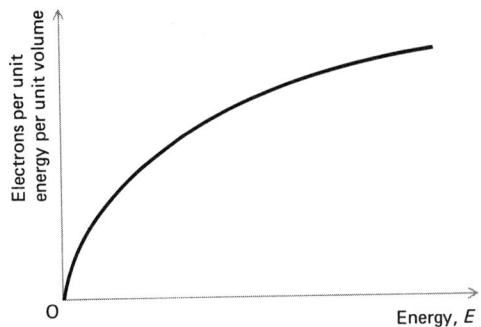

Figure 35-6 The number of electrons per unit energy per unit volume plotted against the energy.

SOLUTION: The molecular weight of sodium is 23 kg/kmole, and it has a density of 970 kg/m³. One kilomole has a volume

$$V = \frac{M}{\rho} = \frac{23 \text{ kg/kmole}}{970 \text{ kg/m}^3} = 0.0237 \text{ m}^3/\text{kmole}$$

and each atom has a volume

$$v = \frac{V}{N_0} = \frac{0.0237 \text{ m}^3/\text{kmole}}{6.02 \times 10^{26} \text{ atoms/kmole}} = 3.94 \times 10^{-29} \text{ m}^3/\text{atom}$$

On the assumption that there is one free electron per atom, the number of electrons per unit volume is $N/L^3 = 1/v = 2.54 \times 10^{28}$ electrons/m³; this number is called the Loschmidt number.

Equation (35-8) becomes

$$E = \frac{(1.05 \times 10^{-34} \text{ J-sec})^2 \pi^2}{2 \times 9.11 \times 10^{-31} \text{ kg}} \times \left(\frac{3 \times 2.54 \times 10^{28} \text{ electrons/m}^3}{\pi}\right)^{2/3}$$

$$= 4.98 \times 10^{-19} \text{ J} = 3.12 \text{ eV}$$

Table 35-1 compares the Fermi energy for free electrons in several metals.

Table 35-1

METAL	ELECTRON CONCENTRATION (electrons/m³)	ENERGY, E_f (eV)	ELECTRON VELOCITY (m/sec)
Li	4.6×10^{28}	4.7	1.3×10^6
Ba	2.5	3.1	1.1
Cs	0.86	1.5	0.73
Cu	8.50	7.0	1.56
Ag	5.76	5.5	1.38

35–3 BAND THEORY

Previously, we assumed that the electrons within a metal are free to move about within the crystal. This simple picture led to some insight into the electronic properties of solids. However, a more realistic picture of the environment of an electron within a metal must include the effect of atoms making up the periodic lattice within the crystal. When this is included, it is found that there are certain regions of energy in which there is no solution of the Schrödinger equation. These gaps in the distribution of energy among quantum states separate the allowed energies into bands. The character of these bands determines the electronic properties of solids, classifying them as either conductors or nonconductors.

In reality, electrons move about within the metal under the influence of other electrons and nearby atoms. A complete mathematical description of these electrons is a formidable problem, and for this reason only a qualitative picture will be presented.

In an individual atom, the orbiting electrons are bound to the nucleus by well-defined energy states, and the quantum theory forbids energies intermediate between any two given states. In a solid, the situation becomes more complex because atoms are densely packed, and they influence neighboring atoms significantly. As an example, the 11 electrons in a neutral sodium atom each occupy a specific energy level as indicated in Figure 35–7. When the atoms are part of a solid, they

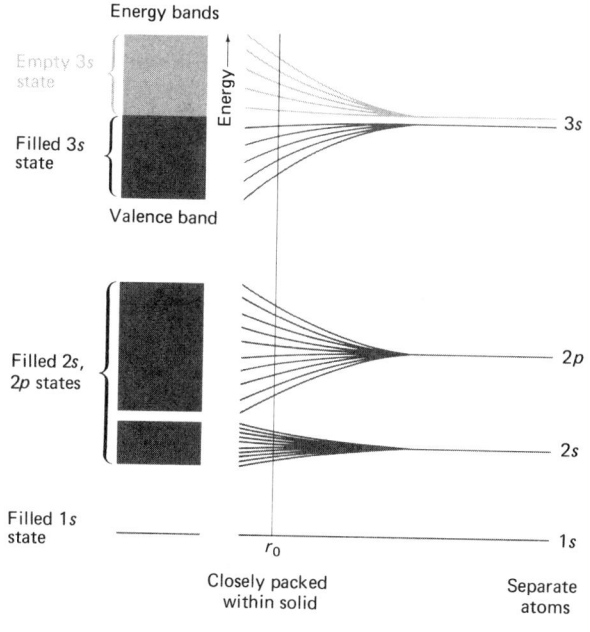

Figure 35–7 The energy levels of sodium become bands when the atoms are close together. In the figure, r_0 represents the spacing between atoms in solid sodium.

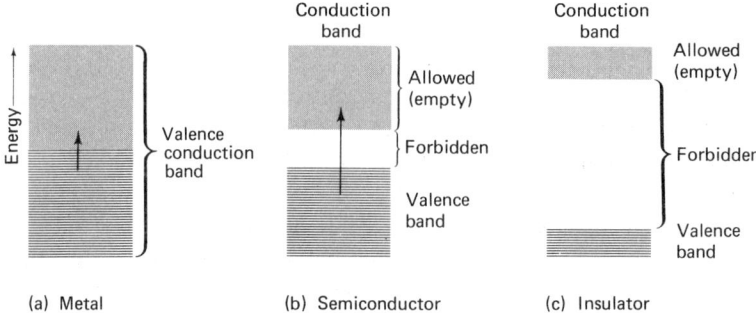

Figure 35–8 Schematic representation of the distribution of energy bands in a metal, a semiconductor, and an insulator.

interact with each other; and the electrons have slightly different energies since, according to the Pauli exclusion principle, no two electrons can occupy the same quantum state. When a crystal contains 10^{20} atoms, there are so many quantum states that they blend into what amounts to a continuum of energies that form bands of the allowed energies.

Whether a solid is an insulator, conductor, or semiconductor, is determined by the energy band structure, and the arrangement of the electrons with respect to these bands. Schematic representations of the energy bands for an insulator, a metal conductor, and a semiconductor are shown in Figure 35–8. In the metal the valence band is not full, and electrons can be readily lifted into the conductor band. On the other hand, if a band filled with electrons is well isolated from a conductor band by a forbidden energy gap, the solid will be an insulator since it is difficult to put electrons into the conduction band. Finally, in a semiconductor the forbidden gap is small, but some electrons can acquire enough energy to jump the gap to the conduction band. A small electromotive force or thermal energy is sufficient to raise electrons from the valence band through the energy gap to the conduction band.

PROBLEMS

35–1 The molecular weight of aluminum is 27 kg/mole, and it has a density of 2700 kg/m³. On the assumption that there will be three free electrons with each atom, how many free electrons will be contained in 1 m³ of aluminum? What is the linear separation of the aluminum atoms in an aluminum crystal?

35–2 Measure the dimensions of a copper penny and determine the total number of free electrons in the penny. Assume one free electron per atom. The molecular weight of copper is 64 kg/kmole, and the density is 8290 kg/m³.

35-3 Assume that there is one free electron per atom in lithium and that it is restricted to move in one dimension. What is the maximum kinetic energy of these electrons? How does this compare with Table 35-1 and with the thermal energy?

35-4 What is the maximum velocity of a free electron in sodium? In gold?

35-5 The total kinetic energy of all N electrons in a volume V can be obtained from

$$\text{total kinetic energy} = \int_0^{E_F} P(E)E\, dE$$

where $P(E)\, dE$ is found in Equation (35-10). Show that this total kinetic energy is equal to $\tfrac{3}{5}NE_F$.

35-6 Assume that the free electrons within a metal are free to move in a plane with sides L. What are the eigenfunctions and energy eigenvalues associated with these electrons?

35-7 From the *Handbook of Chemistry and Physics*, find the ratio of the resistivity of sodium to glass. How is this explained in terms of the band theory?

35-8 Assume that there are two free electrons per atom in magnesium and find the maximum velocity of free electrons.

35-9 How many electrons per unit volume will there be in a sodium crystal that will have an energy less than thermal energy (0.0025 eV)?

RECOMMENDED READING

BRAGG, Lawrence, "X-ray Crystallography," *Sci. Am.*, July 1968.

KITTEL, Charles, *Introduction to Solid State Physics*, 3rd ed., Wiley, New York, 1968.
A good, standard text on the subject. The third edition is much improved and a more readable text than the previous ones.

MOTT, N. F., and JONES, H., *The Theory of the Properties of Metals and Alloys*, Dover, New York, 1958.

PAULING, Linus, and HAYWARD, Roger, *The Architecture of Molecules*, Freeman, San Francisco, 1964.
Beautifully illustrated book showing the structure of many important molecules.

Scientific American, September 1967 issue.
The entire issue is devoted to materials.

36
Solid State II

Walter Houser Brattain
(1902–)

Brattain is a native of Amoy, China, and studied at Whitman College, the University of Oregon, and the University of Minnesota. He was with the technical staff of Bell Telephone Laboratories from 1929 to 1967. Presently, he teaches at Whitman College in Walla Walla, Wash. The invention of the point-contact transistor and his research on the surface properties of semiconductors are his chief contributions to solid state physics. In 1956 he, with William Shockley and John Bardeen, was awarded the Nobel Prize in physics for research on semiconductors.

36–1 FERMI–DIRAC DISTRIBUTION
36–2 SEMICONDUCTORS
36–3 TRANSISTORS

36–1 FERMI–DIRAC DISTRIBUTION

In the previous chapter, it was pointed out that although the electrons are relatively free to move about within a metal, the Pauli exclusion principle applies to these electrons just as it does to electrons in restricted orbits of atoms. No two of the free electrons within the metal can occupy the same energy state. Particles that obey the Pauli exclusion principle are represented statistically by Fermi–Dirac statistics.

If $f(E)$ is the probability that a given quantum state of energy is occupied by an electron, then at zero temperature the following conditions will be imposed:

$f(E) = 1 \quad$ if $E < E_f$
$f(E) = 0 \quad$ if $E > E_f$

At a finite temperature, the probability of a given quantum state being occupied is given by the Fermi–Dirac distribution law:

$$f(E) = \frac{1}{e^{(E-E_f)/kT} + 1} \tag{36–1}$$

In this equation, E is the energy of a given quantum state, E_f is the Fermi energy (the energy of the topmost filled level), k is the Boltzmann constant, and T is the absolute temperature.

Figure 36–1 shows the distribution function $f(E)$ plotted against the energy E at absolute zero and at a few temperatures $T > 0$. Notice that when $E = E_f$, $f(E) = \frac{1}{2}$ at any temperature. The number of electrons with energies in the range E to $E + dE$ at a temperature T can now be determined from Equations (35–10) and (36–1), where

$n(E) \, dE = P(E) \, dE \, f(E)$

and

$$n(E) \, dE = \frac{V}{2\pi^2} \left(\frac{2m}{\hbar^2}\right)^{3/2} \frac{E^{1/2} \, dE}{e^{(E-E_f)/kT} + 1} \tag{36–2}$$

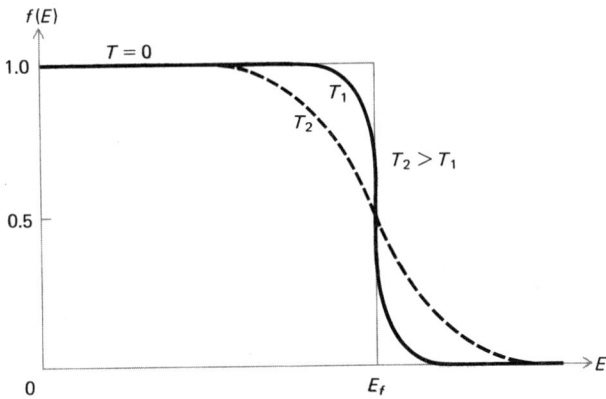

Figure 36-1 The Fermi–Dirac distribution plotted against the energy.

At temperature $T = 0°K$, $f(E) = 1$ if $E < E_f$, since all the energy states are filled up to E_f and consequently $\exp[(E - E_f)/kT] = 0$. The number of electrons with energies less than or equal to the Fermi energy is

$$N = \int_0^{E_f} n(E)\, dE = \frac{V}{2\pi^2}\left(\frac{2m}{\hbar^2}\right)^{3/2} \int_0^{E_f} E^{1/2}\, dE$$

or

$$N = \frac{2}{3}\left[\frac{V}{2\pi^2}\left(\frac{2m}{\hbar^2}\right)^{3/2}\right] E_f^{3/2} \tag{36-3}$$

At $T = 0°K$ then, the Fermi energy is, from Equation (36-3),

$$\boxed{E_f(0) = \frac{\hbar^2 \pi^2}{2m}\left(\frac{3N}{\pi V}\right)^{2/3}} \tag{36-4}$$

which is in agreement with Equation (35-8).

The classical theory requires that at $T = 0°K$ the electrons all have zero energy, but according to Equations (36-3) and (36-4) and Table 35-1 there will be a range of energies up to a maximum E_f. An effective temperature of the electron gas, known as the *Fermi temperature*, is defined by

$$T_f = \frac{E_f}{k} \tag{36-5}$$

Thus, from the Fermi energy of sodium listed in Table 35-1, the electron gas has enough energy to give it an effective Fermi temperature of

$$T_f = \frac{3.1 \text{ eV} \times 1.6 \times 10^{-19} \text{ J/eV}}{1.38 \times 10^{-23} \text{ J/°K}} = 3.7 \times 10^{4}°K$$

Classically, if heat were added to the electrons, it would be expected that each would increase in energy by about kT. The Pauli exclusion principle prevents electrons from acquiring energy and moving to a higher energy level unless it is unoccupied. Upon heating, only those electrons that have energy near the Fermi energy can accept an increase in energy of the order kT.

36–2 SEMICONDUCTORS

In Figure 35–8, the semiconductor was pictured as having a small forbidden energy gap between the filled valence band and an empty conduction band. At low temperatures, all of the electrons in these crystals are in the valence band and there is no electrical conduction. The crystal is effectively an insulator. However, at room temperature, some electrons have received sufficient thermal energy to jump the forbidden gap and enter the valence band where, when an electrical field is applied, they become conducting electrons. Because of these characteristics, these crystals are known as semiconductors. Because of their simple structure, silicon and germanium are the best-known semiconductors.

Small amounts of impurities added to semiconductor crystals change the character of the forbidden energy gap. When traces of phosphorus, with a valence configuration $3s^2 3p^3$, replace some silicon atoms, which have a valence configuration $4s^2 4p^2$, in a crystal, an extra electron not involved in covalent bonding is available. The "extra" electron from phosphorus, as shown in Figure 36–2, can now become a conduction electron within the crystal.

When an impurity such as phosphorus in Figure 36–2 provides an extra electron that can then be used as a conduction electron, it is known as a *donor*. The semi-

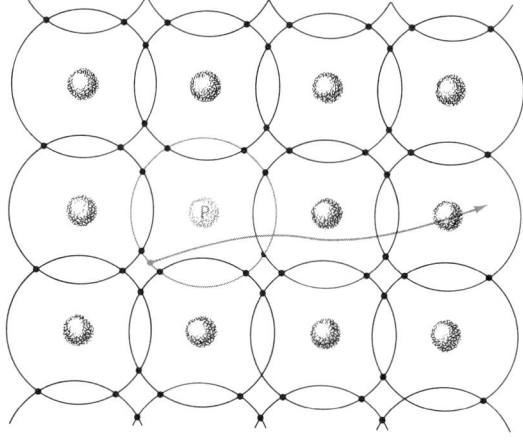

n-type silicon

Figure 36–2 Phosphorus impurities added to a silicon crystal provide an extra electron for conduction.

conductor that has had a donor impurity added is an *n-type semiconductor*, because it is the negative electrical charge that carries the current.

An impurity such as aluminum, with a valence electron configuration $3s^2 3p^1$, enhances the current by a somewhat different mechanism. When traces of aluminum are added to a silicon crystal as shown in Figure 36-3, the covalent bonding is one electron shy and electrons are readily accepted by the hole. Other electrons in turn fill the hole left by the previous electron and so on, providing a mechanism for conduction. In this case, the impurity atoms are designated as *acceptors* because they accept electrons. This semiconducting crystal is a *p-type semiconductor*, because the flow of current can be described in terms of the holes. This is effectively the same as having a current composed of moving positive charges.

The effect of donor and acceptor impurities on the forbidden region of the energy is represented schematically in Figure 36-4.

Crystals of silicon and germanium can be produced such that one portion of the crystals is an *n*-type semiconductor and the remainder is a *p*-type, with a thin intermediate region known as the *p–n* junction between them. This semiconductor has the important electrical property that it has a greater tendency for current to travel in one direction than another, and it thus acts as a rectifier.

Because there are free electrons in one region of the crystal and positive "holes" in the other [Figure 36–5(a)], a difference in potential is established between the regions. In the absence of an applied electrical field, the conduction electrons from the *n* region tend to diffuse into the *p* region and seek out the positive holes. This is effectively the same as saying that positive holes from the *p* region move into the *n* region. As the electrons and holes leave their respective regions, the *n* region becomes more positive, the *p* region becomes more negative, and an

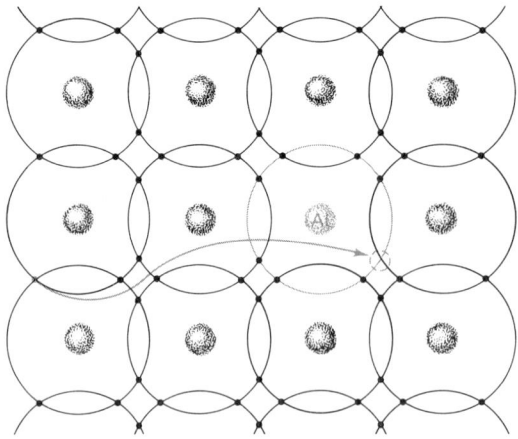

p-type silicon

Figure 36–3   Aluminum is shy one electron, and other electrons move to fill the vacancy.

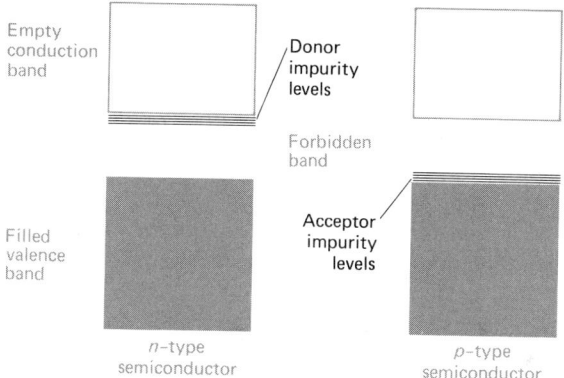

Figure 36–4 The addition of donor and acceptor impurities provide additional energy levels in which conduction may take place.

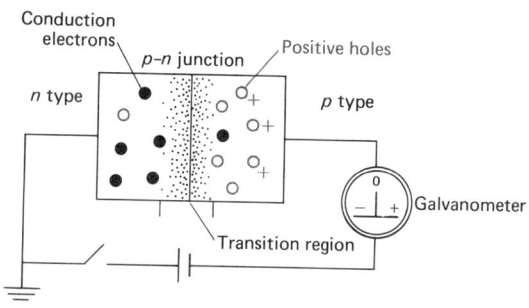

(a) No applied voltage

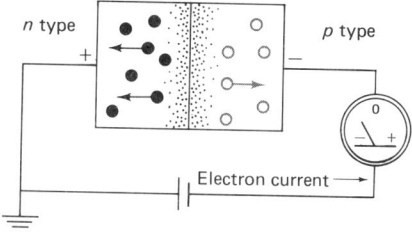

(b) Reverse applied voltage

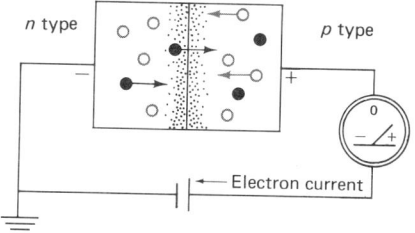

(c) Forward applied voltage

Figure 36–5 The *p–n* junction in a semiconductor acts as a rectifier of the current. (a) No applied voltage. (b) Reverse applied voltage. (c) Forward applied voltage.

electrical field is established that then attracts some electrons back into the *n* region and some holes back into the *p* region. Eventually, an equilibrium condition is reached where the current from the conducting electrons equals the current from the holes.

Now let an electrical field be applied across the semiconductor. When the *n* region is made positive and the *p* region negative as it is in Figure 36–5(b), there is only a small current because only a few electrons can be pulled through the transition region from the *p* region. When the field is reversed [Figure 36–6(c)] so that the *n* region is negative and the *p* region is positive, the current rises sharply because the electrons are attracted toward the region in which they tend to move anyway.

36–3 TRANSISTORS

An important use of the junction-type semiconductor is the *transistor*. A transistor shown in Figure 36–6 is made up of three distinct regions: one is an *n*-type region called the *emitter*, containing a large concentration of donors; one is a weak *p*-type region called the *base*; and the third, the *collector*, is another *n* region. The transistor shown in Figure 36–6 is used in a power amplifier circuit.

With a small bias voltage between the emitter and the base, a rather large electron current flows in the direction from the emitter to the base. The *p–n* junction between the emitter and the base has the same rectifying action as described previously. On the other hand, the electron current in the direction from the collector to the base is very small because of the strong reverse bias established by V_2.

The base region is made so thin (of the order of 500 Å) that almost all of the electrons pass through the base without encountering a positive hole and enter the collector. The rectifying action of the junction between collector and the base prevents any appreciable electron current by diffusion of electrons from the collector to the base. Thus, almost all of the current from the emitter enters the collector.

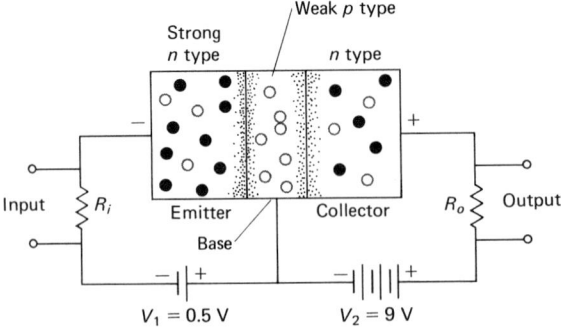

Figure 36–6   Schematic diagram of an *n–p–n* transistor used in an amplifier circuit.

The efficiency of a transistor is measured by the α of the transistor, which is defined as

$$\alpha = \frac{I_c}{I_e} \tag{36-6}$$

where I_c is the collector current and I_e is the emitter current. Values for α for transistors are of the order of 0.95 to 0.995.

EXAMPLE 36–1: What is the amplification of the circuit shown in Figure 36-6 if the transistor has a value of $I_c/I_e = 0.95$, and $R_i = 50\ \Omega$ and $R_o = 1000\ \Omega$.

SOLUTION

$$A = \frac{P_o}{P_i} = \frac{I_c^2 R_o}{I_e^2 R_i}$$

From the values given above, then, $I_c/I_e = 0.95$ and

$$A = (0.95)^2 (\tfrac{1000}{50}) = 18$$

PROBLEMS

36–1 Use the energy listed in Table 35–1 and determine the probability of finding an electron in lithium at room temperature and with energy 0.10 eV above the Fermi energy.

36–2 Show that the probability of finding a particle in an energy state equal to the Fermi energy is $\frac{1}{2}$.

36–3 If the Fermi energy for sodium is 3.1 eV, what is the probability of finding an electron with an energy of 3.2 eV at room temperature.

36–4 The Fermi energy for silver is 5.5 eV. At a temperature just below the melting point 1230°K, what is the probability of finding an electron with an energy equal to 5.0 eV? 5.6 eV?

36–5 Cesium has a Fermi energy of 1.5 eV and, based on one electron per atom, an electron concentration of 8.6×10^{27} electrons/m^3. How does the de Broglie wavelength of these free electrons with Fermi energy compare with the average distance between electrons?

36–6 Check the electron configuration of the atoms in Table 24–1, and determine several atoms that would be donors and several that would be acceptors for a silicon semiconductor.

36-7 (a) For cesium at room temperature, how many electrons per unit volume will have an energy within $\pm 0.01 E_f$?
(b) What fraction of the total number of electrons does this represent? See Table 35-1.

36-8 By comparing the description of the n–p–n transistor in Figure 36-6, describe qualitatively the action of a p–n–p transistor.

36-9 For a transistor circuit as shown in Figure 36-6, the amplification is found to be 27 when the input resistance is 50 Ω and the output resistance is 1500 Ω. What is the α of the transistor in this case?

36-10 What is the Fermi temperature for free electrons in copper? Use this temperature and determine the root-mean-square velocity by assuming that the free electrons are an ideal gas. Compare this velocity with that found in Table 35-1.

RECOMMENDED READING

KITTEL, Charles, *Elementary Solid State Physics, A Short Course*, Wiley, New York, 1962.

PAULING, Linus, *The Nature of the Chemical Bond and the Structure of Molecules and Crystals*, Cornell University Press, Ithaca, N.Y., 1966.

"Resource Letter Ser-1 on Semiconductors," *Am. J. Phys.* **32**, 329 (1964).

WEST, Charles A., and THOMSON, Robb M., *Physics of Solids*, McGraw-Hill, New York, 1964.

FIVE
Elementary Particles

> ... nature seems to take advantage of the simple mathematical representation of the symmetry laws. The intrinsic elegance and beautiful perfection of the mathematical reasoning involved and the complexity and depth of the physical consequences are great sources of encouragement to physicists. One learns to hope that nature possesses an order that one may aspire to comprehend.
>
> CHEN NING YANG
> *Elementary Particles*, 1959

At the beginning of the twentieth century, electrons, protons, and neutrons were the elementary particles of nature. However, by the 1960s, with the advent of huge accelerating machines, a plethora of elementary particles—hyperons, kaons, leptons, and so on—had been uncovered. There were then so many fundamental units of nature that confusion existed rather than order. Modern physicists look to mathematical reasoning to provide a thread of continuity and symmetry to all of these elementary particles.

37
Elementary Particles

Sir Joseph John Thomson
(1856–1940)

Thomson was born in Manchester, England, and received his education from Owens College, Manchester College, and Trinity College at the University of Cambridge. From 1905 to 1918, he taught at the Royal Institute of Great Britain. Thomson is considered the discoverer of the electron and the pioneer of the study of elementary particles. Recipient of the Hughes medal and the Copley medal, in 1902 he was awarded the Hodgkins medal by the Smithsonian Institution. For his investigations on the transmission of electricity by gases, he was awarded the Nobel Prize for physics in 1906.

37–1 CHARGES AND FORCES
37–2 QUANTUM NUMBERS OF ELEMENTARY PARTICLES

37-1 CHARGES AND FORCES

As the last century drew to a close, physicists as we have noted earlier, pictured the material universe in terms of atoms, which were unique and distinct for each element. It was conjectured that the various atoms could account for all chemical reactions and for the vast complexity of nature. However, the reasons for differences between the atoms of different elements and the mechanism of chemical reactions were not understood. The first breakthrough toward understanding the atom was in 1897, when J. J. THOMSON discovered by experimentation the electron and reasoned that it was a part of every normal atom. Later work demonstrated, in fact, that electrons can be obtained from all elements and that each electron has a mass approximately equal to 1/1837 the mass of a hydrogen atom. This pioneering work with electrons led physicists to establish two new concepts:

1. Distinct and unique particles exist in nature, which are smaller than the previous known chemical atoms; these particles are part of every atom.
2. Definite properties, such as charge and mass, are associated with these particles to identify them.

This discovery marked the birth of the branch of physics devoted to the discovery and study of the elementary particles. Cosmic rays provided the first rich source of new particles, followed by those produced in giant accelerators. Techniques for the discovery and study of the particles now range from cloud chambers and blocks of photographic emulsion to very large bubble chambers filled with liquid hydrogen.

Some 35 years after the detection of the electron, C. D. ANDERSON discovered experimentally another particle that has the same mass as the electron but that has electrical charge equal in magnitude but opposite in sign to Thomson's electron. This was obviously a distinctly different particle, the positron. The existence of this particle had been predicted in the theoretical work of P. A. M. DIRAC when he applied the principles of relativistic mechanics to the wave equation of the electron. Dirac had been forced by logic to account for the symmetrical properties of the theory of relativity—in this case, symmetry with respect to reversal of an algebraic sign. When the predicted particle was found in nature, physicists concluded that nature itself must exhibit this symmetry as well. Thus, the discovery of the positron was the first example to demonstrate that some of the mathematical symbols and terms used to identify and describe elementary particles may be reversed in algebraic sign to yield valid descriptions of other elementary particles. Even more subtle symmetries have since been found in the family of elementary particles to assist in arranging them in an orderly table and to point the way to more particles in nature. Thus, the discovery of elementary particles has been furthered by both theoretical predictions and experimental results. There is now quite a long list of elementary particles, and it has become a rather serious problem in recent years to even define what is meant by elementary in view of this large number. It is, perhaps, too early in our understanding of nature to know which of these particles to call truly elementary in the most fundamental sense. The list of properties for any given elementary particle might begin with the rest mass of the particle and then continue with the electrical charge, the value of an intrinsic angular momentum that is always associated with the particle (its spin), the values of other sorts of "charge" that determine how the particle interacts with other particles, and so on. These numbers, along with their algebraic signs, are the "quantum numbers" of the elementary particles, and the list of quantum numbers constitutes the true name (or description) of each of the particles.

We have implied that each of the quantum numbers in the list for an individual particle is a physically observable quantity. A certain physical reality is ascribed to each of these numbers, since each has a definite physical consequence when a particle interacts with another particle. We should keep in mind, however, that the philosophical foundations on which every physical observation are based are only intervals in space and in time. Quantities are derived from these raw data, which we name mass, charge, spin, and so on. We should also remind ourselves that these limitations are true for any branch of physics. In the case of the elementary particles the space and time intervals that are measured are in fact usually separated from the actual particle itself by a rather long and complex chain of physical events. In other words, what is measured is the amplification of a local event on the surroundings. This method provides length and time intervals that are large enough to observe. Thus, experiments having to do with elementary particles have their own limitations imposed on them.

To measure lengths and times associated with successive positions of an elementary particle, it is necessary first to locate the particle accurately within some measuring apparatus. The particle may, however, be of the order of 10^{-15} m in diameter and so be quite unobservable by any ordinary means, including the most powerful of microscopes. One way to overcome this limitation is to make the particle affect a region much larger than its own volume in such a way that changes in the larger region can be observed. In other words, the "amplification" of a local event is made to take place. Thus, an electron with dimensions of perhaps 10^{-10} m may pass through an undeveloped photographic plate. In doing so, its path will take it through many grains of silver halide, each of which may have dimensions of the order of 10^{-6} m. Excitation of tiny regions in each such grain penetrated then render the entire grain developable, resulting in a black spot easily observed using the proper microscope. A string of such grains in the developed plate then defines the trajectory of the electron in passing through the plate. (See Figure 37–1.)

In the above example, it was the electrical charge on the electron that excited the silver halide grains. Neutral particles, such as neutrons and neutrinos, interact with the material of a detector through nonelectrical charges. Neutral particles do this through nuclear charges or weak interaction charges that produce forces

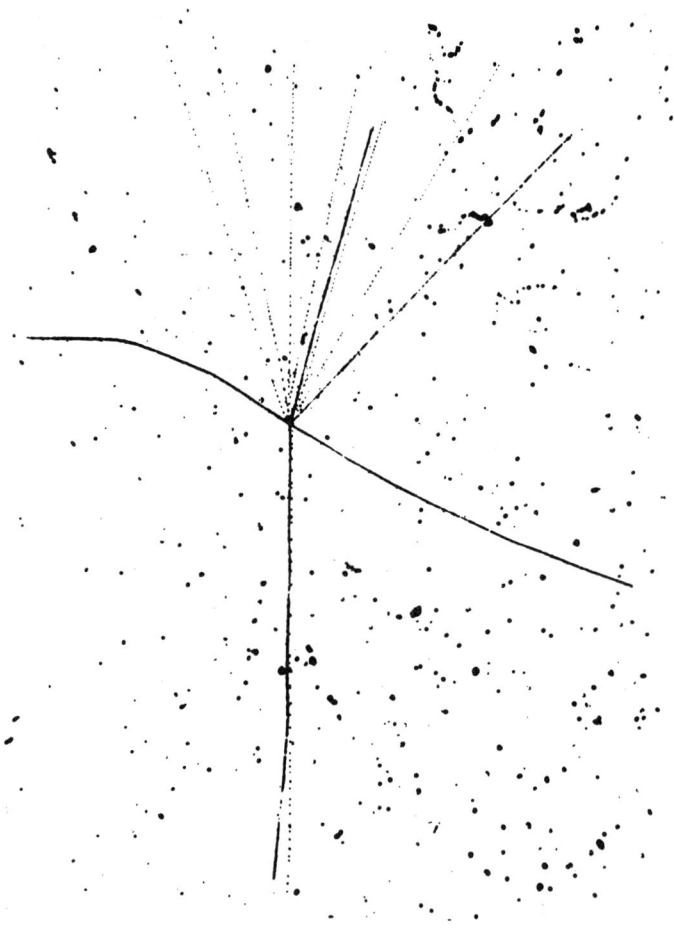

Figure 37–1   Tracks produced in a nuclear emulsion plate by the interaction of high-energy cosmic-ray particles with emulsion grains. Photographic plates were carried to high altitude by balloon in a Brookhaven National Laboratory experiment. (Courtesy of Brookhaven National Laboratory.)

between the colliding particles. Interactions between particles are due to the various sorts of force that exist, and arise, in turn, from various sorts of charge carried by the particles. The word *charge* is used here in its general sense to mean the source of a given force field. An example is mass, which is one kind of "charge" because it gives rise to gravitational force fields and produces gravitational interactions with other particles. A physical particle may be defined by the energy carried by these various "charges" and the momentum associated with each of them. The list of quantum numbers that specifies the signs and magnitudes of these "charges" and momenta therefore becomes the identity of the particle.

The quantum numbers are determined by experiment. Electrical charge, for instance, is often determined by the direction of curvature that a particle undergoes in traversing a magnetic field. When a particle with a new set of quantum numbers is found, we know that we have discovered yet another particle in nature.

37-2 QUANTUM NUMBERS OF ELEMENTARY PARTICLES

We may now speculate that the "elementary" particles of present-day physics are not truly fundamental, and also that the list of quantum numbers describing each particle is not yet complete. However, we shall study and examine the various quantum numbers that are now known to define elementary particles. These include rest mass, electrical charge, spin and magnetic moment, statistics, isospin, particle number, strangeness, parity, and hypercharge. Even more fundamental entities are beginning to appear both through current experimental work and theoretical concepts today—some of these are called quarks, some triplets, some partons; but we shall leave the discussion of these concepts to the next chapter. In this section, we shall define the various quantum numbers that are now known and have been investigated.

Rest Mass. Physicists claim that all particles, by their very existence, represent a localized finite concentration of energy, so by the Einstein mass-energy relation,

$$E = mc^2 = \sqrt{(m_0 c^2)^2 + p^2 c^2} \tag{37-1}$$

every particle exhibits the property of mass. This is very general, as we have seen before, and holds for any system under the special theory of relativity. For a particle at rest, we note that $p = 0$, and Equation (37-1) reduces to $E = m_0 c^2$. Some particles are known, such as the photon and the neutrino, that are never to be found at rest, and there are excellent theoretical grounds for considering that their rest masses m_0 are identically zero. They have no rest mass, and for them, Equation (37-1) reduces to

$$E = mc^2 = pc \tag{37-2}$$

The rest mass of an elementary particle provides one means of classification. In fact, no elementary particle has been isolated truly at rest, so the rest masses must be found from dynamical methods. In these measurements, particles are scattered from one another, and the resulting energies and angles of scattering are measured. This is an example of the amplification of a local event. By applying the kinematics of special relativity, we can obtain the mass of one particle as a ratio with the mass of another particle (see Figure 37-2). We then correlate the system of masses built up in this fashion with the gross mass of matter in bulk through Avogadro's constant. This constant is obtained, in turn, by macroscopic measurements utilizing electrolysis and Faraday's law, or by the analysis of Brownian motion as done by Perrin. These procedures for obtaining the masses of the elementary particles are analogous to those used to find the masses of the chemical atoms. The difference in these procedures lies in the types of measurements that are made. For elementary particles, kinematical quantities from collisions are measured or their trajectories through known magnetic fields. Typical bubble chamber photographs of the trajectories of electrons and positrons in a magnetic field are shown in Figure 37-3. For chemical atoms, the relative combining weights of the atoms are measured.

The rest masses of the elementary particles are often quoted as multiples of the rest mass of the electron m_e or of the proton m_p, or they are given in units of energy such as electron volts. Classification by mass is generally accomplished as shown in

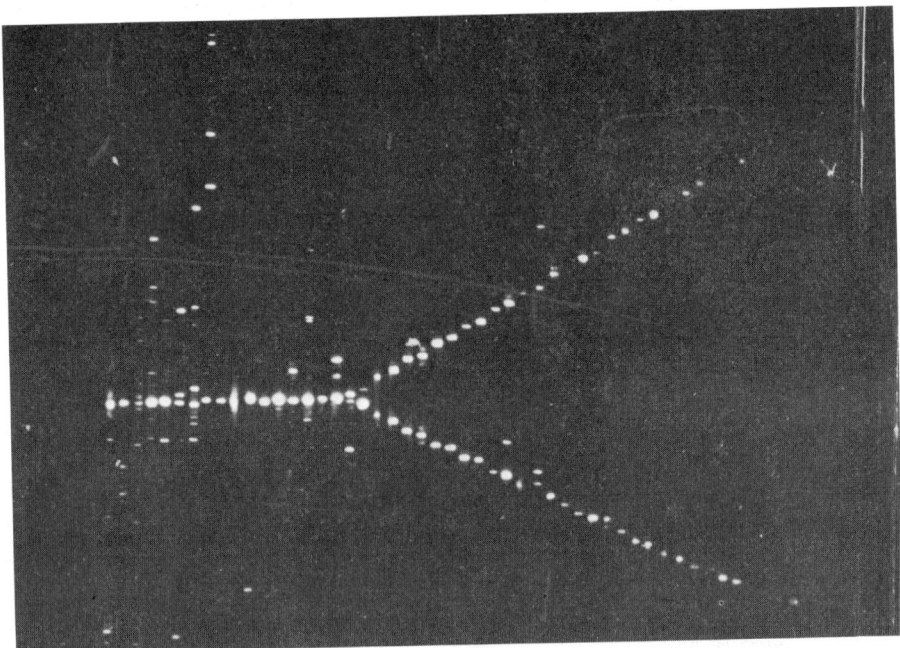

Figure 37-2 Spark chamber picture of the scattering of a 1.5-Gev π^- meson, which enters from the left. (From Harold Enge, *Introduction to Modern Physics*, Addison-Wesley, Reading, Mass., 1966. Used with permission.)

CHAPTER 37: ELEMENTARY PARTICLES · 485

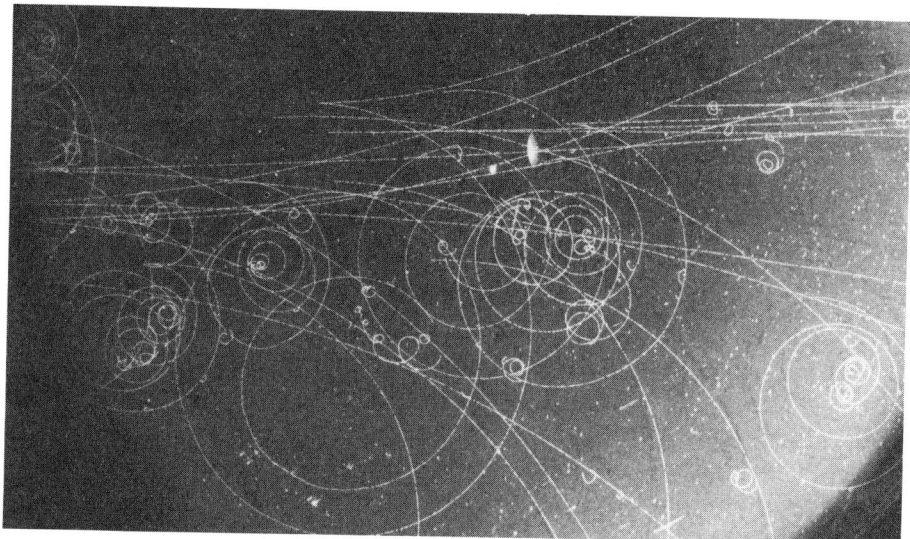

Figure 37–3 Hydrogen chamber picture showing Compton-scattered electrons and several examples of pair production. (From Harold Enge, *Introduction to Modern Physics*, Addison-Wesley, Reading, Mass., 1966. Used with permission.)

Table 37–1. These limits are approximate and encompass those particles presently known. We note that the striking feature of this classification is that the particles within each class have definite and characteristic values of rest mass. That is, there is no continuous spectrum of rest masses.

As we have noted previously, there are two apparently independent physical bases for defining the mass of any object: gravitational and inertial. The first basis requires that we treat mass as the source of a gravitational field. Then mass is defined in terms of the forces that exist between the masses due to these fields via Newton's universal gravitation equation,

$$F_G = G \frac{M_1 M_2}{r^2} \tag{37-3}$$

Table 37–1 Classification of the elementary particles

CLASS	REST MASS (MeV)
photons	0 (usually placed in a class of its own)
leptons	0 to 106 inclusive
mesons	135–888
baryons	938 and above

The second basis treats mass as the cause of the inertial forces that arise when an object of given mass undergoes an acceleration. This inertial force is given by another of Newton's equations,

$$F_I = \frac{d}{dt}(Mv) = Ma \qquad (37\text{-}4)$$

These "masses," whether gravitationally or inertially defined, are presumably identical in nature as well as in magnitude, for any given object. Delicate experiments which have been performed tend to confirm this view. It should be pointed out, however, that this philosophical question has not yet been settled definitively by experiment.

Both conceptions of mass are encompassed by Einstein's mass-energy relation, given by Equation (37-1), and are included in a unified way as one physical entity in the general theory of relativity. This theory dispenses with the Newtonian type of forces altogether and treats mass as the cause of a sort of bending or distortion of space-time in its near vicinity (refer to Chapter 4).

It is also presumed, on rather good but not exact experimental grounds, that the mass associated with antiparticles is identical for each particle in the pair.

Electrical Charge. By the close of the eighteenth century, Coulomb had proposed the quantization of electrical charge. His treatment was based on the accurate demonstration of the inverse-square law of force between two point charges *in vacuo*. Almost a century later, Thomson measured the charge-to-mass ratio for the electrons in cathode rays. Finally, the work of Millikan in determining the value of the electronic charge began the era of the study of elementary particles. Since that time, Millikan's law of multiple proportions has held fast. This law states that charges carried by atoms, nuclei, and the elementary particles are all found to be either zero or integral multiples, positive or negative, of the basic electronic charge. Only in recent years has there been a basis for suspecting the existence of non-integral electrical charges in nature. Such a set of particles was first suggested by Gell-Mann and Zweig in 1969;* they are popularly called *quarks*, and are presumed to have one-third and two-third multiples of electronic charge for each member. These quarks are further hypothesized to be elementary components of baryons and mesons; they will be discussed in the next chapter.

A number of searches have been made for these particles in accelerators, in nature by physicochemical methods, and in cosmic rays. Most results have been negative, only two experiments reporting possible success in finding the particles.†

---

* See M. Gell-Mann, *Phys. Lett.* **8,** 214 (1964) and G. Zweig, *CERN (Geneva) Report* No. 8419/TH **412** (1964) for the first discussions of these ideas.

† For accounts of the various experiments searching for quarks and the results of each to date, see Yu. M. Antipov et al., *Phys. Lett.* **30B,** 576 (1969); D. D. Cook et al., *Phys. Rev.* **188,** 2092 (1969); F. Ashton et al., *J. Phys. A: Proc. Phys. Soc. (London)* **1,** 569 (1968); C. B. A. McCusker and I. Cairns, *Phys. Rev. Lett.* **23,** 658 (1969); I. Cairns et al., *Phys. Rev.* **186,** 1394 (1969); T. Massam, *CERN (Geneva) Report* No. 68-24 (1968); and H. Faissner et al., *Phys. Rev. Lett.* **24,** 1357 (1970).

A positive demonstration of the existence of fractionally charged particles would open an important new field in the study of the elementary particles and would constitute one of the most startling and important discoveries of these times.

Spin and Magnetic Moment. Spin and magnetic moment are distinct, physically, from one another. One is a purely mechanical entity, while the second is an electromagnetic one. They are closely related to one another, however, so we shall discuss them under one heading. In 1921 Compton suggested that the charged electron might be spinning and so constitute a natural electromagnet which could be the unit of magnetism in atomic physics. This concept was later used quantitatively by Uhlenbeck and Goudsmit, and independently by Bichowsky and Urey, to explain the fine structure splitting of atomic spectral lines.

To account for these effects as well as some of the peculiar splitting of lines observed in the Zeeman effect, it is necessary to assume that the electron constitutes a magnet of strength equal to 1 Bohr magneton:

$$\mu_B = \frac{e\hbar}{2m_e} \tag{37-5}$$

This magnetic effect of the spinning electron may be stated in terms of the mechanical angular momentum of the spinning mass by considering both effects of spin as vectors. Then Equation (37-5) may be written

$$\boldsymbol{\mu}_s = -\frac{e}{m_e}\mathbf{L}_s \tag{37-6}$$

where $\boldsymbol{\mu}_s$ is the spin magnetic moment of the electron and $\mathbf{L}_s$ is its spin angular momentum.

A striking effect is now found for this spin angular momentum, understandable only as a quantum mechanical effect. The spin magnetic moment vector may, of course, be measured relative to any arbitrary axis. This axis, say the z axis, then specifies the corresponding component μ_{sz} of the magnetic moment vector $\boldsymbol{\mu}$. When this is done for a given electron, regardless of the direction of the axis being used, the result is always $\mu_{sz} = (e/m)L_{sz} = 2m_s\mu_B$. The value of L_s, the spin angular momentum, is always $(\sqrt{3}/2)\hbar$, and the value of the component of this vector in the direction of any axis defined is always $L_{sz} = L_s \cos\theta = m_s\hbar = \pm\frac{1}{2}\hbar$.

Thus, relative to any axis, there are only two directions available for the spin vector to point, instead of many such directions as exhibited by a child's spinning top. This spin vector, although defined in a three-dimensional (ordinary) space, has only two observable components, and relative to any defined axis, one of these is always $\frac{1}{2}\hbar$ and the other is $-\frac{1}{2}\hbar$. This sort of vector is called a *spinor*, and the statement is abbreviated by saying that the electron has "spin $\frac{1}{2}$."

We may obtain some physical feeling for this perpetually spinning object, the electron, by considering it in the light of a tiny, irreducible packet of energy. As such, it occupies a finite volume, and thus we have an elementary region of energy density. But energy density has the dimensions of pressure, and we have learned in mechanics that pressure implies a rate of change of momentum. So our elementary volume of energy density has an associated momentum that is continually changing. If it is to remain constant in magnitude, and if it is not to go anywhere else, then it must be changing by rotation around a fixed axis. This sort of momentum is angular momentum, or spin. We cannot push this model too far, but it does help in visualizing the physical situation.

Statistics. Statistics treats the relative probabilities of finding various possible distributions of n objects among m different containers. This statement becomes relevant to the discussion of elementary particles by defining as "containers" the physical states available to the particles. A most striking property of the elementary particles is that among any given type each is identical. That is, every electron is, apparently, identical to every other electron in the universe. Every proton is identical to every other proton. And so on down the list. Thus, when we visualize a system containing several electrons, it does not matter *which* electron occupies this position or that energy level; any one of those present will do as well. We must be careful to see that the statistical equations describing such a system have the same property.

The description of a system of electrons is written in terms of the wave functions of each of the electrons, putting together a linear combination of the wave functions for each of the particles. Each particle occupies a given region of space or a given band of energy or a given range of momentum, so these quantities must be in the wave functions of the different particles. But we have said that the particles are *not* different. To resolve this problem, we pretend, at first, that each particle may be labeled with a unique number, such as 1, 2, 3, and so on. Then a linear combination is formed from all possible combinations of each particle and each available state. The equation for the system must be insensitive to *which* electron goes *where* in the system. We require then that the total wave equation, when it is squared, be insensitive in magnitude to interchanges among the various electrons that have been labeled. Recall that it is not the wave function ψ but rather its square $\psi^*\psi = |\psi|^2$, that gives the physically observable system. When all this is done with care, it is found that any system may be described by one of only two kinds of equations: a wave function, which is a product of the individual wave functions, such as

$$\psi_i(r_i)\psi_k(r_k)\psi_l(r_l) \cdots \psi_n(r_n)$$

taken over all permutations among the various values of r, and one that is a determinant, such as

$$\begin{vmatrix} \psi_1(r_1) & \psi_1(r_2) & \psi_1(r_3) & \cdots & \psi_1(r_n) \\ \psi_2(r_1) & \psi_2(r_2) & \psi_2(r_3) & \cdots & \psi_2(r_n) \\ \psi_3(r_1) & \psi_3(r_2) & \psi_3(r_3) & \cdots & \psi_3(r_n) \\ \vdots & \vdots & \vdots & & \vdots \\ \psi_n(r_1) & \psi_n(r_2) & \psi_n(r_3) & \cdots & \psi_n(r_n) \end{vmatrix}$$

These two forms are insensitive in magnitude to interchanges among the ψ_i. The first one does not even change sign, so it is called *symmetric*. The determinant, however, changes its algebraic sign when two terms are interchanged, so it is called *antisymmetric*. It is found that every particle in nature obeys either the symmetric or the antisymmetric forms of wave function. The first is called a *boson* and obeys the Bose-Einstein statistics. If, on the other hand, a particle obeys the second form (the determinant), then it is called a *fermion* and obeys the Fermi-Dirac statistics. Bosons have integral intrinsic spin; that is, their spin is zero or an integer multiple of $\hbar$. The fermions have half-odd integer spin, that is, $\frac{1}{2}$, $\frac{3}{2}$, $\frac{5}{2}$, and so on multiples of $\hbar$.

Furthermore, the determinant for the fermions vanishes if two of the r's are the same. This is an expression of the Pauli exclusion principle, which in this case states that no two particles may be at the same point at the same time. The fermions are the "hard" particles of nature. They prevent the universe from collapsing into a point under the forces of attraction such as gravity, nuclear forces, electromagnetic forces, and so on.

Bosons, on the other hand, may occupy the same position. Photons may be superposed, for instance. The bosons are the quanta of the force fields in nature. Photons are the quanta of the electromagnetic field, gravitons are the quanta of the gravitational field, pions are the quanta of the strong nuclear fields and so on.

Isospin. Fermions have spin $\frac{1}{2}$, and the angular momentum vector may point either way along any given axis. It has two possible values, $+\frac{1}{2}$ and $-\frac{1}{2}$. Imagine, for a moment, an abstract space called *spin space*. This space has only two dimensions. A given particle, say an electron, may be described in this space by the symbols + or −, and in this space we have these two sorts of electron. A beautiful system of mathematics related to ordinary vector analysis has been developed around the properties of the *spinors*, the vectors in this spin space.

Now consider the neutron and the proton. They may be considered as being two different states of the same particle: a nucleon. Again, we may imagine an abstract space containing the nucleon. There is a vector attached to the nucleon in this space; if it points "up," the nucleon is a proton, while if the vector points "down," the nucleon is a neutron. There are no other possibilities in this restricted example we are considering. So the space is two dimensional, just as is the spin space above. The vector hooked to the nucleon is a spinor, and all the mathematics of spinors apply. This space is called isotopic spin space or, just *isospin space*. For convenience, the nucleon is said to have an isospin of $+\frac{1}{2}$ if it is in the proton state and an isospin of $-\frac{1}{2}$ if it is in the neutron state. Another way of saying the same thing is to say that the proton has isospin $+\frac{1}{2}$, and the neutron has isospin of $-\frac{1}{2}$.

These ideas may be extended to the bosons as well. The pion, for instance, exists as neutral, negative, or positive in electrical charge. These may be considered as merely three states of the same particle in isospin space. A vector of magnitude $+1$ describes the positive pion, of magnitude -1 describes a negative pion, and of magnitude zero describes the neutral pion. In other words, a positive pion has isospin +1, a negative pion has isospin −1, and a neutral pion has isospin zero. These three states of the pion constitute a *triplet* in isospin space. The two states of the nucleon above constitute a *doublet* in isospin space.

Rotations of a given vector in isospin space imply a transition, say from a neutron to a proton. Such a transition by a nucleon that is a member of a nucleus would be an example of ordinary β decay.

Particle Number. The fermions, the "hard" particles of nature, are conserved in a very particular way. They may interact with one another and with other particles, but *the total number of baryons remains constant*. For this reason, every baryon is given a baryon number +1, and a count may be kept using this number. In a similar fashion, *the number of leptons remains constant*, so each lepton (neutrino, electron, and negative muon) is given a lepton number of +1. The bosons do not possess this property, so all bosons (photons and mesons, for instance) are given a particle number zero.

Strangeness. The quantum numbers we assign are derived from observable properties of the particles in their interactions with other particles. The quantum number S, called *strangeness*, is a particularly transparent example of this fact. A number of baryons are known, but all appear to be merely excited states of the nucleon. Thus, all known baryons decay spontaneously to nucleons, preserving their baryon number along the way. They produce other particles in the process, but we ignore these for the present. Another fact concerning these decays is that they are all "slow" in terms of a nuclear time scale. Their half-lives are, typically, of the order of 10^{-10} or 10^{-9} sec. They are reluctant to decay, and the force driving them to do so is rather weak in overcoming this.

It was found, however, that (given sufficient energy) a proton–proton collision produces these excited states millions of times faster, and always in the company of a particular sort of meson, the K meson. Thus, production of excited baryons is always associated with the production of K mesons as well. The strong force that causes this excitation obviously induces a new property in the excited baryons. This property is called strangeness, and it gives the baryons numbers ranging from zero (for the nucleons) upwards in positive integers for the excited states. We assign negative values of S to the K mesons produced in association with the baryon excited states. Strangeness is thus conserved in the production of baryon states (the strong interactions). It is not conserved in their weak decays.

Parity. In the discussion of statistics, the effects on a state function of the interchange of coordinates between identical particles was considered. Let us consider a different sort of change in coordinates—a change in the coordinate system itself. If $-r_k$ is substituted for every r_k appearing in the argument of a state function, the coordinate system is merely reflected through its origin. This reflection through the origin is called a *parity transformation*. If the corresponding wave function changes sign, it is said to possess *odd parity*, while if the sign remains the same, it has *even parity*.

The physical world should not be sensitive to the sort of coordinate system that is imposed on it in order to measure it. This property then means that the parity of a particle or system of particles should remain the same following an interaction as it was before the interaction.

The parity of a scalar quantity is even, since it does not change sign under parity transformation. Some one-component quantities that do change sign under space inversion are known, however; these are called *pseudoscalars*. A three-component vector such as linear momentum, which does change sign under space inversion, has odd parity. A three-component vector such as angular momentum does not change sign under the parity transformation and is called a pseudovector. It has an even parity.

It has been found that *the parity of a system is conserved (it remains even or odd) under all types of interactions except the weak ones, where it is always changed*.

Hypercharge. It has been found possible to correlate the electrical charge (q), the isospin component (I_z), the baryon number (A), and the strangeness (S) for all the strongly interacting particles (the mesons and the baryons) by the relation

$$q = e\left(I_z + \frac{A}{2} + \frac{S}{2}\right) \qquad (37\text{-}7)$$

where e is the electronic charge.

The numbers A and S are commonly grouped as their sum and called the *hypercharge*:

$$Y = A + S \qquad (37\text{-}8)$$

so that Equation (37-7) reads

$$q = e\left(I_z + \frac{Y}{2}\right) \qquad (37\text{-}9)$$

The quantum numbers of all the strongly interacting particles satisfy this relation. At the present time, physicists do not know why.

QUESTIONS

37–1 What specific lessons did the discovery of the electron and the positron carry for the field of elementary particles?

37–2 What is meant by the term "quantum number" in reference to the elementary particles?

37–3 How is an elementary particle "observed"?

37–4 Discuss the difference between gravitational mass and inertial mass.

37–5 How does the inverse-square law of force define the concept of electrical charge?

37–6 How are the spin and the magnetic moment of a particle related?

37–7 What is meant by the "statistics" of an elementary particle?

37–8 In what sense does the Pauli exclusion principle apply to the elementary particles?

37–9 Discuss the role of symmetry in the concepts of isospin, strangeness, and parity.

37–10 In what sense is hypercharge a "charge"?

RECOMMENDED READING

ADAIR, R. K., and FOWLER, E. C., *Strange Particles*, Wiley, New York, 1963.
   A well-written book at the senior undergraduate or junior graduate level.

FRISCH, D. H., and THORNDIKE, A. M., *Elementary Particles*, Van Nostrand Reinhold, New York, 1964.
   Undergraduate level, strongly recommended reading by the Commission on College Physics.

SWARTZ, C. E., *The Fundamental Particles*, Addison-Wesley, Reading, Mass., 1965.

YANG, C. N., *Elementary Particles*, Princeton University Press, Princeton, N.J., 1963.
   This amazing book contains the Vanuxem Lectures given by Prof. Yang at Princeton University (1959). The lectures were intended for university people interested in physics. The clarity, simplicity, and insight of the author in this field have made this book a must in the library of every physics student, graduate or undergraduate, as well as for scientists.

38
Interactions of Elementary Particles

Luis Walter Alvarez
(1911–)

Born in San Francisco, Alvarez received his Ph.D. in physics from the University of Chicago in 1936. He was a member of the faculty of the University of California from 1936 to 1945. At present, Alvarez is Senior Physicist at the Lawrence Radiation Laboratory. His varied research in high-energy physics includes work on linear accelerators, the theory of elementary particles, and hydrogen bubble chambers. In 1968, for his contribution to the physics of subatomic particles and techniques for their detection, he was awarded the Nobel Prize.

38–1 ANTIPARTICLES
38–2 CLASSES OF INTERACTIONS
38–3 INTERACTIONS AND CONSERVATION LAWS

38–1 ANTIPARTICLES

We have seen in the last chapter that a particle is "named" by means of a set of quantum numbers. Included in this list is the spin of the particle. The spin angular momentum was hypothesized to explain the splitting of spectral lines, and the first formal theory of spin (that of Wolfgang Pauli) was based on nonrelativistic mechanics. This was quickly followed, however, by a relativistic theory formulated by P. A. M. Dirac. This theory predicted, among other things, that there should exist among the particles of nature positively charged electrons. It was easily extended to cover all fermions (spin $\frac{1}{2}$), and so it predicted negatively charged protons as well. These were quickly named as the *antiparticles* of the normally seen negative electron and the positively charged proton. Both have now been observed to exist experimentally, and the expected antiparticles of most other particles have been observed as well. Modern theory regards all particles as having antiparticles, but in some cases the particle is its own antiparticle!

An antiparticle may be thought of as an image produced by a kind of reflection in space-time. In this concept, all of the properties of the "object" are contained in the "image," except that some are reversed. The properties are the quantum numbers of the last chapter. Thus, if the object particle has a negative charge, a strangeness equal to $+1$, and a positive magnetic moment, its antiparticle might have a positive charge, strangeness equal to -1, and a negative magnetic moment. The masses would be the same. The parity and particle number, might be likewise reversed. Notice that we use the word "might." This is to cover the cases of bosons, which are their own antiparticles, and in which, therefore, these quantum numbers are not reversed except for the case of strangeness in the K mesons. The reversal is complete for the fermions except for the mass and spin. These are the same for both particles and antiparticles, and this holds for both fermions and bosons.

Consider the case of an object that is reflected in a plane, highly polished sheet of metal, say, copper. Let the copper sheet be grounded, and let the object be charged and have some charge distribution such that there are more negative charges near its top than near its bottom. Now consider the image. The optical image will have many of the features of the object, but it will be reversed, left-to-right. The light rays falling on the mirror, which come from the object, will be reversed in phase as they leave the image. The image will be charged, but *positively*. There will be a higher concentration of positive charge near its top than near its bottom, as is the case of the negative charges in the image. Thus, several of the properties of the object are contained in the image, but are reversed. Not all are reversed, for "up" at the object is still "up" at the image. Much may be learned from contemplation of this situation; the copper sheet plays the part (in a very limited sense) of the space-time "mirror." Space-time, however, reflects *all* properties into the image, reversing some, so that the image is just as "real" as is the object and lays claim to being a proper particle along with the rest. Which one is particle and which is antiparticle is merely a matter of common definition. Each is "anti" to the other, just as the two hands, right and left, are anti to one another in form.

Antiparticles bear another and very special relationship to one another: If particle and antiparticle of the same species (say an electron and a positron) are brought together, they combine with one another in one of nature's most violent interactions. Both disappear, and their entire mass-energy is converted to photons or to other particles. This has been termed "annihilation" and is so called in common terminology. It is *not* annihilation into nothingness, however, since energy and angular momentum are preserved and appear in the product particles, even if these product particles are photons. Bosons, which are their own antiparticles, may be thought of, in a sense, as continuously annihilating themselves and reappearing as the same bosons. The world of the elementary particles does, indeed, seem strange at times!

One concept of the particle–antiparticle relationship was provided by Dirac shortly after his development of the theory that predicts their existence. This is that all space is completely filled, jam-packed with particles in *negative energy states*. This "Dirac sea" filling all of space contains, in this view, both electrons and nucleons, as well as neutrinos and muons (another particle, a lepton, which we have not mentioned before). Because they are in a "negative energy state," none of these particles is visible or detectable in any way. Each is real, however, and each is characterized by its list of quantum numbers; but *its mass is negative*, and this makes it real but undetectable by experiment.

If the picture of the situation as drawn by Dirac were complete, then the world would be a void. But Dirac later postulated that some of these particles have been lifted out of the negative energy sea and that they have been given *positive* mass. They are now detectable and are "ordinary" particles in every way. A sufficient number of them have been thus lifted out of the negative sea so that we have the present observable universe. The "holes" left behind do not just fill up as would holes in water, but they remain and are the antiparticles we observe. These holes would have all the properties of the particles that occupied them, but the quantum numbers would be reversed. The mass, being a hole in a negative sea, would be positive.

This early picture drawn by Dirac was able to account for many of the observed phenomena, but it has now been incorporated in much more sophisticated form in the modern relativistic "field theory" used to describe the elementary particles today.

38-2 CLASSES OF INTERACTIONS

The interactions of the elementary particles may be considered as the direct source of all natural phenomena. The explosion of a star or the growth of a flower, the flight of a bee, or of a neutrino into the void—all are controlled from instant to instant through particle interactions. A number of interactions are considered in modern physics. We shall discuss the five most firmly based in theory—gravitation, weak interactions, electromagnetic interactions, strong interactions, and very strong interactions.

Gravitation. Gravitation was the first interaction to be known and is usually discussed in terms of the forces between macroscopic objects or in terms of the general relativity. It operates quite certainly on the elementary particle level, but it is so weak there that if any other forces are present, it is overwhelmed and (so far) is then undetectable. Its tremendous effect on the macroscopic and astronomical levels is due to the vast number of elementary particles that go to make up the earth or a star. As the gravitational field of each particle is added to that of all others in any body, the total force generated is sufficient to cause extreme pressures at the centers of stars and so create the conditions for other forces to come into play and generate stellar energy.

The strength of gravitation is commonly stated in terms of Cavendish's gravitational constant:

$$G_m = 6.7 \times 10^{-11} \text{ N-m}^2/\text{kg}^2$$

This may be put into dimensionless form for comparison with the other interactions by use of the mass of the nucleon M, Dirac's constant $\hbar$, and the speed of light c:

$$\boxed{g_m = G_m \left(\frac{M^2}{\hbar c}\right) = 6 \times 10^{-39}} \tag{38-1}$$

The quantum of the gravitational field is the *graviton*, undoubtedly a boson. It has not been observed experimentally as a unique particle. Gravitational waves have been observed, however, by J. Weber at the University of Maryland.* For this purpose, Weber uses an "antenna" made of a large bar of aluminum 2 m long by 1 m in diameter. The bar is suspended by a strong wire from its midpoint in a vacuum chamber, and exceedingly sensitive strain gauges are fixed to its surface. Any vibrations that occur in the bar are converted to electrical signals by the gauges and are recorded. A similar bar is suspended in a vacuum chamber at the Argonne National Laboratory near Chicago. As gravitational waves pass over the pair of

---

* See Joseph Weber, "The Detection of Gravity Waves," *Sci. Am.*, May 1971.

CHAPTER 38: INTERACTIONS OF ELEMENTARY PARTICLES · 497

bars, each is set into vibration. Careful analysis of the timing between the two sets of vibrations and of the vibrations themselves indicates that the gravity waves are coming from the center of our galaxy where, apparently, large amounts of matter are being explosively converted into gravitational energy (as well as into other forms at the same time).

The extremely small value of the gravitational interaction given by Equation (38-1) has, so far, kept it out of the detailed theory of the elementary particles, although any complete theory obtained in the future must include it.

Weak Interactions. The diameter of an atomic nucleus is of the order 10^{-15} m. When a very energetic cosmic-ray proton, or one traveling in the beam of a giant accelerator, strikes the nucleus and penetrates it, it takes about 10^{-23} sec for the proton to travel across the nucleus. The high-energy proton is traveling at close to the speed of light. The particle has, therefore, about this time to produce an interaction with the nucleus, and often a very large shower of particles emerges from the nucleus along with the original proton. A characteristic time for nuclear reactions may be taken as 10^{-23} sec.

Radioactive decay of nuclei and of the unstable particles, however, take billions of times longer than this. The shortest-lived β decays, for instance, require a time of the order 10^{-10} sec, while many are known that take seconds, minutes, or hours. Some are on record that take thousands of years. Evidently, the forces that drive β decay are much weaker than those that drive nuclear reactions in collisions.

A striking property of the weak interaction is that a single type of force seems to underlie this vast range of lifetimes. It is typified by the point interaction of four fermions, in this sense: Two fermions come together at a point in space and at an instant in time, and two other fermions emerge from this point instantaneously. The interaction, apparently, has no range as do the gravitational interaction and the electromagnetic interaction. If it does have a range over which it can act, that is, if the particles can actually be separated by some small distance during the interaction, then this range must be very small. It has not been measured as yet.

Let Δx equal the as yet undetected range of the weak interaction. If this is greater than zero, then some quantum of the force field must carry the interaction from one point to another. This is the function of the graviton in gravitational force and of the photon in electromagnetic force. The weak interaction quantum must lie within Δx, so that, by the uncertainty principle, .

$$\Delta x \, \Delta p \geqq h$$

where Δp is the uncertainty in the momentum of the particle. If we let the total momentum p of the quantum equal its uncertainty Δp, then we have set a limiting value on the momentum. In this case it is a lower limit. Since Δx is extremely small (much smaller than the diameter of a nucleus), the momentum p must be quite large. Assuming that the speed of the quantum is nearly the velocity of light c, we can write the momentum as Mc, and so the uncertainty relation becomes

$$M \geqq \frac{h}{(\Delta x)c}$$

Experimentally, the quantum should remain "virtual," so long as a given interaction occurs at too low an energy for the production of "real" quanta, that is, with kinetic energy in the center-of-mass system below the rest mass energy of the particle Mc^2. At energies above this, however, the quantum should be produced in a given interaction and should be observable traveling away from the point of the collision. The maximum energies available at the large accelerators have set a lower limit on Mc^2 of about 2 GeV, although much larger accelerators are being built and will be used to search for the quantum at higher rest mass values. If found, it is expected that the quantum will be a boson, as are the quanta of the other known force fields.

The strength of the weak interaction may also be expressed in terms of the universal weak interaction coupling constant G_w, although this constant is not dimensionless as are the others. It is about 1.4×10^{-62} J-m³. To make a comparison of this with other couplings, one may make use of the Compton wavelength of the pi meson to obtain an approximate value for a dimensionless constant:

$$\boxed{g_w = G_w^2 \left(\frac{1}{\hbar c}\right)^2 \left(\frac{\hbar}{Mc}\right)^{-4} \cong 5 \times 10^{-17}} \quad (38\text{-}2)$$

where M is the rest mass of the pion. Obviously, if one uses the rest mass of a presumed heavy quantum (the "intermediate boson") mentioned above, it is much smaller than this.

At present, there is discussion among physicists of the possible existence of a *superweak* interaction exhibited by a resonant binding state of muons with protons and with pions. It is too early, however, to say that this exists with certainty. Much more experimental and theoretical work must be done on this problem.

Electromagnetic Interactions. Electromagnetic interaction is the classical interaction of the Maxwell–Lorentz field between electrically charged particles. The quantum of the field is the photon. The interaction is responsible for an astonishing range of phenomena common to the observable world, including the emission of γ radiation from nuclei, all chemical reactions, and the emission of bremsstrahlung by accelerated charged particles. It is the prototype of the relativistically invariant field and is characterized in strength by the dimensionless fine structure constant:

$$\boxed{g_e = \frac{e^2}{4\pi\varepsilon_0 \hbar c} = \frac{1}{137}} \quad (38\text{-}3)$$

Strong Interactions. The strong interaction is the dominant nuclear interaction responsible for most nuclear phenomena, nuclear energy levels, and so on. It is the interaction between baryons at low energy (low relative momenta), such as between the nucleons in a nucleus. The quantum of the strong interaction field is the pi meson, or pion. The strength of the interaction is characterized by the constant

$$g_s = \frac{(g\mu/2M)^2}{4\pi\varepsilon_0 \hbar c} \cong 8 \times 10^{-2} \qquad (38\text{-}4)$$

where g is the pion–nucleon coupling constant, μ is the pion rest mass, and M is the nucleon rest mass. The constant g is the nuclear analog of the electrical charge e, and so might be considered as the "pionic charge" of the nucleon.

Very Strong Interactions. The very strong interaction between nucleons at relativistic energies loses the factor $(\mu/2M)^2$ and becomes typical of interactions between mesons, between mesons and baryons, and between baryons and their antiparticles. In these cases, the coupling constant is the strong interaction analog of the electromagnetic fine structure constant. It is dimensionless and may be written

$$g_{vs} = \frac{g^2}{4\pi\varepsilon_0 \hbar c} \cong 15 \qquad (38\text{-}5)$$

This interaction is responsible for the formation of the so-called meson and baryon resonances, or excited states of the mesons and baryon particles. It is also the coupling that drives their decay, and for this reason their lifetimes are very short. They appear in experimental apparatus as transient states of other particles, reminiscent of resonant phenomena or excited atomic states. A term sometimes used for the strongly interacting particles is *hadron*.

38–3 INTERACTIONS AND CONSERVATION LAWS

We have seen in earlier chapters that symmetries in various sorts of physical "spaces" result in conservation laws. Physical four-space, for instance (x, y, z, t), is characterized by a number of symmetries and corresponding conservation laws. By symmetry, we mean that a physical happening remains unchanged in the space even though a transformation of coordinates is carried out. In quantum mechanics, this is interpreted to mean that all transition probabilities of a given system remain the same under a given coordinate transformation. Each distinct symmetry operation on the coordinates is the source of a correspondingly distinct conserved quantity.

A familiar symmetry of four-space is translation in time, t. Dynamical events are unchanged by relocating the origin of the t axis. Merely changing the setting of a clock does not affect the outcome of a dynamical event, so long as the *rate* of the clock is not changed. This is one approach to the law of conservation of energy. Similarly, the fact that the origin of the (x, y, z) coordinate system may be located arbitrarily expresses the symmetry of space under linear translations. This symmetry produces the conservation of linear momentum.

Physical three-space (x, y, z) is *isotropic*; that is, it is symmetric under rotations about the origin. Dynamical happenings in physical three-space are independent of the particular orientation of the coordinate axes. The result is the law of conservation of angular momentum.

These three laws find complete satisfaction in all of the interactions known. Processes that are clearly dominated by any of the interactions we have discussed, from gravitational to the very strong interactions, are found to obey these three conservation laws. This situation changes, however, as we proceed to other symmetries and other conservation laws.

Reflection of the physical three-space (x, y, z) through the origin, like that shown in Figure 38–1, is known as the parity transformation. It carries the coordinates (x, y, z) into $(-x, -y, -z)$, and is found to affect the algebraic signs of vectors in different ways. A linear momentum vector, for instance, changes sign. An angular momentum vector does not change sign. This property of changing signs or not under the parity operation is expressed as *even parity* if the sign remains the same, and *odd parity* if it changes. All physical systems are found to conserve this property, except for those that undergo a change due to the weak interaction. If a given process is dominated by the weak interaction, then the parity of the system changes as the system goes from an initial to a final state. A striking result of this circumstance is that the neutrino, which is affected primarily by the weak interaction, is absolutely polarized in intrinsic spin, relative to its direction of travel. That is, if a neutrino is viewed along the direction in which it is traveling, it is always spinning in a counterclockwise direction. If an antineutrino is viewed along its direction of motion, it is always spinning clockwise. Nature seems to distinguish absolutely between right- and left-handed coordinate systems when she operates in the weak interactions. The parity operation is indicated by P.

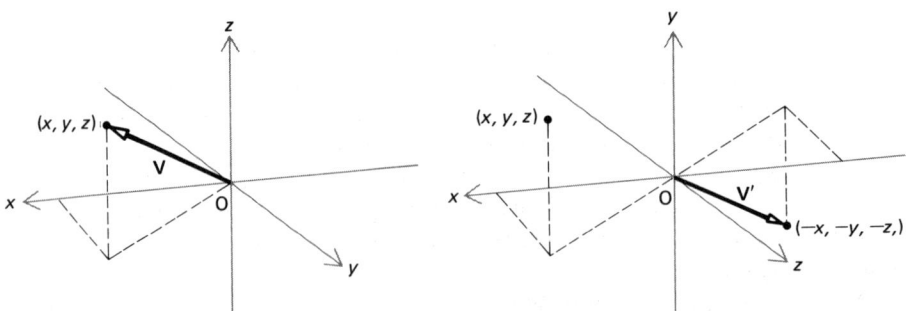

Figure 38–1 Reflection of coordinates x, y, z of the vector **v** through the origin into $-x$, $-y$, $-z$ is a parity transformation.

CHAPTER 38: INTERACTIONS OF ELEMENTARY PARTICLES · 501

There is another symmetry operation of interest in physical four-space. This is the reflection of the time axis t through the origin. This operation, indicated by the symbol T, describes physical happenings but in the backward direction, timewise. It describes the reversibility properties of thermodynamic systems and embodies the law of detailed balance. The reversibility properties of a system remain unchanged by application of the time-reversal operator T. Again, the weak interaction fails to preserve this law when it is in control, and so does the very strong interaction There are certain interactions involving the K mesons (kaons) that seem to violate the conservation law.

Moving from physical space into a more abstract one, we can use the positive, zero, and negative values of electrical charge as "axes" in a "charge space." The rotation of a vector pointing along the positive axis into a direction pointing along the negative axis would describe the change of the charge on a system from positive to negative. This is never found to happen if the vector describes the total charge on an interacting system of particles. The symmetry in charge space gives rise to the law of conservation of electrical charge, and of isospin at the same time. All interactions preserve this conservation law, except perhaps, for the gravitational interaction. Rotation in charge space is an operation named by the symbol C.

Let us proceed to another abstract sort of space, particle–antiparticle space (called G space), whose axes are four: the electrical charges $+$ and $-$, and 0 and the time axis t. A rotation in this space may be thought of as the simultaneous application of the operators C and T, or of the G operator CT. It is found that symmetry in G space under this rotation (called G conjugation) gives rise to the conservation of particle number. Lepton and baryon numbers, in particular, are conserved by this symmetry. It seems that the weak, the strong, and the very strong interactions preserve this symmetry, while the gravitational and electromagnetic ones violate it.

Yet another abstract space is strangeness space. Here, reflections through the origin are the symmetry operations that change the sign of the strangeness of a particle or of a system of particles. It is the cause of the conservation of the strangeness of a system of particles and is symbolized by S. The electromagnetic, the strong, and the very strong interactions preserve strangeness under this operation.

This brings us to a very special three-dimensional space composed of axes made up of isospin and strangeness and the rotations in this three-space. A particular group of rotations called the special unitary or unimodular unitary rotations in this space, given the symbol SU_3, have been most fruitful in classifying the mesons and baryons and in understanding them as families of particles. The operator of SU_3 causes rotations in this space while preserving the length of a vector (unimodular); it is called unitary because the successive application of the operator followed by its "adjoint" results in the original vector. For more details, we must refer the reader to texts on the theory of groups and of linear operators. Suffice it to say that the three-space of SU_3 may be constructed of three base vectors that may (or may not) have their counterparts in physical reality. These physical possibilities are particles given the names of *triplets*, *aces*, or *quarks*.

More about them will be discussed in the next chapter. We can define a quark charge, however, for the mesons and baryons. This quark charge is conserved by the strong and the very strong interactions only.

Finally, we may add the ordinary spin of the particles to those of isospin and strangeness to obtain yet another space (a six-dimensional one), in which we have the operation SU_6. This operation in the six-space conserves quark number and is preserved only by the very strong interaction.

Thus, we find that not only are there a variety of interactions in nature but also a variety of symmetries, charges, and conservation laws. We have not detailed all of them here, but we have included the most frequently discussed ones. Nature, on the other hand, is obviously *not* divided into compartments in this manner but is rather a single, all-encompassing unit of natural law. It is a weakness in our understanding, an incompleteness (and possibly error) in our theories, that forces us to divide the realm of dynamic happenings into these categories. Physicists are searching for a single theory that encompasses all of these aspects under a single interaction and that contains in itself all of the symmetries we observe in nature.

PROBLEMS

38-1 Compile a table of the properties of an electron, using those given in Chapter 37 as you best understand them, and then obtain the corresponding table for the antielectron. Do the same for the proton.

38-2 Show that the units of G_m are understandable in terms of its definition by Cavendish.

38-3 Compile a table showing the various interactions, the symmetry operations proper to the various conservation laws, and the quantity conserved by each, and indicate which conservation laws are obeyed by each interaction.

38-4 If the intermediate boson of the weak interaction has a rest mass equal to 5 GeV (5×10^9 eV), determine the range Δx of the weak interaction from the uncertainty principle.

RECOMMENDED READING

CHEW, Geoffrey F., GELL-MANN, Murray, and ROSENFELD, Arthur H., "Strongly Interacting Particles," *Sci. Am.*, February 1964.

GARDNER, M., *The Ambidextrous Universe*, Basic Books, New York, 1964.
  A popular and yet deep analysis of subjects such as the law of parity and its downfall, the origin of life, the Ozma problem, galaxies, suns, and planets.

MORRISON, P., "The Overthrow of Parity," *Sci. Am.*, April 1957.

TREIMAN, S. B., "The Weak Interactions," *Sci. Am.*, March 1961.

39
The Family of Elementary Particles

Murray Gell-Mann
(1929–)

Born in New York City, Gell-Mann received his Ph.D. from the Massachusetts Institute of Technology in 1951. He was appointed professor of physics at the California Institute of Technology in 1956. In the field of elementary particles, Gell-Mann has conducted very active research; he introduced the strangeness concept, developed the "eightfold way" of symmetry, and predicted the existence of the Ω-particle. He was awarded the Nobel Prize for physics in 1969.

39–1 PHOTONS
39–2 LEPTONS
39–3 HADRONS

39-1 PHOTONS

In the family of elementary particles, the *photon* is among the lightest and is usually listed separately from the rest. Its rest mass m_0 is zero, so that the general equation for the energy of a particle,

$$E^2 = m_0^2 c^4 + p^2 c^2 = m^2 c^4 \qquad (39\text{-}1)$$

reduces to

$$E = pc = mc^2 \qquad (39\text{-}2)$$

which yields a momentum for the photon of

$$p = \frac{E}{c} = mc \qquad (39\text{-}3)$$

Since the rest mass is zero but its energy is finite, the velocity of the photon is c, the velocity of light in a vacuum, and the momentum of the photon is

$$p = mc \qquad (39\text{-}4)$$

in agreement with Equation (39-3). Since the energy of a photon is also given by

$$E = h\nu \qquad (39\text{-}5)$$

where h is Planck's constant and ν is the frequency of the electromagnetic wave associated with the photon, we have for the mass of the photon

$$m = \frac{h\nu}{c^2} \qquad (39\text{-}6)$$

The photon, given the symbol γ, is the quantum of the electromagnetic field. As such it is a *boson* (it obeys Bose–Einstein statistics). All bosons have an intrinsic spin, which is either zero or an integral multiple of $\hbar$. The photon has "spin 1." The photon is itself uncharged, although it carries the Coulomb force field between electrical charges, either as a *virtual* particle in the static field or as a *real* particle in the radiation field.

The photon is stable and, therefore, does not decay spontaneously into any other particle. Its lifetime is thus infinite so long as it does not interact with other particles, and so photons continuously reach the earth from the farthest distances of the universe. These have traveled for a time comparable with the age of the universe itself. Such photons have, of course, partaken of the adiabatic cooling of the universe as it has expanded from an initially small, extremely dense fireball at the beginning of time. The fireball presumably contained radiation at a temperature of $10^{14}\,°K$ or more. In the ten or so billion years of expansion of the universe from this initial state, the primordial photons have cooled to about $3\,°K$ and are now observed as an isotropic radiation filling space.

Most of our information concerning the natural world is carried by photons. Not only physiologically, in terms of our natural sight, but also technologically, photons carry most of the data. Molecular and atomic energy states are known through the colors of the light that they emit and absorb; nuclear spins are known through their interactions with external electromagnetic fields, and excited nuclear states are often translated into γ radiation. Essentially all of our information concerning astronomical matter in the planets, stars, and galaxies comes to us as either radio or optical radiation. The photon may be thought of as the most commonly experienced elementary particle in its free state. The photon is its own antiparticle. It carries only the electromagnetic interaction, and so interacts with other particles only through this and through the gravitational field.

39-2 LEPTONS

Neutrinos. The lightest family of fermions (those particles obeying the Fermi-Dirac statistics) is known as the family of *leptons* and includes four particles and four antiparticles. The lightest of the four are the neutrinos, which have rest mass zero and so obey Equations (39-1) through (39-6), just as do the photons. The two kinds of neutrino known are the *electron-neutrino* (v_e) and the *muon-neutrino* (v_μ). Each is associated with its corresponding antiparticle. Both have spin $\frac{1}{2}$, and neither has an electrical charge. The two are differentiated from one another by their couplings, one to the electron field and the other to the muon field. These couplings are through the weak interaction. When an electron is involved in a weak interaction, so is the corresponding electron-neutrino. When a muon is involved in a weak interaction, the corresponding muon-neutrino is also involved.

A typical reaction involving electrons and neutrinos is that of nucleon β decay,

$$n \to p^+ + e^- + \bar{v}_e$$
$$\text{and}$$
$$p^+ \to n + e^+ + v_e$$

(39-7)

and its inverse,

$$\bar{\nu}_e + p^+ \to n + e^+$$

and (39–8)

$$\nu_e + n \to p^+ + e^-$$

Typical reactions involving muons and neutrinos are those of pion decay,

$$\pi^- \to \mu^- + \bar{\nu}_\mu$$

and (39–9)

$$\pi^+ \to \mu^+ + \nu_\mu$$

and the muon production reaction,

$$\bar{\nu}_\mu + p^+ \to n + \mu^+ \tag{39–10}$$

Note that in all of these reactions an electron is associated with its antineutrino when both appear on the same side of the arrow, but is associated with its neutrino when they appear on opposite sides. The same rule holds for the muon and its neutrino. This is a consequence of the law of conservation of particle number, in this case the conservation of lepton number. If particles are given the number $+1$ and antiparticles (those with bars over them) are assigned -1, then the total particle number remains constant throughout the reaction.

Neutrinos carry only the weak interaction and so interact only very weakly with other particles. This is expressed through an exceedingly small interaction cross section and a correspondingly long mean free path through matter. A typical neutrino interaction cross section is of the order 10^{-49} m^2 or less, while the maximum known is about 10^{-42} m^2. A typical neutrino mean free path is thousands of light years long through condensed matter such as lead.

Neutrinos are the only known particles that appear to be absolutely polarized in spin with reference to their momentum vectors. Neutrinos spin counterclockwise (with *negative helicity*), and antineutrinos spin clockwise (with *positive helicity*) relative to their directions of travel. Neutrinos are stable against decay into other particles.

Electrons and Muons. The two charged members of the lepton family are the electron, e, and the muon, μ. These two particles act very much alike, although they differ in rest mass. The electron has a rest mass of 0.510 MeV, while the muon has a rest mass more than 200 times this: 105.66 MeV. Both the electron and the muon have spin $\frac{1}{2}$, and the particle of each has a negative electrical charge. The antiparticle of each is positively charged.

The electron is stable against decay into other particles, while the muon decays into an electron and two neutrinos:

$$\mu^- \to e^- + \bar{\nu}_e + \nu_\mu \tag{39-11}$$

The positive muon (the antiparticle) decays into the charge conjugate states of the same three products:

$$\mu^+ \to e^+ + \nu_e + \bar{\nu}_\mu \tag{39-12}$$

In these decays, the electron-neutrino is associated with the presence of an electron and the muon-neutrino with the muon as in the preceding reactions. Again, note that lepton number is conserved in these decays. The mean lifetime of the muon in these decays is 2.2×10^{-6} sec.

The electron was the first of the elementary particles to be found in nature (by J. J. Thomson), while the muon was the first cosmic-ray particle heavier than the electron to be found. It was named the "mu meson" at the time, but because of its properties has been included in the lepton family rather than the meson family in recent years. The muon has presented a rather deep puzzle to modern physics in terms of its classification, because it seems to be no more than a heavy (perhaps excited) electron. Some slim experimental evidence appeared in 1970 that the muon forms resonant states with the proton and with the pion (the pi meson), but much more work remains to be done in confirming this. If true, then the question is raised concerning the possible existence of the same property in the electron. To find such resonant states might well constitute the discovery of yet another interaction in nature intermediate between the weak and the strong interactions.

A most fascinating effect caused by the heavy mass of the muon has been discovered by L. Alvarez and co-workers. Since the muon is more than 200 times as massive as the electron but carries a negative electrical charge, it can be bound to a proton as a heavy atomic "electron" to form a kind of heavy hydrogen atom (*muonic hydrogen*). The binding energy is proportional to the mass, however, so it is bound by some 200 times as much as the 13.6 eV that binds the electron in hydrogen. The Bohr orbit is, moreover, inversely proportional in radius to the mass of the particle and so is less than $\frac{1}{200}$ times the radius of the normal Bohr orbit in hydrogen. This is more exactly expressed as the mean radius of the wave function of the muon in its *K* shell.

Now, both ordinary hydrogen and this muonic hydrogen can bind an additional proton to form a *positive hydrogen-ion molecule*. The high binding energy of the muonic hydrogen brings the two protons so close together that their wave functions overlap appreciably. This situation causes a proton–proton reaction, with the formation of a deuteron and the ejection of a fast positron. Actually, the reaction was observed to occur in liquid deuterium, producing a $^3_2$He nucleus and a neutron along with the release of considerable energy. The muon causing this reaction was then shaken loose from the $^3_2$He and was free to cause another such event. Thus, an elementary particle was found to catalyze a nuclear reaction and start a chain of such reactions. The chain is broken when the muon decays or fails to shake free of the $^3_2$He.

The positive muon can form another sort of "hydrogen" by binding an electron to it in a Bohr orbit. The binding energy in the *muonium* atom is less than $\frac{1}{200}$ that in ordinary hydrogen. Muonium lives as a stable atom until the muon nucleus decays by Equation (39–12). Negative muons are capable, of course, of forming the charge conjugate state of muonium with positive electrons.

The electron is the most venerable of the elementary particles. It is the common carrier of electrical charge in the practical world as well as being responsible for essentially all chemical reactions and compounds.

39–3 HADRONS

Hadrons carry the electromagnetic, the weak, and the strong interactions. They may be grouped into two great families: the mesons (spin 0, 1, etc.) and the baryons (spin $\frac{1}{2}$, $\frac{3}{2}$, etc.). The name hadron means strongly interacting particle. The hadrons have been found to fit into a classification scheme that organizes them into subfamilies (called supermultiplets) that have common spins and parities among all the particles. This classification scheme is based on the symmetrical states reached by the group of rotations in three-space, which are unimodular and unitary. This group, symbolized by SU_3, gives rise to the concept of three fundamental particles from which all hadrons may be built.

Quarks. It is not within the scope of this text to discuss the symmetry groups such as SU_3, but we may start with the three fundamental particles and build from there. They have been named *quarks*, from a coined word in James Joyce's *Finnigan's Wake*. The theory is a development of an earlier, partially successful one by Sakata, which used the proton (p), the neutron (n), and the lambda particle (Λ) as three fundamental particles from which to build the hadrons. For this reason, the same symbols are used in the modern theory for the three quarks. We here name them the *park* (p), the *nark* (n), and the *lark* (λ). The quarks should not be confused with the hadrons, which bear the same symbols.

Referring to Figure 39–1(a), it is seen that the three quarks may be plotted on a Y, I_z diagram to form an equilateral triangle with 120° symmetry. Rotation of the triangle around the axis by 180° brings the system into the antiquark configuration. The two triangles are indicated in Figure 39–1(b). It is a fascinating fact that all known hadrons may be composed of these two triangles to yield the proper set of quantum numbers for each.

Recall that the charge Q is given by

$$Q = e\left(I_z + \frac{Y}{2}\right)$$

The values of I_z and Y in Figure 39–1(a) result in fractional electrical charges on the particles. These, along with the other quantum numbers for the quarks, are

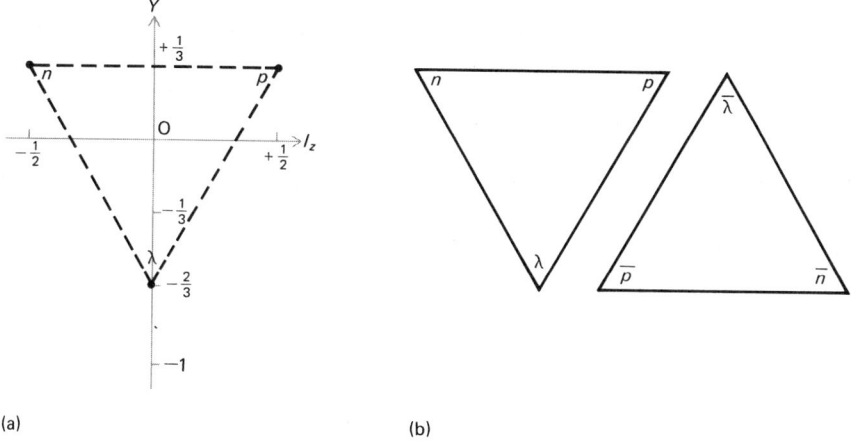

Figure 39–1 (a) The three quarks plotted as hypercharge Y versus z component of isospin I_z. The nark (n) and the park (p) form an "isospin doublet," as do the neutron and the proton. Both have the same value of hypercharge. The lark (λ) constitutes an isospin singlet. (b) The quark triangle and the antiquark triangle. The symbols stand for the nark (n), the park (p), and the lark (λ). All the known mesons and baryons may be composed of these two triangles, and the resulting diagrams have simple symmetries as appear in the following figures. Note that they are obtained from one another by rotation around the origin of (a).

given in Table 39–1. A similar table may be obtained by using the rotated quark diagram. In that case, the baryons numbers B and the spins σ remain the same.

Whether the quarks constitute real, observable particles or are just mathematical aids in classification of the hadrons remains to be seen. In the early part of the 1970s, several possible cases of quark observation in cosmic rays have been reported, but they remain to be confirmed by further work. If they do exist phys-

Table 39–1 The three quarks and their respective quantum numbers

QUARK NAME	SYMBOL	Q	S	Y	I_z	B	σ
park	p	$+\frac{2}{3}$	0	$\frac{1}{3}$	$\frac{1}{2}$	$\frac{1}{3}$	$\frac{1}{2}$
nark	n	$-\frac{1}{3}$	0	$\frac{1}{3}$	$-\frac{1}{2}$	$\frac{1}{3}$	$\frac{1}{2}$
lark	λ	$-\frac{1}{3}$	-1	$-\frac{2}{3}$	0	$\frac{1}{3}$	$\frac{1}{2}$

NOTES

Q = electrical charge in units of an electron charge
S = strangeness quantum number
Y = hypercharge quantum number ($= B + S$)
I_z = z component of isospin quantum number
B = baryon number
σ = intrinsic angular momentum quantum number (spin)

Note that the park and the nark have the same symbols as are commonly assigned to the nucleons proton and neutron. This circumstance stems from historical usage in the Sakata model.

ically, then a number of very interesting consequences result. Theories of cosmology and of stellar energy production will have to be altered. Quark catalysis of nuclear reactions will occur, and so on. It is probable that at least one of the three quarks is stable against disintegration, and perhaps all three are stable.

Mesons. The lightest of the two families of hadron is that of the *mesons*. They are all of either zero spin or of integral spin and may be considered as the quanta of the strong nuclear force field. The mesons may be formed of combinations of a quark and an antiquark, as shown in Figure 39–2. Here the triangles of Figure 39–1(b)

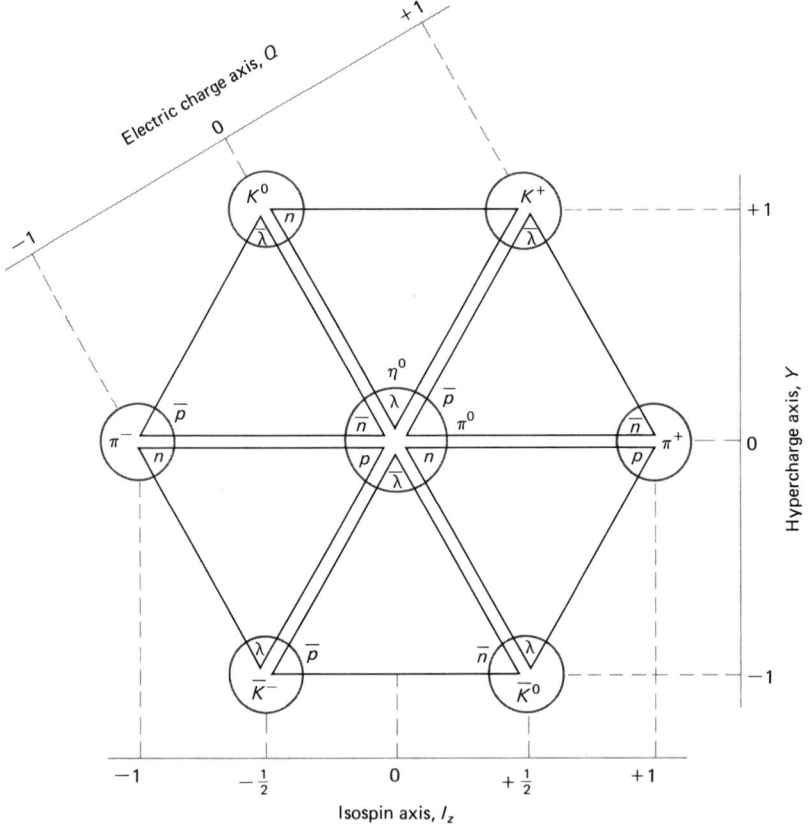

Figure 39–2 Composition of an octet of mesons (all those of zero spin and odd parity) from the basic quark and antiquark triangles. Each peripheral vertex of the hexagon is composed of a pair of quarks. These pairs combine to form the corresponding mesons as marked (K^0, K^+, etc.). Note the threefold symmetry of the diagram in terms of the three axes drawn. The central vertices combine to form a nonstrange meson π^0 and a strange one η^0. There are eight mesons in all accounted for in this diagram. See Table 39–2.

are used to compose a hexagon, each vertex of which is a quark–antiquark pair and which constitutes a particular meson as marked. The central six vertices combine to form a *nonstrange meson* π^0, which constitutes an *isospin triplet* with the π^+ and the π^-. They also form a *strange singlet*, the η^0. Taken all together, these eight mesons form the spin zero, odd parity octet. (See Table 39–2.)

Table 39–2 The spin-zero octet (family-of-eight) mesons and their corresponding quark charges

NAME OF MESON	SYMBOL	QUARK CHARGE	QUARK SYMBOLS
K zero	K^0	antilark + nark	$\bar{\lambda}n$
K plus	K^+	antilark + park	$\bar{\lambda}p$
K minus	K^-	lark + antipark	$\lambda\bar{p}$
anti K zero	$\bar{K}^0$	lark + antinark	$\lambda\bar{n}$
pi minus	π^-	nark + antipark	$n\bar{p}$
pi plus	π^+	park + antinark	$p\bar{n}$
pi zero	π^0	$\left\{\begin{array}{c}\text{nark + antinark}\\ +\\ \text{park + antipark}\end{array}\right\}$	$n\bar{n} + p\bar{p}$
eta zero	η^0	$\left\{\begin{array}{c}\text{nark + antinark}\\ +\\ \text{park + antipark}\\ +\\ \text{lark + antilark}\end{array}\right\}$	$n\bar{n} + p\bar{p} + \lambda\bar{\lambda}$

NOTES: Each meson may be considered as being composed of a pair of quarks as listed, except for the π^0 and the η^0, which contain more. The octet is represented in this composition by Figure 39–2. This family of mesons may be considered the ground states of the meson family for decays via the strong interactions. They all decay to leptons via weak interactions.

These eight mesons constitute the ground states of a whole spectrum of mesons. The first excited levels of this spectrum also form an octet with the same symmetry (Figure 39–3). Many other levels exist as excited particles that decay via strong interactions to the ground state octet. These are shown in Figure 39–4, where the dominant decays are indicated. The ground state mesons decay via weak interactions to leptons as listed in Table 39–3.

Baryons. The baryons are the heavy "hard" particles of nature. They have spin $\frac{1}{2}$ so they are fermions and obey the Pauli exclusion principle. Without this property, all of nature would quickly collapse into a point! Like mesons, the baryons may also be composed of quarks, but now of three quarks rather than a quark–antiquark pair. They may, therefore, be arranged in SU_3 representation diagrams.

512 · PART FIVE: ELEMENTARY PARTICLES

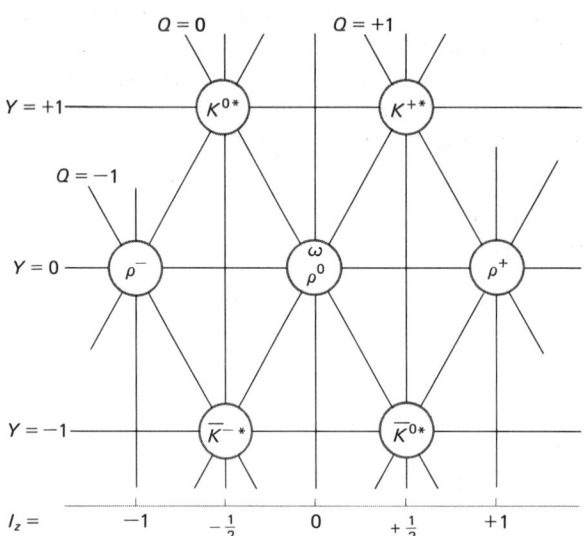

Figure 39–3 A second octet of mesons, those of spin 1 and odd parity, can be arranged in the same manner as are those of Figure 39–2. They may be thought of as excited states of the mesons in Figure 39–2, since they decay to those states as shown in Figure 39–4. A ninth meson belongs to the family, the ϕ, and constitutes a singlet as companion to the above octet.

Figure 39–4 Energy levels of the meson, after the scheme of V. F. Weisskopf. The known and confirmed particle resonance states are shown as energy levels, using their rest masses as reported in the "Particle Properties Tables," *Rev. Mod. Phys.* **41**, 109 (1969). Since the field is still actively developing, some of those shown are probably incorrect, while others not shown remain to be discovered and confirmed. The particle name corresponding to each level is indicated at that level.

The dominant strong decay modes are shown as lines connecting the levels. Two meson octets are indicated by underscoring the levels, the lowest being that of Figure 39–2 and the higher that of the spin-1 mesons (Figure 39–3). Notice the symmetry in the level pattern and decay modes of the $\bar{K}$ and the K mesons. The ground state mesons all decay either to γ rays through the electromagnetic interaction or, ultimately, to leptons through weak interaction decays as shown in Table 39–3.

The multiplicity of each level is given by the value of $(2I + 1)$, as indicated under each column. Thus, there is one particle corresponding to each singlet level, two to each doublet, and three to each triplet level.

Many alternate decay modes are seen in low relative abundance or are suspected.

CHAPTER 39: THE FAMILY OF ELEMENTARY PARTICLES · 513

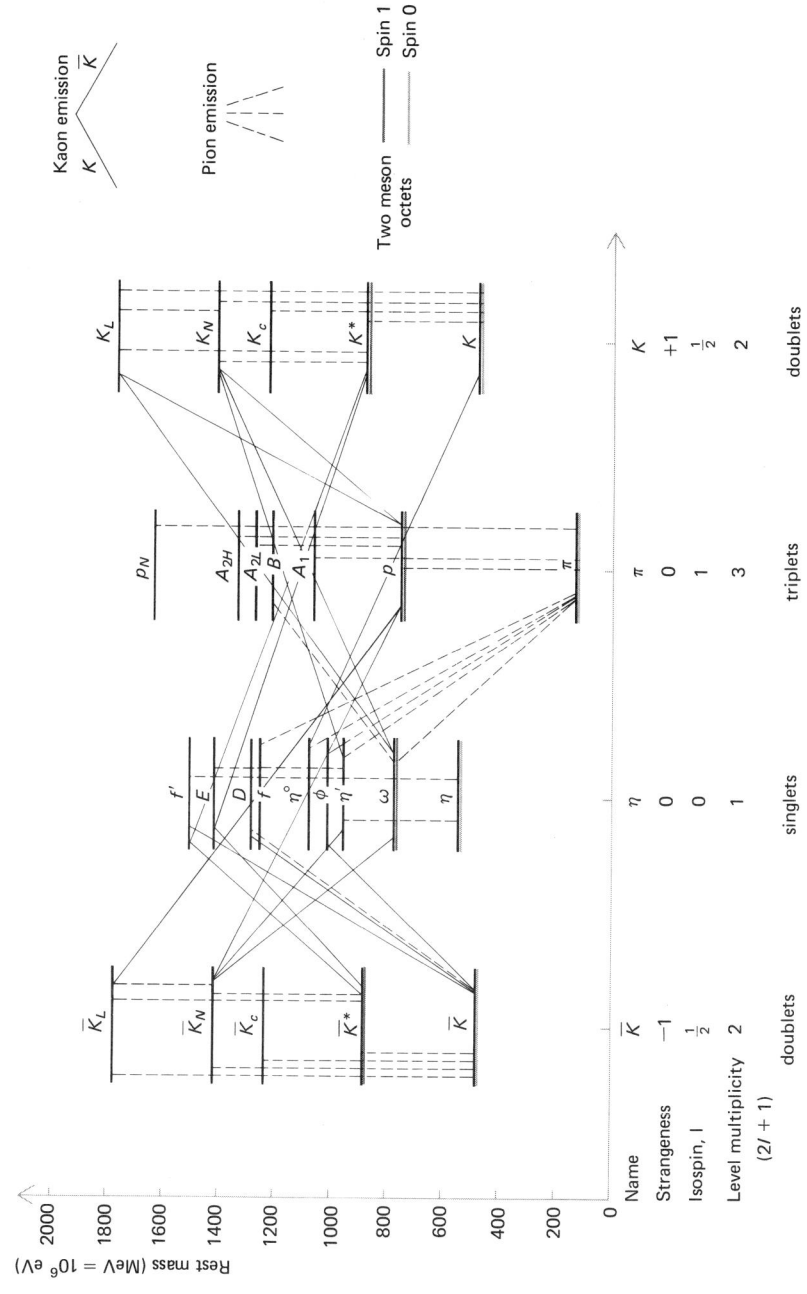

Table 39-3 Ground states of the elementary particles with their quantum numbers

SYMBOL	MASS, M (MeV)	SPIN, σ	PARITY, P	ISOTOPIC SPIN I	I_z	PARTICLE NUMBER BARYON, B	LEPTON, L	STRANGENESS, S	HALF-LIFE, T (sec)	PRINCIPAL DECAY MODES
PHOTONS										
γ	0	1	-1	—	—	0	0	0	—	stable
LEPTONS										
$\nu_e, \overline{\nu}_e$	0	$\tfrac{1}{2}$	—	—	—	0	1, -1	—	—	stable
$\nu_\mu, \overline{\nu}_\mu$	0	$\tfrac{1}{2}$	—	—	—	0	1, -1	—	—	stable
e^-, e^+	0.51098	$\tfrac{1}{2}$	1	—	—	0	1, -1	—	—	stable
μ^-, μ^+	105.658	$\tfrac{1}{2}$	1	—	—	0	1, -1	—	1.52×10^{-6}	$\mu^\pm \to e^\pm + \nu_e + \nu_\mu$
HADRONS										
MESONS										
π^+, π^-	139.58	0	-1	1	1, -1	0	0	0, 0	1.78×10^{-2}	$\pi^\pm \to \mu^\pm + \nu_\mu$
π^0	134.99	0	-1	1	0	0	0	0	0.7×10^{-16}	$\pi^0 \to \gamma + \gamma$
K^+, K^-	493.8	0	-1	$\tfrac{1}{2}$	$\tfrac{1}{2}, -\tfrac{1}{2}$	0	0	1, -1	0.8×10^{-8}	$K^\pm \to \pi^\pm + \pi^0$ $K^\pm \to \pi^\pm + \pi^+ = \pi^-$
$K^0, \overline{K^0}$	497.8	0	-1	$\tfrac{1}{2}$	$\tfrac{1}{2}, -\tfrac{1}{2}$	0	0	1, -1	0.7×10^{-10} 4×10^{-8}	$K_1^0 \to \pi^+ + \pi^-$ $K_2^0 \to \pi^+ + \pi^- + \pi^0$
η^0	548	0	-1	0	0	0	0	0	($<10^{-16}$)	$\eta^0 \to \pi^+ + \pi^- + \pi^0$

CHAPTER 39: THE FAMILY OF ELEMENTARY PARTICLES · 515

BARYONS

Particle	Mass	J	I	I_z	B	S	τ (s)	Decay modes
$p^+, \overline{p^-}$	938.21	$\tfrac{1}{2}$	$\tfrac{1}{2}$	$\tfrac{1}{2}, -\tfrac{1}{2}$	$1, -1$	0	—	stable
$n, \bar{n}$	939.50	$\tfrac{1}{2}$	$\tfrac{1}{2}$	$-\tfrac{1}{2}, +\tfrac{1}{2}$	$1, -1$	0	0.71×10^{3}	$n \to p^+ + e^- + \nu_e$
$\Lambda^0, \overline{\Lambda^0}$	1115.4	$\tfrac{1}{2}$	0	0	$1, -1$	$-1, +1$	1.8×10^{-10}	$\Lambda^0 \to p^+ + \pi^-$
								$\Lambda^0 \to n + \pi^0$
$\Sigma^+, \overline{\Sigma^-}$	1189.2	$\tfrac{1}{2}$	1	$1, -1$	$1, -1$	$-1, +1$	0.6×10^{-10}	$\Sigma^+ \to p^+ + \pi^0$
								$\Sigma^+ \to n + \pi^+$
$\Sigma^-, \overline{\Sigma^+}$	1197.6	$\tfrac{1}{2}$	1	$-1, +1$	$1, -1$	$-1, +1$	1.2×10^{-10}	$\Sigma^- \to n + \pi^-$
$\Sigma^0, \overline{\Sigma^0}$	1193.2	$\tfrac{1}{2}$	1	$0, 0$	$1, -1$	$-1, +1$	$<10^{-14}$	$\Sigma^0 \to \Lambda^0 + \gamma$
$\Xi^-, \overline{\Xi^+}$	1321.0	$\tfrac{1}{2}$	$\tfrac{1}{2}$	$-\tfrac{1}{2}, +\tfrac{1}{2}$	$1, -1$	$-2, +2$	0.9×10^{-10}	$\Xi^- \to \Lambda^0 + \pi^-$
$\Xi^0, \overline{\Xi^0}$	1310	$\tfrac{1}{2}$	$\tfrac{1}{2}$	$\tfrac{1}{2}, -\tfrac{1}{2}$	$1, -1$	$-2, +2$	1.0×10^{-10}	$\Xi^0 \to \Lambda^0 + \pi^0$
$\Omega^-, \overline{\Omega^+}$	1675	$\tfrac{3}{2}$	0	$0, 0$	$1, -1$	$-3, +3$	$\sim 10^{-10}$	$\Omega^- \to \Xi^- + \pi^0 + K^0$

NOTES: Those listed as stable are "absolutely stable," while those shown as decaying are unstable against decay by the weak interaction. All are stable against decay via strong interactions. Note that the only stable hadron is the proton. Both particle and antiparticle states are listed. The double values listed under I_z, B, and S refer to these states, respectively.

SOURCE: Adapted from M. Stanley Livingston, *Particle Physics*, McGraw-Hill, New York, 1968.

The ground states of the baryon family are presented in such a diagram in Figure 39–5. It is seen that the nucleons n and p form an isospin doublet, the three Σ's form a triplet, the two Ξ's form a doublet, and the Λ forms a singlet.

The baryons also have a plethora of excited levels. The first set of such levels forms a family of ten, the baryon decuplet shown in Figure 39–6. Here the Δ's form an isospin quartet, the Σ's form a triplet, the Ξ's form a doublet, and the Ω constitutes a singlet.

These and the other known excited states are shown in the baryon spectrum diagram of Figure 39–7. The neutron and the proton are the lowest of the baryons and constitute the ultimate ground state for the system. These two particles are the constituents of atomic nuclei in their ground states, and as such are absolutely stable against decay into particles. They may, of course, change into one another in the nucleus via the weak interaction in nuclear β decay. The free neutron is radioactive and decays into a proton by β decay. The proton seems to be absolutely stable in all circumstances.

The excited states of the baryons are components of excited nuclei, and so play important roles in the systematics of nuclear transitions. While considerable progress has been made in the classification of the hadrons using the symmetry group SU_3,

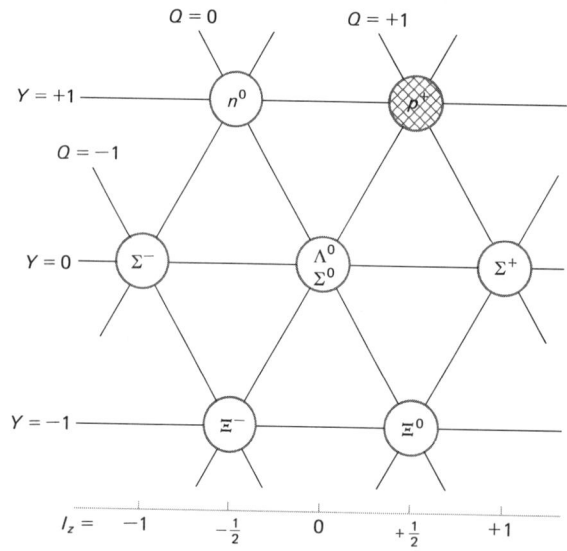

Figure 39–5 Octet of baryons comprising the "ground" states of the baryon family reached by decay via strong interactions (emission of pions or kaons). All are stable against decay via strong interactions. They decay via weak interactions to the proton p^+, which appears to be absolutely stable against all decays of every sort. Unlike the meson octets, which include antiparticles, this diagram shows only the particle states. A similar diagram with opposite charges describes the antiparticles.

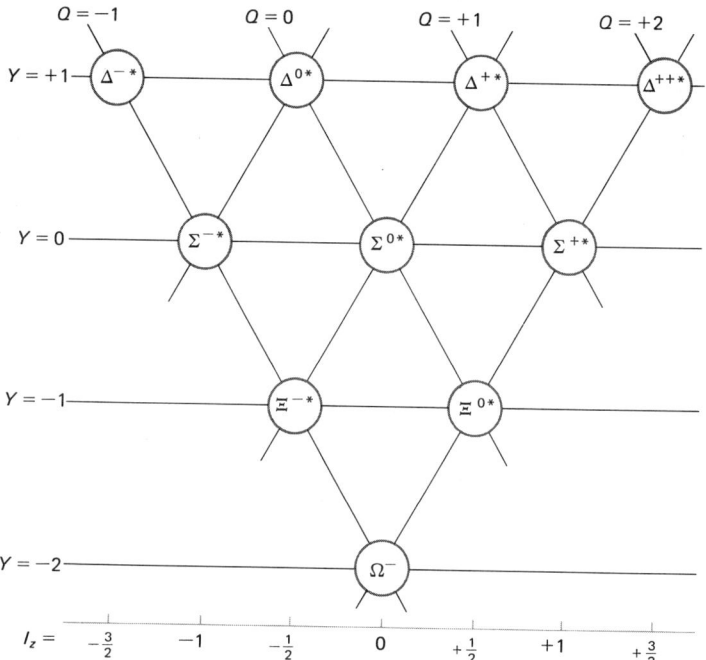

Figure 39-6 Baryon decuplet (family-of-ten) of baryon resonance particles, first excited states of the baryons of Figure 39-4. These decay via strong interactions to the particles of Figure 39-5. Again, like those, only particle states are included. A similar diagram with charges reversed describes the antiparticles.

and some insight may be developing into the structure of the hadrons in terms of quark components, much remains to be done in understanding the basic reasons for these regularities and in classifying the excited states of the hadrons according to some general principles.

Recent years have seen a new approach to the problem of the elementary particle commonly called the "bootstrap theory." In this view, no particle is more fundamental than any other, and each results from the existence of the rest.

Internal Structure of the Hadrons. The principal frontier in elementary particle physics is in determining the details of the structure of the various particles. Several models for the baryon may be conceived. One might be simply a hard particle with a distributed electrical charge throughout. A more sophisticated model is that of a particle associated with a cloud of pions. These pions carry the electromagnetic and nuclear force fields. Another model visualizes the baryon as consisting of a group of smaller particles bound together exceedingly strongly.

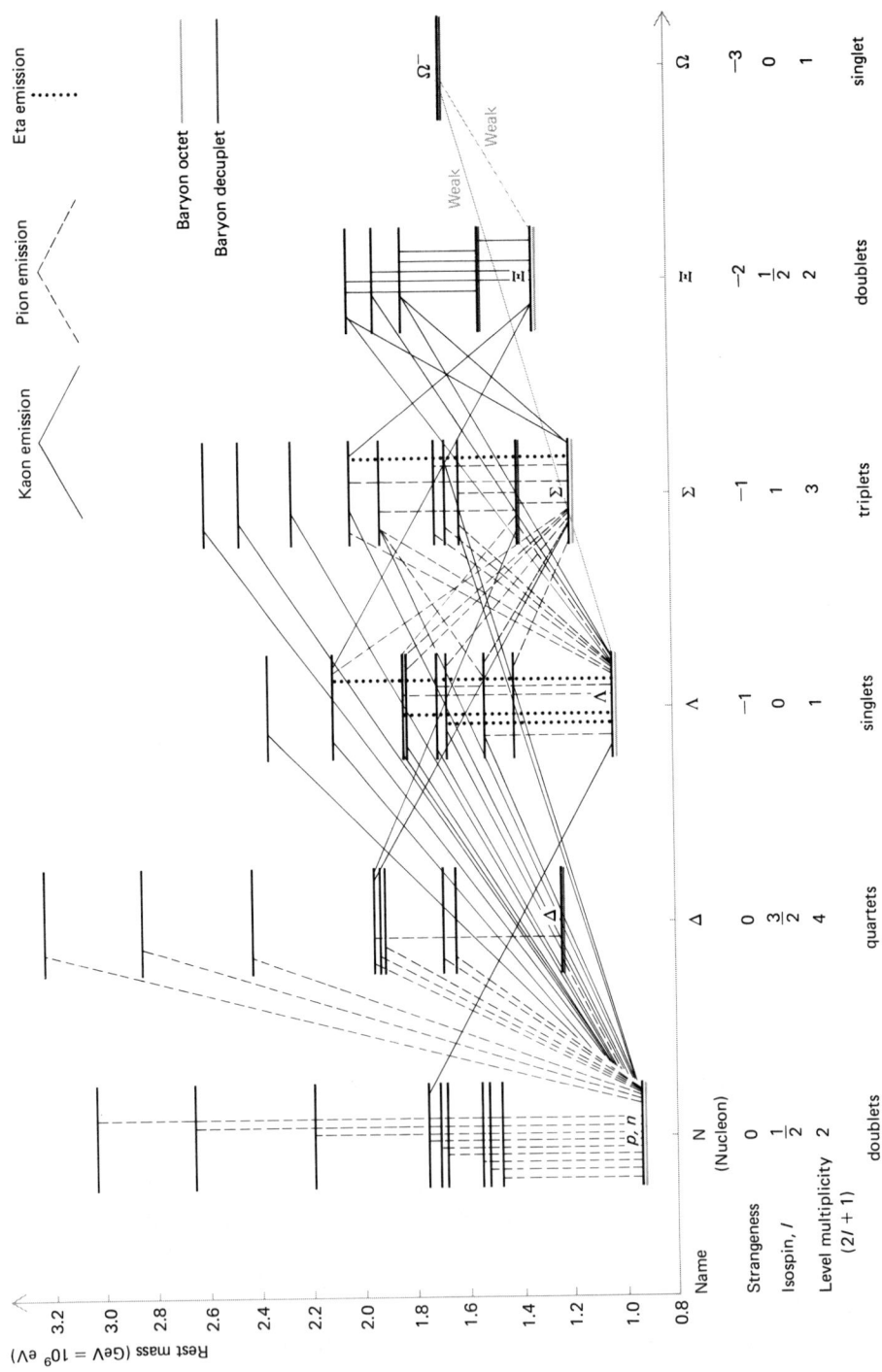

These may be quarks or quarklike particles, or there may be a large number of even smaller particles, recently named *partons*.

Experimental tests of these models are provided by several means. That of longest standing has been the use of electrons with energies in the hundreds of MeV to scatter from protons and neutrons in nuclei. The de Broglie wavelengths of these electrons are much smaller than the sizes of the nucleons, so they may be used to investigate the electromagnetic structure of the nucleon. Much work has been done using this technique,* and the size and shape of the nucleon as evidenced by the Coulomb field is now rather well known. The proton, for instance, has an apparent root-mean-square radius of $(0.77 \pm 0.10) \times 10^{-15}$ m, and the shape of the charge density function throughout the particle is not far from a Gaussian. The nature of the pion cloud is also explored using fast electrons, and it is explored for atomic nuclei as well as for the free proton.

Another experimental test of the structure of nucleons is provided by the exceedingly energetic cosmic rays and the very-high-energy beams in the large accelerators. Here again, the de Broglie wavelengths are small compared with the size of the nucleon, but these particles "feel" the strong interaction forces as well as the electromagnetic ones.

Chia Ping Wang† has applied statistical analysis techniques to the showers of secondary particles produced in such high-energy collisions and has arrived at a picture of the nucleon that consists of many small cells or a grainy sort of structure.

* See Robert Hofstadter, *Electron Scattering and Nuclear and Nucleon Structure*, W. A. Benjamin, Menlo Park, Calif., 1963 (a collection of reprints from the literature).

† See Chia Ping Wang, "Regularity of Multiplicity Distribution in NN and πN Collisions and the Structure of the Nucleon," *Phys. Rev.* **180**, 1463 (1969).

Figure 39–7 Energy levels of the baryon, after the scheme of V. F. Weisskopf. The known and confirmed baryon particle resonance states are shown as energy levels, using their rest masses as reported in the "Particle Properties Tables," *Rev. Mod. Phys.* **41**, 109 (1969). Some of those shown are probably incorrect, while others not shown remain to be discovered and confirmed.

The system used here names the various particles corresponding to the levels by a symbol and the rest mass in million electron volts, for instance, $N(1688)$, $\Delta(2850)$, $\Lambda(1815)$, $\Sigma(1660)$, and $\Xi(2030)$.

The ground state mesons underlined as the baryon octet are those of Figure 39–5 while the baryon decuplet is that of Figure 39–6. Notice how the level multiplicities shown here become evident in those diagrams.

The dominant strong decay modes are shown as lines connecting the levels. All are strong except for those shown with short dashes. Many other decay modes are seen in lower abundance or are suspected. The lowest level of the N is that of p, n, or the "common" proton and neutron.

Since each particle shown here has an antiparticle, a similar diagram exists for those.

PROBLEMS

39-1 Compile a table of the "light" particles, including photons with leptons, and show for each particle its symbol, spin, rest mass, charge, and decay modes, if any.

39-2 Estimate the binding energies of muonic hydrogen and of muonium by calculating the reduced mass in each case. The mass of the proton is about 1836 times as heavy as the electron.

39-3 Show how the conservation of lepton number is obeyed in Equations (39–7) through (39–12). Assign a lepton number $+1$ to particles and -1 to antiparticles.

39-4 The charge conjugate muonium has not yet been observed experimentally. Give a reason for this circumstance.

39-5 Write down the series of reactions that constitute muon catalysis of the D-D (deuterium, deuterium) reaction.

39-6 By rotating the quark triangle of Figure 39–1(a) through 180° about the origin, one obtains the antiquark triangle. Using this fact, write a table of quantum numbers for the antiquark system analogous to those of Table 39–1 for the quarks.

39-7 How many mesons are known and confirmed to exist? How many baryons, including antibaryons, are known and confirmed to exist?

RECOMMENDED READING

BAE, A. N., and OKUM, L. B., "On the Λ-Hyperon Creation Cross Section Near the Σ-Hyperon Creation Threshold," *Sov. Phys.—JETP* (English translation) **8**, 525 (1959).

BURBIDGE, Geoffrey, and HOYLE, Fred, "Anti-matter," *Sci. Am.*, April 1958.

FEYNMAN, R., and GELL-MANN, M., *Phys. Rev.* **109**, 193 (1958).

FRAZER, W. R., *Elementary Particles*, Prentice-Hall, Englewood Cliffs, N.J., 1966.
  Written to provide physics graduate students with a solid background in the subject.

GELL-MANN, M., "Model of the Strong Couplings," *Phys. Rev.* **106**, 1296 (1957).

GELL-MANN, M., and ROSENBAUM, E. P., "Elementary Particles," *Sci. Am.*, July 1957.

HUANG, K., LEE, T. D., and YANG, C. N., "Capture of μ^- Mesons by Protons," *Phys. Rev.* **108**, 1353 (1957).

LEDERMAN, L., "Two Neutrino Experiment," *Sci. Am.*, March 1963.

LEE, T. D., and YANG, C. N., "Possible Determination of the Spin of Λ^0 from its Large Decay Angular Asymmetry," *Phys. Rev.* **109**, 1755 (1958).

LEE, T. D., and YANG, C. N., "Implications of the Intermediate Boson Basis of the Weak Interactions: Existence of a Quartet of Intermediate Bosons and Their Dual Isotopic Spin Transformation Properties," *Phys. Rev.* **119**, 1410 (1960).

LIVINGSTON, M. S., *Particle Physics*, McGraw-Hill, New York, 1968.
  An introductory account in the field of elementary particles for those interested in this subject.

MARSHAK, R., "Pions," *Sci. Am.*, January 1957.

PENMAN, S., "The Muon," *Sci. Am.*, July 1961.

REINES, F., et al., *Phys. Rev.* **117**, 159 (1960).

RICCORDO, Levi Seth, *Elementary Particles*, University of Chicago Press, Chicago, 1966.

ROSSI, B., *High Energy Particles*, Prentice-Hall, Englewood Cliffs, N.J., 1965.
  Can be considered one of the most fundamental references in the subject of high-energy particles.

WIGNER, E. P., "On the Behavior of Cross Section Near Thresholds," *Phys. Rev.* **73**, 1002 (1948).

40
Origin of the Elements

Hans Albrecht Bethe
(1906–)
Born in Strasbourg, Germany, Bethe received his education at the universities of Frankfurt and Munich. In 1935 he came to the United States and joined the faculty of Cornell University. In 1937 Bethe proposed a new theory concerning the energy production of the sun and stars through the fusion of hydrogen to form helium. Among his prolific contributions to physics are the development of atomic power reactors, intercontinental ballistics missiles, and antimissiles. He was awarded the Fermi medal, the Planck medal, and the 1967 Nobel Prize in physics.

40–1 THE PUZZLE OF THE ELEMENTS
40–2 PRESENT DISTRIBUTION OF THE ELEMENTS
40–3 PRIMORDIAL NUCLEOSYNTHESIS
40–4 ELEMENT BUILDING IN THE STARS
40–5 SUPERNOVAE AND THE r PROCESS
40–6 EXPLOSIONS OF GALACTIC NUCLEI
40–7 SUMMARY

40–1 THE PUZZLE OF THE ELEMENTS

One of the most fundamental questions that man may ask of nature concerns the origin of nature herself—how the physical universe has evolved from its beginnings to become what we now observe to exist around us. Nature consists of free elementary particles and chemical elements, with the latter predominating heavily in abundance. How have the chemical elements come into existence and in the proportions in which they are now found? Were they always as they are now, or have they somehow evolved from other states? If they have evolved, then where and when did this happen?

The puzzles presented to us by the chemical elements may be classified as follows:

1. What is the present abundance distribution of the elements; that is, how much hydrogen is there compared to helium, lithium, and so on? How much iron is there compared to copper, bismuth, and so on throughout the periodic table.
2. How does this abundance distribution vary from place to place in the universe? Is it the same on the earth as in the sun, and is it the same in all of the stars? If not, then why not?
3. If the elements have evolved from some primordial state, then what was that state, and what are the processes by which they have evolved?
4. How have the elements become distributed from the places where they were made to the places where they are now found?
5. What is the age of the elements, and are they continuing to evolve somewhere in the universe? If so, where?

The answers to these questions as this text is being written are surprisingly detailed, although they are far from being complete. The subject is a very dynamic one and is developing rapidly. It is important to realize that, while there is much that depends on theoretical conjecture, the details of the processes that are involved have been observed in the laboratory. They are well known. It may fairly be said that the processes that have built the elements, starting with the beginning of the universe, have now been reproduced and studied experimentally. The theoretical picture of the origin of the elements, then, is based on solid fact. It may develop, however, that future theory and new facts will change our picture radically.

The outline of the theory of the origin of the elements as presented in this chapter is associated with the names of Hans Bethe, Harold Urey, George Gamow, Robert Alpher, Robert Herman, William Fowler, Fred Hoyle, Geoffrey and Margaret Burbidge, Jesse Greenstein, and many others. It will be seen that the theory depends almost wholly on nuclear physics at the present time. It remains for the future to see how much the extensive developments in the field of elementary particles will affect it.

40–2 PRESENT DISTRIBUTION OF THE ELEMENTS

A total of 105 elements are presently known. The 81 lightest of these have radioactively stable isotopes. These, plus nine radioactive elements, are found in the earth, and technetium and promethium have been found in the stars. Thirteen elements have been produced artificially, from element 93 through 105. These elements have some 1512 isotopes, 280 of which are stable, 67 of which are naturally radioactive, and 1165 of which are radioactive and are produced artificially.

The artificial production of isotopes is accomplished through the fission of uranium, through exposure of target elements to the intense neutron fluxes found in reactors and nuclear explosions, and by exposing targets to γ-ray and charged particle beams from powerful accelerators. They are identified through their chemistry and their modes of radioactive decay, both of which may be predicted quite firmly from the properties of the known elements and isotopes.

All mass numbers from 1 to 238 are found naturally occurring in the earth and stars except for mass 5 and mass 8. The decay lifetimes of these elements, He-5, Li-5, and Be-8, range from 10^{-21} sec to 10^{-16} sec.

Regarding the abundance distribution of these elements, data are obtained from terrestrial rocks, oceans, and gases, from the analysis of meteorites and solar winds, from the analysis of lunar rocks brought back to earth by the Apollo missions, and from the analysis of the solar optical spectrum. These sources yield a picture of the abundance distribution in the solar system as given by Table 40–1.

Notice that hydrogen and helium together make up about 98% of the material of the solar system. The planets and meteorites (with the exception of Jupiter) do not have nearly this amount of these two gases, because they are too small to have retained them by gravitational forces. The solar atmosphere, however, is a better sample of the elemental distribution. Assuming that little mixing has taken place between the surface of the sun and its core where nuclear transmutations are occurring, Table 40–1 may be taken as representative of the material from which the sun and its planets were formed originally.

Many stars similar to the sun exhibit similar abundances in their spectra, while a great many others have distributions that differ widely from these values. If we take Table 40–1 to represent fairly the general universal distribution of the elements and try to understand it along with those cases that are found to deviate from it, we then have a basis for proceeding. The general cases of stars observable

Table 40–1 Relative abundance of the elements by mass

ELEMENT	FRACTIONAL ABUNDANCE BY MASS
hydrogen: ^1_1H	0.71
^2_1D	0.0001
helium: ^3_2He	0.00006
^4_2He	0.27
Li, Be, B	0.00000001
C, N, O, Ne	0.018
silicon group: Na–Ti	0.002
iron group: $50 \leq A \leq 62$	0.0002
middleweight group: $63 \leq A \leq 100$	0.000001
heavyweight group: $A > 100$	0.0000001

SOURCE: After W. A. Fowler, *Nuclear Astrophysics*, Memoirs of the American Philosophical Society, Philadelphia, Vol. 67 (1967).

in nearby galaxies, as well as the bulk of those in our own galaxy, the Milky Way, follow this distribution quite well. The same holds true for the light of the more distant galaxies and for the (still not understood) quasars, except for a remarkable lack of helium in the latter.

40–3 PRIMORDIAL NUCLEOSYNTHESIS

There are several theories as to the original state of the universe which have led to the presently observed expanding system, and we shall discuss these in more detail in Chapter 41. We shall follow the line of reasoning here that starts with an explosive origin, or "big bang" theory, as suggested by Gamow and his collaborators. It is for this model that the theory is best established and most fully developed at the present time.

In this theory, the original state of the universe was essentially a great fireball consisting of very dense radiation in equilibrium with elementary particles. Beginning with "time zero," when there was a radiation fireball of exceedingly high temperature, we assume that the ball expanded and cooled as $T \cong t^{1/2}$, where T is the temperature in MeV and t is the age of the universe in seconds. At $t = 10^{-4}$ sec, the temperature was about 100 MeV ($10^{12}\,°K$). At age 1 sec, the temperature t was about 1 MeV ($10^{10}\,°K$). At times shorter than 1 sec, when temperatures exceeded 1 MeV, the radiation was in equilibrium with particle–antiparticle pairs. At times much earlier than this (10^{-6} sec and less), the creation and annihilation of all the elementary particles assisted in the cooling of the fireball and the maintainance of equilibrium with matter. Since nuclear process times are of the order of 10^{-23} sec, we may assume the fireball to be in equilibrium with matter down to times approaching this value.

All elementary particles are unstable against decay except neutrinos, protons, and electrons. As the fireball expanded and cooled, therefore, the various particles decayed to these three, and these were successively frozen out of the process as time went on. We may, then, look for the synthesis of nuclei to begin as the temperature falls to values below nuclear binding energies but still high enough to trigger nuclear reactions between the protons left behind. This occurs at temperatures of about $10^{10} \, °K$ down to about $10^{7} \, °K$. The fusion of protons begins a chain of reactions called *hydrogen burning*, which results in the production of helium:

$$p^+ + p^+ \rightarrow {}^{2}_{1}D + e^+ + \nu$$
$$^{2}_{1}D + p^+ \rightarrow {}^{3}_{2}He + \gamma \qquad (40\text{--}1)$$
$$^{3}_{2}He + {}^{3}_{2}He \rightarrow {}^{4}_{2}He + p^+ + p^+$$

Another particle, the neutron, which is radioactive is also frozen out of the fireball due to its long lifetime against radioactive decay (12 min). This now begins a second chain of synthesis, which lasts to much lower temperatures and also produces helium:

$$p^+ + n \rightarrow {}^{2}_{1}D + \gamma$$
$$^{2}_{1}D + n \rightarrow {}^{3}_{1}H + \gamma \qquad (40\text{--}2)$$
$$^{3}_{1}H \rightarrow {}^{3}_{2}He + e^- + \bar{\nu}$$
$$^{3}_{2}He + n \rightarrow {}^{4}_{2}He + \gamma$$

It is thus seen that He-4 is produced rather readily in the fireball by the *p–p* reaction during the first few seconds following the initial cooling to less than 2.2 MeV, the binding energy of the deuteron, and proceeds via a chain of neutron captures for a matter of some minutes. Reactions (40–2) actually proceed with higher cross section at lower temperature, so that chain (40–2) is terminated by the expansion and consequent dilution of the hydrogen gas combined with complete capture of the neutrons or their radioactive decay:

$$n \rightarrow p^+ + e^- + \bar{\nu} \qquad (40\text{--}3)$$

Why, then, does not the chain of neutron capture proceed beyond helium? The answer lies in the zero cross section of He-4 for thermal neutron capture combined with the exceedingly short lifetime of the mass-5 elements Li-5 and He-5. Should the gap at mass-5 be bridged by a few reactions, there is another at mass 8 to cross. The result of the primordial fireball is thus to produce a hot gas consisting of hydrogen and helium, with a small admixture of deuterium, He-3, Li-7, and B-11. The percentages of these is sensitive to the details of the expansion that is assumed. If the density during the first few minutes is too low, then most of the neutrons will decay before being captured by protons, while if the density is too high, far too much helium will be produced.

Based on some rather simple assumptions in the solution of the equations of general relativity and concerning the numbers of neutrinos and antineutrinos and their effects, one is led to calculate the primordial gas following fireball expansion and prior to formation of stars as about 30% He-4 and 70% H-1, with traces of H-2 and He-3, in surprising agreement with the abundances listed in Table 40–1 for the lightest elements. The heavier elements are not produced at all.

The effects of particle–antiparticle annihilation and the reactions between remnant mesons and hyperons are not considered in this treatment. These must change the picture only slightly, however, in view of the results of the simpler picture.

One of the many fascinating current puzzles regarding the origin of the elements now enters the picture. It arises in observations of spectra of the oldest stars in the Milky Way. For all stars, it is a good bet that there is practically no mixing between the central core where energy is being produced through nuclear processes and the surface layers of the star. This means that the surface layers are accurately representative of the material from which the star formed originally. And in the oldest of the stars, we find little or no helium! Some take this to mean that the details of the model of the primordial fireball are wrong and that very little helium was produced there—perhaps the explosion occurred much more rapidly than is thought, or perhaps the neglected matter–antimatter annihilation reactions destroyed the helium by breaking it down into nucleons. At any rate, some astrophysicists hold that the primordial gas was purely hydrogen, while others maintain that the solar system abundances are representative of the whole universe in spite of the old stars. The presence of helium is very difficult to detect spectroscopically in the stars, so this might be one answer.

In analyzing the abundance distributions of the elements in stars in our galaxy, as well as in the stars of nearby galaxies, one finds distributions that are common to all and in which the particular types are functions of position in the galactic discs relative to the nuclei and in the spiral arms and galactic halos. These distributions suggest that explosive events have occurred in the galactic nuclei. We shall return to this point later.

40–4 ELEMENT BUILDING IN THE STARS

If the process of nucleosynthesis ends in the primordial fireball with the production of only the lightest of the elements up to helium, then we must look elsewhere—and obviously at later times—for the processes that will produce the heavier elements. The right conditions are found in the stars. The expansion and cooling of the primordial fireball stopped the p–p and neutron-capture chains, but just the reverse situation is found near the center of the forming protostar. There are no free neutrons, however, so only the p–p chain may be ignited. As the star forms from the cloud of gas and dust in its vicinity, it begins to contract under its own gravitational field, and the central temperature and density rise to high values.

When the temperature reaches some $10^7 °K$ and the density approaches 10^5 kg/m³, hydrogen burning begins again.

Hydrogen Burning. We must now consider this process, the p–p chain in somewhat more detail. For one thing, there are more branches in the chain than the single one indicated in reactions (40–1). The more important of these are given in Table 40–2. It is seen that the p–p fusion chain results in the consumption of four protons to form He-4 in each branch. The positrons slow down and combine with negatrons in an annihilation reaction, and the neutrinos escape from the star with the velocity of light. The He-4 on the left side of the second branch, the one that combines with He-3, reappears at the end of the chain and may be considered as a catalyst for the branch.

Table 40–2 The principal branches in the p–p (hydrogen burning) process

$$p^+ + p^+ \to {}_1^2D + e^+ + \nu$$
$${}_1^2D + p^+ \to {}_2^3He + \gamma$$
$$\phantom{p^+ + p^+ \to {}_1^2D + p^+ \to}{}_2^3He + {}_2^3He \to {}_2^4He + p^+ + p^+$$

or

$$\phantom{p^+ + p^+ \to {}_1^2D + p^+ \to}{}_2^3He + {}_2^4He \to {}_4^7Be + \gamma$$
$$\phantom{p^+ + p^+ \to {}_1^2D + p^+ \to {}_2^3He + {}_2^4He \to}{}_4^7Be + e^- \to {}_3^7Li + \nu$$
$$\phantom{p^+ + p^+ \to {}_1^2D + p^+ \to {}_2^3He + {}_2^4He \to {}_4^7Be}{}_3^7Li + p^+ \to {}_2^4He + {}_2^4He$$

or

$$\phantom{p^+ + p^+ \to {}_1^2D + p^+ \to {}_2^3He + {}_2^4He \to {}_4^7Be}{}_3^7Li + p^+ \to {}_4^8Be + \gamma$$
$$\phantom{p^+ + p^+ \to {}_1^2D + p^+ \to {}_2^3He + {}_2^4He \to {}_4^7Be + e^-}{}_4^8Be \to {}_3^8Li^* + e^+ + \nu$$
$$\phantom{p^+ + p^+ \to {}_1^2D + p^+ \to {}_2^3He + {}_2^4He \to {}_4^7Be + e^- \to}{}_4^8Be^* \to {}_2^4He + {}_2^4He$$

NOTE: ${}_4^8Be^*$ indicates an excited state of ${}_4^8Be$.

A most important aspect of the p–p chain is the fact that it is exoergic, releasing a large amount of energy as it proceeds. The four protons have a greater mass in the free H-1 state than they do when combined in the nucleus of He-4 as two protons and two neutrons (the conversion of a proton to a neutron is signaled by the appearance of a positron on the right-hand side of a reaction). The excess mass—the difference between the total mass of the four free protons and that of the He-4 nucleus—appears as kinetic heat energy through Einstein's relation,

$$E = \Delta m\, c^2 \qquad (40\text{–}4)$$

where c is the speed of light in a vacuum, 3×10^8 m/sec. Now, the joule may be expressed as kg-m²/sec², so that $c^2 = 9 \times 10^{16}$ J/kg. The four protons consumed have a total mass some 0.7% greater than does the He-4 nucleus, so the conversion of 1 g of hydrogen into helium yields some 6×10^{11} J. In other words, 1 lb of hydrogen becomes 0.993 lb of helium, releasing about 100 million kW-h of energy as it does so. This is the source of energy for a main-sequence star such as the sun until its supply of hydrogen in the core is consumed.

Helium Burning. As the hydrogen supply in the core of a star is converted to He-4, the helium accumulates in the core to higher and higher concentration until it so dilutes the hydrogen that the *p–p* reaction "goes out" in the center. Hydrogen burning continues in a thin shell around this helium core, and the helium core slowly grows larger and larger. Eventually, the diameter of the hydrogen burning shell is such that its temperatures and pressure fall too far, and it effectively goes out as well. At this point, there is insufficient radiation energy to support the great mass of the star, so it begins to contract under the force of its own gravitation. The gravitational energy released at the center of the collapsing star begins to heat the core to even higher temperatures, and increases its density greatly. As the core contracts, however, the envelope of the star *expands* due to the higher temperature inside it. Thus, the atmosphere at the surface of the star has an increased area from which to radiate what energy reaches it from below. The greater surface results in a cooling of the top layer, and it turns from the bright yellow to red. These expanded stars are the *red giants*.

As the core continues to heat up due to gravitational contraction, its density approaches 10^8 kg/m^3 and its temperature becomes about 10^{8}°K. At this time, the helium in the core begins to fuse in the chain called *helium burning* (Table 40–3).

Table 40–3 The helium burning chain

$${}^{4}_{2}\text{He} + {}^{4}_{2}\text{He} \rightarrow {}^{8}_{4}\text{Be} \rightarrow {}^{4}_{2}\text{He} + {}^{4}_{2}\text{He}$$

or

$${}^{8}_{4}\text{Be} + {}^{4}_{2}\text{He} \rightarrow {}^{12}_{6}\text{C}^* \rightarrow 3\ {}^{4}_{2}\text{He}$$

or

$${}^{12}_{6}\text{C}^* \rightarrow {}^{12}_{6}\text{C} + 2\gamma$$
$${}^{12}_{6}\text{C} + {}^{4}_{2}\text{He} \rightarrow {}^{16}_{8}\text{O}$$
$${}^{16}_{8}\text{O} + {}^{4}_{2}\text{He} \rightarrow {}^{20}_{10}\text{Ne}$$

We have seen in the foregoing that the lifetime of Be-8 against breakup into two He-4 nuclei is extremely short. At a sufficiently high temperature and pressure, a small amount of the Be-8 captures a He-4 before it can break up, going to an excited state (7.656 MeV) of the carbon nucleus in a very rapid, or "resonant" reaction. Most of the C-12* decays back into three He-4 nuclei, but a small amount of it decays to stable C-12 with the emission of two γ rays. It is seen that some O-16 and Ne-20 are also formed, once the gap at mass 8 has been successfully bridged. This gap presented a difficult problem for a long time in developing a theory of nucleosynthesis in the stars and was solved only when the excited state of carbon was found with exactly the correct energy to permit the rapid resonant capture of a He-4 by the Be-8 before it decayed. The experimental work that made it possible to construct this chain of events was done at the Los Alamos Scientific Laboratory, California Institute of Technology, Stanford University, and Brookhaven National Laboratory.

Only a small fraction (0.07%) of the mass of the He-4 is converted into energy in these processes, but this is sufficient to reverse the collapse phase of the star. Red giants are not stabilized for long periods after the onset of helium burning. Stars that become unstable during this phase may eject large quantities of gas into interstellar space containing hydrogen, helium, carbon, oxygen, and neon. New stars that form from these gases will then contain appreciable quantities of these heavier elements in their cores.

The CNO Bi-cycle. If a star contains these processed gases, which include carbon, nitrogen, and oxygen, then when its central temperature reaches a sufficiently high value, a new set of catalyzed reactions begins to process hydrogen into helium, the *CNO bi-cycle* (Table 40–4).

Table 40–4 The CNO bi-cycle

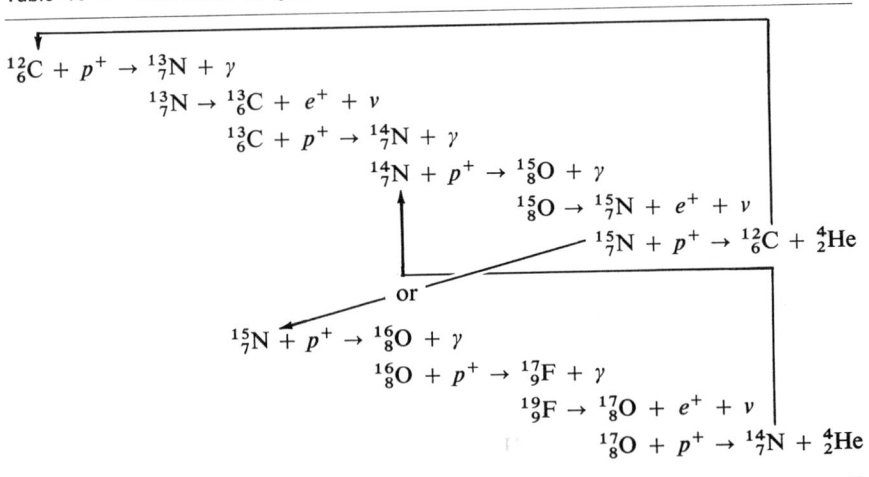

NOTE: The total number CNO nuclei remains unchanged.

Carbon and Oxygen Burning. We have seen how hydrogen is burned to helium in the first-generation stars, and that helium is used to produce carbon and oxygen in the first-generation red giants. Gases ejected from these stars through explosive processes then form second-generation stars that contain hydrogen, helium, carbon, and oxygen. Hydrogen burning can now proceed through the *p–p* chain and through the CNO bi-cycle as well, with the production of C-13, N-14, and N-15. If explosive instabilities arise in the star during this phase, then these elements are ejected into the interstellar medium for the formation of *third-generation stars*.

How about the heavier elements? Red giants need not become unstable at early times. This is a function of their size and the details of their structure. If a red giant remains stable, then the supply of helium that burns at its core also becomes exhausted, accumulating the carbon and oxygen "ash" in its core. The helium burning reaction goes out, and contraction of the core begins again with consequent rise in density and heating. Remember that a collapsing star generates heat at its center through conversion of its potential gravitational energy.

As the temperature and density rise farther, the C-12 and the O-16 begin to burn in a series of fusion reactions accompanied by decays that produce many new elements. The most abundant among these are isotopes that are integral multiples of the α particle (He-4 nucleus): Ne-20, Mg-24, Si-28, and S-32. Thus, stars that form from the gases and dust resulting from the explosions of earlier generations of stars can synthesize elements farther up into the mass scale.

The Alpha Process (α Process). In the range neon–sulphur, the resistance to simple fusion supplied by the Coulomb repulsions of the bare nuclei stops this process. As the star continues to contract in its core, the intense radiation field generated in the very dense material begins to photodisintegrate the heavier nuclei such as silicon. In this process, the γ rays that lie in the high-energy tail of the Planck distribution of the radiation have sufficient energy to tear the nuclei apart. A typical case is that of Si-28, which can be broken down into seven free α particles as the temperature approaches 3 billion °K.

With the production of free, high-energy α particles, a new process takes place—the direct capture of α particles by those nuclei that happen not to be broken down by photodisintegration. For example, a second Si-28 nucleus may successively capture seven α particles to form Ni-56. The process is complex, obviously, and the path of any given nucleus to a higher mass element may be accompanied by intermediate photodisintegrations followed by α-particle capture. Nickel-56 is radioactive, capturing an electron from the plasma to become Co-56. This is also radioactive, capturing a second electron to form the stable Fe-56. A neutrino is emitted with each capture. Many nuclei may be so produced through α capture, photodisintegration, and radioactive decay. Those near Fe-56, from V-50 to Ni-62, are produced in this fashion.

Thus, we see that the stable red giant is an important factory for the production of the elements up to the iron group and, by becoming unstable to explosive emission of these elements, supplies the interstellar space with them for the formation of later generation stars enriched in these elements.

The Equilibrium Process (e Process). As the star continues to collapse in the core region and so to heat up there, the temperature approaches 4 billion °K and the core density approaches 3 billion kg/m³. The star is now in a rapidly evolving state approaching the catastrophic explosion that will blow the newly made elements into space. All possible nuclear reactions are now occurring in an equilibrium state: Fusion, α capture, photodisintigration, and neutron capture, as well as weak interaction processes such as electron capture by nuclei, and neutrino emission in electron–positron pair annihilation, are all happening to keep the isotopic mixture in a *quasi equilibrium state*. The collapse is hastened by the refrigeration action of the breakup of the iron group elements into lighter ones. It is at this stage that the iron group elements—titanium, vanadium, chromium, manganese, iron, cobalt, and nickel—are produced.

Neutron Capture in Red Giants (s Process). The high temperatures attained in the core of the red giant star results in the production of free neutrons through such reactions as

$$^{13}_{6}C + \alpha \rightarrow {}^{16}_{8}O + n \tag{40-5}$$

Thus, alpha-in-neutron-out (α, n) reactions, accompanied by (γ, n), (p, n) and $(n, 2n)$ reactions supply free neutrons. These are captured in hydrogen or in the heavier iron-group elements, in general, in (n, γ) reactions. Very often, the capturing nucleus becomes radioactive in the process. As the next neutron capture for that nucleus may be many years away, it has sufficient time to decay to a stable isotope. The next neutron capture eventually occurs, and then decay back to another stable isotope occurs. Thus, the nucleus—via successive neutron captures and β decays—becomes heavier and heavier but never goes very far from the stability valley.

An example of a part of this chain might start with Fe-56 and proceed as shown in Table 40-5.

Table 40-5 A part of the slow (s-process) neutron capture chain

$$^{56}_{26}Fe + n \rightarrow {}^{57}_{26}Fe + \gamma$$
$$^{57}_{26}Fe + n \rightarrow {}^{58}_{26}Fe + \gamma$$
$$^{58}_{26}Fe + n \rightarrow {}^{59}_{26}Fe + \gamma$$
$$^{59}_{26}Fe \rightarrow {}^{59}_{27}Co + e^- + \bar{\nu}$$
$$^{59}_{27}Co + n \rightarrow {}^{60}_{27}Co + \gamma$$
$$^{60}_{27}Co \rightarrow {}^{60}_{28}Ni + e^- + \bar{\nu}$$
$$^{60}_{28}Ni + n \rightarrow {}^{61}_{28}Ni + \gamma$$
etc.

This process can account for many but not all of the stable isotopes above iron as well as many below, and for those isotopes with very long half-lives against β decay. The s process ends at Bi-209. Above this point, a chain of α-radioactive isotopes is made that cycles back to Bi-209 as shown in Table 40–6.

Table 40–6 The terminal chain in the s process

$$\begin{aligned}
&^{209}_{83}\text{Bi} + n \rightarrow {}^{210}_{83}\text{Bi} + \gamma \\
&^{210}_{83}\text{Bi} \rightarrow {}^{210}_{84}\text{Po} + e^- + \bar{\nu} \\
&^{210}_{84}\text{Po} \rightarrow {}^{206}_{82}\text{Pb} + \alpha \\
&^{206}_{82}\text{Pb} + n \rightarrow {}^{207}_{82}\text{Pb} + \gamma \\
&^{207}_{82}\text{Pb} + n \rightarrow {}^{208}_{82}\text{Pb} \\
&^{208}_{82}\text{Pb} + n \rightarrow {}^{209}_{82}\text{Pb} + \gamma \\
&^{209}_{82}\text{Pb} \rightarrow {}^{209}_{83}\text{Bi} + e^- + \bar{\nu}
\end{aligned}$$

Note how this chain synthesizes ^4_2He from four neutrons.

The discovery of technetium in the spectra of some stars supports the neutron capture hypothesis in the s process, since those isotopes of technetium that can be made in this fashion would have decayed away had the technetium been produced more than several hundred thousand years ago. The recent discovery of promethium in a star is even more surprising; it has a half-life of only 18 yr and has no stable isotopes.

40–5 SUPERNOVAE AND THE r PROCESS

We have been mentioning the explosive instabilities that may set in during the red giant phase of stellar life. The most catastrophic of these is known as a *supernova*, in which a true explosion blows away all of the stellar envelope and a large fraction of its core material as well. The explosion is such that the average density of the star may drop by more than a factor of eight in 15 min. It is due to the final accelerating collapse of the core of the star when it has so exhausted its nuclear fuel that it can no longer support the great mass of the envelope and core against gravitational forces. The sudden heating transfers a large amount of energy to the envelope and outer core layers. Nuclear burning of fuel in the envelope, triggered by the sudden heat, contributes to the explosion. Temperatures reach billions of degrees Kelvin, and the velocity of the gases becomes of the order of 10^6 m/sec.

Under these conditions, neutron densities in the core may reach as high as 1000 kg/m^3. Multiple neutron captures occur with extreme rapidity, but so also

Figure 40–1 A star is seen exploding in the upper spiral arm of galaxy NGC 4725. The supernova was found on May 5, 1940, when its brightness was estimated as being greater than 100 million times that of our sun. The picture here was taken 5 days later when it had faded appreciably. Galaxy NGC 4725 is in Ursa Major and is some 10 million light years from the earth. (Photograph from the Hale Observatories.)

do the reverse (γ, n) reactions occur in the extreme high radiation flux. The result is a sort of transient equilibrium between the (n, γ) and the (γ, n) reactions.

All of the elements are synthesized in such an environment, and especially those neutron-rich isotopes that were bypassed by the s process and the elements beyond Bi-209, the thorium and uranium isotopes in particular. Those α-radioactive isotopes that are bypassed by the direct neutron capture are made in turn as the heavier transuranic elements are made and decay. The process ends at $A = 275$. This mass is so unstable against neutron-induced fission that it breaks up very quickly to distribute elements throughout the periodic table. The r process thus produces the parents of the presently observed α-radioactive series.

As the supernova expands and cools, its "light curve" dies away with a luminosity that is decreasing exponentially, with a half-life of 55 days. This matches the spontaneous fission curve of californium-254, discovered at Eniwetok in the clouds produced by nuclear explosions. It is believed that this isotope is responsible for the slow decay of supernova light curves.

Figure 40-2 An explosion of the nucleus galaxy Messier 82. The picture was taken in the light of hydrogen-α with the 200-in. Hale telescope at Mount Palomar. Note the extensive flamelike structure extending from the nucleus of the galaxy. The rest of the galaxy is a myriad of stars composing the galaxy. It has been estimated by R. Lynds and A. Sandage that the total mass of the expanding material is about 5.6 million solar masses. There are many other galaxies that show effects of explosions in their centers. Such events may have occurred in our own galaxy since its formation. (U.S. Naval Observatory Photograph.)

40-6 EXPLOSIONS OF GALACTIC NUCLEI

The galaxies, in general, possess central regions where the stellar population is exceedingly high—the galactic nuclei. Processes are apparently going on in these regions that are as yet little understood but that produce unusual optical spectra, vast quantities of radio-frequency and infrared energy and, on occasion, violent explosions that eject vast quantities of mass from them. An outstanding example of such explosions is given by the galaxy M82, where the material ejected from the galactic nucleus approaches that of 10 million solar masses. A vast flame may be seen as originating in the galactic core and sweeping through the galactic regions.

40–7 SUMMARY

It is seen how the primordial explosion produced hydrogen and helium and how the p–p chain of hydrogen burning heats the main-sequence stars while synthesizing more helium from the hydrogen at its core. The sequence is carried still further in the red giant phase of stellar life, as helium burning and burning of heavier elements occur with higher internal temperatures and densities. Ejection of gases during novae instabilities and smaller explosions supplies the interstellar medium with these lighter elements from which succeeding generations of stars may form.

As red giant life continues, other processes set in with increasing temperatures and densities until, finally, a catastrophic explosion such as a supernova occurs. Here, the remaining elements and isotopes are produced and are supplied to the interstellar space for yet further star formation and the production of planets such as the Earth.

In concluding this chapter, it is apropos to quote from the Jayne Lecture of William A. Fowler to the American Philosophical Society in 1965:

"My major theme has been that all of the elements heavier than helium, and perhaps the helium too, have been synthesized in stars. Let me remind you that your bodies consist for the most part of these heavier elements. Thus it is possible to say that you and your neighbor and I, each one of us and all of us, are truly and literally a little bit of stardust."

PROBLEMS

40–1 Plot a graph of an abundance distribution of the elements as a function of their mass.

40–2 Plot a graph of the temperature of the primordial fireball from time zero to age 10 sec.

40–3 Why is it that as the nuclear fuel becomes exhausted in the center of a star, the temperature begins to rise there?

40–4 Referring to the hydrogen burning processes of Table 40–2, state in what sense the protons may be regarded both as fuel and as catalysts for the chain.

40-5 How do you account for the fact that although the core temperature of a red giant is very high, its color is red compared to "normal" stars?

40-6 How does it happen that a main-sequence star may have heavier elements in its core and atmosphere than helium?

40-7 Sketch the life history of a red giant star that proceeds to the supernova stage, indicating the onset and termination of the various processes of nucleosynthesis at the various stages.

40-8 Why is it that the processes of fusion and the e process build elements only so far as the iron group?

40-9 Why is the process of neutron capture interspersed with β decay in a red giant star called a "slow" process?

40-10 Why is the discovery of promethium in the spectrum of a star surprising?

RECOMMENDED READING

ALLER, L. H., *Abundance of the Elements*, Wiley, New York, 1961.

ALPHER, R. A., BETHE, H. A., and GAMOW, G., "The Origin of the Chemical Elements," *Phys. Rev.* **73**, 803 (1948).

ALPHER, R. A., and HERMAN, R. C., "Theory of the Origin and Relative Abundance Distribution of the Elements," *Rev. Mod. Phys.* **22**, 153 (1950).

BURBIDGE, E. M., et al., "Synthesis of the Elements in Stars," *Rev. Mod. Phys.* **29**, 547 (1957).

HACK, M., "The Hertzsprung–Russel Diagram Today," *Sky and Telescope*, May 1966 and June 1966.
This article outlines the basis for the classification of main-sequence and red giant stars by their luminosity and color.

SCHWARZCHILD, M., *Structure and Evolution of the Stars*, Princeton University Press, Princeton, N.J., 1958.

41
Origin of the Universe

George Gamow
(1904–1968)

Born in Odessa, Russia, Gamow received his Ph.D. in physics from the University of Leningrad in 1928. He came to the United States and was a naturalized citizen. His early nuclear fluid hypothesis of atomic nuclei led to the present theory of nuclear fission and fusion. Gamow is the author of the "big bang" theory of the origin of the universe, and by writing many popular books such as, One, Two, Three . . . Infinity, Creation of the Universe, *and* Mr. Thompkins Learns the Facts of Life, *he has helped to educate the public in many scientific concepts.*

41–1 AGE OF THE UNIVERSE
41–2 SIZE OF THE UNIVERSE
41–3 EXPANDING UNIVERSE
41–4 BIRTH OF THE UNIVERSE

41-1 AGE OF THE UNIVERSE

Since the dawn of reason in mankind, the question has been asked "From whence has the universe come?" The extent of the universe may have been radically different for different ages of man, and the answer itself was often based on myth or mysticism. It was always based on man's observation of nature, however—he looked around him and tried to explain how this has all come about. We, both scientists and nonscientists, today, are like all men of past ages in this respect as we consider this question. We are able to call on a vast extent of modern knowledge, however, in formulating our ideas concerning the universe. Let us take into account all that we have discussed in this text—the facts that we have learned and the questions we have posed—to wonder about the universe at large and to attempt an answer to the question as to how it came about. Our solution will be very incomplete and most probably incorrect in some aspects, but let us build on our present knowledge to discuss the origin of the universe and pass our results on to future students of this question.

In the last chapter, we considered the origin of the chemical elements, and in so doing we came a long way in a discussion of the origin of the universe itself, for it is from the elements that all matter is built. It remains now to consider the universe at large and to ask how it has evolved as a function of time and how it looked at various times. Let us begin by investigating the present evidence on the age of the universe.

Age of Meteorites. With the discovery that there are radioactive elements in nature came the question as to when and where they were formed. They have a finite age in the sense that if they were all formed at the same time, then they are all decaying with their various characteristic half-lives and should disappear from detectability when a sufficient number of half-lives have passed. We notice that certain of these elements, thorium, and uranium for instance, are about as abundant in nature as are the nonradioactive heavy elements. The most abundant isotope of uranium is U-238 with a half-life of 4.5×10^9 yr. This is found in company with the more rare isotope U-235, which is only about 0.7% as abundant as is U-238. The rare isotope U-235 has a half-life of about 0.7×10^9 yr, so it decays more than seven times faster than U-238. If these isotopes were formed at about the same times and in the same amounts—a most reasonable sort of assumption—then we may use the above data to calculate how long they have been decaying.

It is believed that meteorites are true members of the solar system along with the earth, although the details of how they came to be is still a matter of much argument. Uranium-238 decays, ultimately, to the stable isotope Pb-206, while U-235 yields Pb-207, and Th-232 yields Pb-208. It is found that meteorites that are relatively rich in uranium are also rich in lead. Measurements of the ratio U-235/U-238 and Pb-207/Pb-206, combined with the known lifetimes for decay to these isotopes, yields ages for the meteorites that range from 4.5 to 5.0 billion yr.

Age of the Earth. The above method using radioactive methods can be applied to determining the age of the rocks that form the crust of the earth. By age, we mean how long it has been since they were in a molten state along with the rest of the earth, and so we search for the mineral that yields the greatest value. The method, first used by Lord Rutherford, investigates the ores of uranium and thorium as well as other rocks that contain these elements. So long as those rocks were in the melted state, the lead formed from radioactive decay could be separated from the parent elements, but when the rocks solidify, all components are frozen into place, and the lead produced from that time on is found along with its parent thorium or uranium which is still decaying to produce lead.

The ratios Pb-208/Th-232, Pb-207/U-235, and Pb-206/U-238 found in the oldest mineral, monazite (found in Rhodesia), yield a value of 2.7 billion yr. A technique that employs the isotopic ratios of uranium and of lead as described for the meteorites yields an age for the earth of 3.35 billion yr. These values are in excellent agreement with those obtained for the meteorites, taking into account uncertainties associated with various ways that the elements may be lost from the rocks and ores.

The age of the oceans may be estimated by a method proposed by the astronomer Edmund Halley (for whom the famous comet is named). This uses the present salinity of the ocean water, about 3%, and estimating the rate at which salt has been carried into the oceans by rivers, geologists arrive at an age for the oceans of about 3 billion yr, in agreement with the age obtained for the oldest rocks.

Age of the Moon. An estimate of the time that has elapsed since the moon was effectively in contact with the earth has been made by George Darwin, son of the biologist Charles Darwin. The moon is known to be receding from the earth at a rate of about 5 in./yr. This effect, caused by the friction of the ocean tides raised on the earth by the moon, also causes a gradual increase in the length of the day and in the length of the month. The value for the time elapsed since the moon was effectively in contact with the earth is about 4 billion yr.

A more modern method for estimating the age of the moon depends on the analysis of radioactive elements found in the samples of the moon's rocks brought back to earth by the Apollo missions of NASA. As this book is being written, the results of these measurements are not yet available, but we leave a space here where they may be written in when they are found and published:

Age of moon as determined by analysis of rocks = _____
Date _____

Age of the Stars. In Chapter 40, we saw that the primordial fireball distributed gases of helium and hydrogen into space, and from these gases the first stars were born. We have also seen that stars age by using up their hydrogen and helium contents, becoming red giants, and ultimately exploding or otherwise ejecting their contents into interstellar space. It is from these gases and dust that new stars then form. Is there a means of estimating the age of the oldest stars in the galaxy—those stars that are approaching their periods of instability?

It is known from theoretical models of stars that stellar brightness increases with the cube of the star's mass. A star twice as heavy as another burns its fuel at a rate eight times as fast. Dividing this factor of eight by the factor of two in mass, the lifetime of the lighter star will be some four times that of the heavier one. Calculating the masses of the stars observed in the Galaxy and measuring their brightness yields a value of about 5 billion yr for the oldest stars, a value that agrees logically with the results obtained for the ages of the earth, the moon, and meteorites.

Age of Stellar Clusters. A few hundred closely clumped groups of stars are known that all move with nearly the same velocity and in nearly the same direction through the Galaxy. It is presumed that the stars in such a cluster were all formed at about the same time from a common cloud of gas and dust. It has been calculated by B. J. Bok that the average life of such a cluster against dispersal by gravitational interactions with the rest of the Galaxy is between 1 and 10 billion yr. Thus, the lifetime of the Galaxy cannot be more than a few billion years.

Age of the Galaxy. If we look upon the stars of the Milky Way as molecules of a gas, then we may assume that in their assembly into the Galaxy, some stars had much more kinetic energy of motion than did other stars. Through gravitational interactions, however, the distribution of kinetic energy among the stars should finally be reached in an equilibrium where the velocity of each star through the Galaxy is inversely proportional to its mass.

The astronomer F. Gondolatsch has analyzed the velocity distribution of the stars in the neighborhood of the sun and has found that such equipartition of energy has reached about 98% of its final value. Based on this theory, he arrives at an age for the Galaxy between 2 and 5 billion yr.

Thus, we see that every effort made to determine the age of the solar system and the Galaxy (the Milky Way) results in an answer in the range of several billion years. These findings impress us as being more significant than mere coincidence, so they provide a basis for building a model of the universe which had its beginnings at a time in the billions of years ago. We now ask, what evidence can be gained from a consideration of the universe at large?

41-2 SIZE OF THE UNIVERSE

In discussing the present size and mass of the total universe, we must call on definite theoretical models for its description. These models are based on particular solutions (approximate ones) of a set of field equations derived from a theory of general relativity. While there are several such theories to choose from, the most common one is also the first of these theories: the general relativity of ALBERT EINSTEIN.

Einstein's equations are a set that ascribe an intrinsic geometry to space-time and that then predict the effects of the presence of matter on this geometry. It has as an ultimate aim the description of *all* of physics in terms of this intrinsic space-time geometry. In its development, the Einstein theory calls on the algebra of mathematical objects called *tensors*—constructs that are related to the generalized idea of a vector—but in four-dimensional space. These four dimensions have three "spacelike" dimensions and one "timelike" dimension, so the space is called space-time for short. It is related in some deep sense to the "physical vacuum" of which we spoke in Chapter 1.

In using the equations of general relativity, we begin by considering the universe in what may be—at first sight—an unusual way: We think of it as an ideal gas in which the individual molecules are galaxies! To treat the Milky Way as a single molecule of some gigantic gas may seem incongruous, but let us realize that there are probably more galaxies in the universe than there are grains of sand on all of the beaches of the world. In fact, we may go further and consider the clusters of galaxies that are seen as the molecules. Our universe then becomes a fluid continuum of a highly idealized sort.

As with any gas, the fluid may be described by an average density ρ and an average internal pressure p, both of which are functions of time but not of position within the universe. The internal energy of the gas is composed of the mass-energy and the pressure, which is the same as an energy density p/c^2.

The relative values of ρ and p/c^2 can be estimated by assuming that the pressure is due to the random motion of the molecules (as is true of any gas), that is, local deviations from the average state. It is a well-known result of the kinetic theory of gases that the pressure and density are related by

$$\frac{p}{\rho} = \frac{1}{3} v_{\rm rms}^2 \tag{41-1}$$

where v_{rms} is the root-mean-square random velocity of the gas molecules. Thus, for our galactic-cluster gas molecules, we have

$$\frac{p/c^2}{\rho} = \frac{1}{3}\frac{v^2}{c^2} \qquad (41\text{-}2)$$

Now, the observed random motions of the galaxies yield velocities much less than c, so that we expect the pressure to contribute much less energy than does the density ρ.

We must now refer to the Einstein field equations for guidance, and the first step in this process is to assume some particular form for the "line element." This is merely the name for the particular generalized law in four-dimensional space-time that plays the part of the familiar Pythagorean relation for the length of the hypotenuse of a right triangle in two-dimensional space. The line element in this case, however, sets the particular geometry of the space. Without going further into the algebra of the situation, we merely say that we shall now use the line element used by astronomers in considering the universe at large, the so-called *Robertson–Walker metric*. To describe it a bit further, however, we now make a simple comparison: It is easy to see that an ordinary sphere is a two-dimensional surface embedded in ordinary three-dimensional space. The "ordinary" space we refer to is the "flat" Euclidean space without any intrinsic curvature. Now, the Robertson–Walker metric describes a universe that has the geometry of a three-dimensional "hypersphere" embedded in a four-dimensional Euclidean space.

The radius of this sphere is given by

$$\frac{1}{R_0^2} = \frac{k}{R^2(t)} \qquad (41\text{-}3)$$

where $R(t)$ is the "radius of the universe," R_0 is the uniform curvature of the three-dimensional sphere, and k is a constant. For $k = 0$, the curvature is zero, and the hypersphere is itself an ordinary Euclidean subspace of the four-dimensional space. For $k = 1$, R_0 equals $R(t)$, while for $k = -1$, the R_0 becomes imaginary, corresponding to a hypersphere of imaginary radius, or a "pseudo hypersphere."

We have mentioned the Robertson–Walker metric because it yields a shift of the wavelengths of light emitted by distant objects—the red shift which is actually seen. It also relates the magnitude of the red shift to the distance, and so gives a theoretical explanation of Hubble's law, which we shall see in the next section.

The geometrical picture of a three-dimensional hypersphere (representing the universe) embedded in a four-dimensional space-time is a construct to assist us in visualizing the generalized idea of "curved space." In this construct, all of the universe lies on the surface of the hypersphere. George Gamow has described this picture of the universe in terms of a large balloon that has billions of tiny dots on its surface, each of which represents a galaxy or a cluster of galaxies. The balloon is steadily expanding through being blown up, so the distances between the dots are steadily increasing. There is no "edge" to such a universe, although it is "closed" in that a light ray that starts at any point will travel along the surface until it eventually comes back to the starting point. This can happen only, of course, if the rate of expansion is slow enough compared to the velocity of the light ray.

An element of the surface of a three-dimensional sphere is proportional to R^2, where R is the radius. In like manner, an element of the surface of our four-dimensional sphere is proportional to R^3. The result of solving the Einstein equations with this metric is

$$\frac{d}{dt}(\rho R^3) + \frac{p}{c^2}\frac{dR^3}{dt} = 0 \qquad (41\text{-}4)$$

Thus, if $V(t)$ is the actual volume content of the surface element, then $M = \rho V$ measures the mass content, and Equation (41-4) can be written

$$dM + \frac{1}{c^2}p\,dV = 0 \qquad (41\text{-}5)$$

Now Mc^2 measures the energy content of the element and $p\,dV$ the work done against the pressure, so the energy balance under the cosmic expansion is preserved.

If the equation of state of the ideal fluid is

$$p = p(\rho) \qquad (41\text{-}6)$$

then we can obtain from Equation (41-4) the dependence of R on ρ. Equation (41-4) may be written in the alternative form

$$\frac{dR}{R} + \frac{1}{3}\left[\frac{d\rho}{\rho + (p(\rho)/c^2)}\right] = 0 \qquad (41\text{-}7)$$

Simple integration then gives the relationship of R and ρ. In the case of the pressure being so small as to be effectively zero, we can write

$$R^3\rho = \text{constant} \qquad (41\text{-}8)$$

For an ideal gas, $p = \alpha p$, where α is a constant, so we obtain

$$R^{3[1 + (\alpha/c^2)]}\rho = \text{constant} \qquad (41\text{-}9)$$

In particular, if the pressure is entirely due to radiation, we have

$$p = \left(\frac{c^3}{3}\right)\rho \qquad (41\text{-}10)$$

with the result that $R^4\rho = \text{constant}$.

Referring to Equation (41–3), we see that different models for the universe may be obtained by assigning various values to k and various specific forms to the function of time $R(t)$. Now it happens that the values of k are restricted to 0 and ± 1. If $k = +1$, then one obtains, from the Einstein equations, a spherical universe of radius

$$R = c\left(\frac{3}{8\pi G\rho}\right)^{1/2} \tag{41-11}$$

where G is the gravitational constant. Observational results of astronomy place the mean density of space somewhere between 10^{-24} and 10^{-27} kg/m^3, resulting in a radius for the universe that lies in the range 10^9–10^{11} light years. (A light year is the distance that light travels in a vacuum in 1 yr.)

The total mass of the universe is now given by

$$M = \tfrac{4}{3}\pi R^3 \rho \tag{41-12}$$

The total mass of the universe is preserved during its cosmic evolution.

41–3 EXPANDING UNIVERSE

There are several simple sorts of solution to the equations that result from particular assignments of the form for $R(t)$. One of these results in the radius $R(t)$ that describes a cycloid. R goes through 0 at $t = 0$. It then expands, goes through a maximum, and contracts again. The trouble with this model is that it requires the matter to be more condensed than is allowed for in the homogeneous fluid sphere model, and the pressure term would have to be larger to counterbalance this. It is considered to be too unphysical in these respects.

If we take the value of $k = 0$, then we have a nonstatic *exploding model* of the universe proposed by Friedman. The radius is given by

$$R = At^{2/3} \tag{41-13}$$

where A is a constant. It describes a universe in continuous expansion starting abruptly at the origin of time, $t = 0$. The total time elapsed from $t = 0$ to the present may be taken as T and is given by

$$8\pi G\rho = \frac{3}{T^2} \tag{41-14}$$

Equation (41–13) shows that $R(t)$ increases monotonically from zero time to infinity. This expansion results in a red shift of the spectra of the stars that increases with distance from any given point, and the value of this red shift may be used to compute the velocity of a distant point relative to a local one as a function of its distance. This velocity–distance relation is called *Hubble's law*. Based on the local (or laboratory) value of a spectral line λ, the red shift is $\Delta\lambda$. EDWIN HUBBLE (1936) discovered that the red shifts of the stars can be described by

$$\frac{\Delta\lambda}{\lambda} = \frac{L}{cT} \qquad (41\text{--}15)$$

where L is the astronomical distance to the star and T is *Hubble's constant*.

The distance L is measured for the locality of the Galaxy and nearby galaxies by using parallax (the apparent shift of a star in position as seen from different points on the Earth's orbit at different times of year) and the periodic brightening of certain stars called cepheid variables. Such stars possess an absolute luminosity that varies periodically in time, and the period has been shown to be directly related to the absolute luminosity. Cepheid stars in distant galaxies thus yield a measure of the distance if their apparent brightness is taken as proportional to L^{-2}.

The value of Hubble's constant obtained by astronomers is

$$\boxed{T = (4.1 \pm 2) \times 10^{17} \text{ sec}} \qquad (41\text{--}16)$$

or about 13 billion yr. Equation (41–14) states that this same constant is the elapsed time since $t = 0$, or the age of the universe since its initial expansion. Equation (41–14) then yields a present density for the universe of

$$\rho = (1.1 \pm 0.5) \times 10^{-26} \text{ kg/m}^3 \qquad (41\text{--}17)$$

in agreement with experimental limits quoted above.

Not only does the radius R increase monotonically with time, but at times very early in the expansion the radius is increasing *faster* than the velocity of light. This is a result of the general theory of relativity and would be forbidden by the inapplicable special theory of relativity. For large R, the velocity approaches the velocity of light. The birth of the universe is explosive, since dR/dt is infinite at time $= 0$.

The value of 13 billion yr for the age of the universe is in agreement with the results quoted for the various ages in the foregoing, because these represent the development of the universe at much later times, so their ages should be somewhat less. We have arrived at a picture of the universe as a system of galaxies that is in constant expansion. The average distance between galaxies is increasing in a way that makes the farthest galaxies travel faster relative to a local point than do the nearer galaxies. We have a picture that we can turn around and extrapolate back in time. We know the age rather well, and we know the total mass of the universe. We shall now build a picture of the development of the universe from time zero that is consistent with this.

41–4 BIRTH OF THE UNIVERSE

In order to build a logical model of the universe, it is necessary first to have a physical vacuum in which to contain the developing fireball. Thus, the prerequisite to our explosive model is that state of three-space called the physical vacuum. As we have pointed out before, such a vacuum—while containing no material particles or radiation—does contain all the particles of nature in virtual states. The density of such a vacuum is extremely high in terms of these virtual states. Furthermore, such a vacuum contains all the laws of nature. Along with the virtual particle states, there exist in the vacuum all the requisite symmetries and quantum numbers needed to describe the particles.

In order to contain the physical vacuum we must, of course, have three-space itself. The intrinsic geometry of this space must be such as to produce the symmetries and laws of nature before they can come into play by influencing the actions of material particles. Thus, it is seen that the initial conditions necessary for a universe is that it be first provided with three-space and the physical vacuum contained in it.

Time Zero. We have been speaking about the explosive model of the expanding universe. At time $t = 0$, the entire mass energy plus pressure energy was produced in a singularity and proceeded to expand from that point with the velocity of light. The temperature of this fireball was, of course, exceedingly high, being of the order of $10^{14}\,°K$ at times of about 0.1 sec. The simplest choice for the initial composition of such a fireball is pure radiation.

The Production of Matter and Antimatter. In the first few moments of the expanding ball of radiation, we may expect many photon–photon collisions, with the occasional production of material particles such as electrons, mesons, and nucleons. As soon as these are produced, photon–particle collisions rapidly produce many more particles. In all cases, such production is in particle–antiparticle pairs, so both matter and antimatter are produced. An equilibrium is soon reached between the expanding radiation and the matter it contains. Particle–antiparticle annihilations consume the matter and turn it into radiation in parallel with the photon collisions that produce it. Reactions between particles produce all possible elements, but the expansion is so rapid and the temperature so high that, effectively, only hydrogen and helium are built in any consequence, as we have seen previously.

It is a characteristic of the fireball in its early stages that it has infinite heat capacity. Thus, if one tried to raise the temperature by adding radiation, one would only succeed in producing more particles rather than raising the temperature. It is a characteristic of systems with exceedingly high heat capacity that they undergo strong density fluctuations. Thus, as the fireball expands, it rapidly undergoes exceedingly strong density fluctuations, which are at first only transient—appearing and disappearing with great rapidity. As the ball expands further, however, fluctuations become "frozen out"; that is, regions of high density live longer and become interspersed with regions of low density. Thus, a fragmentation occurs which we may regard as being the early galactic clusters (long before the formation of stars and galaxies, of course). The radiation is by now cooled to a point where it is no longer effective in the production of particle–antiparticle pairs.

The Formation of Protogalaxies. The picture we have at this stage is a fireball that has expanded and cooled to produce an expanding cloud of regions that are dense in matter interspersed with regions that are devoid of matter to a great extent. All regions have, of course, an equal mixture of matter and antimatter. The progress of annihilations proceeds, therefore, long after the production of particle–antiparticle pairs has ceased. If there were no other process to intervene, one would now expect the annihilation process to consume all the remaining matter, leaving only radiation and a neutrino sea behind.

The condensed regions, however, possess their own gravitational fields. It is suggested by Hannes Alfvén that this gravitational field now pulls at the heavier components of the plasma—the nucleons—and causes a separation of these from the lighter components—the electrons. Alfvén describes the plasma of matter-antimatter radiation as an "ambiplasma." There are, in effect, two gases mixed in the ambiplasma: one consisting of electrons and positrons and one some 1840 times heavier, the proton–antiproton gas. Thus, there is a separation of these two gases in a gravitational field.

Magnetic fields created by the early expansion and trapped in the ambiplasma now move with the plasma and produce electrical fields. Those electrical fields in the direction of motion of the gravitating plasma now produce a separation of the positive electrons from the negative ones and a similar separation of the protons from the antiprotons. These currents are such as to concentrate the antimatter in a layer in the center of the gravitating column of ambiplasma, leaving the matter to form on both sides of it. (Opposite currents would concentrate the matter in the center, leaving the antimatter behind.) Thus, we now have regions where matter predominates over antimatter and other regions where the opposite is true.

Annihilations now continue until each region is either purely matter or purely antimatter. The regions are, of course, separated by space in which annihilations have gone to completion, so intergalactic space should be exceedingly empty of matter. This is found to be so. The general expansion of the universe has continued, of course, during this process.

Stars and Planets. With the separation of the matter and antimatter regions, we now have the conditions for the gravitational contraction of these clouds and the continued condensation of them until galaxies of stars may form from them as described in Chapter 40. The formation of stars, along with their manufacture of the elements and their explosive ejection of heavy gases and dust into the interstellar medium, provides the conditions for the formation of the planetary rings of dust and gas around stars like the sun. Continued gathering together of these regions into clumps under gravitational attraction then forms planets.

Thus, according to one theoretical model, the universe was born.

QUESTIONS

41-1 Make a table listing the various ages of the components of the universe so that they may be compared with one another.

41-2 In what sense may the universe be described as an ideal gas?

41-3 Derive Equation (41-7) from Equation (41-4).

41-4 Derive Equations (41-8), (41-9), and (41-10).

41-5 How many meters is 1 light year? How many miles is this?

41-6 Using Equation (41-17), calculate the total mass of the universe.

RECOMMENDED READING

ADLER, Ronald, BAXIN, Maurice, and SCHIFFER, Menahem, *Introduction to General Relativity*, McGraw-Hill, New York, 1965.

ALFVÉN, Hannes, *Worlds-Antiworlds*, Freeman, San Francisco, 1966.

EDDINGTON, Arthur, *Space, Time and Gravitation*, Harper & Row, New York, 1959; Cambridge University Press, London, 1920.

GAMOW, George, *The Creation of the Universe*, Viking, New York, 1952.

GAMOW, George, *A Planet Called Earth*, Bantam, New York, 1970.

GAMOW, George, "The Physics of the Expanding Universe," *Vistas in Astronomy* **2**, 1726 (1956).

NUNITZ, Milton K. (Ed.), *Theories of the Universe*, Free Press, NewYork, 1957.

Appendix

Table 1 Periodic table of the elements

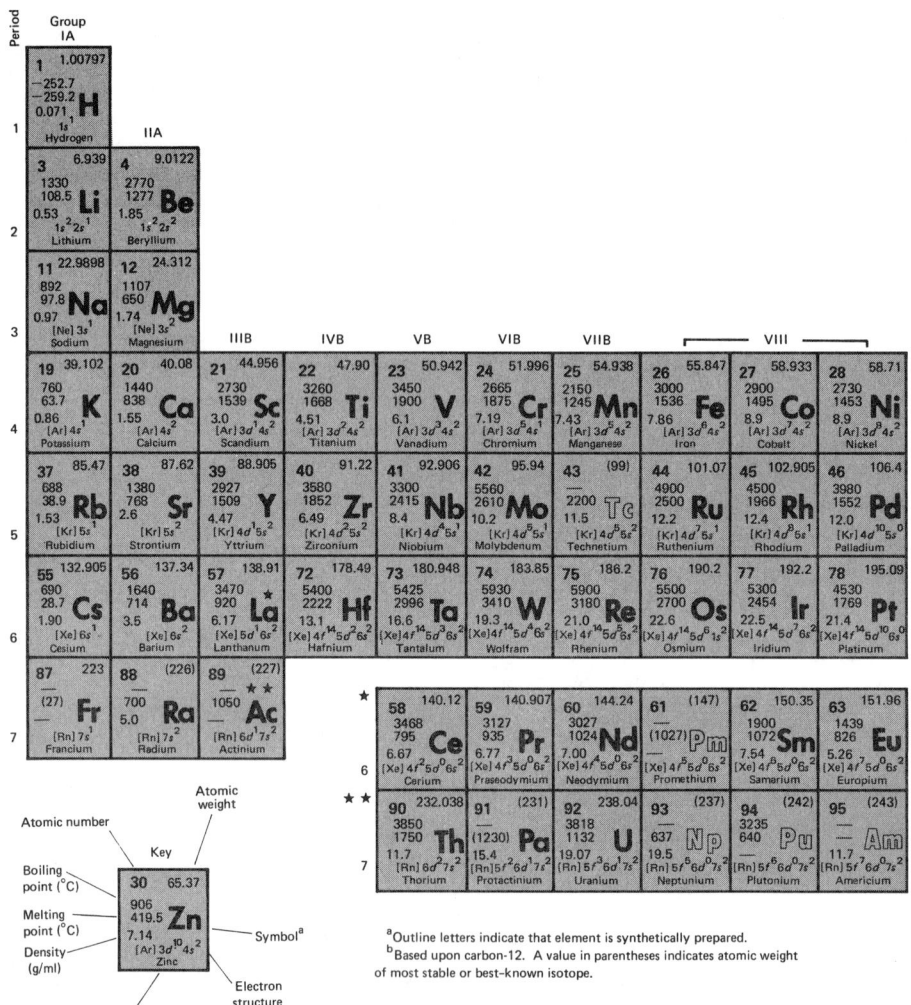

[a] Outline letters indicate that element is synthetically prepared.
[b] Based upon carbon-12. A value in parentheses indicates atomic weight of most stable or best-known isotope.

Table 1 (*Continued*)

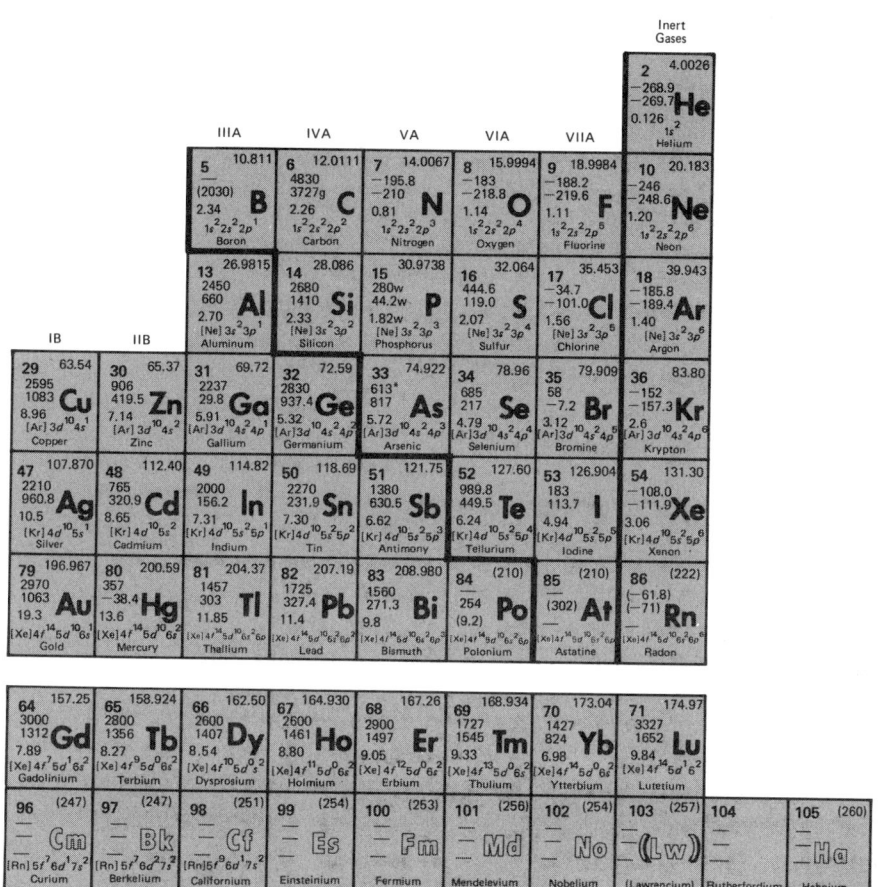

Table 2 Useful mathematical aids

TRIGNOMETRIC IDENTITIES

(1) $\sin^2 \alpha + \cos^2 \alpha = 1$
(2) $\sin 2\alpha = 2 \sin \alpha \cos \alpha$
(3) $\cos 2\alpha = \cos^2 \alpha - \sin^2 \alpha$
(4) $\sin^2 \dfrac{\alpha}{2} = \dfrac{1 - \cos \alpha}{2}$
(5) $\cos^2 \dfrac{\alpha}{2} = \dfrac{1 + \cos \alpha}{2}$
(6) $\sin (\alpha \pm \beta) = \sin \alpha \cos \beta \pm \cos \alpha \sin \beta$
(7) $\cos (\alpha \pm \beta) = \cos \alpha \cos \beta \mp \sin \alpha \sin \beta$
(8) $\sec^2 \alpha = 1 + \tan^2 \alpha$
(9) $\csc^2 \alpha = 1 + \cot^2 \alpha$
(10) $\sin \alpha + \sin \beta = 2 \sin \left(\dfrac{\alpha + \beta}{2}\right) \cos \left(\dfrac{\alpha - \beta}{2}\right)$
(11) $\sin \alpha - \sin \beta = 2 \sin \left(\dfrac{\alpha - \beta}{2}\right) \cos \left(\dfrac{\alpha + \beta}{2}\right)$
(12) $\cos \alpha + \cos \beta = 2 \cos \left(\dfrac{\alpha + \beta}{2}\right) \cos \left(\dfrac{\alpha - \beta}{2}\right)$
(13) $\cos \alpha - \cos \beta = -2 \sin \left(\dfrac{\alpha + \beta}{2}\right) \sin \left(\dfrac{\alpha - \beta}{2}\right)$

SERIES EXPANSION

(1) $(a + b)^n = a^n + na^{n-1}b + \dfrac{n(n-1)}{2!} a^{n-2}b^2 + \dfrac{n(n-1)(n-2)}{3!} a^{n-3}b^3 + \cdots$

(2) $(1 + x)^n = 1 + nx + \dfrac{n(n-1)}{2!} x^2 + \dfrac{n(n-1)(n-2)}{3!} x^3 + \cdots$

(3) $e^x = 1 + x + \dfrac{x^2}{2!} + \dfrac{x^3}{3!} + \cdots$

(4) $\dfrac{1}{1 - x} = 1 + x + x^2 + \cdots$ with $|x| < 1$

(5) $\sin x = x - \dfrac{x^3}{3!} + \dfrac{x^5}{5!} + \cdots$

(6) $\cos x = 1 - \dfrac{x^2}{2!} + \dfrac{x^4}{4!} - \cdots$

(7) $\tan x = x + \dfrac{x^3}{3} + \dfrac{2x^5}{15} + \cdots$

(8) $\ln (1 + x) = x - \dfrac{x^2}{2} + \dfrac{x^3}{3} - \cdots$

Table 2 (*Continued*)

MOST COMMON INDEFINITE INTEGRALS

(1) $\int u^n \, du = \dfrac{u^{n+1}}{n+1} + C \quad \text{except} \quad n = -1$

(2) $\int \dfrac{du}{u} = \ln u + C$

(3) $\int e^u \, du = e^u + C$

(4) $\int b^u \, du = \dfrac{b^u}{\ln b} + C \quad b > 0, b \neq 1$

(5) $\int \sin u \, du = -\cos u + C$

(6) $\int \cos u \, du = \sin u + C$

(7) $\int \tan u \, du = \ln \sec u + C$

(8) $\int \cot u \, du = \ln \sin u + C$

(9) $\int \sec u \, du = \ln(\sec u + \tan u) + C$

(10) $\int \csc u \, du = \ln(\csc u - \cot u) + C$

(11) $\int \sec^2 u \, du = \tan u + C$

(12) $\int \csc^2 u \, du = -\cot u + C$

(13) $\int \dfrac{du}{a^2 + u^2} = \dfrac{1}{a} \tan^{-1} \dfrac{u}{a} + C$

(14) $\int \dfrac{du}{u^2 - a^2} = \dfrac{1}{2a} \ln \dfrac{u-a}{u+a} + C$

(15) $\int \dfrac{du}{\sqrt{a^2 - u^2}} = \sin^{-1} \dfrac{u}{a} + C$

(16) $\int \dfrac{du}{\sqrt{u^2 \pm a^2}} = \ln(u + \sqrt{u^2 \pm a^2}) + C$

(17) $\int \sqrt{a^2 - u^2} \, du = \dfrac{1}{2}\left(u\sqrt{a^2 - u^2} + a^2 \sin^{-1} \dfrac{u}{a}\right) + C$

(18) $\int \sqrt{u^2 \pm a^2} \, du = \dfrac{1}{2}\left[u\sqrt{u^2 \pm a^2} \pm a^2 \ln(u + \sqrt{u^2 \pm a^2})\right] + C$

INTEGRATION BY PARTS

$\int u \, dv = uv - \int v \, du$

Table 3 Natural trigonometric functions

ANGLE					ANGLE				
DEGREE	RADIAN	SINE	COSINE	TANGENT	DEGREE	RADIAN	SINE	COSINE	TANGENT
0°	0.000	0.000	1.000	0.000					
1°	0.017	0.018	1.000	0.018	46°	0.803	0.719	0.695	1.036
2°	0.035	0.035	0.999	0.035	47°	0.820	0.731	0.682	1.072
3°	0.052	0.052	0.999	0.052	48°	0.838	0.743	0.669	1.111
4°	0.070	0.070	0.998	0.070	49°	0.855	0.755	0.656	1.150
5°	0.087	0.087	0.996	0.088	50°	0.873	0.766	0.643	1.192
6°	0.105	0.105	0.995	0.105	51°	0.890	0.777	0.629	1.235
7°	0.122	0.122	0.993	0.123	52°	0.908	0.788	0.616	1.280
8°	0.140	0.139	0.990	0.141	53°	0.925	0.799	0.602	1.327
9°	0.157	0.156	0.988	0.158	54°	0.942	0.809	0.588	1.376
10°	0.175	0.174	0.985	0.176	55°	0.960	0.819	0.574	1.428
11°	0.192	0.191	0.982	0.194	56°	0.977	0.829	0.559	1.483
12°	0.209	0.208	0.978	0.213	57°	0.995	0.839	0.545	1.540
13°	0.227	0.225	0.974	0.231	58°	1.012	0.848	0.530	1.600
14°	0.244	0.242	0.970	0.249	59°	1.030	0.857	0.515	1.664
15°	0.262	0.259	0.966	0.268	60°	1.047	0.866	0.500	1.732
16°	0.279	0.276	0.961	0.287	61°	1.065	0.875	0.485	1.804
17°	0.297	0.292	0.956	0.306	62°	1.082	0.883	0.470	1.881
18°	0.314	0.309	0.951	0.325	63°	1.100	0.891	0.454	1.963
19°	0.332	0.326	0.946	0.344	64°	1.117	0.899	0.438	2.050
20°	0.349	0.342	0.940	0.364	65°	1.134	0.906	0.423	2.145
21°	0.367	0.358	0.934	0.384	66°	1.152	0.914	0.407	2.246
22°	0.384	0.375	0.927	0.404	67°	1.169	0.921	0.391	2.356
23°	0.401	0.391	0.921	0.425	68°	1.187	0.927	0.375	2.475
24°	0.419	0.407	0.914	0.445	69°	1.204	0.934	0.358	2.605
25°	0.436	0.423	0.906	0.466	70°	1.222	0.940	0.342	2.747
26°	0.454	0.438	0.899	0.488	71°	1.239	0.946	0.326	2.904
27°	0.471	0.454	0.891	0.510	72°	1.257	0.951	0.309	3.078
28°	0.489	0.470	0.883	0.532	73°	1.274	0.956	0.292	3.271
29°	0.506	0.485	0.875	0.554	74°	1.292	0.961	0.276	3.487
30°	0.524	0.500	0.866	0.577	75°	1.309	0.966	0.259	3.732
31°	0.541	0.515	0.857	0.601	76°	1.326	0.970	0.242	4.011
32°	0.559	0.530	0.848	0.625	77°	1.344	0.974	0.225	4.331
33°	0.576	0.545	0.839	0.649	78°	1.361	0.978	0.208	4.705
34°	0.593	0.559	0.829	0.675	79°	1.379	0.982	0.191	5.145
35°	0.611	0.574	0.819	0.700	80°	1.396	0.985	0.174	5.671
36°	0.628	0.588	0.809	0.727	81°	1.414	0.988	0.156	6.314
37°	0.646	0.602	0.799	0.754	82°	1.431	0.990	0.139	7.115
38°	0.663	0.616	0.788	0.781	83°	1.449	0.993	0.122	8.144
39°	0.681	0.629	0.777	0.810	84°	1.466	0.995	0.105	9.514
40°	0.698	0.643	0.766	0.839	85°	1.484	0.996	0.087	11.43
41°	0.716	0.658	0.755	0.869	86°	1.501	0.998	0.070	14.30
42°	0.733	0.669	0.743	0.900	87°	1.518	0.999	0.052	19.08
43°	0.751	0.682	0.731	0.933	88°	1.536	0.999	0.035	28.64
44°	0.768	0.695	0.719	0.966	89°	1.553	1.000	0.018	57.29
45°	0.785	0.707	0.707	1.000	90°	1.571	1.000	0.000	∞

Table 4 Exponential functions

x	e^x	e^{-x}	x	e^x	e^{-x}
0.00	1.0000	1.0000	2.5	12.182	0.0821
0.05	1.0513	0.9512	2.6	13.464	0.0743
0.10	1.1052	0.9048	2.7	14.880	0.0672
0.15	1.1618	0.8607	2.8	16.445	0.0608
0.20	1.2214	0.8187	2.9	18.174	0.0550
0.25	1.2840	0.7788	3.0	20.086	0.0498
0.30	1.3499	0.7408	3.1	22.198	0.0450
0.35	1.4191	0.7047	3.2	24.533	0.0408
0.40	1.4918	0.6703	3.3	27.113	0.0369
0.45	1.5683	0.6376	3.4	29.964	0.0334
0.50	1.6487	0.6065	3.5	33.115	0.0302
0.55	1.7333	0.5769	3.6	36.598	0.0273
0.60	1.8221	0.5488	3.7	40.447	0.0247
0.65	1.9155	0.5220	3.8	44.701	0.0224
0.70	2.0138	0.4966	3.9	49.402	0.0202
0.75	2.1170	0.4724	4.0	54.598	0.0183
0.80	2.2255	0.4493	4.1	60.340	0.0166
0.85	2.3396	0.4274	4.2	66.686	0.0150
0.90	2.4596	0.4066	4.3	73.700	0.0136
0.95	2.5857	0.3867	4.4	81.451	0.0123
1.0	2.7183	0.3679	4.5	90.017	0.0111
1.1	3.0042	0.3329	4.6	99.484	0.0101
1.2	3.3201	0.3012	4.7	109.95	0.0091
1.3	3.6693	0.2725	4.8	121.51	0.0082
1.4	4.0552	0.2466	4.9	134.29	0.0074
1.5	4.4817	0.2231	5	148.41	0.0067
1.6	4.9530	0.2019	6	403.43	0.0025
1.7	5.4739	0.1827	7	1096.6	0.0009
1.8	6.0496	0.1653	8	2981.0	0.0003
1.9	6.6859	0.1496	9	8103.1	0.0001
2.0	7.3891	0.1353	10	22026	0.00005
2.1	8.1662	0.1225			
2.2	9.0250	0.1108			
2.3	9.9742	0.1003			
2.4	11.023	0.0907			

Table 5 Nobel Prizes in physics

Year	Laureate(s)
1901	W. C. Roentgen (Germany)
1902	H. A. Lorentz (Netherlands) Pieter Zeeman (Netherlands)
1903	Henri Becquerel (France) Pierre Curie (France) Marie S. Curie (France)
1904	J. W. Strutt (Lord Rayleigh) (England)
1905	Philipp Lenard (Germany)
1906	J. J. Thomson (England)
1907	A. A. Michelson (U.S.)
1908	Gabriel Lippman (France)
1909	Guglielmo Marconi (Italy) K. F. Braun (Germany)
1910	J. D. van der Waals (Netherlands)
1911	Wilhelm Wien (Germany)
1912	N. G. Dalen (Sweden)
1913	Kamerlingh Onnes (Netherlands)
1914	Max von Laue (Germany)
1915	W. H. Bragg (England) W. L. Bragg (England)
1916	no award
1917	C. G. Barkla (England)
1918	Max Planck (Germany)
1919	Johannes Stark (Germany)
1920	C. E. Guillaume (Switzerland)
1921	Albert Einstein (Germany)
1922	Niels Bohr (Denmark)
1923	R. A. Millikan (U.S.)
1924	K. M. G. Siegbahn (Sweden)
1925	James Franck (Germany) Gustav Hertz (Germany)
1926	Jean Perrin (France)
1927	A. H. Compton (U.S.) C. T. R. Wilson (England)
1928	O. W. Richardson (England)
1929	Louis de Broglie (France)
1930	C. V. Raman (India)
1931	no award
1932	Werner Heisenberg (Germany)
1933	Erwin Schrödinger (Austria) P. A. M. Dirac (England)
1934	no award
1935	James Chadwick (England)
1936	V. F. Hess (Austria) C. D. Anderson (U.S.)
1937	C. J. Davisson (U.S.) G. P. Thomson (England)
1938	Enrico Fermi (Italy)
1939	E. O. Lawrence (U.S.)
1940	no award
1941	no award
1942	no award
1943	Otto Stern (Germany)
1944	I. I. Rabi (U.S.)
1945	Wolfgang Pauli (Austria)
1946	P. W. Bridgman (U.S.)
1947	E. V. Appleton (England)
1948	P. M. S. Blackett (England)
1949	Hideki Yukawa (Japan)
1950	C. F. Powell (England)
1951	J. D. Cockcroft (England) E. T. S. Walton (Ireland)
1952	Felix Bloch (U.S.) E. M. Purcell (U.S.)
1953	Fritz Zernike (Netherlands)
1954	Max Born (Germany) Walter Bothe (Germany)
1955	Polykarp Kusch (U.S.) W. E. Lamb (U.S.)
1956	John Bardeen (U.S.) W. H. Brattain (U.S.) William Shockley (U.S.)
1957	C. N. Yang (China, U.S.) T. D. Lee (China, U.S.)

Table 5 (*Continued*)

Year	Laureate(s)	Year	Laureate(s)
1958	P. A. Čerenkov (U.S.S.R.) I. Y. Tamm (U.S.S.R.) I. M. Frank (U.S.S.R.)	1965	R. P. Feynman (U.S.) J. S. Schwinger (U.S.) Sin-itiro Tomonaga (Japan)
1959	Owen Chamberlain (U.S.) E. G. Segrè (U.S.)	1967	Hans Bethe (Germany)
		1968	Luis Alvarez (U.S.)
1960	D. A. Glaser (U.S.)	1969	Murray Gell-Mann (U.S.)
1961	Robert Hofstadter (U.S.) R. L. Mössbauer (Germany)	1970	Hannes Alfven (Sweden)
1962	L. D. Landau (U.S.S.R.)	1971	Dennis Gabor (U.S.)
1963	J. H. D. Jensen (Germany) Maria G. Mayer (U.S.) E. B. Wigner (U.S.)	1972	John Bardeen (U.S.) Leon Cooper (U.S.) John Schrieffer (U.S.)
1964	Nikolai Basov (U.S.S.R.) Alexander Prokhorv (U.S.S.R.) C. H. Townes (U.S.)		

Table 6 Table of isotopes

Z	ELEMENT SYMBOL	A	ATOMIC MASS (amu)	RELATIVE ABUNDANCE (%) OR TYPE OF DECAY[a] AND HALF-LIFE
0	n	1	1.008665	β^-, 13 min
1	H	1	1.007825	99.985
	D	2	2.014102	0.015
	T	3	3.016049	β^-, 12.2 yr
2	He	3	3.016030	0.00013
		4	4.002603	99.9999
		5	5.012296	α, n; 2×10^{-21} sec
		6	6.018900	β^-, 0.81 sec
3	Li	5	5.012541	p, α; $\sim 10^{-21}$ sec
		6	6.015126	7.42
		7	7.016005	92.58
		8	8.022488	β^-, 0.86 sec
		9	9.027300	β^-, 0.17 sec
4	Be	6	6.019780	β^+, 0.4 sec
		7	7.016931	EC, 53.6 days
		8	8.005308	2α, $\sim 3 \times 10^{-16}$ sec
		9	9.012186	100
		10	10.013535	β^-, 2.7×10^6 yr
		11	11.021660	β^-, 13.7 sec
5	B	8	8.024612	β^+, 0.78 sec
		9	9.013335	$2\alpha + p$, $\sim 3 \times 10^{-19}$ sec
		10	10.012939	19.6
		11	11.009305	80.4
		12	12.014353	β^-, 0.019 sec
		13	13.017779	β^-, 0.035 sec
6	C	10	10.016830	β^+, 19.0 sec
		11	11.011433	β^+, 20.5 min
		12	12.000000	98.89
		13	13.003354	1.11
		14	14.003241	β^-, 5770 yr
		15	15.010600	β^-, 2.25 sec
7	N	12	12.018709	β^+, 0.011 sec
		13	13.005739	β^+, 10.0 min
		14	14.003242	99.63
		15	15.000108	0.37
		16	16.006089	β^-, 7.35 sec
		17	17.008449	β^-, 4.14 sec
8	O	14	14.008597	β^+, 73 sec
		15	15.003072	β^+, 2.03 sec
		16	15.994915	99.759
		17	16.999134	0.037
		18	17.999160	0.204
		19	19.003577	β^-, 29.4 sec
		20	20.004071	β^-, 14 sec

Table 6 (*Continued*)

Z	ELEMENT SYMBOL	A	ATOMIC MASS (amu)	RELATIVE ABUNDANCE (%) OR TYPE OF DECAY[a] AND HALF-LIFE
9	F	16	16.011707	β^+, ~10^{-19} sec
		17	17.002098	β^+, 66 sec
		18	18.000950	β^+, 1.87 h
		19	18.998405	100
		20	19.999985	β^-, 11 sec
		21	20.999972	β^-, 5 sec
10	Ne	18	18.005715	β^+, 1.46 sec
		19	19.001892	β^+, 18 sec
		20	19.992442	90.92
		21	20.993849	0.257
		22	21.991385	8.82
		23	22.994475	β^-, 38 sec
		24	23.993597	β^-, 3.38 min
11	Na	20	20.008890	β^+, 0.4 sec
		21	20.997638	β^+, 23 sec
		22	21.994435	β^+, 2.58 yr
		23	22.989773	100
		24	23.990967	β^-, 15.0 h
		25	24.989920	β^-, 60 sec
12	Mg	23	22.994135	β^+, 12 sec
		24	23.985045	78.70
		25	24.985840	10.13
		26	25.982591	11.17
		27	26.984346	β^-, 9.5 min
		28	27.983880	β^-, 21.3 h
13	Al	24	24.000090	β^+, 2.1 sec
		25	24.990414	β^+, 7.2 sec
		26	25.986900	β^+, 7.4 × 10^5 yr
		27	26.981535	100
		28	27.981908	β^-, 2.30 h
		29	28.904420	β^-, 6.6 min
14	Si	26	25.992320	β^+, 2 sec
		27	26.986701	β^+, 42 sec
		28	27.976927	92.21
		29	28.976491	4.70
		30	29.973761	3.09
		31	30.975349	β^-, 2.62 h
		32	31.974020	β^-, ~700 yr
15	P	28	27.991740	β^+, 0.28 sec
		29	28.981816	β^+, 4.4 sec
		30	29.978320	β^+, 2.6 min
		31	30.973763	100
		32	31.973908	β^-, 14.3 days
		33	32.971728	β^-, 25 days
		34	33.973340	β^-, 12.4 days

Table 6 (*Continued*)

Z	ELEMENT SYMBOL	A	ATOMIC MASS (amu)	RELATIVE ABUNDANCE (%) OR TYPE OF DECAY[a] AND HALF-LIFE
16	S	31	30.979599	β^+, 2.6 sec
		32	31.972074	95.0
		33	32.971461	0.76
		34	33.967865	4.22
		35	34.969034	β^-, 86.7 days
		36	35.967091	0.014
		37	36.971040	β^-, 5.1 min
		38	37.971220	β^-, 2.87 h
17	Cl	32	31.986030	β^+, 0.31 sec
		33	32.977446	β^+, 2.5 sec
		34	33.973764	β^+, 1.5 sec
		35	34.968855	75.53
		36	35.968312	β^-, 3×10^5 yr
		37	36.965896	24.47
		38	37.968002	β^-, 37.3 min
		39	38.968010	β^-, 55.5 min
		40	39.970400	β^-, 1.4 min
18	Ar	35	34.975275	β^+, 1.83 sec
		36	35.967548	0.337
		37	36.966772	EC, 34.3 days
		38	37.962725	0.063
		39	38.964321	β^-, 260 yr
		40	39.962384	99.60
		41	40.964508	β^-, 1.83 h
		42	41.963043	β^-, >3.5 yr
19	K	38	37.969090	β^+, 7.7 min
		39	38.963714	93.10
		40	39.964008	0.0118
		41	40.961835	6.88
		42	41.962417	β^-, 12.4 h
		43	42.960731	β^-, 22.4 h
		44	43.962040	β^-, 22 min
20	Ca	39	38.970706	β^+, 0.9 sec
		40	39.962589	96.97
		41	40.962279	EC, 1.1×10^5 yr
		42	41.958628	0.64
		43	42.958780	0.145
		44	43.955490	2.06
		45	44.956189	β^-, 165 days
		46	45.953689	0.0033
		47	46.954512	β^-, 4.7 days
		48	47.952363	0.18
		49	48.955662	β^-, 8.7 min

Table 6 (*Continued*)

Z	ELEMENT SYMBOL	A	ATOMIC MASS (amu)	RELATIVE ABUNDANCE (%) OR TYPE OF DECAY[a] AND HALF-LIFE
21	Sc	43	42.961163	β^+, 3.9 h
		44	43.959406	β^+, 4.0 h
		45	44.955919	100
		46	45.955173	β^-, 84.0 days
		47	46.952402	β^-, 3.4 days
		48	47.952231	β^-, 44 h
		49	48.950025	β^-, 57.5 min
		50	49.951600	β^-, 1.8 min
22	Ti	44	43.959573	EC, $\sim 10^3$ yr
		45	44.958129	β^-, 3.08 h
		46	45.952633	7.93
		47	46.951758	7.28
		48	47.947948	73.94
		49	48.947867	5.51
		50	49.944790	5.34
23	V	46	45.960226	β^+, 0.4 sec
		47	46.954884	β^+, 32 min
		48	47.952260	β^+, 1.61 days
		49	48.948523	EC, 330 days
		50	49.947165	0.24
		51	50.943978	99.76
		52	51.944802	β^-, 3.77 min
		53	52.943370	β^-, 2.0 min
24	Cr	48	47.956930	EC, 23 h
		49	48.951271	β^+, 42 min
		50	49.946051	4.31
		51	50.944786	EC, 27.8 days
		52	51.940514	83.76
		53	52.940651	9.55
		54	53.938879	2.38
		55	54.941080	β^-, 3.5 min
25	Mn	50	49.953990	β^+, 0.29 sec
		51	50.948200	β^+, 45 min
		52	51.945563	EC, 5.6 days
		53	52.941293	EC, 10^6 yr
		54	53.940360	EC, 291 days
		55	54.938054	100
		56	55.938914	β^-, 2.58 h
26	Fe	53	52.945578	β^+, 9 min
		54	53.939621	5.82
		55	54.938302	EC, 2.7 yr
		56	55.934932	91.66
		57	56.935394	2.19
		58	57.933272	0.33
		59	58.934867	β^-, 45 days

Table 6 (*Continued*)

Z	ELEMENT SYMBOL	A	ATOMIC MASS (amu)	RELATIVE ABUNDANCE (%) OR TYPE OF DECAY[a] AND HALF-LIFE
27	Co	56	55.939870	EC, 77.3 days
		57	56.936292	EC, 270 days
		58	57.935754	EC, 71 days
		59	58.933190	100
		60	59.933806	β^-, 5.27 yr
		61	60.932434	β^-, 9.90 min
		62	61.933949	β^-, 1.9 min
28	Ni	57	56.939765	EC, 36 h
		58	57.935342	67.88
		59	58.934344	EC, 8×10^4 yr
		60	59.930783	26.23
		61	60.931049	1.19
		62	61.928345	3.66
		63	62.929666	β^-, 92 yr
		64	63.927959	1.08
		65	64.930041	β^-, 2.56 h
29	Cu	58	57.944468	β^+, 3.3 sec
		59	58.939456	β^+, 81 sec
		60	59.937382	β^+, 2.4 min
		61	60.933444	β^+, 3.3 h
		62	61.932564	β^+, 9.73 min
		63	62.929595	69.09
		64	63.929761	EC, 12.8 h
		65	64.927786	30.91
		66	65.928871	β^-, 5.1 min
30	Zn	61	60.939240	β^+, 89 sec
		62	61.934379	EC, 9.3 h
		63	62.933208	EC, 38.3 min
		64	63.929145	48.89
		65	64.929234	EC, 245 days
		66	65.926048	27.81
		67	66.927149	4.11
		68	67.924865	18.57
		69	68.926653	β^-, 55 min
		70	69.925348	0.62
31	Ga	64	63.936738	β^+, 2.6 min
		65	64.932733	β^+, 15 min
		66	65.931599	β^+, 9.5 min
		67	66.928221	EC, 78 h
		68	67.927997	β^+, 68 min
		69	68.925682	60.4
		70	69.926048	β^-, 21 min
		71	70.924840	39.6
		72	71.926030	β^-, 14.1 h
		73	72.925020	β^-, 4.8 h
		74	73.927220	β^-, 7.8

Table 6 (*Continued*)

Z	ELEMENT SYMBOL	A	ATOMIC MASS (amu)	RELATIVE ABUNDANCE (%) OR TYPE OF DECAY[a] AND HALF-LIFE
32	Ge	67	66.932940	β^+, 19 min
		68	67.928700	EC, 280 days
		69	68.928983	EC, 40 h
		70	69.924277	20.52
		71	70.925090	EC, 11 days
		72	71.922082	27.43
		73	72.923360	7.76
		74	73.921150	36.54
		75	74.922840	β^-, 82 min
		76	75.921360	7.67
33	As	70	69.931300	β^+, 50 min
		71	70.927250	EC, 62 h
		72	71.926430	β^+, 26 h
		73	72.923760	EC, 76 days
		74	73.923910	EC, 18 days
		75	74.921580	100
		76	75.922417	β^-, 26.5 h
		77	76.920668	β^-, 39 h
		78	77.921750	β^-, 91 min
		79	78.920990	β^-, 9 min
		80	79.922950	β^-, 15 min
34	Se	71	70.931970	β^+, 45 min
		73	72.926710	β^+, 7.1 h
		74	73.922450	0.87
		75	74.922510	EC, 120 days
		76	75.919229	9.02
		77	76.919934	7.58
		78	77.917348	23.52
		79	78.918521	β^-, 7×10^4 yr
		80	79.916512	49.82
		81	80.917984	β^-, 18 min
35	Br	76	75.924200	β^+, 16.5 h
		77	76.921399	EC, 58 h
		78	77.921070	β^+, 6.5 min
		79	78.918348	50.54
		80	79.918541	EC, 18 min
		81	80.916344	49.46
		82	81.916802	β^-, 35.7 h
		83	82.915205	β^-, 24 h
		84	83.916595	β^-, 32 min
36	Kr	77	76.924490	β^+, 1.2 h
		78	77.920368	0.35
		79	78.920089	EC, 34.5 h
		80	79.916388	2.27
		81	80.916610	EC, 2×10^5 yr
		82	81.913483	11.56
		83	82.914131	11.55
		84	83.911504	56.90
		85	84.912430	β^-, 10.4 yr
		86	85.910617	17.37
		87	86.913370	β^-, 78 min

Table 6 (Continued)

Z	ELEMENT SYMBOL	A	ATOMIC MASS (amu)	RELATIVE ABUNDANCE (%) OR TYPE OF DECAY[a] AND HALF-LIFE
37	Rb	81	80.919010	EC, 4.7 h
		82	81.917959	β^+, 75 sec
		84	83.914352	EC, 33 days
		85	84.911710	72.15
		86	85.911160	β^-, 18.77 days
		87	86.909180	27.85
		88	87.911190	β^-, 18 min
		89	88.911220	β^-, 15 min
38	Sr	84	83.913376	0.56
		85	84.912900	EC, 64 days
		86	85.909260	9.86
		87	86.908890	7.02
		88	87.905610	82.56
		89	88.907010	β^-, 50.4 days
		90	89.907330	β^-, 28 yr
		91	90.909780	β^-, 9.7 h
		92	91.910520	β^-, 2.7 h
39	Y	85	84.916120	β^+, 5 h
		86	85.914840	β^+, 14.6 h
		87	86.910700	EC, 80 h
		88	87.909500	EC, 108 days
		89	88.905430	100
		90	89.906740	β^-, 64.2 h
		91	90.906910	β^-, 57.5 days
		92	91.904590	β^-, 3.60 h
		93	92.909190	β^-, 10.4 h
		94	93.911510	β^-, 20 min
40	Zr	87	86.914470	β^+, 1.6 h
		89	88.908480	EC, 79 h
		90	89.904320	51.46
		91	90.905250	11.23
		92	91.904590	17.11
		93	92.906080	β^-, 9.5×10^5 yr
		94	93.906140	17.40
		95	94.907920	β^-, 65 days
		96	95.908200	2.80
41	Nb	89	88.912650	β^+, 1.9 h
		90	89.910890	β^+, 14.6 h
		91	90.906960	EC, long
		92	91.906820	EC, 10.1 days
		93	92.906020	100
		94	93.906960	β^-, 2.0×10^4 yr
		95	94.906720	β^-, 35 days
		96	95.907910	β^-, 23.35 h
		97	96.907820	β^-, 72 min

Table 6 (*Continued*)

Z	ELEMENT SYMBOL	A	ATOMIC MASS (amu)	RELATIVE ABUNDANCE (%) OR TYPE OF DECAY[a] AND HALF-LIFE
42	Mo	90	89.913610	β^+, 5.7 h
		91	90.911730	β^+, 15.6 min
		92	91.906290	15.84
		93	92.906530	$EC, \approx 10^4$ yr
		94	93.904740	9.04
		95	94.905720	15.72
		96	95.904550	16.53
		97	96.905750	9.46
		98	97.905510	23.78
		99	98.907870	β^-, 66 h
		100	99.907570	9.13
43	Tc	92	91.913200	β^+, 4.3 min
		93	92.909930	EC, 2.7 h
		94	93.909380	β^-, 52 min
		95	94.907500	EC, 20 h
		96	95.907750	EC, 4.3 days
44	Ru	95	94.909860	EC, 99 min
		96	95.907600	5.51
		98	97.905500	1.87
		99	98.906080	12.72
		100	99.903020	12.62
		101	100.904120	17.07
		102	101.903720	31.61
		103	102.905610	β^-, 40 days
		104	103.905530	18.58
		105	104.907290	β^-, 4.45 h
		106	105.907030	β^-, 1.0 yr
45	Rh	98	97.910000	β^+, 8.7 min
		99	98.908330	EC, 16 days
		100	99.906930	EC, 21 h
		102	101.906150	EC, 206 days
		103	102.904800	100
		104	103.906180	β^-, 42 sec
		105	104.905250	β^-, 36 h
		106	105.906990	β^-, 30 sec
		107	106.906620	β^-, 21.7 min
46	Pd	99	98.912400	β^+, 22 min
		102	101.904940	0.96
		103	102.905410	EC, 17 days
		104	103.903560	10.97
		105	104.904640	22.23
		106	105.903200	27.33
		107	106.905010	β^-, 7×10^6 yr
		108	107.903920	26.71
		109	108.905900	β^-, 13.6 h
		110	109.904500	11.81
		111	110.907640	β^-, 22 min
		112	111.907490	β^-, 21.0 h

Table 6 (*Continued*)

Z	ELEMENT SYMBOL	A	ATOMIC MASS (amu)	RELATIVE ABUNDANCE (%) OR TYPE OF DECAY[a] AND HALF-LIFE
47	Ag	103	102.907770	β^+, 1.0 h
		104	103.908150	EC, 69 min
		105	104.906800	EC, 40 days
		106	105.906390	β^+, 24 min
		107	106,904970	51.82
		108	107.905890	β^-, 2.4 min
		109	108.904700	48.18
		110	109.906050	β^-, 24 sec
		111	110.905280	β^-, 7.5 days
		112	111.907170	β^-, 3.2 h
		113	112.906760	β^-, 5.3 h
		114	113.908500	β^-, 5 sec
48	Cd	106	105.905950	1.22
		107	106.906520	EC, 6.7 h
		108	107.904000	0.88
		109	108.904870	EC, 470 days
		110	109.902970	12.39
		111	110.904150	12.75
		112	111.902840	24.07
		113	112.904610	12.26
		114	113.903570	28.86
		115	114.905620	β^-, 2.3 days
		116	115.905010	7.58
		117	116.907360	β^-, 2.8 h
49	In	108	107.909470	β^+, 40 min
		109	108.907040	EC, 4.3 h
		110	109.907220	β^+, 66 min
		111	110.905480	EC, 2.81 days
		112	111.905640	β^-, 14 min
		113	112.904280	4.28
		114	113.905090	EC, 72 sec
		115	114.904070	95.72
		116	115.905640	β^-, 14 sec
		117	116.904640	β^-, 45 min
		118	117.906300	β^-, 5.1 sec
		119	118.905930	β^-, 2.0 min
50	Sn	111	110.908180	EC, 35 min
		112	111.904940	0.96
		113	112.905010	EC, 118 days
		114	113.902960	0.66
		115	114j903530	0.35
		116	115.902110	14.30
		117	116.903060	7.61
		118	117.901790	24.03
		119	118.903390	8.58
		120	119.902130	32.85
		121	120.904170	β^-, ~25 h
		122	121.903410	4.92
		123	122.905670	β^-, 40 min
		124	123.905240	5.94
		125	124.907750	β^-, 9.4 days

Table 6 (*Continued*)

Z	ELEMENT SYMBOL	A	ATOMIC MASS (amu)	RELATIVE ABUNDANCE (%) OR TYPE OF DECAY[a] AND HALF-LIFE
51	Sb	114	113.909700	β^+, 3.4 min
		115	114.906780	EC, 31 min
		116	115.906990	β^+, 15 min
		117	116.905010	EC, 2.8 h
		118	117.905970	EC, 5.1 h
		119	118.904010	EC, 38 h
		120	119.905060	β^+, 16.4 min
		121	120.903750	57.25
		122	121.905120	β^-, 2.8 days
		123	122.904150	42.75
		124	123.905890	β^-, 60.9 days
		125	124.905230	β^-, 2.0 yr
		127	126.906810	β^-, 12 days
52	Te	116	115.908670	EC, 2.5 h
		117	116.908750	EC, 61 min
		119	118.906470	EC, 16 h
		120	119.904510	0.089
		122	121.903000	2.46
		123	122.904180	0.87
		124	123.902760	4.61
		125	124.904420	6.99
		126	125.903242	18.71
		127	126.905092	β^-, 9.3 h
		128	127.904710	31.79
		129	128.906576	β^-, 67.3 min
		130	129.906700	34.48
		131	130.908576	β^-, 25 min
		132	131.908537	β^-, 78 h
53	I	120	119.909880	β^+, 1.4 h
		122	121.907450	β^+, 3.5 min
		124	123.906180	EC, 4.2 days
		125	124.904580	EC, 60.0 days
		126	125.905512	EC, 13.2 days
		127	126.904352	100
		128	127.905818	β^-, 25.0 min
		129	128.904987	β^-, 1.72×10^7 yr
		130	129.906685	β^-, 12.5 h
		131	130.906128	β^-, 8.05 days
		132	131.907995	β^-, 2.3 h
		133	132.907460	β^-, 21 h
		134	133.909840	β^-, 53 min
		136	135.914740	β^-, 86 sec

Table 6 (*Continued*)

Z	ELEMENT SYMBOL	A	ATOMIC MASS (amu)	RELATIVE ABUNDANCE (%) OR TYPE OF DECAY^a AND HALF-LIFE
54	Xe	124	123.906120	0.0961
		126	125.904169	0.090
		127	126.905100	EC, 36.4 days
		128	127.903538	1.92
		129	128.904784	26.44
		130	129.903510	4.08
		131	130.905087	21.18
		132	131.904162	26.89
		133	132.905550	β^-, 5.27 days
		134	133.905398	10.44
		135	134.907040	β^-, 9.13 h
		136	135.907221	8.87
55	Cs	126	125.909320	EC, 1.6 min
		127	126.907340	EC, 6.2 h
		128	127.907732	β^+, 3.8 min
		130	129.906721	EC, 30 min
		131	130.905468	EC, 9.7 h
		132	131.906110	EC, 6.5 days
		133	132.905090	100
		134	133.906520	β^-, 2.19 yr
		135	134.905800	β^-, 2.0×10^6 yr
		136	135.907130	β^-, 13 days
		137	136.906820	β^-, 30 yr
		138	137.910200	β^-, 32.2 min
		139	138.913230	β^-, 9.5 min
56	Ba	130	129.906200	0.101
		132	131.905120	0.097
		133	132.905610	EC, 7.5 yr
		134	133.904310	2.42
		135	134.905570	6.59
		136	135.904360	7.81
		137	136.905560	11.32
		138	137.905010	71.66
		139	138.908610	β^-, 83 min
		140	139.910460	β^-, 12.8 days
		141	140.913740	β^-, 18 min
57	La	134	133.908290	EC, 6.5 min
		135	134.906700	EC, 19.8 h
		136	135.907440	EC, 9.5 min
		138	137.906810	0.089
		139	138.906060	99.9111
		140	139.909300	β^-, 40.2 h
		141	140.910620	β^-, 3.9 h
		143	142.915720	β^-, 14 min

Table 6 (*Continued*)

Z	ELEMENT SYMBOL	A	ATOMIC MASS (amu)	RELATIVE ABUNDANCE (%) OR TYPE OF DECAY[a] AND HALF-LIFE
58	Ce	136	135.907100	0.193
		138	137.905720	0.250
		139	138.906350	EC, 140 days
		140	139.905280	88.48
		141	140.908013	β^-, 32.5 days
		142	141.909040	11.07
		143	142.912170	β^-, 33 h
		144	143.913430	β^-, 285 days
		145	144.916240	β^-, 3.0 min
		146	145.918270	β^-, 1.4 min
59	Pr	140	139.908782	β^+, 3.4 min
		141	140.907390	100
		142	141.909790	β^-, 19.2 h
		143	142.910630	β^-, 13.7 days
		144	143.913100	β^-, 17.3 min
		145	144.914100	β^-, 5.9 h
		146	145.917200	β^-, 24 min
60	Nd	141	140.909322	EC, 2.4 h
		142	141.907478	27.11
		143	142.909620	12.17
		144	143.909900	23.85
		145	144.912160	8.30
		146	145.912690	17.22
		147	146.915830	β^-, 11.1 days
		148	147.916480	5.73
		149	148.919830	β^-, 1.8 h
		150	149.920710	5.62
		151	150.924220	β^-, 12 min
61	Pm	142	141.912630	β^+, 34 sec
		143	142.910800	EC, 265 days
		145	144.912310	EC, 18 yr
		146	145.914540	EC, 710 days
		147	146.914860	β^-, 2.5 yr
		148	147.917140	β^-, 5.39 days
		149	148.918070	β^-, 54.4 h
		150	149.921090	β^-, 2.7 h
		151	150.921640	β^-, 28.4 h
62	Sm	143	142.914460	EC, 1.0 min
		144	143.911650	3.09
		145	144.913000	EC, 340 days
		146	145.912900	α, 5×10^7 yr
		147	146.914620	14.97
		148	147.914560	11.24
		149	148.916930	13.83
		150	149.917010	7.44
		151	150.919710	β^-, ~ 93 yr
		152	151.919490	76.72
		153	152.921720	β^-, 47.1 h
		154	153.922010	22.71
		155	154.924720	β^-, 22 min
		156	155.925710	β^-, 9 h

Table 6 (*Continued*)

Z	ELEMENT SYMBOL	A	ATOMIC MASS (amu)	RELATIVE ABUNDANCE (%) OR TYPE OF DECAY[a] AND HALF-LIFE
63	Eu	150	149.919610	EC, ~5 yr
		151	150.919630	47.82
		152	151.921480	EC, 13 yr
		153	152.920860	52.18
		154	153.922840	β^-, 16 yr
		155	154.922850	β^-, 1.7 yr
		156	155.924740	β^-, 15 days
		157	156.925300	β^-, 15 h
64	Gd	149	148.918920	EC, 9 days
		150	149.918460	α, ~3 × 10^5 yr
		152	151.919530	0.200
		153	152.921090	EC, 200 days
		154	153.920720	2.15
		155	154.922590	14.73
		156	155.922100	20.47
		157	156.923940	15.68
		158	157.924100	24.87
		159	158.925970	β^-, 18 h
		160	159.927120	21.90
		161	160.929320	β^-, 3.7 min
65	Tb	158	157.925030	EC, ~1000 yr
		159	158.924950	100
		160	159.926760	β^-, 73 days
		161	160.927170	β^-, 7.1 days
66	Dy	152	151.924380	β^+, 2.5 h
		153	152.925370	EC, 5 h
		154	153.924780	α, ~10^6 yr
		156	155.923760	0.052
		158	157.923960	0.090
		159	158.925360	EC, 144 days
		160	159.924830	2.29
		161	160.926600	18.88
		162	161.926470	25.53
		163	162.928370	24.97
		164	163.928830	28.18
		165	164.931700	β^-, 2.3 h
		166	165.932900	β^-, 82 h
67	Ho	162	161.928790	β^+, 12 min
		163	162.928380	EC, 10^3 yr
		164	163.930350	β^-, 37 min
		165	164.930300	100
		166	165.932380	β^-, 26.8 h
		167	166.933120	β^-, 3.0 h

Table 6 (*Continued*)

Z	ELEMENT SYMBOL	A	ATOMIC MASS (amu)	RELATIVE ABUNDANCE (%) OR TYPE OF DECAY[a] AND HALF-LIFE
68	Er	162	161.928800	0.136
		164	163.929300	1.56
		166	165.930400	33.41
		167	166.932050	22.94
		168	167.932380	27.07
		169	168.934710	β^-, 9.4 days
		170	169.935510	14.88
		171	170.938160	β^-, 7.8 h
		172	171.939560	β^-, 50 h
69	Tm	168	167.934330	*EC*, 85 days
		169	168.934350	100
		170	169.935920	β^-, 127 days
		171	170.936570	β^-, 1.9 yr
		172	171.938580	β^-, 64 h
70	Yb	170	169.934880	3.03
		171	170.936460	14.31
		172	171.936560	21.82
		173	172.938300	16.13
		174	173.939020	31.84
		175	174.941390	β^-, 4.2 days
		176	175.942740	12.73
		177	176.945500	β^-, 1.9 h
71	Lu	170	169.938680	*EC*, 2.0 days
		173	172.939040	*EC*, ~1.3 yr
		174	173.940600	*EC*, short
		175	174.940890	97.41
		176	175.942740	2.59
		177	176.944020	β^-, 6.8 days
72	Hf	174	173.940260	0.18
		176	175.941650	5.20
		177	176.943480	18.50
		178	177.943870	27.14
		179	178.946020	13.75
		180	179.946810	35.24
		181	180.949080	β^-, 44.6 days
		182	181.950680	β^-, 9×10^6 yr
		183	182.953800	β^-, 1.1 h
73	Ta	177	176.944740	*EC*, 53 h
		178	177.945910	*EC*, 2.2 h
		179	178.946120	*EC*, 1.6 yr
		180	179.947520	0.0123
		181	180.947980	99.988
		182	181.950140	β^-, 115 days
		183	182.951440	β^-, 5.0 days
		184	183.953850	β^-, 8.7 h
		185	184.955520	β^-, 50 min
		186	185.958310	β^-, 10.5 min

Table 6 (*Continued*)

Z	ELEMENT SYMBOL	A	ATOMIC MASS (amu)	RELATIVE ABUNDANCE (%) OR TYPE OF DECAY[a] AND HALF-LIFE
74	W	180	179.946980	0.14
		181	180.948190	EC, 130 days
		182	181.948270	26.41
		183	182.950290	14.40
		184	183.950990	30.64
		185	184.953480	β^-, 75.8 days
		186	185.954340	28.41
		187	186.957370	β^-, 24 h
75	Re	185	184.953020	37.07
		186	185.955090	β^-, 90 h
		187	186.955960	62.93
		188	187.958240	β^-, 17 h
		190	189.962150	β^-, 2.8 min
76	Os	184	183.952560	0.018
		185	184.954070	EC, 94 days
		186	185.953940	1.59
		187	186.955960	1.64
		188	187.955970	13.3
		189	188.958250	16.1
		190	189.958600	26.4
		191	190.961190	β^-, 15 days
		192	191.961410	41.0
		193	192.964500	β^-, 32 h
77	Ir	186	185.958030	EC, 5 h
		188	187.959010	EC, 11 h
		190	189.960800	EC, 11 days
		191	190.960850	37.3
		192	191.962990	β^-, 74 days
		193	192.963280	62.7
		194	193.965210	β^-, 19 hr
78	Pt	188	187.959570	EC, 10 days
		190	189.959950	0.0127
		192	191.961430	0.78
		193	192.963330	EC, < 500 yr
		194	193.962810	32.9
		195	194.964820	33.8
		196	195.964981	25.3
		197	196.967357	β^-, 20 h
		198	197.967530	7.21
		199	198.970660	β^-, 30 min
79	Au	192	191.964900	EC, 4.8 h
		194	193.965510	EC, 39 h
		195	194.965110	EC, 200 days
		196	195.966554	EC, 5.55 days
		197	196.966552	100
		198	197.968242	β^-, 2.7 days
		199	198.968745	β^-, 3.15 days
		200	199.970810	β^-, 48 min
		201	200.971930	β^-, 26 min

Table 6 (*Continued*)

Z	ELEMENT SYMBOL	A	ATOMIC MASS (amu)	RELATIVE ABUNDANCE (%) OR TYPE OF DECAY[a] AND HALF-LIFE
80	Hg	196	195.965822	0.146
		197	196.966769	EC, 65 h
		198	197.966800	10.02
		199	198.968256	16.84
		200	199.968344	23.13
		201	200.970315	13.22
		202	201.970630	29.80
		203	202.972853	β^-, 47 days
		204	203.973482	6.85
		205	204.976230	β^-, 5.2 min
81	Tl	198	197.970530	ED, 5.3 h
		200	199.970974	EC, 26 h
		201	200.970760	EC, 73 h
		202	201.972130	EC, 12 days
		203	202.972331	29.50
		204	203.973890	β^-, 3.9 yr
		205	204.974462	70.50
		206	205.976080	β^-, 4.20 min
		207	206.977446	β^-, 4.78 min
		208	207.982006	β^-, 3.1 min
		209	208.985295	β^-, 2.2 min
		210	209.990002	β^-, 1.3 min
82	Pb	202	201.972190	EC, $\sim 3 \times 10^5$ yr
		203	202.973400	EC, 52 h
		204	203.973069	1.48
		205	204.974516	EC, 3×10^7 yr
		206	205.974446	23.6
		207	206.975898	22.6
		208	207.976644	52.3
		209	208.981094	β^-, 3.3 h
		210	209.984177	β^-, 21 yr
		211	210.988803	β^-, 36.1 min
		212	211.991896	β^-, 10.64 h
83	Bi	203	202.976830	EC, 12.3 h
		204	203.977700	EC, 11.6 h
		205	204.977360	EC, 15.3 days
		206	205.978320	EC, 6.3 days
		207	206.978474	EC, 30 yr
		208	207.979731	EC, 7.5×10^5 yr
		209	208.980417	100
		210	209.984110	β^-, 5.0 days
		211	210.987294	α, 2.15 min
		212	211.991271	β^-, 60.6 min
		213	212.994329	β^-, 47 min
		214	213.998634	β^-, 19.7 min
		215	215.019000	β^-, 8 min

Table 6 (*Continued*)

Z	ELEMENT SYMBOL	A	ATOMIC MASS (amu)	RELATIVE ABUNDANCE (%) OR TYPE OF DECAY[a] AND HALF-LIFE
84	Po	206	205.980500	EC, 8.8 days
		207	206.981594	EC, 5.7 h
		208	207.981264	α, 2.9 yr
		209	208.982457	α, 103 yr
		210	209.982866	α, 138.40 days
		211	210.986649	α, 0.52 sec
		212	211.988859	α, 0.30 μsec
		213	212.928370	α, 4 μsec
		214	213.951920	α, 164 μsec
		215	214.999469	α, 0.0018 sec
		216	216.001917	α, 0.16 sec
		218	218.008930	α, 3.05 min
85	At	207	206.985720	EC, 1.79 h
		208	207.986500	EC, 1.6 h
		209	208.986140	EC, 5.5 h
		210	209.986970	EC, 8.3 h
		211	210.987496	EC, 7.2 h
		213	212.993090	α, <2 sec
		214	213.996330	α, <5 sec
		215	214.998658	α, ~100 μsec
		216	216.002405	α, ~300 μsec
		217	217.004647	α, 0.018 sec
		218	218.008554	α, 1.35 sec
		219	219.011360	α, 54.0 sec
86	Rn	210	209.989720	EC, 2.7 h
		211	210.990600	EC, 16 h
		212	211.990726	α, 23 min
		215	214.998670	α, <1 min
		216	216.000234	α, 45 μsec
		217	217.003917	α, 500 μsec
		218	218.005592	α, 0.030 sec
		219	219.009523	α, 4.0 sec
		220	220.011396	α, 51.5 sec
		222	222.017530	α, 3.823 days
87	Fr	212	211.996100	EC, 19 min
		217	217.004780	α, <2 sec
		218	217.007520	α, <5 sec
		219	219.009249	α, 0.02 sec
		220	220.012330	α, 28 sec
		221	221.014176	α, 4.8 min
		223	223.019802	β^-, 22 min
88	Ra	219	219.010030	α, <1 min
		220	220.010972	α, 0.025 sec
		222	222.015365	α, 38 sec
		223	223.018565	α, 11.7 days
		224	224.020216	α, 3.64 days
		225	225.023518	β^-, 14.8 days
		226	226.025360	α, 1622 yr
		227	227.029220	β^-, 41 min
		228	228.031228	β^-, 6.7 yr

Table 6 (*Continued*)

Z	ELEMENT SYMBOL	A	ATOMIC MASS (amu)	RELATIVE ABUNDANCE (%) OR TYPE OF DECAY[a] AND HALF-LIFE
89	Ac	221	221.015690	α, <2 sec
		222	222.017750	α, 5 sec
		223	223.019119	α, 2.2 min
		224	224.021690	EC, 2.9 h
		225	225.023143	β^-, 10.0 days
		226	226.026180	β^-, 29 h
		227	227.027814	β^-, 21.6 yr
		228	228.031169	β^-, 6.13 h
		231	231.038600	β^-, 15 min
90	Th	223	223.020890	α, 0.9 sec
		224	224.021379	α, ~1 sec
		225	225.023660	EC, 8 min
		226	226.024890	α, 31 min
		227	227.027768	α, 18.17 days
		228	228.028749	α, 1.91 yr
		229	229.031629	α, 7340 yr
		230	230.033080	α, 80.00 yr
		231	231.036350	β^-, 25.6 h
		232	232.038211	α, 1.39×10^{10} yr
		233	233.041428	β^-, 22.1 min
		234	234.043570	β^-, 24.10 days
91	Pa	226	226.027800	α, 1.8 min
		227	227.028854	α, 38.3 min
		228	228.031000	EC, 22h
		229	229.031952	EC, 1.5 h
		230	230.034366	β^-, 17.7 days
		231	231.035936	α, 3.43×10^4 yr
		232	232.038611	β^-, 1.32 days
		233	233.040108	β^-, 27.4 days
		234	234.043370	β^-, 6.66 h
		235	235.045440	β^-, 24 min
		237	237.051050	β^-, 39 min
92	U	227	227.030920	α, 1.3 min
		228	228.031278	α, 9.3 min
		229	229.033200	EC, 58 min
		230	230.033926	α, 20.8 days
		231	231.036330	EC, 4.2 days
		232	232.037167	α, 73.6 yr
		233	233.039498	α, 1.62×10^5 yr
		234	234.040900	α, 2.48×10^5 yr
		235	235.043933	α, 7.13×10^8 yr
		236	236.045733	α, 2.39×10^7 yr
		237	237.048581	α, 6.75 days
		238	238.050760	α, 4.51×10^9 yr
		239	239.054320	β^-, 23.5 min
		240	240.056700	β^-, 14.1 hr

Table 6 (*Continued*)

Z	ELEMENT SYMBOL	A	ATOMIC MASS (amu)	RELATIVE ABUNDANCE (%) OR TYPE OF DECAY[a] AND HALF-LIFE
93	Np	231	231.038330	EC, ~50 min
		233	233.040600	EC, 35 min
		234	234.042830	EC, 4.4 days
		235	235.044069	α, 410 days
		236	236.046625	β^-, >5000 yr
		237	237.048030	α, 2.20×10^6 yr
		238	238.050930	β^-, 2.10 days
		239	239.054320	β^-, 2.35 days
		240	240.056700	β^-, 60 min
		241	241.058170	β^-, 16 min
94	Pu	232	232.041080	EC, 36 min
		233	233.042690	EC, 20 min
		234	234.043290	EC, 9 h
		235	235.045330	EC, 26 min
		236	236.046072	α, 2.85 min
		237	237.048277	α, 45.6 days
		238	238.049520	α, 89 yr
		239	239.052161	α, 24,360 yr
		240	240.053974	α, 6.58×10^3 yr
		241	241.056711	β^-, 13 yr
		242	242.058710	α, 3.79×10^5 yr
		243	243.061990	β^-, 4.98 h
95	Am	237	237.049780	EC, 1.3 h
		241	241.056689	α, 433 yr
		242	242.059480	β^-, 16 h
		243	243.061382	α, 7.95×10^3 yr
		244	244.064520	β^-, 25 min
		245	245.066313	β^-, 207 h
		246	246.069830	β^-, 25 min
96	Cm	238	238.053010	EC, 2.5 h
		240	240.055503	α, 26.8 days
		241	241.057510	EC, 35 days
		242	242.058800	α, 163 days
		243	243.061377	α, 35 yr
		244	244.062910	α, 17.6 yr
		245	245.065342	α, 8×10^3 yr
		246	246.067370	α, 5480 yr
		249	249.075800	β^-, 65 min
97	Bk	243	243.062920	EC, 4.5 h
		245	245.066240	EC, 4.95 days
		247	247.070180	α, ~10^4 yr
		248	248.073050	β^-, 23 h
		249	249.074838	β^-, 314 days
		250	250.078490	β^-, 3.2 h

Table 6 (*Continued*)

Z	ELEMENT SYMBOL	A	ATOMIC MASS (amu)	RELATIVE ABUNDANCE (%) OR TYPE OF DECAY[a] AND HALF-LIFE
98	Cf	244	244.065933	α, 25 min
		245	245.067890	*EC*, 44 min
		246	246.068780	α, 35.7 h
		248	248.072350	α, 350 days
		249	249.074704	α, 360 days
		250	250.076550	α, 10 yr
		253	253.084980	β^-, 19 days
99	Es	249	249.076220	*EC*, 2 h
		251	251.079850	*EC*, 1.5 days
		252	252.082900	α, 140 days
		253	253.084685	α, 20.5 days
100	Fm	248	248.077240	α, 0.6 min
		250	250.079480	α, 30 min
		252	252.082650	α, 23 h
		254	254.087000	α, 3.24 h
101	Md	255	255.090570	*EC*, 0.5 h
102	No	253	253.091000	α, ~10 min
		254	254.091000	α, ~3 sec
		255	255.090000	α, ~15 sec
103	Lr	257	257.100000	α, 8 sec
104	Rf[b]	257		α, 4.5 sec
		259		α, ~3 sec
105	Ha[b]	260		α, > 10 msec

[a] The following notation for modes of decay has been used: β^- = beta decay, β^+ = position emission, α = alpha emission, and *EC* = orbital electron capture.

[b] Proposed symbols for rutherfordium and hahnium.

SOURCE: L. A. Konig, J. H. E. Mattauch, and A. H. Wapstra, *Nucl. Phys.* **23,** 28 (1962); General Electric Company chart of nuclides, December, 1968; *Handbook of Chemistry and Physics,* 49th ed., Chemical Rubber Co., Cleveland, Ohio, 1968, table of isotopes, pp. (B-4)–(B-92).

Answers to Odd-Numbered Problems

ANSWERS TO ODD-NUMBERED PROBLEMS · 581

Chapter 2

2-1 $30\hat{i} + 22\hat{j}$ kg-m/sec
2-5 $-57.6\hat{i}$ kg-m/sec
2-7 $-400\hat{k}$ kg-m²/sec; $-400\hat{k}$ kg-m²/sec

Chapter 3

3-1 $1.4c$
3-5 $-10 + 15t - 4.9t^2$ m; $15 - 9.8t$ m/sec; -9.8 m/sec²
3-7   4 km/h
3-11 (a) 8.35×10^{-14} J; 3.34×10^{-20} kg-m/sec
      (b) 7.38×10^{-14} J; 3.14×10^{-20} kg-m/sec

Chapter 4

4-5   (a) $1.5c$
      (b) $0.96c$
4-7   (a) 140 km/h
      (b) ≈ 140 km/h
4-11 $-0.125c\hat{i} + 0.3c\hat{j}$; $67.5°$ relative to x axis

Chapter 5

5-1 $0.0447c$; $0.865c$
5-3   0.2 m
5-5   3.33 sec
5-7   5.3 sec
5-9 $0.975c$
5-11 $0.66c$
5-13 The observer must be moving at the speed of light in the same direction as the pulse.

Chapter 6

6-1 2.5×10^{-10} J or 1.560 GeV
6-3 1.34×10^{-21} kg-m/sec
6-5 $(\sqrt{E^2 - E'^2}/E)c$
6-7   (a) $0.99999345c$; 66.666229 kg-m/sec; 10^{-7} Å
      (b) 8.2 mm
6-9   (a) 45.5×10^{-6} eV
      (b) 34.4×10^{-4} eV
      (c) 8.9×10^6 eV
6-11 5.33×10^{-6} kg-m/sec
6-15 (a) 7.3×10^{-20} kg-m/sec; 4.38×10^7 m/sec; 1.71×10^{-21} kg-m/sec; 4.38×10^7 m/sec
      (b) 5.04×10^{-20} kg-m/sec; $0.1c$; 5.33×10^{-5} kg-m/sec; $0.9986c$

(c) A 10-MeV electron is relativistic, while a 10-MeV proton is almost non-relativistic.

6–17 (a) $0.116c$
(b) 3.2×10^{-23} kg-m/sec; 0.0511 MeV

6–21 (a) 1.5 mm
(b) 107 eV

Chapter 7

7–1 0.6 eV

7–3 (a) 1.175×10^6 m/sec
(b) 3.92 V

7–5 1.91 eV; 1.91 V

Chapter 8

8–3 $0.1 E_0$

8–5 1.24×10^4 V

8–7 0.0354 MeV; 0.0248 MeV; 0.0207 MeV; 0.0151 MeV

8–9 2.82 Å

8–11 29.6°

8–13 40 lines/mm

8–17 (a) 0.036 Å
(b) 40.9°; 0.141 MeV

8–19 1.32×10^{-5} Å; 201.32×10^{-5} Å

Chapter 9

9–3 4 MeV

9–5 ratio = 2

9–7 (a) 0.0183 Å
(b) 0.0146 Å
(c) 4.52×10^{-26} kg-m/sec

9–9 (a) 0.46 cm
(b) 3.06 cm

9–13 1.73×10^{20} photons/pulse

Chapter 10

10–1 (a) 1.25 Å
(b) 2.39×10^{-24} Å

10–3 (a) 6.6×10^{-24} kg-m/sec; 1.24×10^4 eV
(b) 6.6×10^{-24} kg-m/sec; 151 eV

10–5 (a) 5.25×10^{-30} kg
(b) 5.24×10^{-31} kg

10–7 23.25°

10–9 (b) 0.00714 rad

10–15 (a) ≥ 115 m
   (b) 0
10–17 (a) 1.05×10^{-25} kg-m/sec
   (b) 1.15×10^5 m/sec
   (c) 6.02×10^{-21} J

Chapter 11

11–1 1.93×10^7 m/sec
11–3 (a) 2.6×10^{-13} m
   (b) 2120 barns
   (c) 3.74×10^{-3}
11–5 (a) 1.25×10^5 particles
   (b) 6.25×10^4 particles
11–7 (a) 1.06×10^5 particles/sec
   (b) 8.94×10^5 particles/sec
   (c) 4.27×10^5 particles/sec
11–9   0.25
11–13 (a) 3.43×10^{-14} m
   (b) 8.75×10^{-5}
   (c) 0.9999125
11–15 (a) 6.05×10^{-5}
   (b) 7.6×10^{-6}
   (c) 1.51×10^5 particles/m²
   (d) 19 particles

Chapter 12

12–1 (a) 4.12×10^{16} rad/sec
   (b) 2.19×10^6 m/sec
   (c) 13.6 eV
   (d) -27.2 eV
   (e) 13.6 eV
12–3 (a) 9.15 m/sec²
   (b) 8.33×10^{-30} N
   (c) 8.2×10^{-8} N
   (d) electron will spiral into nucleus
12–5 (b) 8.2×10^4 cm^{-1}; 9.75×10^4 cm^{-1}; 1.03×10^5 cm^{-1}; 1.53×10^3 cm^{-1}
12–7 (a) 6560 Å
   (b) $36/5\lambda$
12–11 1.04×10^7 revolutions
12–13 (a) 7×10^{15} Hz
   (b) 1.13 mA
   (c) 13.3 Wb/m²

12–15 (a) 5; 2
　　　(b) 4340 Å
　　　(c) Balmer series
12–17　10 Å; x-ray region
12–19　3

Chapter 13
13–1　4.39×10^{-3} Å^{-1}
13–3　For $n = 1$:
　　　(a) 0.265 Å
　　　(b) 1.65×10^{17} Hz
　　　(c) 4.37×10^6 m/sec
　　　(d) -54.4 eV
　　　(e) 1.05×10^{-34} kg-m^2/sec
　　　(f) 0.015
13–5　(a) 304 Å
　　　(b) 256 Å
13–7　1890 Å; 1216 Å
13–9　(a) 4.5×10^{-31} kg
　　　(b) 5.45×10^{-2} Å^{-1}
　　　(c) series limit 1835 Å
13–11　54.4 eV; 6.04 eV

Chapter 14
14–1　$A_n = \sqrt{2/L}$
14–3　$\frac{1}{4} - (-1)^{n-1/2}/2n\pi$ for $n = 1, 3, 5, \ldots$; $\frac{1}{4}$ for $n = 2, 4, 6, \ldots$
　　　(b) $p/(L/4)$

Chapter 15
15–1　(a) 0.939 MeV
　　　(b) 2.82 MeV
　　　(c) 4.4×10^{-3} Å
15–3　(a) 3.42×10^{-29} eV
　　　(b) 10.3 eV
15–5　$4.95449E - 2$; 0.495449; 0.181638; 1.81638; etc.

Chapter 16
16–1　3.35×10^{-34} J; 2.09×10^{-15} eV

16–3　$\psi_4(\alpha x) = \left(\dfrac{\alpha}{24\sqrt{\pi}}\right)^{1/2} (4\alpha^4 x^4 - 9\alpha^2 x^2 + 4) e^{(-1/2)\alpha^2 x^2}$

16–5　2.19×10^{-6}

16-7 (a) 4.28×10^{-4}
   (b) 8.2×10^{-3}
   (c) 0.0805
16-9 (a) 0.124
   (b) 0.0423

Chapter 17

17-1 (a) $(-26\hat{i} - 78/5\hat{j})$ kg-m/sec; $(6\hat{i} + 18/5\hat{j})$ kg-m/sec
   (b) $(-36\hat{i} - 78/5\hat{j})$ kg-m/sec; $(-12\hat{i} + 18/5\hat{j})$ kg-m/sec
17-3 (a) 7.44×10^{-27} kg
   (b) 4.424 tons
17-5 (a) $0.718c$
   (b) $0.86\ m_0 c^2$
17-7 1.05×10^{-23} J
17-9 (a) 0.145 Å
   (b) 0.0106 Å

Chapter 18

18-1 $E_{100} = -\dfrac{me^4}{32\pi^2 \varepsilon_0^2 \hbar^2}$; $E_{200} = -\dfrac{me^4}{32\pi^2 \varepsilon_0^2 \hbar^2}\left(\dfrac{1}{2^2}\right)$; $E_{210} = -\dfrac{me^4}{32\pi^2 \varepsilon_0^2 \hbar^2}\left(\dfrac{1}{2^2}\right)$

18-3 $n = 4$
   $l = 0;\ m_l = 0$
   $l = 1;\ m_l = -1, 0, +1$
   $l = 2;\ m_l = -2, -1, 0, +1, +2$
   $l = 3;\ m_l = -3, -2, -1, 0, +1, +2, +3$

Chapter 19

19-1 1.51 eV; 1.48×10^{-34} J-sec; 0.85 eV; 1.48×10^{-34} J-sec
19-3 1.2×10^{-14} A-m²
19-5 1.31×10^{-9} Wb/m²
19-7 3.28×10^{-24} J; 2.05×10^{-5} eV
19-9 (a) $\approx 10^{26}$
   (b) 6.1×10^{-13} A
   (c) 0.126 A-m²
   (d) 1.4×10^{22}

Chapter 20

20-3 (a) five possible values
   (b) yes
   (c) five possible orientations
20-9 0.0941 Å

Chapter 21
21-5 (a) $\pi/4$
(b) $\frac{3}{8}$
(c) $\frac{15}{16}$
(d) $\frac{15}{64}$

Chapter 22
22-5 (b) 5.08×10^{-5} eV
(c) 1216.58 Å
22-7 0.0073

Chapter 23
23-3 135°
23-7 2.23×10^{-6} eV
23-9 105.6 amu; 7.7×10^{-34} kg-m^2

Chapter 25
25-1 7.56 F
25-3 (a) 30.1 MeV
(b) 2.96×10^{-14} m
25-5 2.22 MeV
25-7 2.8 MeV
25-9 20.58 MeV
25-11 0.112356 amu; 0.111484

Chapter 26
26-1 4.08×10^{21} Hz
26-3 0.182 MeV
26-5 (b) 13
26-7 0.893 MeV; 79 MeV; 803 MeV
26-9 300 eV (max)

Chapter 27
27-1 1.34×10^{-20} sec
27-3 (a) 8.48 MeV
(b) 6.26 MeV
27-5 3.65 MeV
27-7 7.41 MeV
27-9 (a) 0.576 MeV
(b) 4.65×10^{-31} eV

Chapter 28

28–1 (a) 3.27 MeV
(b) 4.03 MeV
(c) 17.59 MeV
28–3 (a) 1.88 MeV
(b) 1.828 MeV
28–5 1.23 m
28–7 −206 MeV
28–11 (a) 0.627 MeV
(b) −105.3 MeV
(c) stable

Chapter 29

29–1 (a) −1.87 MeV
(b) 2.10 MeV
(c) 0.23 MeV
(d) 1.1×10^6 m/sec
29–3 24.4°
29–5 (b) 2.35 MeV
29–7 (a) 17.35 MeV
(b) $\frac{1}{8}$ MeV

Chapter 30

30–1 6337 yr; 1581 yr
30–3 4.6×10^9 yr
30–5 12,300 sec^{-1}
30–7 13.6×10^{19} Ra-226 nuclei; 8.86×10^{14} Rn-222 nuclei; 4.92×10^{11} Pa-218 nuclei
30–9 $1 - 1/e$
30–13 4.5×10^9 yr
30–17 2.61×10^5 yr

Chapter 31

31–1 (a) 5.48 MeV; 1.08×10^{-19} kg-m/sec
(b) 0.099 MeV; 1.08×10^{-19} kg-m/sec
(c) 3×10^5 m/sec
31–3 0.15 MeV
31–5 β^- decay; β^+ decay
31–7 0.26 MeV; 0.36 MeV
31–9 unstable to α decay; stable to β^+ decay; unstable to β^- decay; unstable to positron (β^+) decay and K capture
31–11 2.00 MeV; 0.784 MeV; 2.32 MeV

31–13 (a) 1.37 MeV; unstable
(b) 7.15×10^{-20} kg-m/sec
31–15 5 cm

Chapter 32
32–1 207.3 MeV
32–3 8.47×10^{-6} kg/h
32.5 3.13 million tons/kg
32–7 (a) 112.1 collisions
(b) 501.5 collisons
(c) 1749 collisions
32–9 0.122; 508 collisions
32–11 -24.15 MeV
32–13 0.89 MeV; 5.49 MeV; 12.86 MeV

Chapter 33
33–3 (a) 1.4×10^{-3} cm
(b) 6.28×10^{-8} N
33–5 (a) 0.225 MeV/cm
(b) 0.675 MeV
33–7 (a) 6.8×10^{-10} sec
(b) 3.4×10^{-9} sec

Chapter 34
34–1 (a) 5.4×10^6 m/sec
(b) 13.8×10^6 m/sec
(c) 299.3×10^6 m/sec
34–3 (a) 6000 W
(b) 6.19×10^4 m/sec
(c) 3.2×10^{-19} A
34–5 (a) $6\pi \times 10^7$ m/sec
(b) 0.148 MeV
(c) 3.52×10^{-22} kg-m/sec; yes
34–7 (a) 1.25×10^{21} crossings
(b) 2.29×10^7 rev/sec
(c) $0.196c$
(d) 0.408 m
34–9 0.11 eV
(b) 2.4×10^{-23} eV

Chapter 35
35–1 2.56 Å
35–3 4.65 eV
35–9 9.34×10^{25} electrons/m^3

ANSWERS TO ODD-NUMBERED PROBLEMS · 589

Chapter 36

36–1 0.013
36–3 0.01
36–5 4.9 Å
36–7 (a) 1.25×10^{26} electrons
      (b) 1.5%
36–9 0.95

Chapter 41

41–5 9.45×10^{15} m; 5.87×10^{12} miles

Index

Absorption, 122
Absorption coefficient, 122
Absorption spectrum, 162
Accelerators, 443
   betatron, 450
   Cockcroft-Walton generator, 443
   cyclotron, 446
   linac, 452
   linear, 452
   Van de Graaff, 445
Acceptor, 472
Actinide series, 321
Activity, radioactive, 386
Age
   of earth, 540
   of galaxy, 541
   of meteorites, 539
   of moon, 540
   of stars, 541
   of stellar clusters, 541
   of universe, 539
Alfvén, H., 548
Alkali metal, 318
Allowed transitions, 270
Alpha decay, 400
Alpha process, 531
Alpher, R., 524
Alvarez, L., 443, 452
Anderson, C. D., 116, 117, 480
Angular momentum, 261
   barrier, 258
   conservation of, 16, 303
   diatomic molecule, 307
   of hydrogen atom, 165
   molecular, 307
   of neutron, 352
   nuclear, 344, 352
   orbital, 302, 344
   of proton, 352
   quantization, 272
   resultant, 344
   spin, 290, 302, 343
   total, 302

Angular probability density, 282
Annihilation, 5
Antilife, 8
Antimatter, 5, 547
Antineutrino, 385, 506, 527
Antiparticles, 118, 494
Antiproton, 120
Apollo mission, 541
Atom, nuclear, 328
Atomic mass unit, 82
Atomic (proton) number, 329
Atomic spectra, 304
Atomic weight, 330
Average life, 388

Back, E., 267
Balmer, J. J., 161
Balmer series, 161
Band spectrum, 161
Band theory of metals, 465
Bardeen, J., 456
Bartlett, J. H., 351
Baryon, 490, 511, 519
Baryon number, 509
BASIC program, 79, 114, 212
Beats, 132
Becker, H., 356
Becquerel, A. H., 384
Beta decay, 450
Betatron, 450
Bethe, H., 416, 522, 524
Big Bang theory, 525
Binding energy, 168
   deuteron, 341
   electron, 336
   nuclear, 335
   per nucleon, 335
   semi-empirical formula, 347
Binomial expansion, 77
Blackbody radiation, 192
Bohr, N. H. D., 49, 157, 345

Bohr magneton, 345
Bohr model of the atom, 163
   correspondence principle, 169
   energy states, 164
   postulates, 163
Bohr radius, 160, 166
Boltzmann constant, 469
Bonds
   chemical, 458
   covalent, 458
   hydrogen, 459
   ionic, 458
   metallic, 458
   molecular, 458
   Van der Waals, 459
Born, M., 196, 202
Bose-Einstein statistics, 321, 489, 504
Bosons, 5, 321, 489, 504
Bothe, W., 356
Bound system, 160
Brackett series, 162
Bragg, W. H., 105, 458
Bragg, W. L., 105, 457
Bragg law of reflection, 107
Bremsstrahlung, 101, 102, 117
Brewster window, 311
Bubble chamber, 427, 429
Burbidge, G., 524
Burbidge, M., 524

Calame, J., 318
Carbon burning, 530
Carbon cycle, 416
Carbon dating, 395
Cepheid stars, 546
Cerenkov detector, 437
Cerenkov effect, 439
Chadwick, J., 345, 355, 358
Characteristic spectrum (x rays), 101
Chemical bond, 458
Cloud chamber, 427

INDEX • 591

CM frame kinetic energy, 374, 376
CNO bi-cycle, 530
Cockcroft, J. D., 373, 443, 444
Cockcroft-Walton generator, 443
Compton, A. H., 98
Compton effect, 109, 117
Compton scattering, 110
Compton wavelength, 111
Conduction band, 466
Conjugate variables, 136, 276, 297
Conservation
 of angular momentum, 366
 of electrical charge, 366
 of laws and interactions, 499
 of linear momentum, 366
 of mass-energy, 238, 366
 of total number of nucleons, 366
Conservation laws, 13
Control rods, 420
Cooper, L., 163
Correspondence principle, 49, 170
Cosmic ray telescope, 433
Covalent bond, 458
Cowan, C., 403, 405, 433, 434, 437
Cross section probability, 362, 380
Crystal spectrometer, 101
Crystals, 105, 457
Curie, M. S., 385, 399
Curie, P., 385
Curie, unit, 386
Curie-Joliot, I., 356
Curie-Joliot experiment, 357
Curve of growth, 389
Cutoff frequency, 103
Cyclotron, 446

Davisson, C. J., 127
Davisson and Germer experiment, 129
De Broglie, L. V., 126, 127
De Broglie wavelength, 128, 195
Decay
 alpha, 400
 beta, 405
 gamma, 407
 positron, 402
Degenerate states, 250, 274
Detectors, 360, 424
Deuterium, 181
Deuteron, 340, 360
Diffraction
 electron, 130
 x rays, crystals, 104
 x rays, ruled grating, 107
Dipole moment, magnetic, 262, 289
 neutron, 344
 proton, 344
 table of nuclides, 346
Dirac, P. A. M., 111, 117
Disintegration constant, 386
Donor, 471
Doublet states of sodium, 306

Eigenfunctions, 206
Eigenvalues, 206
Einstein, Albert, 10, 57, 66, 71, 95, 309
Electromagnetic spectrum, 100
Electron, 506
 capture, 406
 decay, 405
 discovery, 480
 emission, 92
 in nucleus, 345
 pair, 6
Electron volt, unit, 82
Electron-neutrino, 505
Elementary particles, 480
 baryons, 485
 classification, 485
 electric charge, 486
 hypercharge, 491
 interactions, 495
 isospon, 489
 leptons, 485
 magnetic moments, 487
 mesons, 485
 photons, 485
 rest mass, 483
 parity, 490
 particle number, 490
 quantum numbers, 483
 spin, 487
 statistics, 488
 strangeness, 490
 table of, 514, 515

Endoergic reaction, 360, 367
Energy, total, 80
Energy level diagrams
 baryon, 518
 helium, 176
 hydrogen, 167
 hydrogen, corrected, 181
Excitation energy, 168
Excited state, 165
Exclusion principle, 317, 321, 353, 461
Exoergic reaction, 367
Expanding universe, 545

Fermi, E., 412, 413
Fermi-Dirac law, 469
Fermi-Dirac statistics, 321, 461, 469, 489, 505
Fermi energy, 461, 463
Fermi temperature, 470
Fermions, 5, 321, 489
Fields, 21
Fine structure, 288
First excitation potential, 183
Fission, 413
Fission fragments, 413, 415
Fitzgerald, G. F., 58, 66
Forbidden transitions, 270
Force, 75
Fowler, W., 524
Franck, J., 174, 183
Franck-Hertz experiment, 183
Frank, I. M., 438
Frisch, O. R., 351, 352, 413
Fusion, 416

Galactic nuclei, 535
Galaxies, 534
Galaxy, age, 541
Galilei, Galileo, 26
Gamma decay, 407
Gamow, G., 524, 538, 543
Geiger, J. W. (Hans), 144, 423
Geiger-Muller counter, 436
Geiger-Nutall law, 402
Gell-Mann, M., 486, 503
Gerlach, W., 292
Germer, L. G., 127, 129
Glaser, D. A., 427
Goeppert-Mayer, M., 339
Gondolatsch, F., 541
Goudsmit, S., 288, 317

592 · INDEX

Gravitational coupling constant, 496
Graviton, 496
Grazing angle, 107
Ground state, 182
Group velocity, 132, 134
Gyromagnetic ratio, 263

Hadron, 508
  internal structure, 517
Hahn, O., 413
Half-life, 387
Halmitonian, 203
Harmonic oscillator, 193
  classical, 216
  eigenfunctions, 219, 221
  eigenvalues, 219, 221
  probability densities, 222
  quantum mechanical, 218
  Schrödinger equation, 219
  zero point energy, 220
Hathway, D. L., 408
Heisenberg, W. K., 135, 198, 215
Heisenberg's microscope, 137
Helicity, 506
Helium burning, 529
Herman, R., 524
Hermite polynomial, 220
Hertz, G. L., 174, 183
Herzberg, G., 301
Hoyle, F., 524
Hubble, E., 546
Hubble's law, 543, 546
Hydrogen atom, 12
  angular momentum, 165, 257
  bound states, 256
  energy eigenfunctions, 256
  energy states, 166
  ground state, 165
  magnetic moment, 262
  normalized radial functions, 285
  stationary states, 165
Hydrogen bond, 459
Hydrogen burning, 526, 528
Hydrogenlike atoms, 175
Hypercharge, 491

Impact parameter, 146
Impulse, 16
Impurities, 471
Inertial frame of reference, 14

Inertial observer, 34
Integral cross section, 151
Interactions, 23, 496
  electromagnetic, 498
  gravitational, 496
  strong, 330, 499
  very strong, 499
  weak, 497
Interferometer, 41
Intrinsic spin, 489
Ionic bond, 458
Ionization chamber, 432
Ionization energy, 168
Isotone, 329
Isotope, 329

Jauncey, G. E. M., 109
Joliot, F., 356

K capture, 406
Kerst, D. W., 450
Kinetic energy
  in laboratory frame, 376
  relativistic, 77, 238
Kunsman, C. H., 129

Lagrangian function (kinetic potential), 235
Laguerre polynomials, 250
Lanthanide series, 321
Lark, 508
Laser, 309, 311
  CO_2, 313
  pulsed, 312
  ruby, 312
  transitions, 314
Laue x-ray pattern, 107, 108
Lawrence, E. O., 442, 446
Legendre polynomials, 249
Lenard, Philipp, 93
Length contraction, 56, 64, 66, 237
Lepton, 5, 490, 505
Libby, W. F., 395
Linac, 452
Line spectrum, 161
Linear accelerator, 452
Linear momentum, 14
  conservation of, 15, 72, 75, 231
  invariance, 32
  in pair annihilation, 121
Liquid drop model, 328, 347
Livingston, M. S., 325, 446
Lorentz, H. A., 49, 55, 267
Lorentz factor, γ, 48, 58, 236

Lorentz triplets, 267
Loschmidt number, 464
Luminiferous ether, 41
Lyman series, 162

Magic numbers, 351
Magnetic quantum number, m_l, 249, 258
Marsden, E., 144
Maser, 310
Mass, 72, 75
Mass defect, 334
Mass-energy conversion, 81, 238
Mass number, 329
Matter-field, 195
Mean life, 388
Meitner, L., 413
Mendeleeff, D. I., 320
Meson, 510
  nonstrange, 511
Metallic bond, 458
Michelson, A. A., 40
Michelson-Morley experiment, 41, 63, 236
Milky Way, 536, 541
Millikan, R. A., 91
Millikan oil-drop experiment, 92
Million electron volt, unit, 82
Models of mechanics
  classical (Newtonian), 230
  quantum, 230
  special relativistic, 230
Moderating ratio, 419
Moderator, 418
Molecular bond, 458

Nark, 508
Neutrino, 385, 402, 505, 527
Neutron
  capture, 361
  cross-section, 418
  delayed, 414
  detection, 360
  discovery, 356
  fast, 421
  number, 329
  production, 359
  prompt, 414
  source, 359
  thermal, 361
Newton, I., 12

INDEX • 593

Newton's laws, 16, 33, 231
  invariance of, 234
Nondegenerate states, 250, 274
Normalization condition, 259
Nuclear
  density, 334
  energy levels, 362
  forces, 330
    charge independence, 333
    saturation effect, 333
    short range, 330
    strong, 333
Nuclear binding energy, 335, 349
Nuclear reaction, 319, 365, 418
  endoergic, 367, 379
  exoergic, 367
  probability, 381
  Q value, 367
Nuclear reactors
  breeders, 421
  critical, 420
  subcritical, 420
  supercritical, 420
Nucleon, 329
Nucleosynthesis, 525
Nuclide, 329

Operation SU_3, 501
Operation SU_6, 502
Operators
  angular momentum, 258
  Hamiltonian, 204, 205, 239, 246
    rectangular coordinates, 246
    spherical coordinates, 247
  momentum, 205, 257
Optical levels, 183
Orbital quantum number, 1, 250, 256
Orthogonality condition, 240
Oxygen burning, 531

Pair annihilation, 121
Pair production, 117
Parity, 490
  even, 491, 500
  odd, 491, 500
  transformation, 491

Park, 508
Parton, 519
Paschen, F., 267
Paschen series, 162
Pauli, W., 316, 317, 403
Pauli principle, 317, 353, 461
Periodic table, 320, 322, 323
Pfund series, 162
Phase velocity, 132, 134, 197
Photodisintegration, 340, 360
Photoelectric effect, 94, 117
Photoelectron, 93
Photographic emulsion, 425
Photon, 485, 504
Photon flux, 122
Physical four-space, 499
Pickering series, 177
Pile reactor, 420
Pion, 6, 330, 489, 511
Planck, M. K., 193, 255
Planck's constant, h, 96, 193
Planck's distribution law, 193
Planetary model of the atom, 158
Plasma, 416
Plum pudding atom, 144
p-n junction, 473
Population inversion, 310
Positron, 117
  decay, 402
Positronium, 121
Potential barrier, 222
  reflection coefficient, 224
  transmission coefficient, 224
Potential well, 206
  eigenfunctions, 210
  eigenvalues, 210
  energy quantized, 208
  probability density, 210
  zero point energy, 209
Pound, R. V., 82
Powell, C. F., 425
Principal quantum number, n, 166
Principle of complementarity, 131
Probability density, 279
Probability density distribution, 285
Probability per unit length, 196, 239
Proportional counter, 435
Protogalaxies, 548

Proton-proton cycle, 417
Pumping (laser), 312

Q value, 367, 369
Quanta, 127, 193
Quantum numbers
  magnetic, m_l, 249, 291
  magnetic spin, m_s, 291
  orbital, 1, 250, 291
  principal, n, 291, 304
  rotational (molecular), 308
  spin, s, 290
  total angular momentum, j, 302
  total orbital angular momentum, L_T, 319
  total spin angular momentum, S, 319
Quantum numbers (elementary particles), 483
Quark, 501, 508

r process, 533
Rad, 409
Radial probability distribution, 280
Radioactive dating, 393
Radioactive decay, 386
Radioactive series
  actinium, 392
  neptunium, 392
  thorium, 392
  uranium, 392
Radioisotope, 385
Radiological health hazards, 407
Radius of nucleus, 334
Rare earth, 321
Rayleigh-Jeans law, 192
Rebka, G. A., 82
Red giants, 529
Reduced mass, 178
Reines, F., 403, 405, 437
Rem, 409
Relativistic mechanics, 235
Relativity
  classical principle, 28, 234
  general, 47
  special principle of, 68
  special theory, 61
  special theory, schematic review, 84
Rest energy, 83, 238
  electron, 83
  neutron, 83

Rest energy (*continued*)
  proton, 83
Rest mass, 72, 237
Riemann, G. F. B., 3
Robertson-Walker metric, 543
Roentgen, W., 99
Rutherford, E., 143, 144, 328, 345, 365
Rutherford scattering formula, 144, 150, 153
Rydberg, J. R., 162
Rydberg constant, 162, 168, 169
  corrected for reduced mass, 180

Saturation current, 93
Schrödinger, E., 10, 191, 196
Schrödinger equation, 196
  harmonic oscillator, 219
  operator form of, 206
  time-dependent, 197, 239
  time-independent, 198, 200
Scintillation detector, 436
Secular equilibrium, 391
Segre, E., 413
Selection rules, 270, 309
Semiconductors, 466
  n-type, 472
  p-type, 472
Semi-empirical formula for BE
  Coulomb effect, 348
  lack of symmetry effect, 349
  pairing term, 349
  surface effect, 348
  volume effect, 348
Separation energy, 336
Shell model, 328, 351
Solid state, 460
Sommerfeld, A. J. W., 244
Space quantization, spin, 343
Space-time, 4
Spark chamber, 430, 432, 434
Spectra
  atomic, 304, 307
  molecular, 305
  molecular rotational, 308
Spin angular momentum, 288, 290
  spin down, 291
  spin up, 291

Spin gyromagnetic ratio, 295
Spin-orbit interaction, 290
Spinors, 489
Spectroscopic notation, 304
Speed of light, invariance, 44, 58
Spherical harmonics, 257
s process, 532
Stable nuclides, 342
Stars
  age, 541
  element building, 527
  red giants, 529
Stationary states, 165
Stellar clusters, age, 541
Stern, O., 287, 292
Stern-Gerlach experiment, 292
Stimulated emission, 310
Stopping potential, 93
Strangeness, 490
Strassman, F., 413
Strong interaction, 499
  coupling constant, 499
Supernovae, 533
Synchrotron, 453

Tamm, I. M., 438
Thomson, G. P., 127
Thomson, J. J., 144, 479, 480
Threshold energy, 376
Threshold frequency, 96
Time dilation, 59, 61, 237
  pion, 62
Torque, magnetic dipole, 263
Total number of states, 273
Townes, C. H., 310
Transformation
  coordinates, 30
  Galilean, 49, 233
  Galilean group, 31
  Lorentz, 45, 48, 49, 236
Transistors, 474
Tritium dating, 395
Tunnel effect, 221, 333

Uhlenbeck, G., 288, 317
Unbound state, 167
Uncertainty principle, 345
  energy and time, 139, 241, 497
  momentum and position, 135

Universe
  age, 539
  birth, 547
  expansion, 545
  mass, 545
  origin, 539
  radius, 543
  size, 543
Uranium dating, 394
Urey, C. H., 183, 278, 524

Vacuum, physical, 4
Van de Graaff, R. J., 364, 366, 445
Van de Graaff generator, 445
Van der Waals bond, 459
Velocity
  Galilean composition, 31
  Lorentz composition, 49
Very strong interactions, 499
Vibration-rotation band, 308

Walton, E. T., 443, 444
Wang, C. P., 519
Wave equation, 195, 245
  azimuthal, 248
  hydrogen atom, 251, 279
  polar, 249
  radial, 250, 256
Wave functions, 195
Wave packet, 131, 240
Wave-particle duality, 240
Weak interactions, 497, 505
Weisskopf, V. F., 67
Wilson, C. T. R., 427, 428
Work-energy principle, 19
Work function, 96

X rays, 99
  diffraction, 104
  diffraction by ruled grating, 107
  discovery, 99
  properties, 99

Yukawa, H., 327, 330
Yukawa potential, 332

Zeeman, Pieter, 266, 267
Zeeman effect
  anomalous, 267
  normal, 267, 270
Zero point energy
  harmonic oscillator, 220
  potential well, 209

SELECTED CONVERSION FACTORS

1 sec = 1.667×10^{-2} min = 2.778×10^{-4} h = 3.169×10^{-8} yr
1 m = 39.4 in. = 3.28 ft
1 Å (angstrom) = 10^{-10} m = 10^{-4} μ (micron)
1 AU (astronomical unit) = 1.496×10^{11} m
1 light year = 9.499×10^{15} m
1 kg = 2.205 lb
1 amu (atomic mass unit) = 1.6604×10^{-27} kg = 931.48 MeV
1 J = 0.239 cal = 6.242×10^{18} eV
1 cal = 4.186 J = 2.613×10^{19} eV
1 eV (electron volt) = 1.6022×10^{19} J
1 N = 0.225 lbf

MOST COMMON REST MASSES AND ENERGIES

NAME	SYMBOL	REST MASSES (kg)	(amu)	REST ENERGIES (J)	(MeV)
atomic mass unit	amu	1.6604×10^{-27}	1.000000	1.4922×10^{-10}	931.48
proton	m_p	1.6725×10^{-27}	1.007287	1.5031×10^{-10}	938.26
neutron	m_n	1.6748×10^{-27}	1.008665	1.5052×10^{-10}	939.55
electron	m_e	9.1091×10^{-31}	0.000549	0.8186×10^{-13}	0.5110
hydrogen atom	m_H	1.6734×10^{-27}	1.007829	1.5038×10^{-10}	938.72